The Natural Philosophy of En

STUDIES IN THE HISTORY OF PHILOSOPHY OF MIND

Volume 11

Editors

Henrik Lagerlund, *The University of Western Ontario, Canada*

Mikko Yrjönsuuri, *Academy of Finland and University of Jyväskylä, Finland*

Board of Consulting Editors

Lilli Alanen, *Uppsala University, Sweden*

Joël Biard, *University of Tours, France*

Michael Della Rocca, *Yale University, U.S.A.*

Eyjólfur Emilsson, *University of Oslo, Norway*

André Gombay, *University of Toronto, Canada*

Patricia Kitcher, *Columbia University, U.S.A.*

Simo Knuuttila, *University of Helsinki, Finland*

Béatrice M. Longuenesse, *New York University, U.S.A.*

Calvin Normore, *University of California, Los Angeles, U.S.A.*

Aims and Scope

The aim of the series is to foster historical research into the nature of thinking and the workings of the mind. The volumes address topics of intellectual history that would nowadays fall into different disciplines like philosophy of mind, philosophical psychology, artificial intelligence, cognitive science, etc. The monographs and collections of articles in the series are historically reliable as well as congenial to the contemporary reader. They provide original insights into central contemporary problems by looking at them in historical contexts, addressing issues like consciousness, representation and intentionality, mind and body, the self and the emotions. In this way, the books open up new perspectives for research on these topics.

For further volumes:
http://www.springer.com/series/6539

David Dunér

The Natural Philosophy of Emanuel Swedenborg

A Study in the Conceptual Metaphors of the Mechanistic World-View

Translated by Alan Crozier

Springer

David Dunér
Division of History of Ideas and Sciences
Centre for Cognitive Semiotics
Lund University
Biskopsgatan 7, 223 62 Lund
Sweden

ISBN 978-94-007-9821-2   ISBN 978-94-007-4560-5 (eBook)
DOI 10.1007/978-94-007-4560-5
Springer Dordrecht Heidelberg New York London

Springer is part of Springer Science+Business Media (www.springer.com)

Contents

Introduction 1
Prologue on a Grain of Sand 1
Biographical Guide.. 6
Literature About a Phenomenon 8
A Theory of Swedenborg's Brain 13
    Space and Thought.. 18
    Metaphorical Thought 21
    Seeing with the Inner Eye.. 26
    Thinking with Books.. 31
Overview—The World Machine Seen from Above............................ 34

The Space 37
A Blue Camera Obscura 37
The Society of the Curious... 41
Armed Eyes.. 47
Attempts to Find East and West Longitude 50
We Are Educated by Studying, Experiencing, and Thinking.................. 57
Unrest Disturbs My Work.. 61
From Barbarism to Culture 67
The Immutable World 70

The Sign 77
Everything Is Silent, No One Knows Yet the Destination 77
Learned Games with the Number Sixty-Four 81
The Geometrical Number Eight 87
The Useful Number Eight... 94
The Lord Is Wrathful.. 99
A Peripeteia on the Decimal 102
A Million Million 109
Rhetorical Arithmetic 113
Trees, Boxes, and Universal Mathematics 119
To Think Is to Count 125

The Wave 129
The Water Waves in Leiden 129
The Surging of the Sea 133
Sound in the Mountains of Lapland 145
In the Baroque Echo Temple 154
Thunder and Organ Peals 159
Fire and Colours 168
One Membrane Trembles from the Other's Trembling 180
The Beautiful Geometry of Tremulation 182
To Live Is to Tremble 191
The Circles of the Body 193
Hearing the Music from Within 199
Vision Extends into the Invisible 204

The Sphere 207
Hell Upon Earth 207
Flying in the Air 209
The Geometry of War 214
Nature—A Composite Analogy 218
The World Machine and the Little Machine 223
Peas and Cannonballs 226
A Sea of Bubbles 237
The Power of the Water Bubble 246
The Vapours Rising Over the Mountain 250
The Geometry of Heat 255
A Mineral Cabinet Without Stones 257
The Fruits of the Volcano 266
In the Bride-Chamber of the Mineral Kingdom 270
Vanitas Bubbles of Soap and Water 277

The Point 279
The Spider in the Polygonal Web 279
The Point That Delineates the World 285
A Grain of Dust at the Equator 289
Nature's Labyrinth 294
The Janus Face of the Mathematical Point 300

The Spiral 303
Helical Lines 303
The Circle of Time 309
The Force of the Moon 311
Whirls and Voids 313
On the Eternal Spring in the Age of Winter Cold 316
From Centre to Circumference and Back 323
Impossible Figures 329
The Microcosmic Spiral Motion 332

Magnetic Effluvia 335
The Magnetic Sphere and the Sidereal Heaven 337
The Macrocosmic Vortex 341
The Declination of the Magnetic Needle 344
The Membrane Between Body and Soul 347

The Infinite 353
A World That Is Not Even a Point 353
The Limits of the Unknowable 355
The Infinite Is the Ultimate Cause of the Finite 359
The Fantastic Order of the Brain Machine 361
The Limits of the Universe 365
The Nexus Between Infinite and Finite 369
The Last Effect of Creation 370
The Degree of Perfection 372
Escape to the Oracle of Reason 374
The Soul Machine 376
The Philosopher, the Happiest or the Unhappiest of Mortals 383

Conclusion 385
The Convolutions of the Brain 385
From Angular to Perpetuo-Spiritual Form 387
A Blind Man Who Can See, and the Form of Ideas 391
Swedenborg's Euphoria 393
Spiral Dances in Paradise 396
The Primary Metaphors of Correspondences 398
Memorabilia from Earthly Life 401
The Geometry of the Spiritual World 408
The Helical Motions of Emotions 409
The Spiral Forest at Adramandoni 411
Epilogue on a Garden 414

Bibliography 417

Index 459

List of Figures

Introduction

1. Hultkrantz, 60; Photograph, double exposure, by Johann-Martin Bernigeroth the Younger's copperplate after an original drawing by Johann Wilhelm Stör in Swedenborg's *Principia* (1734), and a crayon portrait by an unknown artist, Nordiska Museet; cf. also Henschen; Odelberg, 397–409; Lenhammar (2010).
2. Bromell, 'Lithographiæ svecanæ', *Acta literaria Sveciæ* (1725), 90–102; Copperplate, Johan van den Aveelen; Photo: LUB.
3. Kircher (1646), icon. 11, fol. 188; Copperplate; Photo: LUB.

The Space

1. Zahn, 210f; Copperplate; cf. *Polhems brev*, 76, commentary, 255; Lucretius, 4.225; Photo: LUB.
2. H. Vallerius and Rockman, 40; Woodcut; Photo: LUB.
3. *Dædalus Hyperboreus* IV, 64, 95. Woodcut, J. van den Aveelen?
4. KVA, Codex 86; *Photolith.* II, 66. Drawing, Swedenborg; Photo: KVA.

The Sign

1. *En ny räkenkonst*, § 6; Photo: KB.
2. Swedenborg, *Projekt och uträkning på domkrafter*. LiSB, N 14a, nos. 50–51; *Photolith.* I, 98; cf. Acton, *Reckoning*, 30; Drawing, Swedenborg; Photo: LiSB.
3. *De cupro*, 41, tab. xxvii; Copperplate; cf. *Opera* I, 233; Photo: LUB.

The Wave

1. Photo: author 2001; cit. *Opera* I, 19.
2. Photo: author 2001.
3. LiSB, N 14a; Drawing, Jan Klopper; Photo: LiSB.
4. Tiselius (1730), 46/47; Woodcut; cf. review in *Acta literaria et scientiarum Sveciæ* (1734), 51–56; Photo: LUB.
5. *Acta literaria Sveciæ* (1722), 354; Woodcut; Nordenmark (1933), 53; *Miscellaneous*, 157; Photo: LUB.
6. *Miscellanea*, tab. II; Copperplate.
7. *Miscellanea*, tab. III; Copperplate; cf. *Miscellanea* (ms). LiSB, N 14a.
8. Schott (1657), II, book II, 110/111; Copperplate; Photo: LUB.
9. *Dædalus Hyperboreus* I; Copperplate, J. van den Aveelen; *Letters* I, 86, 100.
10. *Then swenska psalm-boken* (1694), frontispiece; Ps 28:6; Swedberg (1710–1712), II, introduction; cf. Lamm (1915), 6; translation, 8.
11. *Dædalus Hyperboreus* III; Copperplate, J. van den Aveelen; *Letters* I, 100; *Opera* I, 249.
12. Bidloo, tab. V, fig. 1–2; cf. *De cerebro*, translation, II; Copperplate, Abraham Blooteling (?), after a drawing by Gerard de Lairesse; Photo: LUB.
13. *Psychologica*, 35; Drawing, Swedenborg.

The Sphere

1. *Photolith.* I, 21; Drawing, Swedenborg; Photo: LUB.
2. *Dædalus Hyperboreus* III; Copperplate, J. van den Aveelen?
3. Polhem, *De gravitate et compres[s]ione aeris.* KB, X 552:1, 308; Photo: KB.
4. KrA, Rappe, book IV, 123; Photo: KrA.
5. *Prodromus principiorum*, tab. I, 1; Photo: KB.
6. *Prodromus principiorum*, tab. III, 24; Photo: KB.
7. RA, Bergskollegii arkiv, EIV:169, 1725:I, 760; Drawing, Swedenborg; Photo: RA.
8. Vallemont (1693), 106/107; cf. Valentini, III, tab. XXXIII, 68; Copperplate; Photo: LUB.
9. *De victriolo*, 405; Drawing, Swedenborg, after Lister, 93, fig. 1 & 6; cf. Valentini, III, 65; Photo: KVA.
10. Agricola; Woodcut; *De victriolo*, 4/5.
11. *De sulphure et pyrite*; Drawing, Swedenborg; Photo: KVA.
12. *De sale commune*, 291; edition, 131f; translation, xxxvi, *The New Philosophy* (1989:1), 48; Basilius Valentinus (1694); Drawing, Swedenborg after Lister, 93, fig. 3; Photo: KVA.
13. *De magnete*, 263; Drawing, Swedenborg after Hauksbee, pl. VII; cf. Triewald (1735–1736), I, tab. I; Photo: KVA.

14. *De magnete*, 79; Valentini, III, tab. XXXIX, N. 21; Copperplate; Photo: KVA.
15. *Principia*, tab. IX:4; Copperplate.
16. *De ferro*, tab. XXII, 216/217; Copperplate, J. W. Stör, after an engraving by Ph. Simonneau in Réaumur, pl. XI, 406; Photo: LUB.
17. *De cupro*, tab. LXXXV; Copperplate, after Henckel (1725), tab. XI, XII, 157; Photo: LUB.

The Point

1. Kircher (1650), VI, 441; Woodcut; Photo: LUB.
2. KVA, Codex 87; *Photolith.* II, 111f, fig 3–7; *Opera* II, 7–9; *The minor Principia*, 6–8; Drawing, Swedenborg; Photo: KVA.
3. Rudbeck the Elder (1938); Woodcut.
4. Kunckel von Löwenstern (1679), frontispiece; Copperplate; Photo: LUB

The Spiral

1. KVA, Codex 86; *Photolith.* II, 39; Drawing, Swedenborg; Photo: KVA.
2. Photo: author 2000.
3. Burnet (1694), 21; Copperplate; Photo: LUB
4. KVA, Codex 87; *Photolith.* II, 161, fig. 27–28; *Opera* II, 38; *The minor Principia*, 42f; Drawing, Swedenborg; Photo: KVA.
5. KVA, Codex 87; *Photolith.* II, 183, fig. 37; *Opera* II, 50; *The minor Principia*, 58f; Drawing, Swedenborg; Photo: KVA.
6. KVA, Codex 87; *Photolith.* II, 284, fig. 67; *Opera* II, 106; *The minor Principia*, 127; Drawing, Swedenborg; Photo: KVA.
7. KVA, Codex 87; *Photolith.* II, 323, fig. 76; *Opera* II, 127; *The minor Principia*, 152; Drawing, Swedenborg; Photo: KVA.
8. *Principia*, tab. I; Copperplate; Photo: LUB.
9. *Principia*, tab. II:11, 12; Copperplate.
10. *Principia*, tab. V:12; Copperplate.
11. *De magnete*, 147; Valentini, III, tab XXXIX, N. 22; Copperplate; Photo: KVA.
12. Gilbert, translation, 135, 239; Swedenborg, *De magnete*, 11; Woodcut.
13. *Principia*, tab. XXV; Copperplate; Photo: LUB.
14. *Principia*, tab. XXVIII; Copperplate; Photo: LUB.
15. KVA, Codex 88; *Photolith.* III, 108; Drawing, Swedenborg; Photo: KVA.

The Infinite

1. Comenius (1684); Woodcut; cit. Comenius (1716), 88; cf. Comenius (1683), 88.

Conclusion

1. KVA, Codex 3B; *Diarium spirituale*, translation, IV, 368; Drawing, Swedenborg; Photo: KVA.
2. Scapula & Estienne, title page; Woodcut; Photo: Swedenborg Society, London; *De Messia*, translation, 104.

Introduction

Of making many books there is no limit and no end on the earth.
    Lo! Scarcely has a great pile of sand the same number.
And as the birds lack number, like the fish in the sea,
    so the many books on earth lack number, too.
The greater part of them utter empty talk, useless sentences without substance,
    making a display of sonorous words that are nonsense.
Beware lest you are deceived by the cunning fair looks of a book:
    for often there is poison hidden under sweet honey.

Swedenborg, from Swedberg's *Vngdoms regel och ålderdoms spegel* ('Rule of Youth and Mirror of Old Age', 1709).

Prologue on a Grain of Sand

Books are grains of sand, countless as the birds of the air and the fish in the sea. Far too many books are written, Emanuel Swedenborg complained.[1] Yet for his own part, he never managed to overcome the rather severe graphomania that haunted him right up to the end. Through time it resulted in a considerable heap of sand. Altogether his writings, translations of them, and books about him have built up an impressive and not inconsequential sandbank in the ocean of knowledge. Being stranded on this sandbank is not without risks. The quicksand of uncertainty and unreliability can give way beneath one's feet, and waves of scrutiny can wash everything out to sea.

Anyone out there on that sandbank, looking around, cannot avoid realizing that metaphor—thinking about something through something else—occurs abundantly in Swedenborg's thought. Books are sand. The world is a machine. The metaphors impelled new thoughts, allowing him to form ideas about the unknown from what

---

[1] Swedenborg, in Swedberg (1709b), cf. 388–392; *Ludus Heliconius*, ed. Helander, 60f.

D. Dunér, *The Natural Philosophy of Emanuel Swedenborg*, Studies in the History
of Philosophy of Mind 11, DOI 10.1007/978-94-007-4560-5_1,
© Springer Science+Business Media Dordrecht 2013

is known. That is what this particular grain of sand, this book, will be about; it is a study that seeks to provide a description and understanding, not only of what Swedenborg thought, but also of *how* he thought as a mechanistic natural philosopher, up to and including 1734. That was the year in which he published two works about the principles of natural things and about the infinite, *Principia rerum naturalium* and *De infinito*, in which he espoused a mechanistic world-view that stands out in clear contrast to the organic conception that he embraced in the following ten years and the voyages he began from 1745 into an immaterial spiritual world. The early Swedenborg regarded the world as gigantic machine, everything was machines, everything could be explained as geometry in motion. If Swedenborg's world machine were dismantled into its constituent parts—space, signs, waves, the sphere, the point, the spiral, and infinity—one could arrive at a better understanding of how it was designed. It is an investigation that aims to paint the first concerted and complete picture of Swedenborg's early natural philosophy, from geometry and metaphysics to technology and mining science, situated in his time and context. That is the most important contribution to the research on Swedenborg and the history of eighteenth-century science. Additionally, in order to reach a deeper understanding of his natural philosophy and his creative mind, this book aims to penetrate his way of reasoning, with image schemas, metaphors, categories, and other cognitive abilities. In that respect this study is a contribution to the current methodological development of a cognitive history in what has been called the cognitive turn within the humanities. But the approach can also shed new light on this influential visionary's esoteric theology and doctrine of correspondences, which actually is a metaphorical system grounded in his cognitive qualities and experiences.

The verses about the vanity of writing and reading books were written by Swedenborg for his father, Bishop Jesper Swedberg's book *Vngdoms regel och ålderdoms spegel* ('Rule of Youth and Mirror of Old Age', 1709), at the time when he himself was completing his university studies in Uppsala. Perhaps he was tired of everything. Waking nights weaken the body, he continued his lamentation, make a young person resemble the night, an internal disease consumes him and his face increasingly takes on the blue-black colour of death. The words alluding to the meditation in Ecclesiastes about the vanity of vanities encapsulated an edifying theme for the disconsolate reader: 'And further, by these, my son, be admonished: of making many books there is no end.'[2] When Swedberg later sums up his life of toil and diligence in his autobiographical *Lefwernes beskrifning* ('Description of My Life', 1729) he says: 'but I am not as yet afflicted in the way the same Preacher [Ecclesiastes] writes: *much study is a weariness of the flesh*.'[3] If anything, assiduous work enlivened him instead. In any case, a medical student does not need so many books, said the professor of medicine at Uppsala University, Lars Roberg, an acquaintance of Swedenborg's. One book is sufficient, the book of nature. 'Like the

---

[2]Eccles. 12:12; cf. Browallius, 10.
[3]Swedberg (1941), 239, 527.

people with whom one mixes, one should ensure that books are honest, beneficial, and honourable,' he declared, for 'Books are for the Soul what food is for the body. It should be adequate and not a mixture of many different kinds.'[4]

Books can give nourishment, but also stomach-ache. Swedenborg read and wrote indefatigably. In June 1740, at the tercentenary celebrations of Johann Gutenberg and the art of printing, Swedenborg wrote about how printing presses were propelled by the waves flowing from the source of Pallas, goddess of wisdom.[5] A torrent pours forth from the presses. The intensity of the flow from his own pen increased when, as seer of visions, he entered into the geography of the spiritual world. In *De telluribus* (1758) he describes how, during a voyage in space, he got into conversation with some spirits from a distant solar system. On our globe, he explained, there are remarkable sciences and arts that are unparalleled anywhere else in the universe:

> such as astronomy, geometry, mechanics, physics, chemistry, medicine, optics, philosophy. I went on to mention techniques unknown elsewhere, such as ship-building, the casting of metals, writing on paper, the diffusion of writings by printing, thus allowing communication with other people in the world, and the preservation for posterity of written material for thousands of years. I told them that this had happened with the Word given by the Lord, so that there was a revelation permanently operating in our world.[6]

With the aid of printing, people can pass on and store their thoughts and inventions for thousands of years, to future generations, to new creatures on earth. Among the immense numbers of stars and planets in the vast universe there is only one globe where the art of writing is known. Only there can sciences, mathematics, and philosophy be found.

In Swedenborg's ideas about writing, about books as grains of sand, one can— as so often—find threads going back to earlier sources, such as classical authors or the Bible. Counting grains of sand is a classical metaphor for the impossible. Archimedes starts *The Sand Reckoner* from the third century BC by saying that there are some 'who think that the number of the sand is infinite in multitude'.[7] How many grains of sand, he asks, would it take to fill the whole universe with sand? He proceeds from the number of grains of sand in a volume the size of a poppy-seed, and with a new method for expressing very large figures, Archimedes succeeds in stating the figure as 'a myriad-myriad units of the myriad-myriad-th order of the myriad-myriad-th period'.[8] This figure is much larger than the number of grains of sand in the entire spherical universe, which Aristarchus of Samos had calculated to contain a number of less than 10,000,000 units of the eighth order, that is to say, 10^{63}. It would thus be a very large but not infinite number. The quantity

[4]Roberg (1747), 27.

[5]Swedenborg, in *Gepriesenes Andencken*, 93; *Ludus Heliconius*, ed. Helander, 82f, commentary 164.

[6]*De telluribus*, n. 136.

[7]Archimedes, *Psammitēs*, 1.1.1–2; translation, 221; *Ludus Heliconius*, ed. Helander, commentary 149; cf. Horace, 1.28.1.1–4; Sturm (1699), I, 11.

[8]Archimedes, *Psammitēs*, 3.4.27–5.1; translation, 228.

of grains of sand expresses unfathomable numbers, but this can also be invoked in a more comforting way. In an attempt at encouragement, Ovid wrote that there are innumerable possible hunting-grounds for women, and attempting to list them was like trying to count the sand on the beach.[9]

The Bible likewise contains grains of sand. God's thoughts are more in number than the sand, the Psalter says, or as the Swedish bishop Johannes Rudbeckius renders this in his hymnal *Enchiridion* (1622):

> And if to tried to count by hand /
> The thoughts of God to tell the score /
> They would be more than all the sand /
> That lieth on the ocean floor /
> And therefore I shall try no more.[10]

The seed of Abraham will multiply as the sand upon the sea shore, hostile armies attack like a swarm of locusts or the dust of the earth, countless are the sins, and death grinds us down to dust and sand.[11] 'That which is crooked cannot be made straight: and that which is wanting cannot be numbered,' wrote the philosopher Andreas Rydelius, citing Ecclesiastes.[12] The antiquarian Johan Peringskiöld the Elder, in *En book af menniskiones slächt, och Jesu Christi börd* ('A Book about the Human Race, and the Lineage of Jesus Christ', 1713), alludes to the enormous growth of the children of Abraham predicted in the Book of Genesis: 'as the dust of the earth, as the stars of the heaven, and as the sand which is upon the sea shore.'[13] Fertility comes from above! Grains of sand, stars, and locusts represent immeasurable amounts that instil trembling amazement. In *Gudz werk och hwila* ('God's Work and Rest', 1685), Bishop Haquin Spegel writes about the countless sand at the bottom of the sea and the wet shore, like herbs, grass, and spices in the summer fields, the light-winged snow covering the mountains, and the 'blue-starred ceiling' with its glistening carbuncles and jewels, 'In an order that we humans cannot grasp.'[14]

The sand blows through scientific texts as well. The metallurgist Georg Bauer, better known under the name Georgius Agricola, cited Naumachius' assessment of gold and silver as mere dust, like the stones scattered on the beaches and along river banks.[15] Sir Isaac Newton, elected by history as the very model of a sober scientist, saw himself in his own eyes as a boy who had played on the sea shore, taking

---

[9]Ovid, *Ars amatoria*, 1.253f.

[10]Rudbeckius, 166; Ps. 139:18.

[11]Gen. 13:16, 22:17, 28:14, 32:12; Num. 23:10; 1 Kings 4:20; Hos. 1:10; Jth. 2:20; Ecclus. 1:2, 44:21; Rom. 9:27.

[12]Rydelius (1737), 161; Eccles. 1:15.

[13]Peringskiöld, title page.

[14]Spegel (1685), ed. Olsson and Nilsson, I:1, 189.

[15]Agricola, translation, 8; cf. Wisd. of Sol. 7:9.

pleasure in 'finding a smoother pebble or a prettier shell than ordinary, whilst the great ocean of truth lay all undiscovered before me'.[16] The polymath Eric Benzelius the Younger saw with a different eye and searched with his learning in 'our dark gravel', wrote the historian Olof Dalin in a memorial sketch. But if you could not find it there, 'You will now find total knowledge / In eternity's own light.'[17] The grain of sand is the trifle that disappears in the mass. The world is so large, and man so small: 'I, a speck of dust in Paradise, a human', sighed the alchemist and councillor of the realm Gustaf Bonde.[18]

The sand metaphor, like many other metaphors, can be found throughout the history of ideas, and it cannot really be traced to any specific source. Instead it seems as if the metaphors were an integral part of language, of our way of reasoning. One premiss of this study of metaphor in Swedenborg is that it is not always possible to demonstrate influence from a particular quarter; instead there are special cognitive abilities which enable us to think in one way or another. Metaphor is not just poetic adornment or educational parable with its origin in classical authors or the Bible, but is an intrinsic part of the way scientists reason. They imagine the invisible and unknown in terms of what is visible and known, they describe reasoning and research as something spatial, a walk along the sea shore, a voyage over a perilous ocean. An unusually palpable example of this metaphorical way of reasoning is the mechanistic natural philosopher and visionary biblical exegete Swedenborg. Few have taken the metaphorics of the mind to the ultimate limits as he has done. It is also, in line with Carlo Ginzburg, a matter of searching for what was taken for granted, things for which no arguments were given, in other words, not concentrating on the most obvious parts of a work of art, but on the seemingly insignificant details.[19] Like a physician confronted with symptoms, the historian standing before the fragments sees them as signs, clues to the historical context. By studying a small detail, some metaphors, in Swedenborg, we can also see out into the surrounding reality, into the society, the thought, and the science around him, as if discerning the macrocosm reflected in a microcosm. Anyone who points the microscope at a heap of sand, wrote Robert Boyle, will easily see that each grain of sand has its distinctive size and shape, like a rock or a mountain.[20] In the words of William Blake, we see a world in a grain of sand.[21] There are so many books, countless as the grains of sand, and on top of everything—here is another grain of sand on the shore of the immense ocean of knowledge.

---

[16]Brewster, II, 331.

[17]Dalin (1744), 23.

[18]Bonde, dedication; Edenborg (1997), 43.

[19]Ginzburg (1979), 93, 109; Ginzburg (1986), translation, 96–125.

[20]Boyle (1965), 194.

[21]Blake (1989), 589.

Biographical Guide

What was known was just a small part of the great unknown. Swedenborg was in a space that seemed to lack bounds, in a geographical space with blank spots and a cosmos without a discernible end. What he knew about the world was a starting point for understanding the unknown. Movement in space and action in the world are an essential part of reasoning. From the beginning, when he entered this life on 29 January 1688, it was the dirty alleys, cold brick churches, illuminated court halls and aristocratic palaces of Stockholm. He was christened Emanuel, 'God with us'.[22] But where was he to go? 'I do not know my exit or my entrance: O lead me therefore in all innocence', as a children's hymn says in Swedberg's hymnal (1694).[23]

His way took a scientific course, a choice that was not unlikely encouraged by his cousin, the future physician Johan Moræus who taught Latin in the Swedberg home.[24] In a shaky hand, Swedenborg had written his name in a small disputation booklet from 1695 about the cure of diseases.[25] In the following year his mother, Sara Behm, left this earthly life after an illness. At the age of eleven he matriculated at Uppsala University, and for ten years the blocks around the River Fyrisån were the centre of his geography, with short lines between the Cathedral and the main university building, the Gustavianum. Coming from a family of mine owners, he regarded the 'nation' of Västmanland-Dalarna, the union of students from those mining provinces, as his true geographical domicile. When Swedberg left his post as professor of theology and moved to Skara in 1703, where he served as bishop in a diocese that included the Swedish congregations in America within its bounds, Swedenborg remained in the house in Uppsala. His brother-in-law, the university librarian Eric Benzelius, married to Swedenborg's sister Anna, moved in. Also living in the house in 1704 were some other distant relatives, including the teenagers Andreas Rhyzelius and Andreas Kalsenius. 'Our hostel that autumn and the following spring,' wrote Rhyzelius, 'was a dark, wretched, and unhealthy chamber down by the river; yet we throve and studied tolerably well there.'[26] What exactly Swedenborg studied we do not know. It is not unlikely that he received a broad education in science, mathematics, and philosophy.

On 1 June 1709 Swedenborg completed his studies with a viva voce in the great auditorium of the Gustavianum, defending a thesis, written by himself, about various maxims in Seneca and the actor and mime Publilius Syrus, *L. Annæi Senecæ & Pub. Syri Mimi forsan & aliorum selectæ sententiæ cum annotationibus Erasmi &*

[22]Swedberg (1941), 239; SSA, CIa:10.
[23]*Then swenska psalm-boken* (1694), Ps. 389:3.
[24]Linnaeus (1742), 298.
[25]Below and Ribe. KVA.
[26]Rhyzelius (1901), 38; cf. Broberg (1990), 90–104.

græca versione Jos. Scaligeri (1709).[27] His opponents were Andreas Rhyzelius and Jonas Unge. Professor Fabian Törner presided instead of Johan Eenberg, who had recently died. After the disputation, the meadows and fields around Brunsbo, the bishop's residence outside Skara, became his home while he awaited his departure into the outside world. The Danes had cut off the sea routes after the momentous defeat of the Swedes at the hands of the Russian troops of Peter the Great on the Ukrainian fields near Poltava on 28 June 1709, so there was no option but to wait and pass the time studying mathematical sciences.[28] He wanted to get away, out into the world, to escape as quick as he might, far from the dreadful, desolate province of Västergötland.[29]

In the late summer of 1710 his circles widened to include Europe. The encounter with the big world was rapturous, with its foreign languages, sounds, and smells, its palaces, cathedrals, and libraries. Together with some friends, Sven Bredberg, Eric Alstrin, and Jacob Ludenius, he trudged the streets of London, visiting the sights. 'The air is daily laden with the continual smoke of the coal that is burned here,' Bredberg noted in his travel diary, 'probably the cause of the severe chest pains that I have to endure.'[30] Swedenborg discussed the problem of longitude with English astronomers, admired beautiful books in the Bodleian Library in Oxford, and then continued over to the continent to stroll along the canals in Leiden, to immerse himself in mathematics in Paris, and to publish poems in Greifswald. On his return to Sweden, after five years, he had the honour of becoming assistant to the famous engineer Christopher Polhem, which lasted until 1719, and was appointed by King Charles XII as extraordinary assessor in the Royal Board of Mines. This work brought him to damp winter days in Lund, salty bays in Bohuslän, the intractable rapids at Trollhättan, to the front lines at the Norwegian border, and down into the terrifying depths of the Falun mine. On 23 May 1719 the children of Bishop Swedberg were ennobled, assuming the name Swedenborg. As the eldest son, Emanuel took up his seat in the diet, where he was an active participant until his old age. A new trip abroad followed in the years 1721–1722, taking him to Holland and Germany. Then came ten years with neither foreign travel nor published writing. Yet he was far from idle during this time. He had become an ordinary member of the Board of Mines and took an active part in its commissions, with matters concerning mines, ironworks, models of machines, steam engines, and much besides. He was commissioned by the board to make tours of inspection to sulphurous mining districts, and in his free time he wrote thick bundles of paper about the mineral kingdom, about chemical processes and mining. In 1733–1734 he was once again on the move, visiting Germany and Bohemia, to publish his largest

---

[27]Preparatory notes for the dissertation from Cicero, Plautus, and Florus can be found in KVA, cod. 37; cf. *Opera* I, 202; *Letters* I, 4.

[28]*Opera* I, 201; *Letters* I, 3.

[29]Swedenborg, in Palmroot and Unge, 14; *Ludus Heliconius*, ed. Helander, 62f.

[30]Bredberg, 112, cf. 118–120.

work hitherto, *Opera philosophica et mineralia* (1734), three volumes about the principles of natural things, about iron and copper. The plan here is to follow him closely until that year.

After 1734 he abandoned his mechanistic explanation of the world and oriented himself towards an organic world-view. The world was no longer a machine but something living. In ambitious projects he threw himself into the anatomy and physiology of the human body, in search of the abode of the soul. He went to France and Italy, constantly on the trail of the physiological functions of the brain. On the fifth trip to Holland and England (he is said to have spent 22 years of his life abroad), however, he was tormented by strange dreams in a time of anguished inner crisis.[31] His life was never to be the same again. In his nocturnal dreams he began to discern a message. Someone wanted to say something to him. Then in 1745, after a visit to a tavern in London, he met Christ. Two years later he resigned from the Board of Mines, left the geometrical fields of natural science and embarked on research expeditions in a completely different landscape, the immaterial space of the spiritual, non-geometrical world. There he found a breathing space, enjoying a completely new freedom of movement between the cities, societies, gardens, and planets of the spiritual world. In his thoughts and dreams his geography expanded. With his discovery of the correspondences between natural and spiritual meaning, he gained the key to the inner message of the Bible. Finally, on 29 March 1772, he took the irrevocable step over to the other side, the spiritual world.

All cannot have been in vain after the many years that had elapsed: the years of growth, the travels, the almost thirty years as assessor in the Board of Mines, all the intensive studies in mathematics, geology, astronomy, metallurgy, physiology, anatomy, and philosophy. Apart from the world that was known to him, his spatial experience incorporated what friends and relatives had told him about America, the Orient, Tartary ... and all the books that had given him insight into everything from mining in Siberia to imaginary trips to the inhabited moon. He must have carried all this earlier experience with him on his spiritual journeys. When a spirit wondered how a philosopher like him could become a theologian, he replied, 'from early youth [I] had been a spiritual fisherman.' And a 'fisherman', Swedenborg explained, 'in the spiritual sense of the Word, signifies a man who investigates and teaches natural truths, and afterwards spiritual truths rationally'.[32]

Literature About a Phenomenon

Swedenborg's remarkable destiny, moving from natural science to the spiritual doctrine, from rationalism to mysticism, from mathematics to angelic song, has baffled many exegetes. Some lament this 'tragic fate' and say that Swedenborg

---

[31]*Resebeskrifningar*, v.

[32]*De commercio animæ et corporis*, n. 20.

was once a very promising scientist, but was unfortunately afflicted by mental illness. The astronomer Erik Prosperin wrote in 1791 about Swedenborg that 'from the great chasm between the *Algebra* and the new Jerusalem, we had reason to sigh over the delusions of the human mind'.[33] Others, including the German philosopher Immanuel Kant, mounted an attack with a whole book, *Träume eines Geistersehers* (1766), and declared that Swedenborg was a charlatan and a liar and that everything he wrote had its origin in the pure fantasies of a sick brain. There is also a diametrically opposed standpoint. Swedenborg was the second Christ who brought the third testament, a universal genius, the spiritual Columbus, the Buddha of the North, a titanic mental power cast in heroic form.[34] Swedenborg is at once fascinating and terrifying. Many have been amazed at Swedenborg's curious hybrid combining dry-as-dust mathematics with oneiric fantasy. Friedrich von Schelling, describing Swedenborg's doctrine to the author Per Daniel Amadeus Atterbom, wrote of how 'the loveliest comfort, the most devout poetry, the most brilliant depth of thought carry on a strange and wonderful war with abstract dogmatism and poor mathematics'.[35] Schelling's disciple, Gotthilf Heinrich von Schubert, explained Swedenborg's abstruse dream mathematics as 'a higher form of algebra'.[36] Swedenborg's destiny raises questions about the relationship between science and religion, between the truths of reason and the Bible, and the issue of what is meaningful in life. He is an excellent case for studying the interplay and incongruity between science and religion. In him there is existential wonder in the face of God and nature, revelation and rationality, wisdom and love. Swedenborg would see 'That which earthly eyes do not see: / The fierce geometry, the crystal / Labyrinth of God and the sordid / Milling of infernal delights', wrote the Argentinian author Jorge Luis Borges.[37] He saw a line through the fog. Quite simply, he sought to understand the world.

The first task worthy of respect for research into Swedenborg was to organize, catalogue, publish manuscripts, and translate the works of a man who seemed to observe Apelles' maxim 'nulla dies sine linea', not a day without a line. He left almost 42,000 pages behind. It has mainly been representatives of the Swedenborgians' own church, the 'New Church', that have taken charge of the work of publishing all this.[38] The resulting editions may not be scholarly, but they are useful. Among the more important contributions we may mention the photolithographic reproduction of Swedenborg's manuscripts by the American Swedenborgian Rudolph L. Tafel from 1869 to 1870, and James Hyde's *A Bibliography of the Works of*

---

[33]Prosperin, 11.

[34]Spear; Suzuki.

[35]Munich, January 1818; Cited in Horn, translation, 32.

[36]Schubert, 10; Jonsson (1983b), 169.

[37]'Emanuel Swedenborg', Borges (1989), 287; Borges (1982); translation, 16; Báez-Rivera, 73; Kutik, 79.

[38]Tafel, 5–6; Holmquist (1909b); Eby; Söderberg (1989), 53–72; Williams-Hogan (2002), 227–244; Rose (2004), 7; Rose (2005).

Emanuel Swedenborg (1906).[39] Since 1898 the Swedenborg Scientific Association has published a journal of widely varying quality, *The New Philosophy*, for studies of Swedenborg's philosophical, scientific, and theological writings. Another periodical with similar content came later, *Studia Swedenborgiana*. In the first half of the twentieth century, three American New Church members were active. These were Alfred Stroh, who directed the publication of the unfinished collection of Swedenborg's scientific writings, *Opera quædam aut inedita aut obsoleta de rebus naturalibus* (1907–1911); Cyril Odhner Sigstedt, who took great pains to put together a collection of documents concerning Swedenborg, called 'Green Books'; and finally perhaps the foremost of them, Alfred Acton, who translated a long series of Swedenborg's scientific writings and compiled Swedenborg's letters and memorials, with a commentary, in *The Letters and Memorials of Emanuel Swedenborg* (1948–1955).[40]

In this book I will present all the scientific texts by Swedenborg, supplemented with quite a few new discoveries of hitherto unknown texts by and about him. Based on the empirical material, this will give a more balanced and complete analysis of Swedenborg than anything achieved hitherto. For my task here, the New Church studies and interpretations of Swedenborg's writing cannot be used without some difficulty. They have a different purpose, often based on an assumption that everything Swedenborg wrote is true, and that there is an internal coherence, that the contradictions are only apparent in his theological 'canonical works' from *Arcana cælestia* (1749–1756) onwards. Swedenborg's scientific theories chiefly interested his fellow scientists around the last turn of the century.[41] Yet the history of science as written by scientists is tricky to use. In many cases the aim has been either to admit Swedenborg to the success story of science or to expel him from it, to elevate him into a national hero, or to show that his science was as prophetic as his spiritual doctrine. Swedenborg's research on the brain has attracted particular interest and has been rated highly. He has not infrequently been assessed in relation to the scientific level of the exegete's own time, as scholars read into it their personal and contemporary concepts of science and disciplinary boundaries, trying as far as possible to avoid 'extra-scientific', philosophical, and cultural factors. Here I will analyse Swedenborg from a perspective in line with contemporary cognitive history of science, that also reckons with the cognitive processes of the agent and the spatial, cultural, and social environment. In general, it may also be said that many Swedenborg studies have had problems setting him into his own times; the

---

[39]See also Stroh and Ekelöf; Wainscot; Woofenden (2002).

[40]Stroh (1912, 1918); see also Potts; concerning Swedenborg's Latin, Chadwick; Berggren; Chadwick and Rose.

[41]Svante Arrhenius wrote about Swedenborg as an astronomer, Gustaf Retzius and Martin Ramström about his research on the brain, Hjalmar Sjögren about geology, Gustaf Eneström about mathematics, and Wilhelm Oseen published Swedenborg manuscripts. These are only short essays; see Ramström, Retzius; Nordenmark (1933); Stroh (1908); *Transactions of the International Swedenborg Congress 1910*; cf. Broberg (1983), 120f, 127.

Swedish works and the Swedish history of ideas have caused particular difficulty. It is easy to be affected by this rootlessness in time and geography. One of the major contributions of this work is to place Swedenborg within the European, especially the local Swedish, intellectual context and debate.

In academic research there is a strange silence surrounding Swedenborg, the eighteenth-century Swede who—alongside Carl Linnaeus and King Charles XII—has attracted the greatest attention outside Sweden. Only one doctoral dissertation has been written about Swedenborg in Sweden, and that was in the discipline of comparative literature in 1961. In Sweden is it chiefly literary historians with an obvious interest in the history of ideas who have devoted time to Swedenborg. A classic study is the one by the literary historian Martin Lamm from 1915, published much later in English as *Emanuel Swedenborg: The Development of His Thought* (2000). Here Lamm seeks to demonstrate that Swedenborg's theological system can be viewed in the light of his natural philosophy, that there is a link between them. The dream crisis is not a watershed between the two periods. Lamm's method rests in large measure on an implicit idea about intellectual influence, that Swedenborg's theological and scientific thought can be explained on some causal basis as the impact of other thinkers. Lamm was constantly searching for 'the genesis of Swedenborgian theosophy', that is, the influences exerted on Swedenborg by his own scientific studies of Plotinus, Descartes, Locke, Polhem, or others.[42] This kind of quest for sources can easily lead one in the wrong direction. As regards the influence of Milton, Inge Jonsson demonstrates in *A Drama of Creation: Sources and Influences in Swedenborg's Worship and Love of God* (2004, Swedish original 1961) how easy it is to be mistaken concerning influence by proceeding from external similarities of thought.[43] Instead the similarities can be ascribed to older common sources, in this case Ovid. The focus in Jonsson's dissertation is Swedenborg's religiously inspired creation story from 1745. By means of extremely careful studies of Swedenborg's works on natural philosophy, chiefly from the time after 1734, and by searching out many of his sources, Jonsson sheds light on the Swedenborgian creation drama. In *Swedenborgs korrespondenslära* ('Swedenborg's Doctrine of Correspondences', 1968) Jonsson has subsequently published what may be described as the weightiest and most authoritative study ever written about Swedenborg. He has tracked down countless sources and predecessors of the late Swedenborg's most central idea, the doctrine of correspondences between words and concepts. The doctrine of correspondences, he shows, can be traced back to the works on natural philosophy, having arisen from the mathematical spirit and with its roots in, among other things, the universal mathematical tradition.

The only substantial contribution by a historian of ideas is Tore Frängsmyr's examination of Swedenborg's geogony in a chapter of his dissertation *Geologi och*

---

[42]Lamm (1915), translation, xxiii.

[43]Jonsson (1961), translation, 10, 175; cf. Lamm (1915), translation, 184; otherwise there are four dissertations about Swedenborg: Schlieper; Kirven; Calatrello; Woofenden (1970).

skapelsetro ('Geology and the Doctrine of the Creation', 1969). From more recent years we may mention the sympathetic portrait by the author Olof Lagercrantz (1996, published as *Epic of the Afterlife* in 2002), although his method is not entirely without problems, and Lars Bergquist's detailed, sensitive reading of *Swedenborg's Dream Diary* (2001, Swedish original 1988). Bergquist has also produced the latest in the sequence of biographies of Swedenborg.[44] In *Swedenborg's Secret* (2005, Swedish original 1999), he has more skilfully than anyone else succeeded in capturing Swedenborg's life story, his personality and existential quest. In addition to this we have the Latinist Hans Helander's richly commented and exemplary scholarly editions of Swedenborg's Latin poems, an achievement without parallel in Swedenborg studies. Finally, an Italian historian of philosophy, Francesca Maria Crasta, has written a piece of sound scholarship about Swedenborg's natural philosophy, *La filosofia della natura di Emanuel Swedenborg* (1999), although it does not go far beyond what can be found in earlier research.[45] But what has been written *about* Swedenborg is overshadowed by everything that has been written about the reception of Swedenborg, about his successors and readers, about his literary influence on a large number of philosophers and authors, about Swedenborgianism and the history of the Swedenborgian Church.[46]

To sum up, it may be said that there is no comprehensive monographic study of Swedenborg's early mechanistic period up to 1734. In addition, there is a dearth of studies in the history of ideas using contemporary source material, especially the Swedish material. Nor has anyone tried to answer the question not only of what Swedenborg thought, but of *how* he thought. This gap in the research is what the present book aims to fill. The Polish author Czesław Miłosz formulated the challenge: 'the Swedenborg phenomenon, in effect, belongs to those enigmas which, if ever solved, would shed light on the laws of human imagination in general.'[47] I take the reverse approach, of proceeding from what we know about the human imagination to see if it can shed light on the phenomenon that is Swedenborg. This is a book about how the natural philosopher Swedenborg thought.

---

[44]Since Sandels, the ones that have attracted most attention are: Kleen; Toksvig; Benz; Sigstedt; cf. Lenhammar (1988).

[45]Jonsson (2002), 312f; see also Crasta (2002); N. Newton (1999); Holmquist (1909a, 1913); Lindh (1927–1929); Piotrowska, 29–38; Brock; Hoppe; Jonsson (2008).

[46]Works have been written about such readers of Swedenborg as Oetinger, Kant, Goethe, Blake, Coleridge, Schelling, Boström, Balzac, Emerson, Whitman, and Baudelaire. See e.g. Florschütz; McNeilly (2004, 2005); Rix; Stengel; Wilkinson; dissertations about the reception of Swedenborg and about Swedenborgianism are, among others: Lenhammar (1966); Williams-Hogan (1985); Sjödén; Hallengren (1994); Häll; see also Garrett; Hallengren (1989); Hjern; Hanegraaff (1996), 424–429; Sanner; Gabay; Hanegraaff (2007); Williams-Hogan (2008).

[47]Miłosz (2000), 137; Miłosz (2007), 1–16.

A Theory of Swedenborg's Brain

What did Swedenborg think about, how did he think as he walked through the landscape, peered through the microscope, or sat in the library with a book opened in front of him? It was in Swedenborg's brain, in his conscious and unconscious thinking, that his thoughts were formed. In the natural cycle, his brain has gone the way of all flesh, but his thoughts have survived as signs and letters in his writings (Fig. 1). As an author, scientist, and visionary, he has passed on a part of his world, his thoughts and experiences, to posterity. If we proceed from the way people think in general, their mental abilities, reason and cognition, in other words, if we consider *how* people think, not just *what* they think, we could get close to an understanding of how Swedenborg shaped his ideas about nature, man, and God. This is a cognitive history of ideas, a history of thinking. What is called the 'cognitive turn' in the humanities has generated vigorous growth of research into different cognitive explanatory models of human expressions and cultural evolution, for example, in cognitive poetics, neuroaesthetics, and cognitive anthropology.[48]

Fig. 1 Swedenborg's skull. The contours of a cranium drawn on the portrait of Swedenborg in *Principia* from 1734. To the right, a photographic projection of a model of the skull on a later portrait of Swedenborg. Around 1910 Johan Vilhelm Hultkrantz performed a forensic examination of a skull in Swedenborg's coffin and drew the conclusion that, despite suspicions to the contrary, it really was Swedenborg's own skull. Yet Doctor Hultkrantz was mistaken. Swedenborg's true skull was not found until it appeared at an auction at Sotheby's in London in 1978, when it was purchased by the Swedish state for the reasonable price of £1,500 (Hultkrantz, *The Mortal Remains of Emanuel Swedenborg* (1910–1912))

---

[48] Atran; Tomasello (1999); Turner; Richardson and Steen; Tomasello (2005), 203–217; Atran and Medin; Boyd; Dutton.

These approaches are combined in a theory of cognitive science in order to arrive at an understanding of creative processes. In the historical sciences there is also a growing interest in cognitive-historical analyses, particularly in archaeology and history of science.[49] The aim of the cognitive history of science suggested here is to reconstruct scientific thinking on the basis of cognitive theories.[50] Research in cognitive history has generally dealt with the fundamental cognitive practices such as reading and counting, as well as scientific and religious perceptions.[51]

A cognitive history can achieve a deeper understanding of the creative mind, can connect the historical studies to other fields of knowledge production, and can better cover the diversity of human modes of thinking in history. Explanations based on discourse theory and social constructivism are inadequate for explaining the entire breadth of human thought. A cognitive history also considers the things and the environment surrounding man, his perceptions, emotions, and cognitive processes. Here, it is not primarily a matter of which of Swedenborg's ideas are true or false, what he wrote or did not write, from whom he acquired one idea or the other—even though that is important. It is rather a matter of trying to reach an understanding of how he arrived at his conceptions and statements about the world, or in a way how the thoughts moved in his brain. To aid him he had the human capacity to create mental images, to store knowledge, to communicate, to construct idealized models, to categorize, and to use metaphors, metonymies, mythological associations, and images.[52] Swedenborg drew analogies, made comparisons and derivations, performed mathematical and geometrical calculations, drew structural parallels, and interpreted nature's signs. In one sense his natural philosophy rested on a hope that reason is rational and logical, that nature is geometrical, and that the mind and mathematics are transcendent, that is to say, independent of us, objective, and eternally true.

What makes the thought of a different age enigmatic and difficult to understand may have to do with the way in which these mental capacities were used. People had a different spatial perception, focused on other metaphors, divided the world into other categories, drew boundaries differently, had other mental images, made other associations, and had a different pre-understanding when observing things. Approaching an alien mode of thought becomes even more complicated when one considers the communication between minds, that is, how people tried to convey thoughts with the aid of language, orally, or by reading and writing, visually with images, or with musical notes. We are dealing with a cognitive mental universe that no longer exists. The cognitive approach is a 'palaeontology of ideas', an endeavour to progress from a diachronic 'genealogy of ideas', following the threads back in

---

[49] Mithen; Renfrew, Frith and Malafouris.

[50] Nersessian (1992), 4–7, 36–38; Nersessian (1995), 194–211; Nersessian (2005); see also Lawson (1994), 481–495; Gooding; Tweney, 141–173; Carruthers, Stich and Siegal; Heintz, 391–408; Lawson (2004), 1–5; Whitehouse, 307–318.

[51] Olson; Netz (1999); Andersen, Barker and Chen; Martin and Sørensen.

[52] Lakoff, 371.

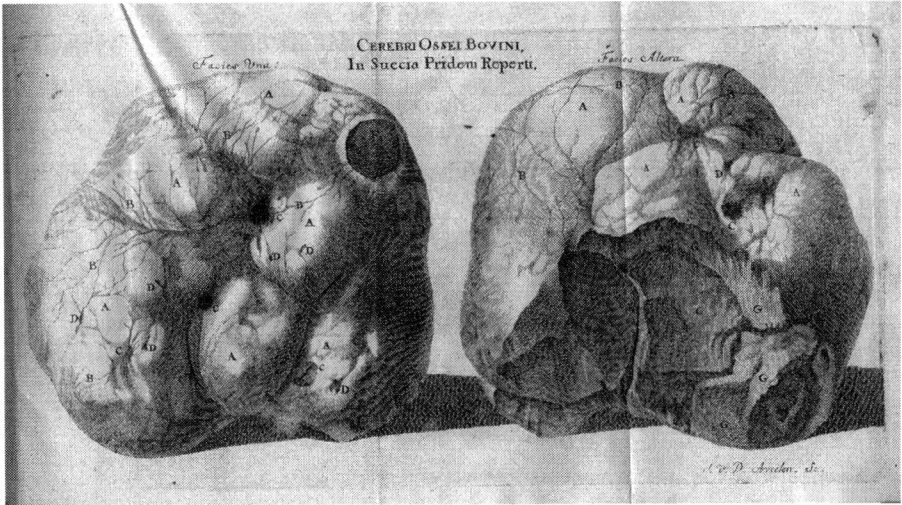

Fig. 2 Ossified and petrified bovine brain, found in Sweden some time previously, belonging to Professor Magnus von Bromell's collection. Swedenborg himself had seen this remarkable collector's item at the professor's home, and he took it as support for his theory that life consisted of tremulations. In a petrified brain there are no tremulations (Bromell, "Lithographiæ svecanæ", *Acta literaria Sveciæ* (1725))

time or down to the present day, to a more synchronic perspective that dwells on what was different in a foreign time, retrieving extinct and forgotten ideas like fossils from another world with a different way of thinking (Fig. 2). These fossils of history can reveal the scope of the human mind, and the differences can tell us what they valued, how their cognitive reality differed from ours. It is possible to think differently from us.

The attempts that have been made to get at bygone people and their world-view, on their own terms, have not infrequently invoked a kind of intuitive empathy for their emotional moods and social life. A cognitive history of ideas should require something more than this. We must imagine the actual sensory impressions, we must see, hear, feel, smell, and taste as they did, try to see things as they did, try to get close to their pre-understanding, perception, mental images, associations, categorizations, and metaphors—quite simply, we must start to think as they did. All this is of course a chimera. Total empathy with a period is rendered difficult by the fact that the historical person and the historian understand a particular situation on different terms, by living in different cognitive, or—if you will—semiotic worlds with meanings and modes of thought that do not quite agree.[53] But to understand a different consciousness, the thought of a different age, as in this case Swedenborg's, may necessitate an attempt to imagine how others *see* the world. It is a matter

---

[53]Lotman (1990), 271.

of searching for the unconscious image schemas, the metaphors, the models, the shared general perceptions of a specific culture, that is to say, what is taken for granted and not consciously reflected on. A cognitive history of ideas could provide an understanding of how other people create meaning in life, why they believe what they believe, and how concepts and experiences guide people in a particular direction in life and thought.

There are at least three assumptions about thought that a cognitive history of ideas can rest on. In cognitive science it has been ascertained, firstly, that our concepts and reason are associated with and structured by the body, the brain, and our everyday action in the world.[54] Mind is embodied, situated and distributed. Space, the environment in which we live, the registration of the senses, and the movement of the body through the physical landscape, all are significant for thought. Secondly, it has been shown that most of our thinking takes place without us being aware of it. There are unconscious cognitive processes to which the conscious mind has no access, such as memories, mental images, conclusions, and perceptions of meanings. The unconscious conceptual system structures our conscious thought. Thirdly, reason is metaphorical, that is, abstract concepts are understood in terms of concrete ones, as conceptual metaphors allow us to think about one thing with the aid of something else. Based on a knowledge of the known, we draw conclusions about the unknown. To these assumptions one can also add that a thinking person feels and is social, belonging to a culture and to history. Reason is thus also shaped by emotions, interpersonal relations and, not least of all, by the surrounding culture and its history. Man is a historical creature, bearing the imprint of his history.

Thought and reason are thus mostly embodied, unconscious, and metaphorical, but also emotionally connected. Our cognitive, conceptual system enables us to fill everyday life with meaning, giving us a kind of embodied quotidian metaphysics. For even the most abstract philosophical, scientific, and mathematical problems we use these unconscious cognitive processes into which we have no insight, mostly unaware that we are thinking in metaphors. Since people have different cognitive capabilities, this also means that a person's thoughts can display individual features; in other words, one can find personal styles in philosophical and scientific theories. Personality permeates choices and the design of theories, arguments, and texts. Individuals each have their own world since they have their own experience. There is no wholly absolute reality that is the same for all human beings. Something of a textbook example of a personally coloured natural philosophy is Swedenborg's. In this work I will give a complete account of his entire natural philosophy and furthermore illuminate his personal style based on image schemas and metaphorical thinking.

These individual cognitive abilities remain more or less the same throughout life. We are, so to speak, prisoners of our thought patterns yet simultaneously unaware

---

[54]Lakoff and Johnson (1999), 3, 7, 10; Johnson; Varela, Thompson and Rosch; Krois et al.; Calvo and Gomila.

of this captivity. It is natural to change one's ideas, to adapt them, but on the cognitive level our freedom of action is limited because most of our thoughts come 'automatically' and unconsciously, in a flow that we cannot control. Swedenborg thought in more or less the same way all through his life. This cogitative inertia in Swedenborg is tellingly captured by Bergquist, when he quotes the saying that an author always writes the same book and a painter always paints the same picture.[55] A great deal of Swedenborg's thought can be understood in terms of this cognitive 'rigidity'. Many, most plainly Lamm, have sought to discern continuity and coherence in the ideas of the early and the late works of Swedenborg. In other words, there have been endeavours to establish some consistency, which seems to be lacking.[56] But it is not possible to gloss over the gaps, the irregularities, and the differences between the ideas. The incoherence is significant. For there are obvious contradictions between the periods, and in many cases he rejected his earlier ideas outright, for example as regards geometry and natural theology. Lamm's thesis has been fruitful for my work too, but an important shift in approach must be made. I will argue that it is not primarily the *ideas* that are shared between the early and the late Swedenborg; instead the common feature can be found at a deeper cognitive level, in his *way of thinking*. The content of the ideas changed, but he went on thinking in more or less the same manner. He was aware of which ideas he abandoned or retained, but he was more or less unaware of his persistent cognitive processes.

The starting point for a cognitive history of ideas that I defend here is that philosophy, science, and mathematics do not really happen just in texts, in language, in laboratories, or in social contexts, but in brains and minds in interaction with the world around the subject, and are thus connected to the body, to perception, thoughts, and feelings. We humans are captured in our brains situated in the world, we are dependent on our thoughts and senses, our prior knowledge, our mental images, when we try to create a picture of the world. Science, in other words, is shaped by our distinctive way of reasoning, in metaphors, with aesthetic, axiological preferences and emotional factors. This bodily foundation means that 'non-scientific' and 'irrational' decisions are a part of scientific and mathematical activity, and thus the embodied-mind thesis, along with the theory of situated and distributed cognition, is difficult to reconcile with an internalistic history of science which presupposed eternal transcendent truths and rationality independent of the context. In that sense it is impossible to distinguish science from non-science. Let us now consider in depth four problems raised by my work with Swedenborg's natural philosophy, which impact on the rest of this book. These are: the significance of spatial perception for thought; metaphorical thinking; conceptual vision; and the problem of reading and writing. These problems have to do with the cognitive capacities of spatiality, metaphors, perception and distributed cognition.

---

[55]Bergquist (2000a), 22; Lakoff and Johnson (1999), 556f.
[56]Skinner, 15–19, cf. 48; Rorty, 73.

Space and Thought

A question to which Swedenborg's natural philosophy often returns is, which path leads into the unknown? How can one arrive at a knowledge of a world beyond the senses, the evanescently tiny world that dodges the most powerful microscope, and the immense, boundless space that disappears into the darkness in the telescope? Swedenborg was not a man to submit to laborious empirical and experimental work. He preferred to think about what others had collected. As a natural philosopher he tried to find, independently of experience, the true internal causal structure and properties of reality. Natural philosophy was supposed to be rational, searching for general principles, how things really are deep down, and not like natural history, searching only to acquire knowledge through experience, by producing an ordered account of knowledge and of what is found in nature. At one and the same time he was in the small world, with empirically known space around him, with towns, churches, and meadows, and also in the big world, where the unknown began, the cosmological macrocosm. Swedenborg's mastery of this unknown space, both in the cosmos and in the world of particles, beyond the limits of vision, is one of the main lines in this book. The premiss is that ideas about the unknown space always go back to experiences of the known and familiar ambient space. That is why the unknown space also often resembles the known.

Our experience of space is significant for thought in that the body is connected to, conditioned by what it walks on and moves through, what it touches, tastes, smells, sees, and breathes, as a part of a larger context. Swedenborg realized how all the senses affected his state when, on a journey through Germany around midsummer 1733, he stopped at a Catholic church. He was spellbound by the powerful effect of the Catholic mass on the senses. Castrati and eunuchs sang in clear voices, he saw beautiful people, inhaled the smell of incense, heard the play of voices. To be sure, he observed, it was very beautiful, but Catholicism had been invented precisely to charm the external, not the internal senses.[57] This implies that perception, the movements of the body, and the manipulation of objects are significant for reasoning and forming concepts.[58] Experiences of the world are a source from which consciousness can draw nourishment. The theory of situated cognition proceeds from the assumption that we also use the world around us in our thinking; cognition thus arises in interaction between the brain, the body, and the world.[59] There is no sharp line dividing the brain from the world. In other words, cognitive activity cannot be separated from the situations in which it takes place. An understanding of a historical situation, or of ideas in history, therefore cannot be geared solely to human consciousness itself, but also to the world around this consciousness.

---

[57]*Resebeskrifningar* 21 June 1733, 17; *Documents* II:1, 22, cf. 67f.

[58]Lakoff and Johnson (1999), 38; cf. Gärdenfors (2000b).

[59]A. Clark; Clark and Chalmers, 7–19; Brinck, 407–431.

The brain creates inner representations of events in the outside world, builds up an internal mental world of perceptions or interpreted sensory impressions and ideas which simultaneously become independent of the actual presence of what they concern.[60] Spatial experience develops in interaction with the surrounding world. Yet thinking does not just take place in the brain, but also in the body, in the hands, in the steps, and thoughts are placed out in the world, outside the head, in the landscape, in pictures, texts, objects, managed by means of pens, books, calendars, maps, as external memory banks and thought processors. According to this theory of distributed cognition, thinking can be said to float out into things.[61] Material culture can be described as an extension of our bodies and our thinking. It is therefore necessary to study material culture in order to understand the thinking of a period. Thinking quite simply needs the outside world if it is to function.

The dependence of human thought on the brain, the body, and physical experiences means that it cannot be transcendental in the sense that thought goes beyond how human beings happen to think. While there may be a world of thought independent of humans, we can never, for natural reasons, know it. We have cognitive limitations as a function of the body, the environment, and the long history of interaction between them. Ultimately, our concepts of the world proceed from the brain and the body in interaction with the external world. The meanings of a word exist in the head and are linked to perception.[62] The world we observe is therefore shaped by our cognitive preconditions. In other words, the world is dependent on our understanding of it, and our knowledge of it is not stable but changing. The subject adapts to the world and also changes it.[63] Swedenborg's and other scientists' theories therefore chiefly concern how they regard and understand reality, more than the external world in itself.

The significance of spatial perception for our concepts, such as spatial relations or orientation in space in relation to gravitation, is revealed in many ways in thought and language. We picture concepts as 'containers' with an inside, an outside, and a boundary between them. We perceive figure and background, part and whole, centre and periphery, straight and curved, cycles, balance, near and far, vertical and horizontal orientation, front and back. The logic of these bodily based 'image schemas' is used in language and in abstract thought, in philosophy, science, and mathematics alike.[64] Image schemas are embodied prelinguistic structures of experience that underpin conceptual metaphor mappings, that is, recurring structures that give patterns for reasoning and understanding. We talk, for example, about *higher* mathematics, we say that we are *near* a solution to the problem, that we have time *ahead of* us . . . Spatial experience is crucial for the symbols we create. Man, according to Ernst Cassirer, has proceeded from organic and perceptual spatial

---

[60]Gärdenfors (2008b), 81.

[61]Giere, 285–299; Giere and Mofatt, 1–10.

[62]Gärdenfors (1992), 95–108.

[63]Gärdenfors (2008a), 28.

[64]Lakoff and Johnson (1999), 36; Johnson, 101–138; Lakoff and Núñez, 34.

experiences to an abstract or symbolic space that has opened up new knowledge, a new direction in his cultural life.[65]

Human consciousness shapes our model of the world on the basis of constants such as the rotation of the earth (the sun's movement over the horizon), the movement of the stars, the cycles of the seasons, and the relations of the human body to the outside world. Gravitation, the vertical position of the body, has resulted in the universal human experience of the opposition of up and down. Swedenborg's teacher, the professor of mathematics Harald Vallerius, was amazed at the paradoxical phenomenon that humans always walk around on the globe at right angles to the horizon, and always on the horizontal plane, and thereby simultaneously walk uphill and downhill.[66] Other pairs of opposites are right-left, alive-dead (which stands for something in movement, warm, and breathing, as opposed to something motionless, cold, and not breathing). The Russian cultural semiotician Yuri Lotman stresses this boundary that humans draw between inside and outside, between the cosmos of culture and the chaos beyond it.[67] 'Our' space, our culture, secure, harmoniously organized, is contrasted with 'their' space, the other, dangerous, chaotic. Cultures moreover organize themselves in the form of a special space-time, with a system of coordinates, temporal divisions into past, present, and future, and spatial categories of inside, outside, and the boundary between them. Territoriality is a fundamental instinct, defining a territory and staking out borders. The differences between inner and outer, above and below, occur repeatedly in Swedenborg. The theme is infinitely varied through the history of ideas, between material reality down here and the world of ideas up there, this world and the other world, man on earth and God in heaven. 'Above' reality is the ideal city, the island, or the country, as in the Utopian geography of the Renaissance, not only Thomas More's *Utopia* but also Tommaso Campanella's *La Città del Sole*, Francis Bacon's *New Atlantis* or, for that matter, in Swedenborg's spiritual world. And mankind is somewhere between Tartarus and the Empyrean.

This exposition on space and thought is intended to underline that human life in space affects our thinking, and that spatial perception and ideas about the world are dependent on certain cognitive and mental factors. Swedenborg's perception and experience of space is thus not unimportant for an understanding of his natural philosophy. In other words, there is an interaction between the inner and the outer world. Day and night, light and darkness, gravitation, the landscape, and the compass points are a part of our thinking. As Lotman writes, 'Thought is within us, but we are within thought, just as language is something engendered by our minds and directly dependent on the mechanisms of the brain, and we are with language. [. . .] And finally the spatial image of the world is both within us and without us.'[68]

---

[65]Cassirer (1953), 65, cf. 286; Cassirer (1923), I, 146–166; an unconvincing attempt to prove similarities between Cassirer's symbol theory and Swedenborg's doctrine of correspondences can be found in Gardiner.

[66]H. Vallerius and Rimmius, 15; see also H. Vallerius and Bohm, 13.

[67]Lotman (1990), 131–133, 136.

[68]Lotman (1990), 273.

Metaphorical Thought

Metaphors ineluctably make their way into thought, as this introduction testifies, with its description of the amount of books being like the sand in the sea and of Swedenborg's life as a journey along a crooked road. A thinking being is a metaphorical thinker, Aristotle said, explaining metaphor as the application of the name of one object to another.[69] The use of metaphors in thinking—using the concrete to capture the abstract, proceeding from the known to learn the unknown— as I will show, has a highly palpable and central role in Swedenborg's thought. He constantly uses metaphors, from the known reality to the unknown, from the visible to the invisible, in both his natural philosophy and his theology. Swedenborg thought in metaphors.

In cognitive semantics, as represented by George Lakoff, Mark Johnson, and others, certain conclusions have been drawn from assumptions in cognitive science about the way humans think. One feature that has been seized on is the fact that humans think metaphorically. Our basic concepts do not function beyond our everyday experiences. To conceptualize non-everyday phenomena or abstract thoughts requires conceptual metaphors. Metaphor can then mean understanding and experiencing something with the aid of something else, or that a structure in one domain is transferred to another, from a source (the sensorimotor domain) to a target (subjective experience) which simultaneously preserves the deductive structure. Metaphors entail conceptualizing something in terms of some other thing, and function in a way as models for less well-known areas. We transfer knowledge about the known to the unknown, from the familiar to the unfamiliar, from the commonplace world, society, human life, engineering and handicraft, to the invisible particle world, to the soul and God. One could say that metaphorical thought means finding similarities between things, but also forgetting dissimilarities, being able to generalize and abstract. The creation and use of metaphors requires creativity and imagination.

Many of our fundamental concepts are organized on the basis of one or more spatial metaphors.[70] There are metaphors that transfer a structure, or proceed from a spatial orientation that arises from the action of the body in physical reality. Our experiences of physical objects give rise to ontological metaphors, that is, seeing events, emotions, ideas, and states as objects, entities, substances, or containers. They can be metaphors such as imagining life as a journey or intellectual influence as a physical force. Time can be understood spatially as something flowing along a line or in a circle. Thinking can be described in terms of movement, moving forward step by step without skipping any stages, or taking the straightest course to the conclusion without going in circles or getting away from the subject. To think is to travel. It is a walk along a path, a voyage on the sea, a journey with

---

[69] Aristotle, *Peri poiētikēs*, 21.1457b6–32.
[70] Lakoff and Johnson (1980), 14, 17, 25, 30; cf. Gärdenfors (2000a), 2, 255.

or without a goal. The researcher can get lost in the labyrinth of reality. He cannot find the narrow trail out of the jungle, he can be driven off course on the ocean of knowledge, or after much searching he may find the straight road towards the goal, 'truth'. The landscape with its settlement, habitability, shifts of light and shade, also gives conceptual patterns. Wilderness and darkness are ignorance and irrationality. Fortified castles and light represent sure knowledge and wisdom. To think is also to see. Knowledge is vision. What is unknown, difficult to comprehend, is obscure darkness. Without knowledge we grope in the dark. To acquire knowledge is to shed light on things, a knowledge that enables us to see and allows new findings to see the light of day. Knowledge brings enlightenment, we see, feel, everything is clear. What is significant and important is of greater weight or size. Similarity is understood as physical nearness, difficulties are burdens, and organizational structures are like physical structures. Swedenborg himself constantly returns to descriptions of thinking in such metaphors of space, travel, and light, and this is far from unusual in philosophy. These metaphors are used unconsciously, automatically in everyday life and arise from our quotidian experience. Language itself contains many dead metaphors based on the movements of the body, in words such as *understand, reflect, grasp, comprehend*. Without metaphors, abstract reasoning would be impossible.[71]

Metaphorical concepts have their origin not just in our physical but also in our cultural experience. The more layers of metaphors we employ, the more abstract and culturally specific the concept becomes.[72] Some metaphors proceed from some special cultural knowledge, for example metaphors based on Euclidean geometry. People who live in cultures with no knowledge of Euclidean geometry would not understand such metaphors. Euclidean geometry gives the world a specific visual metaphorical structure, a world of relations between points, lines, and circles. In many cases, then, scientific theories and concepts about the world are founded on spatial metaphors with a physical and cultural origin. Philosophers and natural scientists use the same conceptual system as ordinary people in their own culture. In philosophical theories they incorporate the concepts available in the historical context and the general theories, models, and metaphors that are common and typical in the culture to which they belong, but they also rework these basic concepts, see new links, and draw new conclusions. It is the shared concepts and ideas that make a specific philosophical theory comprehensible to people within a particular culture. Philosophical theories can be interpreted as attempts to refine, expand, clarify, and make consistent certain common metaphors and 'popular' or 'general' theories shared by people in a culture. What a particular philosophical theory also does is to select the 'right' metaphors. Differences between philosophical views thus depend on different choices of metaphors. Each philosopher's metaphysics has its origin in what he takes as central metaphors. A 'world-view' can therefore be regarded as a consistent constellation of concepts, especially metaphorical concepts, over one or

---

[71]Lakoff and Johnson (1999), 59; Lakoff and Núñez, 41.

[72]Danesi, 73f, 78.

more conceptual domains.[73] The world-view is the reality for the people of its time. The task is thus to find some of Swedenborg's central metaphors, with which he sought to create a consistent natural philosophy.

In philosophical analysis and scientific theory formation, then, metaphors play an important part. Philosophical and scientific texts are more or less strewn with metaphors, analogies, metonymies, similes, and comparisons. In the history of science they have often been dismissed as unscientific and uninteresting adornment.[74] They have mostly been regarded as poetic whims, educational and rhetorical devices, or simply as superfluous linguistic expressions that obscure the view of the true logical structure of the scientific arguments, the purely rational scientific and mathematical. Against this I claim that metaphors, the linguistic form, the tropes that modify the basic meaning of a word, are of crucial importance. They are not mere external ornament, but a major part of creative thought by establishing visual analogies and abstract ideas. For this reason they also provide valuable clues to how scientists think. Scientific reasoning uses metaphors to a great extent as conceptual tools or as theoretical models of the external world. Structural metaphors and process metaphors are particularly common in scientific reasoning, metaphors that try to get away from the emotional and subjective. In science one must form new concepts for the new phenomena one is describing, and this is often done with the aid of metaphors related to what is already known. On the basis of his knowledge about water waves, Swedenborg was able to picture sound waves and light waves; from his experience of peas and cannonballs, Polhem was able to visualize the structure of invisible particles of matter.

In Swedenborg's times there was a keen interest in metaphors. We find a baroque style characterized by a fondness for tropes in which certain semantic units are replaced by others. Throughout the baroque, metaphors and allegorical expressions were popular in art, literature, architecture, on coins, in music and science. Even in scholarly dissertations, metaphors were employed, such as the quest for knowledge and reason as a journey or a wandering, the disputation and the debate as a fencing duel or a struggle on a battlefield, the truth that is brought to light, as when a gardener removes the bark from a tree to arrive at its heart, or studies and writing as a textile craft that weaves knowledge together.[75] But they used similes, analogies, metaphors as something more than a stylistic, aesthetic decoration. These devices could reveal the secrets of creation. Emanuele Tesauro, one of the foremost theorists of literature during the baroque, explained quite simply that 'God wrote the book of nature in metaphors, and so it should be read.'[76] The philosopher of history Giambattista Vico, in his *Scienza nuova* (1725), proceeded from four basic tropes, in which man develops from thinking in anthropomorphic fables or metaphors,

---

[73]Lakoff and Johnson (1999), 338–341, 511.

[74]There are of course exceptions, see Crombie, part IV; Spranzi, 451–483.

[75]Sellberg (2002), 104f; Örneholm, 78–81.

[76]Mazzeo, 54; Helander (1988), 30.

via metonymy and synecdoche to reflective thinking or irony.[77] Metaphor for him was a capacity in the human mind to connect things and events in the world, a way to think about unknown things. In Swedenborg this interest in metaphors is displayed in an uncommonly clear way. His Latin poems, as Helander has shown, are model expressions of a poetic, allegorical baroque world. This baroque metaphorical thought, I would add, can also be expanded to apply to Swedenborg's entire natural philosophy, which in many respects is baroque metaphor taken to its extreme. The poet and the scientist are combined in metaphorical thinking about the world.

Besides metaphorical thought there are other palpable points of contact between the poetry, art, and music of the baroque and the science and mathematics of the age. Firstly, all creative activities are part of the history of human creativity. There is an affinity between the creativity of the mathematician and that of the poet, as Tesauro once declared.[78] In language, art, religion, and science there is an aspiration to build up an ideal world of one's own. Ideas are governed by wishes, a will or desire as to what the world should be like, an effort to arrange the world as it ought to be. This creative imagination can also be found in Swedenborg's natural philosophy, an innovative mode of thought in which the theories, as in all other researchers, are influenced by wishful thinking. Theories are the self-fulfilment of wishes. 'For what a man had rather were true he more readily believes,' wrote Francis Bacon.[79] Secondly, art and science can be said to be united in a 'baroque style', not necessarily understood as a term for an era, but rather as what Ludwik Fleck calls a 'thought style', or what the historian of ideas Gunnar Eriksson has denoted 'baroque science'.[80] The thought style has a specific direction, is dominated by certain aesthetic ideals, and involves a hierarchy between authors. In the specific baroque style, both in the arts and in natural philosophy, people sought for effect, contrasts, and rich symbolism. The baroque gave expression to vigour and passion, a magnificent metaphysics with huge pretensions, monumental buildings, grandiose trumpet fanfares, bulging female bodies, but also personal thoughts, delicate arias, and silent, introverted still lifes. A baroque work of art was supposed to be a universal artwork with internal coherence. In the baroque style there is an aspiration for completeness and variation, an ambition to see things from different perspectives and levels, parallels and antitheses. It is a world composed of correspondences, a structure that always refers to something else. In science and philosophy, the baroque style is manifested in an emphasis on the order and interconnection of reality, and a desire for theories that express simplicity, symmetry, harmony, and 'pure', abstract geometrical forms. Everything is expected to have a meaning, an ultimate purpose, and an internal structure.

---

[77] Vico (1744); Danesi, 62; Marshall.

[78] Jonsson (1983a), 88.

[79] Bacon (1620), § 49.

[80] Fleck, translation 99; Eriksson (1994), 149–162; Choluj and Joerden.

The scientific style underwent some radical changes before Swedenborg entered the stage. There was a switch from anthropomorphic to mechanical metaphors, and a dynamic mobility was introduced as a contrast to the static monumentality of the Renaissance. 'The world machine' became the central metaphor in the natural philosophy of the seventeenth and eighteenth centuries, a basic metaphor that generated new metaphors, particularly spatial, visual, and orientational ones: *space is geometry*, *matter is geometry*, and so on. The world became a machine, a world geometry in constant motion. Nature and reason were mechanized. The concepts and metaphors of geometry and mechanics steered and structured reason in the sense that reason could only be extended in one particular direction but not in others. People focused on geometrical forms but avoided what was inconsistent with the geometrical metaphor. The world was ordered, understood, depicted, and discussed in geometrical and mechanical terms. This metaphorics was not always conscious; it was an inherent part of the way of thinking about nature during the mechanistic period. Virtually all branches of human thought were affected by the metaphor of the world machine.

Machines and the machine metaphor gave people an illusion of power over nature and work. This stood for control, order, and regularity. The world machine worked according to laws and rules, transmitted forces in a continuous causality, but could be controlled and manipulated, dismantled into small parts, and abstracted from its context. The mechanistic philosophy involved a new theory of matter with geometrical particles, a new theory of cause and effect through physical contact, and a new method based on mechanical analogies. The mechanistic world-view brought a new perception of what was real, a reality defined by geometrical particles of matter in motion. Its metaphysical goal was to find the mechanics behind reality, the purely mechanical world underlying experience. With mechanical metaphors, Swedenborg and others were able to describe nature's phenomena and processes, to arrive at the rules of mechanical method in order to gain new knowledge, and to distance themselves from personal interests and emotions. It was this mechanistic, geometrical world-view that he later revolted against when he embarked on the description of a non-material spiritual world liberated from earthly geometry.

Alongside mechanics and dynamics, we notice in this age the suggestive force of infinity. The telescope and the microscope opened a perspective of infinity out into the universe and into matter. As Oswald Spengler wrote: 'The same inspired ordering of an infinite world which manifested itself in the geometrical analysis and projective geometry of the seventeenth century, could vivify, energize, and suffuse contemporary music with the harmony that it developed out of the art of thoroughbass, (which is the geometry of the sound-world) and contemporary painting with the principle of perspective (the felt geometry of the space-world that only the West knows).'[81] In Swedenborg we find all this, the magnificent universal work of art, the variations, the parallels, the correspondences, the order, the harmonics, the mechanical metaphors, the dynamics and the perspective of infinity.

---

[81]Spengler, 89; translation, 61.

In some sense he can be called a thinker of the high or late baroque with a style not unlike that of the inventive minds of Polhem, or the Olof Rudbeck the Elder with his Atlantis myths, or Carl Linnaeus and his systematics of the whole of nature. They are united by their Lutheran Protestantism, classical education, grandiose claims, and metaphorical thought. In this section I emphasized that man has an inherent cognitive ability to think in metaphors, which Swedenborg also used in large measure in his thinking, and this took place at a time with specific metaphors and with a general openness to the metaphorics of reality.

Seeing with the Inner Eye

Swedenborg—literally—saw the world differently. The nature of seeing can to some extent give us the keys to Swedenborg's special mechanistic, geometrical world-view. For there is a conceptual vision which indicates that he actually had other sensory experiences than we would have when faced with the same object, or more correctly, which meant that he made other interpretations of what he saw. In his day there was also a new way of seeing, a powerful visual culture, in which non-verbal thought was strengthened, as reading aloud and storytelling gave way to silent reading of books, thus entailing a shift from sound to visual impressions, and that the linear perspective transformed vision from a subjective to a objective, geometrical perception of the world. The Cartesian concept of space brought a new background knowledge through which Swedenborg was able to interpret what he saw. He and other mechanists had a special way of seeing, a geometrical vision.

With new optics, such as the microscope and telescope, the role of sight was reinforced as a fundamental element in the description of reality.[82] The world had a visible geometrical structure. People had to look around for themselves, not just listen to authorities. Philosophy was not to have auditors but *spectators*, said Professor Johann Christopher Sturm.[83] This ocularcentrism appears plainly in Swedenborg's natural philosophy. In London in 1710 he acquired, among other things, a microscope.[84] Swedenborg may possibly have owned a single microscope of Leeuwenhoek's type with 42-fold magnification. He also had another microscope with accessories, according to his own list of possessions from 1770. Yet he did not perform so many systematic studies with it. At any rate he could smell a powerful odour of urine when he directed the sun's rays through his magnifying glass on to a sample of ordinary water.[85] But what did he see in the microscope? When he drew churches with the aid of a camera obscura, looked at stars in his telescope, or gazed at the plates in anatomical books, he saw something that we no longer see. In purely

---

[82] Wilson; Ratcliff; cf. S. Clark.

[83] Ornstein, 175.

[84] Swedenborg to Benzelius, London, October 1710. *Opera* I, 207; Spaak and Althin, 44, 49.

[85] *Prodromus principiorum*, ed. Opera III, 23, 95; translation, 27, 120.

optical terms we would see similar things, but we would simultaneously see quite different things as well. Vision and perception are not a neutral, objective, faithful registration of reality. There is a conceptual or epistemic vision which means an identification of what we see, and this takes place when we apply our concepts to the visual impression. The concepts affect what we see, and if we lack concepts for a phenomenon, we do not see it. The world is distorted by our concepts, and the concepts are distorted by the world. The world we see around us is in fact a world that is reshaped by our concepts. The world is seen with the inner eye.

Seeing is an activity that aims to create order in the chaos of the senses; it is also conditioned by the observer's emotions and associations. Language, art, myths, and science are endeavours to master existence, to find an order in the world, which is thereby structured by human thoughts and feelings. They are not immediate representations of things but expressions of human ideas about things. Science is therefore in large measure a matter of an orientation in and ordering of the experiences of the human being, not of nature itself. In other words, there are no theory-neutral observations; what we see requires interpretation based on previous experiences, concepts, and prior knowledge. Using the theories we already have, we distinguish forms and patterns in nature. Visual perception is a process through which a person who perceives something goes beyond the given information by organizing and interpreting the visual impressions, by stating the configuration more exactly, adding to and filling out an ambiguous image in order to create an unambiguous perception. The observer is forced to divide sensory impressions into their constituent parts and organize them in terms of figure, background, and foreground. This interpretation of sensory impressions is not determined by the impressions themselves but by the mind.[86]

Perception of space is influenced by topological and spatial factors such as inclusion, proximity, and so on. It is also selective, that is to say, attention is concentrated on certain features, and it is organized. In the same way as the observer's visual experience of an object changes according to his or her position in space, every sensory impression is changed by the concepts, knowledge, wishes, needs, values, or interests of the perceiving person. Swedenborg's prior knowledge, his cultural background, life history, and placing in time and space constitute the perspective from which he observed the world. This perspective, with its special way of discerning and organizing, determined what he perceived and how he perceived it. That is why we are often incapable of seeing what he and others saw in the scientific illustrations, in the microscope, on the firmament, although we have largely identical sensory capacities. Culture and cognition impose pattern on visual impressions.

An occurrence which changed the history of vision, and which I argue was significant for the 'geometrization' of reality, so that the world could be described with exact geometry, was the rise of perspective. The monocular linear perspective with a focal point on the horizon, which had been developed during the Italian Renaissance, differed from subjective experience and binocular, mobile, spherical

---

[86]Reisberg, 57f, 91f, 352.

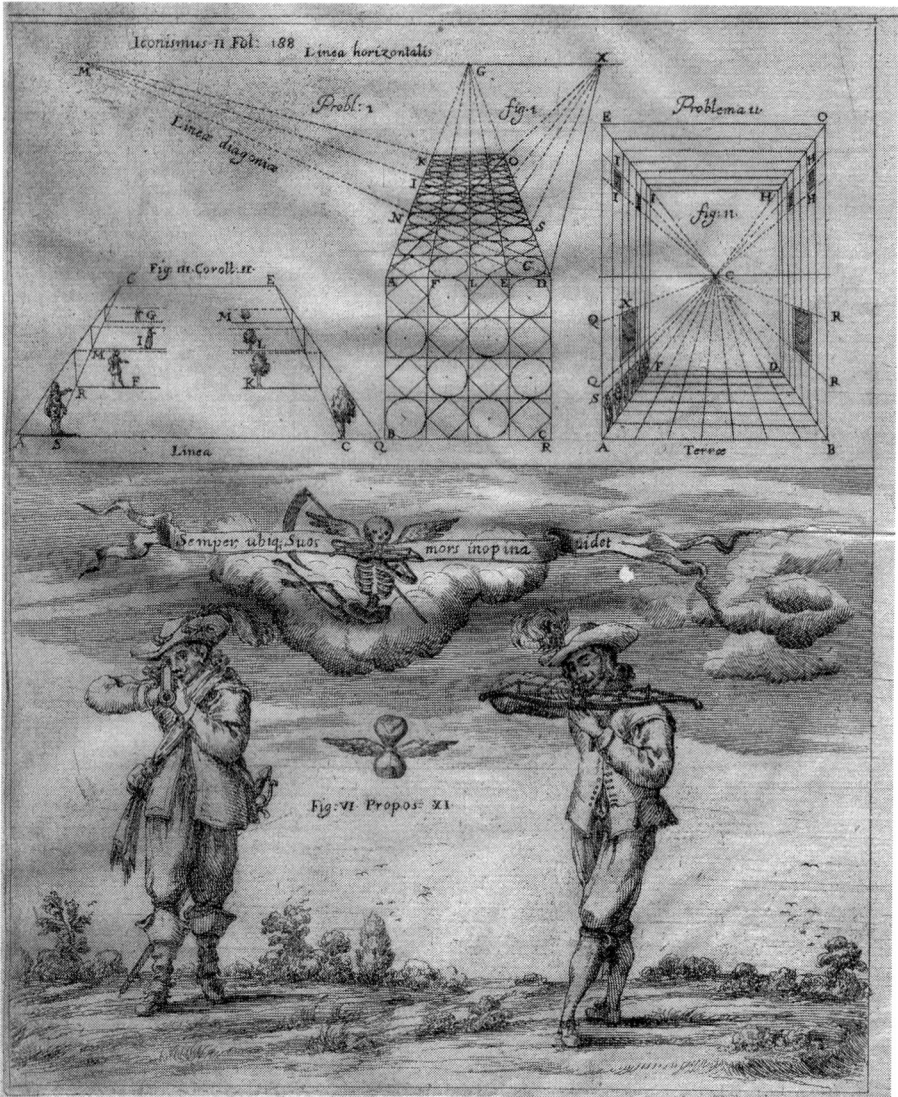

Fig. 3 Unexpected death sees its victims always and everywhere. The crossbowman and the musketeer point their weapons at the observer in the description by the German Jesuit Athanasius Kircher of the dizzying depth effect of linear perspective (Kircher, *Ars magna lucis et vmbræ* (1646))

vision. Sight was coerced into straight lines, and objects, space, and the world became geometrical (Fig. 3). With a ruler and compasses it became possible to draw the world as it is and in agreement with geometry. With straight lines, one could not see round the corner, as in medieval art. From the more 'realistic' depiction of earlier art, which was a more faithful rendering of human cognitive, mental images with their value perspectives—whereby the important things stood out as being biggest and objects were portrayed from their most characteristic side—perspective was used to present that was supposed to be a more objective picture and the same for all observers. The world ended up looking different as a result of the knowledge of perspective. It entailed a mathematical rationalization and abstraction of the psychophysiological perception of space, which led to an experience of space with an infinite extent, free of ambiguity and contradictions.[87]

The theory of perspective had mathematical and scientific consequences. Euclidean geometry did not fully agree with visual geometry. Two parallel lines, or 'track lines' like the traces left by cart wheels on a muddy road, which was Swedenborg's Swedish counterpart, meet at the horizon.[88] Girard Desargue's projective geometry, with a point in infinity where the parallel lines converge, agreed better with our visual perception of reality, a more visual geometry than the tactile Euclidean geometry. Linear perspective and graphics reinforced non-verbal reasoning in science. With perspective one could more efficiently convey one's non-verbal thought to another mind. Architects, engineers, anatomists, and other scientists gained an opportunity to demonstrate what could not be described exactly enough in words. With perspective drawings of landscapes, sections of the human body, and cross-sections of mines, reality was exposed. Interest in the surrounding world, in landscape and still life, increased during the seventeenth century. Topographical renderings, as in Erik Dahlbergh's *Suecia antiqua et hodierna* (1716), became perspectival panoramas from a given fixed point, where the angle of vision became lower, more natural, and the foreground was lined by trees and human onlookers leading the observer's gaze into the depth of the picture, what is known as repoussoir, instead of the steep bird's-eye view and the inconsistent representation of space. Fortification officers learned how to draw landscape, terrain, and bastions according to the rules of perspective, and surveyors were expected to produce correct depictions of fields and towns. The artist became a geometrician. Thanks to perspective, it was assumed, scientific illustrations became more like nature, more lucid and graphic. In fact, however, one could claim that the illustrations, as in Swedenborg's works, in many ways tell us more about the surrounding culture, its perceptions and special vision, than about nature itself. The pictures are not really true to life; they are mental artefacts, an attempt to transfer mental images to paper, they are adapted to a theory and reflect a vision dependent on knowledge. All pictures are interpretations of reality based on experiences, knowledge, and expectations. The pictures in Swedenborg's works are thus not just illustrations but also carry meanings and provide clues to his visualization and mental images.

---

[87]Panofsky, translation, 67–72; Kemp, 41; Edgerton (1980), 179–213; Edgerton (2009); perspective artists are mentioned by Swedenborg, *Resebeskrifningar* 13 July 1733, 28; *Documents* II:1, 35f.

[88]*Regel-konsten*, 4f.

What people saw was not unambiguous. The optical instruments of the day suffered from chromatic and spherical aberration which distorted colours and blurred the focus. Indeterminate images forced the observer to interpret what he had seen, which called for concepts and theories. Harald Vallerius seems to have found himself in that situation. The sensory experience, he wrote in a dissertation on the deceptive property of sight, can lead the intellect to make mistakes about the number, figure, and movement of objects.[89] The laws of perspective cause a circle to appear oval when viewed from the side, a square becomes a rhombus, and with the naked eye, he continues, we cannot distinguish details from a distance because of tremulations in the air, as when one looks at a stone in a river through the waves of the water flowing past. Swedenborg's point of departure was, like that of Vallerius, a mechanistic world-view, a pre-understanding of the world as mechanical and geometrical. He projected the concepts and theories of mechanics on to the world, thus confirming his own mechanistic ideas. He sought and found what he presupposed was there: describable geometrical forms and mechanics. The concepts of vision are shaped by reason, for no one has seen an exactly straight line or a perfect circle. In some sense the mechanists created the world they were out to explain, the geometrical world. By formulating an idea of a rational, geometrical world, they were able to handle what they saw around them. Geometry had the function of organizing visual impressions and memory, like a geometrical net of consciousness through which reality must pass. The mechanistic philosopher interpreted reality in terms of the mechanical and geometrical paradigm, and thereby really did *see* small machines in his microscope, and geometrical structures that others would not have noticed. It was with this geometrical vision that Harald Vallerius saw when, in a dissertation about the meaning of parallel lines in the mathematical sciences, he found parallels in trees, forests, salts, in the cubic, hexahedral, octahedral, dodecahedral forms of the emerald, in the concentric circles of heaven, and in the strings of musical instruments.[90]

There was thus a willingness to see things as geometry, to see with other eyes, to interpret them differently, a preparedness to see something in agreement with their world-view and established thought style. Swedenborg would thus literally see something different in the microscope from what we would see in the same situation. Through his previous knowledge, through his concepts and theories, he interpreted what he saw differently. He would concentrate on other things, interpret in other forms, find other associations, make other evaluations, and receive other mental images. The historical challenge is to try to understand and see reality in the way people did back then, to see the same things, to attempt to interpret and evaluate in similar ways, to see through the microscope in the same manner. What Swedenborg saw in the microscope was a geometrical and mechanical world.

---

[89] H. Vallerius and Rosell, 18, 20; cf. H. Vallerius and Schultin; H. Vallerius and Bredenberg.
[90] H. Vallerius and Swebilius, § XII.

Thinking with Books

The history of reading and writing is a part of the history of distributed cognition. Swedenborg thought with the aid of books and pens. Reading gave him associations, clues, required him to make interpretations, and gave him ideas and matter for continued thinking. In his writing he was forced to make his thoughts concrete in order to find the right words for his mental images. Communication in that sense can be described as an attempt to transfer mental images from one mind to another. A key to Swedenborg's reasoning, how he arrived at his ideas, is the manner in which he communicated, read, and wrote. It is thus in large measure a question of how he interpreted texts, tried to read a meaning into them that he could use in his own thinking. In one way it could be said that the whole of Swedenborg's intellectual activity was about interpreting and creating meaning, about hermeneutic grappling with texts. His doctoral dissertation was about the meanings of the maxims and 'the wisdom of the mimeograph'. Here he is talking about the metaphorical meanings of the concrete words, as in the explanation of a maxim where he demonstrates the rich occurrence of the comparison between a bow and the mind.[91] Both can break. As the bow breaks when it is bent, so can the mind break when loosed. Exercise and meditation are like food to the mind, and if it is not nourished and sustained, it will fall. He also used spatial thinking. Compatibility makes things stronger, while incompatibility makes them weaker. Swedenborg read mineralogical and medical works in search of material for a theoretical synthesis. He made excerpts and drew his own conclusions from them. The dream crisis was an anxious quest for meaning in the chaotic dreams that were so difficult to interpret. He did not find a sure method for interpreting the dreams until he evolved the doctrine of correspondences and his teachings on the realms of spirits and angels. The dreams of the dream crisis and the spiritual world do not differ primarily in their content, but in the method of interpretation. With his journeys to the spiritual world he acquired a narrative framework in which he could understand his visions, and with the doctrine of correspondences he found the key to the interpretation of the Bible's message. In the spiritual doctrine and the doctrine of correspondences he used a reading technique that had its origin in the reading of books, in the work with texts, particularly those on science and mathematics.

Swedenborg was faced with an incalculable amount of books, like the sand in the sea, and this huge quantity of information forced his cognitive self to assume personal responsibility for the organization of all the knowledge. The creation of order could not be left to authorities; it also called for a personal struggle with the texts. Reading for him became a form of self-formation or self-construction. Considering Swedenborg as a reader therefore involves not just looking at *what* he read, but also *how* he used and read other texts. An important first step for understanding his thought is to establish which books and sources he actually may

---

[91]*Selectæ sententiæ*, 24, 28; translation, 17f, 21.

have used, which ones he read or knew of. The next step should be to try to discover his special reading technique, what he was searching for and how he related to the texts. It is a question of how he interpreted, found semantic connections and meaning in them from his point of view and his cultural tradition.

One starting point is that the books, the writings, are something more than just a kind of materialized thoughts. They are not static but possess a semantic mobility, in that their meaning is modified by the outlook of the reader. They are dynamic, changed by new contexts. Moreover, it is not at all certain that the author's message is received intact by the reader, since the two have different codes, that is, different rules for the conveyance of linguistic meanings, based on different linguistic experiences, memories, norms, contexts, cultural traditions, and distinctive individual features.[92] The codes can overlap but are never identical. This applies, of course, not just to Swedenborg as a reader, but also to a modern-day historian. The code or the codes used by the creator of a text must in some sense be reconstructed and then correlated with the researcher's codes. A reader subjects the original material, the text, to cognitive and semiotic manipulations. Reading includes interpretation, perception, and background knowledge. The reader gets certain associations, establishes personal links, receives personal mental images. Reality is therefore deformed not just by the author of the source, but also by the reader, the exegete, and the historian. The reception displaces and distorts, always creating something new.[93]

When Swedenborg read a particular text, it was transformed through his personal comprehension and his special interpretation. It was read in terms of his own norms—aesthetic, religious, scientific—and those of his time, his cognitive ability, metaphors, and mental images. Swedenborg was most things, but not a historian of ideas who tried to be faithful to the original text; instead he had a specific purpose and a special use for the text. He took what he needed for his own thought. What the author actually meant was more or less irrelevant. This also explains why he so often cited his sources incorrectly.[94] He had a reason for this. Correct citation was not an end in itself. The excerpts were rather material on which to build further. When he made his interpretation of the formulations in a book, the ideas no longer belonged to the original author; they now became Swedenborg's own ideas and a part of a new context. The original ideas and Swedenborg's may be very close, but they are not identical. One could thus say that Swedenborg did not read Wolff's Wolff but Swedenborg's Wolff.

The books in his library were not the true source of his ideas but are better regarded as tools for his own thinking. To overstate somewhat, one could say that Swedenborg was not influenced by anyone, if by intellectual influence one means a kind of causal connection, whereby thoughts 'cause' another person's thought, as a

---

[92]Lotman (1990), 13, 218; Lotman (1967); Portis-Winner, 35f.

[93]Chartier, 8f, 21f; Lotman (1996), 64.

[94]Swedenborg's inaccurate quotation technique has been demonstrated by Jonsson (1961), translation, 113.

ball strikes another and changes its direction. The history of reception in that sense is a metaphor based on spatial causality, a genealogy of series of causes and effects: The impact of an idea is a physical force that moves us from one intellectual space to another. Similarities between two thinkers thus seem to presuppose either that one has influenced the other or that they are both influenced by a common source. But I emphasize that there is always an element of interpretation and that there are basic cognitive abilities which entail that ideas cannot be transferred wholly intact. Similarities between Swedenborg's doctrine of correspondences and, for example, Kabbalah, Neoplatonism, or Wolffianism, need not necessarily entail influence or origin in common concrete sources, written or otherwise. In many cases, a cognitive history of ideas can instead explain that the similarities proceed from very similar cognitive premises, that they spring from analogous mental images, or use similar metaphors and categorical divisions quite independently of each other. Not every idea comes from someone else, not all ideas can be traced back to another thinker. The ideas come from minds, from brains. Swedenborg's ideas do not come direct from books and authorities, but from his own human cerebrum in its encounter with the world.

With language one can express things that do not exist, that are not present in time and place. It is a matter of one's own ideas, the internal world and not really about the external world. At the same time, there is a gap between language and cognition. The author cannot express his ideas, cognitive processes, or mental images exactly in words. The text is not capable of directly rendering an author's intention. The documents, the texts, are incomplete, containing lacunae of implicit knowledge and unconscious presuppositions. There are 'non-facts' that were not recorded because they were considered to lack significance. In every culture and genre, in every author, there are selections, conscious or unconscious, of facts that are regarded as significant. In the creation of the work, the author is also subject to intellectual, aesthetic, ethical, religious, and other norms. The work is spatially dependent and part of a chronocentric context. As an author, Swedenborg wanted to say something, aimed at meaning something, and had the intention to express something true that defies time. His writings responded to something and addressed something, contained a meaning, which was intended to be received and was received in a particular way. But his writing was not just about formulating his thoughts in words, putting across a message; not least of all it was a solipsistic act, a personal mental struggle, an articulated introspection through which the inner self was placed in relation to the outside world.[95] When he wrote about the spiritual world, he turned inwards and put his own role and his own action in relation to the world.

An understanding of Swedenborg's thought can be approached by considering how he read, interpreted, and manipulated texts. The books he read were in some sense unfinished and incomplete when he as a reader became an active part in their continuation. The books invited countless readings. Swedenborg carried on writing some of them. In actual fact, the difference between reading and writing is not as

---

[95]Ong, 101f.

great as one might expect; they are interwoven. There is a constant interplay between silent reading and writing. Swedenborg read with a pen in his hand and wrote with a book in front of him.

Overview—The World Machine Seen from Above

This *Introduction* has given a survey of Swedenborg's life and an orientation in the literature about him. It has shown that there is no detailed study of his early mechanistic natural philosophy in the context of intellectual history. The present book can be described as a cognitive history of ideas, examining how Swedenborg thought. In research into his natural philosophy, some theoretical problems arise concerning the concept of space, metaphors, vision, and reading. The rest of the book is thematically structured, with a certain gradual chronological development. Each chapter concentrates on an image schema, a spatial or geometrical figure typical of his natural philosophy, and special disciplines and works of Swedenborg. The idea is also to follow Swedenborg's metaphorics from the known and visible to the unknown and invisible world. If one could for a moment observe *The World Machine* from above, one could see this:

The *Space* is an introductory chapter about the concept of space, about Swedenborg's experiences of space gained through optical instruments and his orientation in the intellectual milieu and in the spatial world with the aid of geometry. The focus here is on Swedenborg's connection with the learned society Collegium Curiosorum, his suggested solution to the problem of longitude, and his manuscripts on geometry dating from the 1710s. The basic standpoint is that human orientation and daily actions in space, together with vision, perception, and experiences of bodily movements, are significant for human reasoning. Swedenborg's natural philosophy is characterized by a geometrization of reality based on the widespread belief that geometry is an ideal method for achieving certainty, and that the ideal objectivity belongs to a transcendent reality.

The Sign concerns the interest that Swedenborg and his contemporaries had in signs, that is, things that stand for something else, that refer to something else. The chapter treats especially some particular sign systems, such as those of arithmetic and algebra, and the division of the world into categories, as in numeral systems, coinage, measures, weights, and volumes. There is an examination of Swedenborg's discussion with the Swedish king, Charles XII, during the years 1716–1718 about numeral systems based on 8 or 64 as an alternative to the traditional decimal system. The signs of mathematics can be used as metaphors to describe reasoning and the structure of the world. This arithmetical study shows that not even mathematics is free from political and rhetorical considerations.

The Wave follows Swedenborg's use of the metaphor of the wave, not only in such scientific disciplines as hydrology, acoustics, optics, and neurology but also in poetry and music. On the basis of his everyday experiences of water waves, Swedenborg was able to use this metaphor to transfer the qualities of these waves to other physical phenomena such as sound waves and light waves. In Swedenborg's time there was an interest in baroque music and the relation of sounds to mathematics. Polhem suggested a number of sound experiments intended to be carried out in the mountains of Lapland, and Swedenborg published these in *Dædalus Hyperboreus*. The two men also debated the mechanical nature of fire and colours. Swedenborg's most original idea was put forward in a manuscript of 1720 about tremulations. He maintained that life consists of waves or tremors of the nerves. The body is like a musical instrument. He was a typical proponent of iatromechanics, discovering circles and waves in the musical body.

The Sphere deals with the sphere as the figure of movement, with the relationship between technology and science, and with analogies, proportions and mental models as important tools for inventing scientific theories, especially in mechanics, physics, and chemistry. War, engineering, and mining gave Swedenborg ideas and metaphors for research into matter. In his early theory of matter, put forward in *Prodromus principiorum rerum naturalium* (1721) and *Miscellanea observata* (1722), he declared, in agreement with Polhem, that the sphere, as in the round form of peas, cannonballs, and bubbles, can give clues to the structure of the invisible world of particles. In one of his first manuscripts about mining he describes effluvia or metallic vapours rising from the rock. His mining studies, such as *De ferro* and *De cupro* from 1734, likewise display metaphors and a geometrical natural philosophy. Behind this analogical reasoning there is an assumption that 'micro-mechanics is macro-mechanics'. The experience of artificial machines made by humans could be transferred to the invisible microcosmic world of particles.

The Point proceeds from the indivisible point of mathematics. In Swedenborg's *Principia rerum naturalium* (1734), mathematical points are given an ontological significance. The world appears when God, like an artist drawing with his pencil, gives motion to the point. The world consists of circulating points. With spider metaphors Swedenborg postulated that the world is built on mathematics, and with labyrinth metaphors he formulated the philosophers' feeling of disorientation in the chaos of nature. Behind this is a conception of the creation of the world as an exercise in practical geometry, as when one draws figures on a sheet of paper with the aid of mathematical points. The world is geometry. A comparison of an early draft of the *Principia* with the printed version shows that in the intervening years he had adopted Wolffian terminology in his description of nature.

The Spiral is the geometrical figure Swedenborg admired most. He wrote about the windings of the spirals in geometry, particle physics, astronomy, and in the nature of the soul. In his mathematical writings he treated the geometry of the logarithmic spiral, and in several astronomical writings he described the eternal spring of the world caused by the spiral movement of the earth. In the *Principia*, particles and planets circulate in perfect spirals. He describes the magnetic force in mechanical terms, as effluvia of particles. Finally, he also made a sketch of a membrane between body and soul in the form of a spiral. Through all this there is a micro- and macrocosmic perspective, in which the world of particles is conceptualized as a small solar system. The experience of water and air whirls is developed into an abstract world of solar vortices and points in spiral motion.

The Infinite is limited to Swedenborg's last mechanistic work, *De infinito* (1734), where he made a strict distinction between the finite and the infinite. Infinity is God in contrast to the finiteness of man and the material world. He puts forward a number of proofs for the existence of God and describes the soul as a machine. His thoughts on infinity give a picture of his metaphysics, of the boundaries of the human mind, and of how he tried to connect science and theology, man and God, reason and revelation, before he turned to organic metaphors.

The book ends with a *Conclusion* surveying the major themes of Swedenborg's anatomical and physiological studies and his spiritual writings from the end of the 1730s to 1772. The survey covers the geometry in his neurological writings, the hierarchies of forms, the agony of the dream crisis, the geometry of the spiritual world, and the metaphorics of the doctrine of correspondences. The analysis shows that a cognitive study of his natural philosophy also sheds light on the late Swedenborg's thought. In his theological system of thought he advanced the ability to think in metaphors further than anyone else has done.

The Space

> If it were possible by some messenger, I would beg to receive the *Camera obscura* which
> had a blue *cylindre* as a covering around it. It lies on the stone ledge in the vault near the
> cupboard.[1]
>
> Swedenborg to Benzelius, 4 March 1716

A Blue Camera Obscura

It was still early in the spring, one day in March 1716, and perhaps time was passing
slowly in the countryside outside Skara. From the episcopal residence of Brunsbo,
Swedenborg wrote to ask Benzelius to send him his camera obscura in the blue
case. With the camera he intended to 'make reflexions on the *perspective* art by
the taking of a number of *vuer* [views] and *prospecter*.' He explained to Benzelius
where the camera could be found by describing the spatial orientation. It is *on* the
stone ledge *in* the vault *near* the cupboard. Benzelius found it. A month later the
camera arrived at Brunsbo, brought by Swedenborg's fellow student from his days
at Uppsala, Master Olof Nordborg, who was in Sweden to collect money for the
Swedish congregation in London. The camera would be a source of pleasure. As
Swedenborg wrote on 12 June, 'I have already learned the drawing of *perspective*,
to my pleasure. I have *exercitium* [practice] from churches, houses, etc.; were I up at
the works in Fhalun or elsewhere, I would draw them as well as any one, *ope hujus
instrumenti* [by the help of this instrument].'[2] One may wonder what the mine in
Falun would look like in it. The vertiginous depths of the mine shaft, all the parts of
the mine machinery would come in the proper relationships to each other, and with

---

[1]Swedenborg to Benzelius, Brunsbo, 4 March 1716. *Opera* I, 242; *Letters* I, 91.
[2]Swedenborg to Benzelius, Brunsbo, 12 June 1716. *Opera* I, 253, cf. 248; *Letters* I, 107; cf. Hyde,
no. 67.

D. Dunér, *The Natural Philosophy of Emanuel Swedenborg*, Studies in the History
of Philosophy of Mind 11, DOI 10.1007/978-94-007-4560-5_2,
© Springer Science+Business Media Dordrecht 2013

the art of perspective they could be exactly depicted in full agreement with the rules of geometry. No drawings from these early summer days are preserved, however, and no writings on the art of perspective, if he ever undertook any. Perhaps they went up in smoke when the episcopal residence of Brunsbo was burnt down once again in 1730.

Seeing reality in a camera obscura makes the world seem geometrical, everything takes on the correct proportions as they appear to the eye. The rays of light pouring into the dark chamber through a tiny hole force vision into the geometry of perspective, everything falls into straight, converging lines like the furrows of the spring ploughing. The upside-down image of the outside world, cast by the light on the opposite wall, could be turned the right way up again with the aid of mirrors, and lenses could be employed to sharpen the blurred image. A work used by Swedenborg, *La physique occulte* (1696) by the French writer Pierre Le Lorrain de Vallemont, describes how the light corpuscles through a laterna magica form colourful phantom images against white walls. Through a hole in a camera obscura one can look out at a beautiful garden: 'One sees the birds flying around and passing by, the people coming and going, the flowers in all their splendid colour, and all this rendered with such a degree of exactitude.'[3] Nature can be copied with its own aid. This chapter is about how space, vision, perception, and orientation functioned as fundamental cognitive points of departure for Swedenborg's thought. In several clear cases one can see how spatial experience, orientation in space, finds expression, as in his proposal for determining longitude. These attempts to grasp space also serve as a foundation for geometry. For him, as for so many others, it was geometry that possessed the secure method. Geometry pointed the way to an objective and transcendent world.

With Giambattista della Porta's *Magiæ naturalis* (1589), a work that Swedenborg would later refer to in passing, the camera became widely known as an aid to artists.[4] Refined by Johannes Kepler, portable cameras soon became popular among artists, scientists, cartographers, and others. They could be used for optical experiments, for investigating the nature of light, for solar observations, for perspective drawings, or—as in the paintings of Vermeer van Delft and Canaletto—to create an exact linear realism, rendering things in a way that was almost true to life. Pictures became so palpable, virtually identical to the real thing. The dividing line between reality and illusion was erased. But the camera also liberated amateurs from the monopoly of professional artists when it came to the exact rendition of reality. A camera obscura, as Harald Vallerius the Younger explained in a dissertation on the instrument from 1700, represents the world more exactly than an artist, in true proportions and dimensions, with vivid colours, optical sharpness, and a wealth of detail, but it can also capture movements. Such is the light in the darkness! In a quarter of an hour, Vallerius proclaimed, anyone can learn the art of depicting a person more perfectly than all artists together are able to do.[5] A few years later Bonde Humerus presided

---

[3]Vallemont (1696), 277.

[4]*Principia*, 240; translation, II, 35.

[5]H. Vallerius and H. Vallerius the Younger; Ellenius (1960), 288f.

in Lund at the public defence of another dissertation about the camera obscura in which the ease of use of the instrument was expressed in a similar manner. Even the most ignorant in the art of perspective can, with a camera obscura, draw 'towns, villages, citadels, palaces, gardens, meadows . . . '[6]

Swedenborg's blue camera obscura was probably of a simple kind bought in London, an easily managed portable camera made of cardboard and leather, as microscopes and telescopes often were. In October 1710 he informed Benzelius, in a letter full of despondency and homesickness, that he had purchased a number of books on mathematical science, as well as some scientific instruments, such as a telescope, various quadrants, prisms, microscopes, and artificial scales, and in addition 'what I admire, and you would also,—a camera obscura.'[7] The main reason for the interest that scientists showed in the camera was optics. It became a standard item in works about optical phenomena. In a number of books that Swedenborg used or had in his library, such as those by Cherubini and Johann Zahn, he would almost certainly have been able to learn about the intriguing optics of the camera[8] (Fig. 1). For the understanding of vision, the camera came in very handy. In fact, Swedenborg, like many before him, as he could read in Sturm and Richter, envisaged the eye as a camera obscura.[9] Vision was of the same nature; it was, so to speak, the body's own perspective drawer. The eye with its membranes, fluids, and retina was a camera obscura interacting perfectly with the fluctuations of the ether, wrote Swedenborg in 1745, but the metaphor was probably familiar to him long before this.[10] Through the little hole, the pupil, light penetrates into the dark chamber, the eyeball, hits the other side, the retina, as an inverted image, and is passed on through the visual nerve by the vital spirits to the brain. 'The human *eye* is a wonderful *Camera obscura*', Roberg observed, so 'that a camera obscura bought in *Venice* or *Amsterdam* is nothing more than botchery in comparison.'[11]

The camera quite simply became a model for vision in the ophthalmology of the time. It could be described as a metaphorical way of thinking, whereby one thing is understood with the aid of something else. If one experimented and learned how to understand the optics of the camera, one could transfer this to human vision. Professor Anders Spole, for example, performed many experiments with a camera obscura in order to illustrate myopia, that is, nearsightedness.[12] The camera obscura as a model for the eye also gave reason to assume the correctness of perception, that

---

[6]Humerus and Gane, 15f.

[7]Swedenborg to Benzelius, London, October 1710. *Opera* I, 207; *Letters* I, 13; cf. *Resebeskrifningar* 25 July 1733, 32; *Documents* II:1, 41.

[8]Kircher (1671), 703ff, 792; Cherubini; Sturm (1676), I, 161–168; Zahn; Valentini, III, 59; Wolff (1721–1723), a. ed. (1737–1738), II, § 143, 149.

[9]Sturm (1676), I, 7–9; C. F. Richter, I, 177; Derham (1760), 131; also e.g. Leonardo da Vinci, Kepler, Scheiner, Descartes, Schott, Molyneux, Willis, and Newton; D. C. Lindberg, 164–168; Pomian, 218.

[10]*De cultu*, part 3, e.

[11]Roberg (1747), 37.

[12]Spole and Kiilberg; Berg, 46.

Sic quasi membranæ volitant simulacra per auras
Quaq; pater quoquq; licet conjuncta feruntur.

Usq; adeò omnibus à rebus   quæque fluenter
fertur et in cunctas dimittitur undiq; partes
                                                                    Lucretius

vision is objective. An unspoiled eye could convey an exact depiction of reality. But the metaphor was taken further to even more abstract phenomena. Wisdom and knowledge are a kind of vision. As the eye needs light to see, thought needs reason to be able to understand. 'The eye is the light of the body,' wrote Jesper Swedberg in *Vngdoms regel och ålderdoms spegel* ('Rule of Youth and Mirror of Old Age'), making comparisons between the human senses and understanding.[13] Spectacles are like two windows with convex glass which help the body's two lights, Swedenborg responded in Latin in a poem in the same book. John Locke compared understanding with a camera obscura and thus also with the eye. Man's outer and inner senses are like windows that admit light from the world outside to the dark room of understanding on the inside.[14] In other words there is a general system of metaphors linking the eye, perception, light, and understanding.

The Society of the Curious

Swedenborg was living in a unique time during his student days in Uppsala. In Sweden time flowed in its own separate channel, 1 day before the Julian calendar and 10 days after the Gregorian. From this isolated existence in time and place, he and others with an interest in natural science in Uppsala tried to forge bonds and build a bridge over the ocean of knowledge. Swedenborg's own thinking is difficult to understand without a grasp of the thoughts that moved within him, all the learned conversation, correspondence, and lectures, when thinking was in constant interaction with its surroundings.

The lectures in Uppsala give us a glimpse of the intellectual milieu in which Swedenborg's thinking germinated. In the year before Swedenborg defended his dissertation, 1708, the lecturers in the large and the small auditoria at the Gustavianum included the professor of anatomy and botany, Olof Rudbeck the Younger, and the professor of medical theory and practice, Lars Roberg. The professor of mathematics, Harald Vallerius, shared his insight into the physical world of optics,

---

Fig. 1 In optics, light rays appear to expand in cones and spheres from the objects or from the eye in a precise geometry. The picture is like a cross-section through the visual pyramid with the base at the object and the eye at the apex. In the *top picture* the images fly like membranes, hither and thither through the air, to be reunited once again. In the *bottom picture* they radiate from different points on a tree trunk towards an eye, with a quotation from Lucretius: 'So from every object flows a stream of matter, spreading out in all directions.' From a work in Swedenborg's library about artificial eyes, that is to say, different kinds of cameras, microscopes, and telescopes (Zahn, *Oculus artificialis teledioptricus sive telescopium* (1702))

---

[13]Swedberg (1709b), 147f; *Ludus Heliconius*, edition, 56f; cf. Matt 6:22, Luk 11:34.
[14]Locke, book II, ch. XI, § 17.

while Johan Upmarck, the Skyttean Professor of eloquence and politics, delivered his orations, and Johan Arent Bellman expounded the text of Cicero's epistles and speeches. Pehr Elvius, professor of mathematics, made excursions in geography, and the Greek language was expounded by Professor Olof Celsius. The lectures of the professor of practical philosophy, Johan Eenberg, dealt with a universal philosophy that proceeded by a dialectic method inspired by Pufendorf, through thesis, antithesis, and examples. The professor of theoretical philosophy, Fabian Törner, who would preside at Swedenborg's disputation the following year, lectured on the peripatetic school, that is to say, Aristotelian philosophy, and on Cartesian logic, and tried to give an answer to the question whether they were opposites or in harmony.[15] Swedenborg may have attended some of these series of lectures.

Two men of particular significance for the young Swedenborg were Benzelius, who was close to him, and Polhem, with whom he cooperated intensively for a couple of years. Benzelius's chief interest was classical and oriental languages, but he encouraged Swedenborg's scientific interest: 'Among all my brothers and sisters, I find none who has wished or does wish me well save d: Brother,' Swedenborg declared to him.[16] As for Polhem, the famous inventor of Stjärnsund in Dalarna, he found a community of thought. 'If it so be that my foreign journey must needs stand over till the Spring [of 1711], then I am quite content to be with him for some time,' wrote Swedenborg to Benzelius in spring 1710.[17] Benzelius and Bishop Swedberg tried to prevail upon Polhem to take on his young admirer as an apprentice in mechanics. He declined at first, but after financial negotiations, brokered by the rector of Husby, Jacob Troilius, Polhem said that he could not undertake this trouble 'for less than four Rixdalers a week'.[18] Polhem was always anxious to be properly paid. In July, however, he confessed, in letters to Benzelius, that he regarded Swedenborg as 'capable of helping me in my current work in mechanics and its experiments'.[19] But there was not to be any study sojourn at Stjärnsund this time. Instead Swedenborg headed off, out into Europe.

In the late summer of 1710 an epidemic of plague swept in over Stockholm and then spread towards central Sweden. In Uppsala the university was forced to close, and those who could do so sought the protection of the sulphurous fumes in the mining districts, Bergslagen. But many remained, including the hospital director Roberg who stayed on to provide care to people suffering from the plague. The whole autumn was one of heavy, oppressive weather. The ground was covered by a suffocating fog, a slight, almost imperceptible wind blew persistently day

---

[15]*Catalogus prælectionum publicarum* (1708); translation, 29–32.

[16]Swedenborg to Benzelius, Brunsbo, 5 October 1718. *Opera* I, 287, cf. 201; *Letters* I, 198; *De infinito*, dedication; translation, 3; Benzelius to Bignon, Uppsala, 5 August 1710. LiSB, B 56; Benzelius (1791), xxii; Liljencrantz (1940), 29.

[17]Swedenborg to Benzelius, Brunsbo, 6 March 1710. *Opera* I, 203, cf. 202; *Letters* I, 7; concerning Polhem, cf. Dunér (2005a), 5–12; Dunér (2005b), 100–118.

[18]Troilius to Benzelius, Husby, 3 June 1710. LiSB, S III, no. 93.

[19]Polhem to Benzelius, Stjärnsund, 16 July 1710. *Polhems brev*, 6.

after day, and the sun was obscured for several months.[20] Plague, famine, the defeat at Poltava, and the strange weather brought a feeling that Doomsday was nigh. It was a sign, God's warning to mankind, said Bishop Swedberg.[21] Some professors withdrew during the ravages of the plague, and towards the end of the year they formed a small learned society, under Benzelius's leadership, called Collegium Curiosorum, the society of the curious. Their conversations concerned natural philosophy and mechanics, but inevitably they also touched on the question of the cause of the pestilence. Did the waxing and waning of the moon have any influence on the plague? The society, which held its first minuted meeting at the start of January 1711, was to be significant in several ways for Swedenborg's early scientific activity. The polymath Benzelius acted as the spider in the web for learned life in Uppsala. He had a far-ranging network of contacts, purchased foreign academic journals for the library, had connections with *Acta Eruditorum* in Leipzig and *Journal des Savants* in Paris, was a good friend of the French philosopher Nicolas Malebranche, and corresponded with the famous German philosopher and mathematician Gottfried Wilhelm von Leibniz about history and linguistics.[22] Benzelius's home or the university library were the meeting places for the natural scientists Elvius, Roberg, Rudbeck the Younger, Harald Vallerius with his sons Johan and Göran, and also the humanist Upmarck, to whom Swedenborg in his dissertation had given the illustrious title 'our light of Athens'.[23] Swedenborg and Polhem, who were elsewhere, were closely connected to the society as corresponding members. The publication of Polhem's manuscript on natural philosophy and mechanics was high on the society's agenda.[24] The Collegium Curiosorum, as a speech in memory of Johan Vallerius put it, 'tested the different hypotheses and axioms of the philosophers which the world hails as oracular utterances, indeed, the very foundations of their doctrine'. And 'our Dædalus Hyperboreus', Christopher Polhem, wandered through nature's impenetrable labyrinths. All this 'was brought, as if with Ariadne's thread, along certain trails and towards certain goals'.[25]

These men were curious, for which there are two Swedish words, *vetgirig* literally 'eager to know', and *nyfiken*, meaning roughly 'greedy for something new'. The French word *curieux* was discussed in a different context in Lund in the company of King Charles XII a few years later. Bishop Swedberg then objected to the suggestion that it should be translated as *nyfiken* and took the example: 'old Ol. Rudbeck was a *curieux* man, but he was not *nyfiken*. For with his own hair, collar, and wide trousers he proved not to be greedy for anything new, but of the

[20]Forssell, 275.

[21]Swedberg (1711), 590f.

[22]LiSB, Benzelius, B 53; Benzelius (1791), xvi; Ekenvall (1950–1951), 141ff; Ekenvall (1953).

[23]*Selectæ sententiæ*, 17; translation, 13; Benzelius (1791), xxii.

[24]*Collegium Curiosorums protokoll* 16 January and 24 January 1711. *Protocoller vid Soc. Scient. i Ups. sammankomster 1711*. LiSB, N 14a, no. 66; edition, 61.

[25]Hermansson, 29f; cited in Liljencrantz (1939), 298.

old world.'[26] The name of the society, Collegium Curiosorum, however, signals—alongside the general curiosity—also a certain shift towards the German culture of learning. In Germany there was already a Collegium or an Academia Naturae Curiosorum with an orientation towards medicine, and of perhaps great significance was the Collegium Curiosum siue Experimentale in Altdorf under the leadership of the mathematician and physicist Johann Christopher Sturm, well known to the Uppsala circle as Polhem and others had met him in Altdorf.[27] During his tour in 1697–1700 Benzelius had also dined with Johann Andreas Schmidt in Helmstedt, which likewise had a Collegium Curiosum Experimentale.

The Collegium Curiosorum in Uppsala may be regarded as an alternative to the university, that is, an organization seeking new knowledge and not merely preserving and commenting on old knowledge, an association that also aspired to unity instead of academic disputes. They also seem to have wanted to link progressive science with society, the economy, and the armed forces—the great utility—instead of with religious and academic purposes. It is obvious that this core of persons, who would be mixed in new constellations over the years, is an important background to Swedenborg's thinking. Conversations with Benzelius, Polhem, and others in this circle aroused many thoughts and ideas in Swedenborg's mind. The oral communication was surely at least as important as the written. Historians of ideas in general tend to ascribe excess weight to the causal influence and significance of written works, at the expense of the spoken word and personal conversation. For natural reasons the printed word with its preservation, in visual spaces on the pages of books, has created this hegemony, while conversations have long since faded away to nothing. Yet letters, minutes, and notes preserve clues to what the conversations may once have been about. Between the lines we hear the sound of the spoken language. Oral conversations, where a great deal was said that escapes us, should give us reason to revalue and not to forget Swedenborg's personal contacts and oral channels of information.

In some sense they formed an intellectual community or a thought collective where Swedenborg was also, for a time, close to the hub, but mostly on the periphery, yet without losing contact with the values and opinions that that were fundamental to them. The circle around the Collegium Curiosorum embraced and exchanged certain ideas and thoughts, developed a special knowledge, a culture, a style of thought. The style of thought focused their attention, assembled them around a method, a way of processing perception, a shared outlook on what was a sound argument or a correct conclusion. They shared many basic beliefs as to what nature is and how natural science should be pursued. They were all faithful Cartesians, united in mechanistic natural philosophy, and they cultivated the geometrical method, the ideal of clarity, objectivity, truth, accuracy, and order. They shared a passion for mathematics, machines, and handicraft; they had a blind

---

[26]Swedberg (1941), 573.

[27]There are also several journals with the name *Relationes curiosæ* (1682–1723); B. Hildebrand, I, 27f; cf. Kenny, 53–57.

faith in practical utility, the blessings of work, engineering, and the ability of science to accumulate prosperity. Utility, which would become even more important during the Swedish Age of Liberty, was not infrequently expressed in a kind of 'manure or fodder' metaphors. *Helpful ideas are nourishing food*, they are useful ideas for fertilizing the country to make the economy grow and flourish.

Apart from the fact that this style of thought had shared ideas and outlooks in the traditional sense, I claim here that there is a more fundamental cognitive community which also explains why they understood each other, what the other person was trying to say, and which could facilitate communication between them. It is that they shared the same metaphors, assessed some metaphors as being more central than others, such as that *the world is a machine, matter is geometry, to think is to see, science is food or energy*. The shared metaphors meant that they shared similar mental images, could communicate them and arouse similar mental images in others. When everyone imagined, for example, sound or the particles of matter, they had similar associations, representations, and mental images.

The self-evident foundation for the curious explorers' community of thought is the Cartesian, mechanistic natural philosophy. Swedenborg rarely referred directly to René Descartes himself. But with great probability he had acquired a substantial dose of the basics of Cartesian natural philosophy during his student years, from *Synopsis Physica* (1678) by the professor of medicine, Petrus Hoffwenius, which his teacher Harald Vallerius had reissued in 1698 with comments by Roberg. *Synopsis Physica* was a very popular series of disputation exercises gathered as a textbook, and was to be used until well into the eighteenth century. It reveals a clear Cartesian world-view which was made comprehensible by being described in Aristotelian terms. In several respects it closely followed Johann Clauberg's *Physica* (1664), and Clauberg is said to have been an important introducer of Cartesian natural philosophy to Sweden. In Uppsala on 20 October 1683 the items left in Master Vallerius's room, listed on the back of a manuscript on arithmetic, were: '2 bookcases. 1 writing case 1 black chest. 1 blackboard. 1 poker. 1. round staff 1 small press. 1 bundle of bedding in which are 2 blankets, 1 small under-bolster, 1 main bolster 1 pillowcase, Claubergius.'[28] Clauberg is as important as pokers and pillowcases.

What they fastened on in Cartesian natural philosophy was the spatial concept, the dualism, the vacuum, the theory of whirls, the particle theory, and that everything could be explained in terms of matter and movement. They rejected Aristotelian teleology, preferring causal explanations. The world consists of two substances, the extended and the thinking substance, that is, body and soul or matter and spirit. Matter, as Descartes explained in *Principia philosophiæ* (1644), is identical with space, in other words nothing other than extent in length, width, and height. Matter is independent of sensory properties, it has no hardness in itself, no power of attraction, colour, or other perceptible properties; it is pure geometry. It entails a quantitative instead of a qualitative way of observing nature. The material is the same as the

---

[28]H. Vallerius, 'Arithmetica figurarum m.m.'. UUB, A 518, fol. 40v.

spatial, and means size, form, and position in relation to other bodies. From this it follows, by definition, that the vacuum cannot exist. All space contains matter. When we refer in everyday speech to a vacuum or an empty space, we really mean a space that lacks the things we think it should contain. A fish pond is empty when there are no fish in it, even though it is full of water. Space became immanent in nature. Since matter was the same as geometrical space, one could hope to achieve the same exactitude in natural philosophy as in geometry. The whole natural world could be explained and described in terms of the two basic concepts of Cartesian natural philosophy, matter and movement; all scientific truths had their foundation in these. In Descartes' philosophy space and time were absolute and objective. In the same way, for Newton space and time were absolute and independent of what was in them.[29] Space remains the same and immobile. Leibniz took a different view, assuming space to be relative to physical objects.

Even the not always so innovative dissertations are of great value for getting at the entrenched, 'self-evident' knowledge and perceptions of the world. The dissertations reflect the scientific culture of the academic environment, the values and the salient ideas. In this Upsaliensian community of thought, as Elvius also expressed it in a dissertation, there is a fusion between Cartesianism and Francis Bacon's empiricism.[30] The aim of the Collegium Curiosorum was not wholly unlike the vision in Bacon's *New Atlantis* (1627) of 'Solomon's House', an order or society the aim of which was 'knowledge of causes, and secret motions of things; and the enlarging of the bounds of human empire, to the effecting of all things possible.'[31] In July 1711 Benzelius noted at the back of a book that he had received as a gift from Swedenborg, namely, Bacon's *De Augmentis Scientiarum* (1652), that he had now reread it for at least the second time.[32] On one important level Descartes and Bacon differed. Descartes' deductive method, based on sure principles, was a contrast to Bacon's empirical method in *Novum Organum* (1620), which 'first lights the lamp, then shows the way by its light, beginning with experience digested and ordered, not backwards or random, and from that it infers axioms, and then new experiments on the basis of the axioms so formed'.[33] It is a matter of inductively collecting facts first, then comparing them and drawing conclusions. This method, Bacon says, is like drawing a straight line or a perfect circle with a ruler and compasses, which does not require the same certainty and practice of the hand as for someone who tries to do it freehand.

When the plague had abated and things had returned to their normal order, the Collegium Curiosorum disintegrated by 1711. Swedenborg, who was still on his travels, wrote to Benzelius in 1714 that he now felt like coming home to tackle Polhem's inventions, make drawings of them and describe them, and set up a

---

[29]Cf. Janiak, 130–162.

[30]Elvius and Bechstadius.

[31]Bacon (1627).

[32]Bacon (1652), 684. LiSB.

[33]Bacon (1620), § 82, cf. § 61; translation, 67.

mathematical society where his inventions would be 'so fine a foundation'.[34] When he came home he breathed new life into the idea of establishing a scientific society on the model of those in foreign countries. In a proposal submitted to King Charles XII in 1716 he takes up Polhem's mechanical laboratory, recommends the creation of an observatory in Uppsala Castle, and suggests the formation of a 'Societas Mathematica' to specialize in mechanics, to invent machines for manufacture, shipbuilding, artillery, mines, field mills, and ballistics—all topics of interest to a warrior king.[35] Swedenborg's most important contribution to the idea of an academy was the publication of Sweden's first scientific journal, *Dædalus Hyperboreus*, of which six issues appeared between 1716 and 1718.[36] Having visited England and being an Anglophile, it is not impossible that he was trying to emulate the Royal Academy's publication *Philosophical Transactions*. More lasting than the society of the curious was the 'guild for book learning', Bokwettsgillet in Uppsala, founded on the initiative of Benzelius on 26 November 1719. It became a far-ranging society, closer to Leibniz's universal academy than Swedenborg's proposal. In Bokwettsgillet the scientific programme comprised both natural science and the humanities. The members met every Friday to discuss a variety of matters and hear essays read. The society also published a journal, *Acta Literaria Sveciæ*. Swedenborg became a member in February 1720, and was to participate actively by sending in manuscripts to be read aloud, but he only attended in person once.[37]

Armed Eyes

The circle around the Collegium Curiosorum took a keen interest in optical instruments. But they were interested in the physical properties of the instruments and the remarkable phenomena that could be observed with them, rather than their practical use in the systematic investigation of nature. In a dissertation on the optical laws for glasses, Elvius compared their properties with the camera obscura and gave good advice to anyone who needed to acquire spectacles.[38] Johan Vallerius lectured about various optical phenomena, about the rules of perspective, anamorphoses,

---

[34]Swedenborg to Benzelius, Rostock, 8 September 1714. *Opera* I, 227, cf. 228f; *Letters* I, 59.

[35]Swedenborg, *Förslag till den nya Soc. Scient. i Upsala* (1716). LiSB, N 14a, no. 12 1/2, fol. 26v; *Photolith.* I, 2; *Opera* I, 241f, 246, 265; There is also a proposal, wrongly ascribed to Swedenborg, intended for Bokwettsgillet in 1726. *Petenda societatis literariæ*. LiSB, N 14a, no. 5, fol. 13; *Photolith.* I, 1.

[36]*Dædalus Hyperboreus* was advertised in *Ordinaire Stockholmiske post tidender* (1716:2), 17 & 36; *Stockholmiske kundgiörelser* (1717:14); *Acta literaria Sveciæ* (1720), 26; Althin.

[37]*Bokwetts Gillets protokoll*, 13, 108, 154, 169f; in 1728 Bokwettsgillet was elevated into the Societas regia literaria et scientarium, and Swedenborg was once again enrolled as a member in autumn 1729, proposed by Polhem. *Polhems brev*, 174; *Opera* I, 321.

[38]Elvius and Walingius; Berg, 52–60.

telescopes, and microscopes.[39] In 1710 Benzelius suggested that the university should purchase microscopes from England and from Johan van Musschenbroek in Holland in order 'to see Circulationem Sanguinis, worms in vinegar, etc.'[40] The task was assigned to Swedenborg. In April 1711 he ordered lenses for a 24-foot telescope from the famous instrument maker John Marshall.[41] With the backing of the Collegium Curiosorum, Elvius wrote to Swedenborg and asked him for information about different astronomical instruments. Elvius, who had defended a dissertation about telescopes, asked Swedenborg to go to the astronomer John Flamsteed to find out, among other things, how the instruments were made, what they cost, and how the telescopes and diopters were used.[42] Just over a year later Swedenborg sent the lenses to Sweden, along with some books and other instruments. On his return visit to Marshall's workshop he was able to view through a microscope the rapid blood movements of a fish (or frog). It reminded him of small rivers, Swedenborg said.[43]

In Leiden in 1713 Swedenborg visited a splendid observatory and saw the most beautiful brass quadrant he had ever seen. He wished he could pursue astronomical observations there for two or three months. During his visit to Holland he also learned glass cutting, a craft for which he acquired all the implements and bowls.[44] The necessity of establishing an astronomical observatory in Sweden was espoused by the circle of the curious, not least by Swedenborg himself. The question was discussed in letters between him and Benzelius. Swedenborg sent him his proposal *Om nyttan och nödvändigheten af ett observatorii inrättande i Sverige* ('On the utility and necessity of the establishment of an observatory in Sweden', 1717). Sweden has one advantage over other countries, he pointed out: the Swedish cold. On the cold winter nights the sky there is much clearer than in warmer southerly countries.[45] And the frozen lakes are suitable for comfortably measuring degrees of latitude and the true size and shape of the earth. The plans and hopes for an observatory, and the interest in optics lived on during the 1720s in Bokwettsgillet. At a meeting in the autumn of 1722 the professor of jurisprudence, Johan Malmström, demonstrated a 'machine' invented by Antonie van Leeuwenhoek and manufactured by Musschenbroek which showed the circulation of the blood and the true figures of its particles.[46] It was, in other words, a microscope. Members could also examine an artificial eye made in Strasburg.

[39]*Catalogus lectionum publicarum* (1714); cf. J. Vallerius, 'Observationes Geometricæ. objectum Geometriæ est Qvantitas'. KB, X 219.

[40]UUB, Academic Consistory minutes, 10 December 1710; cited in Nordenmark (1959), 142.

[41]Swedenborg to Benzelius, London, 30 April 1711. *Opera* I, 208.

[42]Elvius to Swedenborg, Uppsala, 28 July 1711. *Opera* I, 211–213; Elvius, *Astronomiska meddelanden i bref till Svedberg från Ups. 28 jul. 1711.* LiSB, N 14a, no. 67; *Collegium Curiosorums protokoll*, 66f; Spole and Elvius.

[43]Swedenborg to Benzelius, London, 15 August 1712. *Opera* I, 221; *Letters* I, 42.

[44]*Opera* I, 223f.

[45]*Photolith.* I, 4; Nordenmark (1933), 41; in June 1717 the manuscript was sent to Bernhard Cederholm, secretary of the purchasing deputation. *Opera* I, 272.

[46]*Bokwetts Gillets protokoll* 16 November 1722, 75f.

Optical instruments were not mundane tools. They had symbolic meanings, a powerful metaphysical effect on the senses. The microscopes and the telescopes revealed the enormous scope of God's creation. They allowed vision to extend far beyond the commonplace, to see the infinity of existence. Man was created with weak senses. 'We have not been given acute *eyes*, like the *lynx*', said Linnaeus, 'but contemplation has learned through tubes to see spots on the planets and through microscopes to see the arteries in the louse.'[47] Vision was sharpened in Galileo's telescope. The moon turned out not to be a perfectly smooth globe, but was torn and wrinkled with mountains and oceans. The eye could reach further out into the universe, with new stars, new moons around Jupiter, and rings around Saturn. And there was perhaps something that was more important for the time than the scientific use: the telescope revealed troop movements far behind the enemy's front line.

Swedenborg was living in an age when vision became armed, the world was extended with religious wonder. The boundaries of the world are much further away than we can imagine, and God has filled the tiniest drop of water with life. Science and optics revealed *maximus in minimis*, the greatest in the least. The microscope opened a *terra incognita*, an unknown world beyond the bounds of unarmed vision. Leeuwenhoek saw into the realm of small animals in the world of water drops. 'In the large microscope, one can see a flea like a pig,' the Czech educator Jan Amos Komenský (Comenius) explained to young readers.[48] A mite became an elephant for Bernard de Fontenelle. Magnus von Bromell told how Leeuwenhoek investigated the testicles of a flea, and Jonathan Swift wrote sarcastically about how a curious explorer of nature observed that fleas had their own fleas, and they in turn had fleas, and so on to infinity.[49] It became popular, like Christian Ström, who had learned how to make microscopes from Leeuwenhoek at the start of the eighteenth century, to direct the gaze towards spermatozoa. Johan von Hoorn encouraged Ström, punning on his name, which means 'stream, current, flow': 'I shall urge on the Current so that he does not flow away until he … pours forth these vermiculos [small worms] on a sheet of paper.'[50] Robert Hooke, known to Swedenborg and the Collegium Curiosorum as an instrument maker and astronomer, directed his curiosity in *Micrographia* (1665) towards everything in the unknown microscopic world.[51] The world was geometrical, and therefore he saw ovals, pyramids, cones, and prisms among the grains of sand, and the gravel in urine consisted of rhombi, rectangles and squares. It is surely not impossible, Hooke speculated, that one day man will be able to design even better optical aids with which to discover living

---

[47]Linnaeus (1739).

[48]Comenius (1716), 163.

[49]Fontenelle, translation, 95; Lindroth (1989), II, 440; The microscope, 'The Armed Eye', was presented to a Swedish audience in an appendix to the journal *Swenska Mercurius* (1682), entitled 'This World's Greatest Ideas or the So-called Relationes Curiosae', with a picture of Hooke's flea measuring 26 cm. See Happel; Dal (1996), 18.

[50]Cited in Lindroth (1989), II, 441; Pipping (1991), 216.

[51]*Collegium Curiosorums protokoll*, 67; *Opera* I, 212, 216; cf. Purrington.

creatures on the moon or other planets, or to find the figures in particles of matter—
and perhaps even mechanical inventions to refine the other senses and amplify our
perceptions of taste, smell, and feeling.[52]

Attempts to Find East and West Longitude

It was essential to look around in space, but also to orientate oneself in it and
describe it. As a mathematical discipline, geometry deals with spatial magnitudes
and properties. Geometry emerged from a desire to grasp the earth, the landscape,
the sea, and human dwellings. It is linked to people's surroundings, the fields,
meadows, gardens, towns, fortresses. It was about everyone having a patch of land,
and finding his way through geography. The myth of the flooding of the Nile in
Egypt, homeland of geometry, which Herodotus reported in his history, and which
Polhem, Swedberg, and most others were familiar with, forced man to learn how to
measure the earth.[53] Disorder, fraud and violence, murder and war prevailed before
geometry, the perfect science of lines, planes, and solids, gave each man his own,
wrote the English mathematician John Dee.[54]

Surveying was a kind of practical geometry associated with the land and physical
labour. In maps, which were also called 'geometrical', the landscape, the sea, and
the sky were translated into signs and symbols, as an abstraction, a clean drawing
of the world. The signs on the map were supposed to resemble and correspond
to reality, so that the traveller or the landowner could find his way in the world.
They were intended to give the reader of the map a representation, a vision of
what the earth looked like from above. With the aid of the imagination, mountains,
forests, lakes, and villages came to life in the inner eye of the observer, the map was
filled with people, sounds, colours, discoveries, and laborious distances to bridge.
In Swedenborg's Board of Mines the work included interpretations and studies of
geometrical maps. Translating the depths of the mines into signs involved special
problems, since it was important to be able to render every level.

Swedenborg probably learned Euclidean geometry through Martin Gestrinius'
edition from 1637 of the first six books of Euclid.[55] This was used, for instance,
by Harald Vallerius in the autumn of 1709 when he lectured on Euclid's *Elementa*
(*c.* 300 BC), and in his dissertations he often referred to Gestrinius, Clavius, and the
important commentary on Euclid by the neo-Platonist Proclus.[56] Besides the basic
skills in mathematics that he must have acquired during his years in Uppsala, he

---

[52]Hooke, preface, 80f.

[53]Herodotus, 2.109; *Polhems skrifter* III, 461, cf. IV, 231; Swedberg (1716), 273;

[54]Dee, 13f.

[55]Gestrinius; Dahlin, 98.

[56]*Catalogus prælectionum publicarum* 1709; H. Vallerius and Rimmius, 1, 3, 5; H. Vallerius and
Rockman, 11.

started to study science and mathematics on his own before he went to England. In the spring of 1708, together with Kalsenius, he borrowed a surveyor's rules and proportional circle, and in autumn of that year the third part of Claude François Milliet Dechales' *Cursus seu mundus mathematicus* (1674), a book that was also frequently borrowed by Harald Vallerius and Elvius.[57] Milliet Dechales' third book deals exclusively with applied mathematics, such as fortifications, hydrostatics, navigation, optics, perspective, and so on. In the textbooks of the time, as in those by Åke Rålamb or Christian von Wolff, applied mathematics often took up a great deal of space.[58]

Geometry was considered useful and applicable to the whole of material reality, as a key to the solution of problems, to prosperity, happiness, and order in everything. Imperialist geometry could conquer every field. Geometry could be applied in technology, astronomy, architecture, music, optics, fortifications, stage mechanics, cartography, shipbuilding, navigation, surveying, and everything else for that matter. This theoretical mathematical knowledge would improve and promote practical skills, technology, trade, the natural sciences, and even art and philosophy. What happened was instead the reverse: new technology, new economic circumstances, social regroupings, observations, and aesthetic ideals gave rise to new questions about what was seen, how it was to be described, and how people should create order in reality. In many cases geometry was found practically useful; it functioned well for describing the paths of planets and comets, with their ellipses, parabolas, and hyperbolae, or for the manufacture of lenses for spectacles, telescopes, and microscopes in association with the analysis of curves.

Geometry enabled orientation in space, finding one's place in the world, or—as for fishermen in the archipelago—returning to the place where the fish were most abundant (Fig. 2). There were mercantile needs to find the right harbour on the elemental ocean. Navigation thus became an incentive for geometry and astronomy. Longitude, that is, orientation in an east–west direction, caused particular problems. There was no fixed point from which to proceed, as in the case of latitude in a north–south direction, where it was possible in the northern hemisphere to calculate degrees north and south on the basis of the height of the Pole Star above the horizon. The longitude problem engaged many astronomers and mathematicians, such as Borelli, Huygens, Leibniz, and Cassini. The interest that Swedenborg and others took in the matter was not diminished by the fact that the English parliament in 1714 promised a huge sum of money to whoever was able to solve the problem.[59] A general grasp of the different methods that could be used to determine longitude, for instance through the moon, could possibly have been acquired by Swedenborg in a dissertation supervised by Elvius that was defended not long

---

[57]UUB, Bibliotekets arkiv, G 1, fol. 190; The instruments were borrowed on 25 April, returned on 2 June. Milliet Dechales was borrowed on 18 November, returned on 24 March 1709; cf. 155f, 166.

[58]Rålamb, I; Wolff (1716), 378f, 988–997, 1047, 1312f.

[59]*Methodus nova inveniendi longitudines locorum*; ed., *Opera* III, 203; translation, 215.

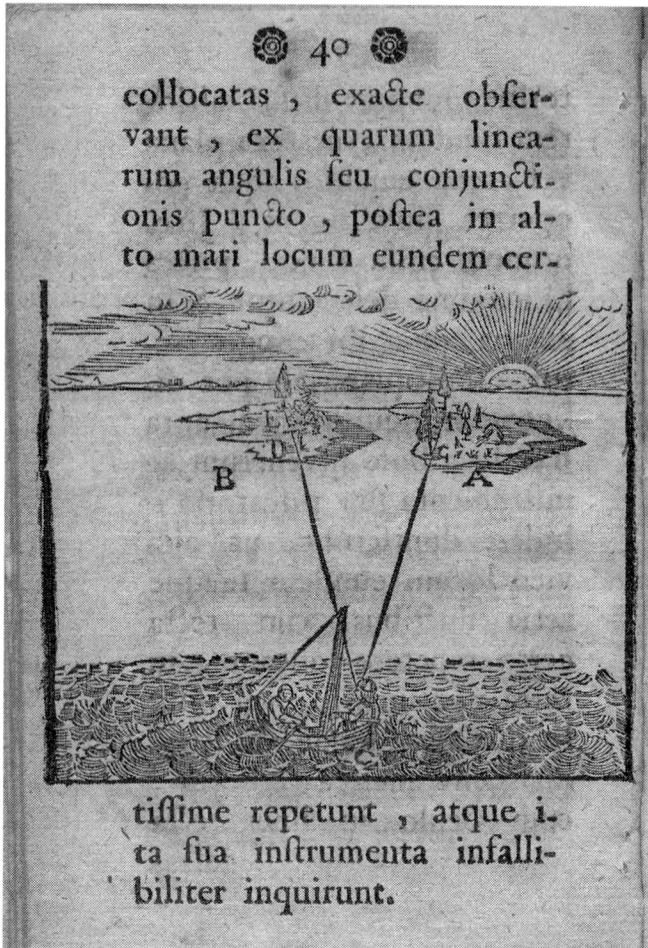

Fig. 2 Two archipelago men fishing for cod can find their way back to a good fishing place by calculating the angles between the boat and a pair of trees on two islands. If *C* is the place where they usually lay out their nets, and there are two islands *A* and *B* which each have two trees, *FG* and *ED*, which they see standing in a straight line, one can easily calculate the exact location of the desired place from the *straight lines EDC* and *FGC* and the resulting angle *ECF* (Harald Vallerius describes the significance of geometry for orientation, how right angles can be used in practical geodesy, in a dissertation about angles, *De angulo* (1698))

before he left Sweden.[60] Swedenborg's own suggested solution, that longitude could be determined by using the position of the moon, was to be one of his most persistent fixed ideas, to which he clung for more than half a century. Swedenborg's view of longitude reveals a characteristic feature of his thinking—an overwhelming

---

[60]Elvius and Duraeus, 27–29.

stubbornness. Once he had made up his mind, he never abandoned an idea no matter what others said or thought. No one could ever disprove the absolute and only possible 'truth' that he had found.

It was in London in the spring of 1711 that Swedenborg was drawn in earnest into the quest for longitude. Here in England there were several mathematicians, such as Wren, Hooke, Newton, Flamsteed, and Halley, who were tackling the difficult task. From the English metropolis Swedenborg wrote home to Benzelius and to say that he paid daily visits to 'the best *mathematicos* in the city here.'[61] Among others, he made the acquaintance of the astronomers Edmond Halley and John Flamsteed. He had visited the latter, who lived at the Royal Observatory in Greenwich, and described him as the best astronomer in England; Flamsteed had drawn up tables of the movements of the moon with which it was possible to determine longitude at sea. Swedenborg was impelled by an insatiable thirst for the mathematical sciences, especially astronomy and mechanics. Just over a year later he declared, with the utmost self-confidence, that he had found the only solution: 'As concerns my *invention de longitudine terrestri invenienda per Lunam* [concerning the finding of the terrestrial longitude by means of the moon], I am sure that it is the only one that can be given, and is the easiest method and in every way the correct one.'[62] All that was needed was better moon tables, which he would obtain from Flamsteed, and 'If this is true, I have won the whole play'! His self-assurance was at its highest. If he could only be encouraged in his mathematical studies, he said, 'I mean to make more *inventiones* therein than any one in our *aetate* [age]'. His letters home exude optimism and reveal a huge capacity for work; nothing is impossible, and there is no time to waste. The rhetoric with which he sought to show how hard he was working, how well he was using his money and time, was no doubt intended to reassure his father who had paid for the trip, and the scholars in Uppsala who must have had great expectations of the information he could provide from the international scientific scene.

He showed his 'only possible' solution for longitude to the learned men of Paris as well, but he was given a rather cool reception.[63] On return to Sweden, however, he tackled the problem with renewed energy. He now hoped to be able to build an observatory on the top of Mount Kinnekulle in order to perform the necessary observations. Bishop Swedberg, who was begging for a post for his son at the academy, got somewhat lost in the terminology when he explained that his good son was thinking of building an observatory on Kinnekulle 'in the belief that he can find Latitudinem in the great ocean'.[64] It was longitude that Swedenborg was out to capture. In a way that was highly characteristic of Swedenborg, his idea for an observatory was to make observations through which he could confirm the

---

[61]Swedenborg to Benzelius, London, 30 April 1711. *Opera* I, 209f, cf. 214f; *Letters* I, 20.

[62]Swedenborg to Benzelius, London, 15 August 1712. *Opera* I, 219; *Letters* I, 39.

[63]*Opera* I, 222, 227.

[64]Swedberg to Feif, Brunsbo, 12 July 1715. RA, Skrivelser till Kansliämbetsmän, 15:38, L–Ö, 3f.

correctness of his method for determining longitude.[65] In other words, *first* a theory, *then* observations to verify the theory.

In the fourth issue for the year 1716 of *Dædalus Hyperboreus* Swedenborg presented his suggestion in print for the first time. The failure of previous attempts, he writes, was not due to the moon itself, since God seems to have placed it there to guide seafarers, but to the way they observe it. He refers to Riccioli, Kircher, Hevelius, takes up the problem of parallax, and mentions that Flamsteed 4 years previously had been compiling a star table and ought now to have 'brought it into the light', but that he had not yet had access to it.[66] Swedenborg's method is rather complicated. Polhem wrote to Swedenborg that he did not understand the bit about longitude as clearly as he ought to, but that it seemed highly possible.[67] In broad outline, his method involves waiting for a point in time when the moon can be seen with the naked eye in a straight line with two fixed stars of the same longitude, as '= ☽ ✳✳ or ✳ ☽ ✳ or ✳ ✳ ☽'.[68] This should then be compared with a clock set according to the latitude (Fig. 3).

Swedenborg published a longer work on the topic in 1718 entitled *Försök at finna östra och westra lengden igen igenom månan, som til the lärdas ompröfwande framstelles* ('An attempt to find eastern and western longitude through the moon, presented for the appraisal of the learned'). In the dedication to Halley he explains that conditions for 'star-art' or astronomy are better in Sweden than in other countries, 'since we have such a pure and clear sky'.[69] The solution to the longitude problem is of great significance; it is both useful and necessary, for otherwise one can end up, as he says, among Hottentots instead of Indians. Without longitude the problem is that, after sailing for some time, one can not only see a different shore, arrive at a different country from what one had hoped; it could also be difficult to find one's way back to the place where one had previously found riches:

> A seafarer often ends up in a country he has never seen before, and unexpectedly finds a large tract full of every bounty, but when he seeks to return there again, he cannot find the place, often goes around it, sometimes travels far to the west and leaves the land to the east that he is actually looking for and sailing in search of. Many such lands and islands, known and unknown, have been discovered, and must now lie in the ocean without being found, which has caused great loss and lack, and still does.[70]

Swedenborg then had this work translated into Latin and published no less than four times, in 1721, 1727, 1754, and as late as 1766. His idea, however, was dismissed quite soon as implausible because of various errors and the fact that he had presupposed what he was looking for. Despite devastating critique from

---

[65]*Opera* I, 231.

[66]*Dædalus Hyperboreus* IV, 87–89, 99.

[67]Polhem to Swedenborg, Stjärnsund, 5 September 1716. *Polhems brev*, 124.

[68]Burman, reviewed in *Acta literaria Sveciæ* (1720), 29; cf. *Opera* I, 301; reviewed in *Neue Zeitungen* 2 June 1721, 345–347.

[69]*Försök at finna östra och westra lengden*, dedication; cf. *Opera* I, 284.

[70]*Försök at finna östra och westra lengden*, 8, cf. 7.

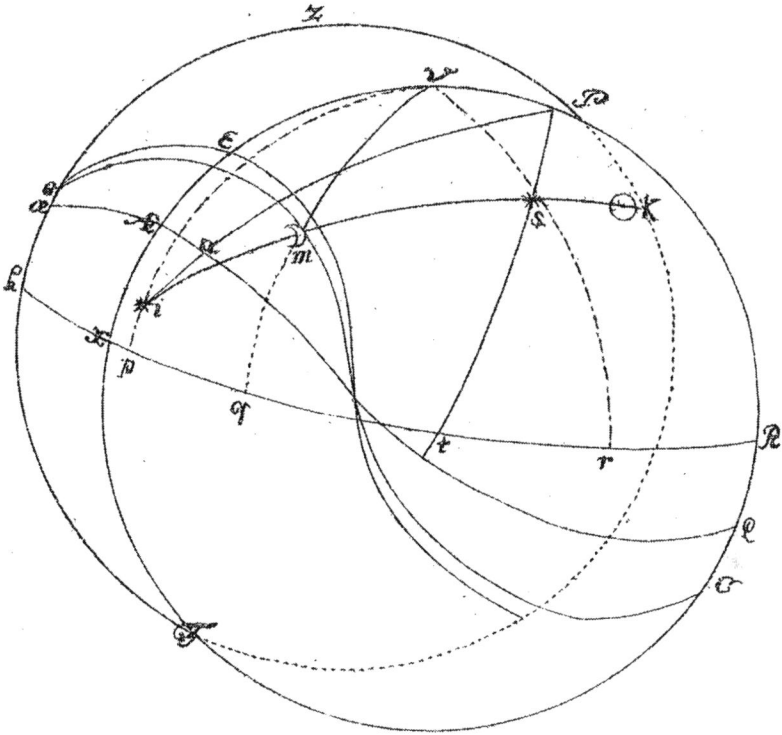

Fig. 3 The meridian where the astronomical tables were compiled, *ZTP*, is compared with the meridian at which the ship and the navigator arrive, *PVEXTR*, where *XR* is the horizon, *V* and *Z* the zenith, *ÆQ* the equator, *P* the pole star, *eEC* the ecliptic or the zodiac, which is the path followed by the moon. Swedenborg, 'A new and certain method for finding eastern and western length, that is, Longitudines Locorum, both at sea and on land, through the moon', *Dædalus Hyperboreus* IV (1716)

astronomers and a rejection by the prize committee in London, he was stubborn in his certainty that he was right. The essay on longitude was read aloud to the members of Bokwettsgillet, and his brother in that society, Eric Burman, wrote a favourably disposed review in *Acta literaria Sveciæ*.[71] In the same journal in 1722, however, Conrad Quensel, professor of mathematics in Lund, demonstrated the impossibility of the method.[72] There are problems with the parallax of the moon, he says, and it is not possible, as Swedenborg believes, to ignore the minutes in the observations and calculations. Quensel ends by noting that even a cloudy day is

[71]*Bokwetts Gillets protokoll* 9 June 1721, 51, cf. 14, 19, 62, 132; also reviewed in *Acta eruditorum*, May 1722, 266–268, and *Neue Zeitungen* 4 June 1722, 436f.

[72]Quensel, 270f; summary of Quensel's article in *Neue Zeitungen* 11 March 1723, 183.

better than the dark of night. Swedenborg was not pleased; he wrote a response that was printed in the journal, but anonymously. 'Swedenborg's Friend' would admit no shortcomings.[73]

Most remarkable of all, however, is that the 78-year-old Swedenborg came with ten copies of his essay to the Board of Longitude on the day when it was to discuss the award of the prize to the instrument maker John Harrison.[74] Although the prize money was now given to a craftsman, Swedenborg the theorist would not concede. Instead he tried in the same year to convince the Swedish astronomers. But once again he was totally refused. That stubborn 'Old Swedenborg', wrote the astronomer Pehr Wilhelm Wargentin to his colleague Fredric Mallet, has 'not read anything more recent than what Riccioli wrote about this, and knows nothing at all about what has been added on this subject in the last 30 years, yet he believes his method is the best, indeed, the only possible one.' And Wargentin concludes: 'One might expect better of a man who knows arcana coelestia and can ask the spirits about everything.'[75] The professor of mathematics in Lund, Nils Schenmark, was also given a copy, and like most scholars before him he demonstrated the insurmountable difficulties of the method.[76] Swedenborg's response to Schenmark reveals that he was unable to accept and unwilling to understand his statement. He refused to let himself be convinced that he was in error.

Alongside this Swedenborgian obstinacy, the longitude problem also reflects man's search to find his bearings in the world, to reach the right harbour and his proper place on earth. Geometry and the longitude problem concern orientation in space, in the world, measuring our planet on land and at sea, ascertaining the distances between places and finding one's way there and back. Time after time Swedenborg returns to a metaphor in which thinking is conceived as navigation on the sea, when one can be cast off course by storms, driven by the currents, washed up on an unknown shore. Life and thought are a hazardous voyage across a turbulent sea. Travellers could appeal to the Lord in prayer to be allowed to survive through uproarious tempests. An official of the Board of Antiquities, Samuel Columbus, depicts life through the image of a sea voyage in *Odæ Sveticæ* (1674), where man sails against the wind, runs aground on sandbanks, is beaten against rocks, and tries to follow the light of a star over the sea.[77] Reason and mind are often regarded in philosophy as the body's navigator or captain, or as Descartes writes, it is not sufficient to say that a rational soul 'has its abode in the human body as the helmsman in his boat'.[78] In his dissertation on the lymphatic vessels, Olof Rudbeck

---

[73]*Amicum Responsum*, 315–317; MS in LiSB, N 14a, no. 149; *Photolith.* I, 203f; *Bokwetts Gillets protokoll* 17 August 1722, 70, cf. 20, 22; *Neue Zeitungen* 27 December 1723, 1012.

[74]Swedenborg to KVA, Stockholm, 10 September 1766. *Opera* I, 339f.

[75]Wargentin to Mallet, Stockholm, 18 September 1766. Nordenmark (1944–1945), 245f; the reference is to Riccioli's anti-Copernican work, Riccioli (1651).

[76]Schenmark to Swedenborg, Lund, 22 March 1767. *Opera* I, 340–344.

[77]Columbus, 20–23; Warnmark, 92–101.

[78]Descartes (1637), V; *Oeuvres* VI, 59.

the Elder writes that he was obliged to put names on the vessels he found in the body, so that he would not suffer the same fate as those who 'head out to sea and then forget the name of the harbour they seek, and thus expose themselves to the winds, without knowing whither they are striving or being carried'.[79] It was essential to find fixed points in thought which one can always have as a point of departure, sure, unchanging, and the same for everyone, just as voyages by sea require fixed harbours, reliable charts, compasses, and astrolabes. One of the few fixed points for a disoriented person—besides God, the central point of everything—was what was regarded as objective and iniquitous and always the same: geometry.

We Are Educated by Studying, Experiencing, and Thinking

Theoretical mathematics had a relatively modest position in Sweden, despite its enormous rhetorical potential, when Swedenborg joined the game. The mathematical inventions of the seventeenth century—logarithms and especially modern symbolic algebra, analytical geometry and infinitesimal calculus—turned mathematics away from physical ontology, that is to say, it left behind the properties of objects, in order to concentrate instead on relationships between objects.[80] In mechanics, for example, geometrical methods were increasingly replaced by mathematical methods such as infinitesimal calculus. Leibniz's calculus had led into hitherto unknown lands, wrote Guillaume-François-Antoine L'Hospital in the first textbook of differential calculus, *Analyse des infiniment petits* (1696).[81] Reality became increasingly abstract.

This was what Swedenborg as a mathematician needed to learn. But it was scarcely from the university lectures on offer that he acquired his knowledge of advanced mathematics. That could instead be achieved in two main ways. One could do as Polhem did, look at the figures in foreign books and teach oneself, or one could seek out someone who knew mathematics and take private lessons.[82] For Swedenborg it was particularly the latter alternative. The purpose of his first trip abroad, to London and especially to Paris, was to visit prominent mathematicians in order to learn the new mathematics. From London he wrote to Benzelius: 'I am now working through *Algebram et Geometriam subtilem*, intending to make such advance in these subjects as, in time, to be able to [*continue Pålhammar's inventions*].'[83] Benzelius's letter of recommendation for Swedenborg was of crucial significance throughout his study tour, granting him access to scholars. In return he was supposed to purchase books and instruments for the university and the

---

[79] Rudbeck the Elder (1653), ch. II.

[80] Mahoney, 703.

[81] L'Hospital, preface.

[82] *Polhems skrifter* IV, 394.

[83] Swedenborg to Benzelius, August 1711. *Opera* I, 215; *Letters* I, 30.

Collegium Curiosorum. For the university library he bought a couple of books in advanced mathematics, including Newton's variant of infinitesimal calculus, *Analysis per quantitatum series fluxiones, ac differentias* (1711) and perhaps the most important book in England about the new calculus, Humphry Ditton's *An Institution of Fluxions* (1706).

From mathematical studies in England and discussions on algebra in Holland with the Swedish diplomat Johan Palmqvist, Swedenborg travelled via Brussels and Valenciennes to Paris. His year's stay in Paris was filled with studies in geometry and algebra, and hour-long conversations about mathematics, above all with the mathematician Pierre Varignon, but also with the astronomer and geometrician Philippe de La Hire. The latter, who was acquainted with Göran Vallerius, was a central figure in French scientific life; he had made himself known for his studies of conical sections and had published several essays on geometry.[84] Swedenborg's conversations with La Hire no doubt concerned astronomy and longitude, since the French savant had previously published astronomical tables of the movements of the sun, the moon, and the planets. Swedenborg's contact with Varignon was through Benzelius, via the latter's good friend and correspondent, Abbé Jean Paul de Bignon, president of the *Journal des savants* and member of the Académie des Sciences. After L'Hospital, Varignon was a vigorous representative in France of the new differential calculus and had used infinitesimal calculus in essays to study various movements. Through these two, Bignon and Varignon, Swedenborg tried to get three of his essays published in the proceedings of the Académie des Sciences: a new algebraic analysis that could perform useful things that ordinary algebra could not do; another new algebraic method whereby an unknown quantity is sought through geometrical and arithmetic proportions; and his essay on longitude.[85] They were never accepted by the reputable journal.

Everything suggests that the acquaintance of these prominent French mathematicians was to be of great significance for Swedenborg's assimilation of the new mathematics. What has evaded the attention of all those who have tried to assess Swedenborg as a mathematician is that in Paris in September 1713 he had bought an advanced textbook in the new calculus based on Leibniz's version, Charles René Reyneau's *Usage de l'analyse* (1708). In an extant copy he has not only written his signature and drawn up his own table of contents, but also made numerous marginal notes at various sections about bomb trajectories, curve lines, parabolas, logarithmic curves, cycloids, differential and integral calculus, tangents, the great and the small, and so on.[86] At the very back of the book, in Swedenborg's own hand, we read: 'Erudimur discendo, experiendo et cogitando', that is, 'We are educated by studying, experiencing, and thinking.'

---

[84]UUB, G. Vallerius; 'G. Wallerii bref till Mr de la Hire Oct. 1712. Om grufvor, bergverk och deras förenade observationer'. LiSB, N 14a, no. 68, fol. 157–161.

[85]Swedenborg to Benzelius, Paris, 9/19 August 1713. *Opera* I, 222.

[86]Reyneau (1708), II; Swedenborg's copy is preserved in Stockholm University Library (SUB); it is the second part of Reyneau's *Analyse demontrée, . . .*, Paris 1708; Swedenborg's table of contents in *Usage de l'analyse* concerns different passages between pages 630 and 824.

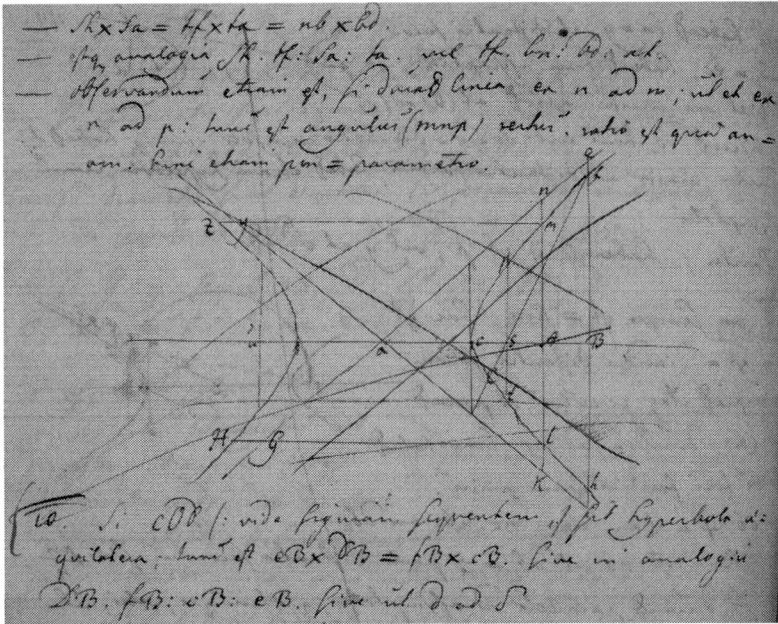

Fig. 4 Analysis of the hyperbola. Swedenborg's manuscript *Geometrica et algebraica* was presumably written during his mathematical studies in Paris in 1713–1714, when he used Reyneau's *Usage de l'analyse*

It also turns out that one of Swedenborg's mathematical manuscripts, *Geometrica et algebraica*, is actually based on Reyneau's *Usage de l'analyse*.[87] In this manuscript Swedenborg considers differential and integral calculus for determining subtangents, curve lengths, areas, volumes, minimum and maximum, and describes different geometrical figures, such as the circle, the ellipse, the parabola, the hyperbola, the cycloid, the spiral, etc (Fig. 4). Several different suggestions have been put forth as regards *Geometrica et algebraica*. First of all, it has been assumed that it was started in 1719, secondly that it was intended as a continuation of his unfinished Swedish textbook in algebra, *Regel-konsten* from 1718, or as a freestanding book in Latin.[88] None of this seems to be correct. Instead my thesis is that *Geometrica et algebraica* was not written in 1719 but in connection with his mathematical studies in Paris in 1713–1714. Moreover, it was not at all intended for publication, but was simply his own study notes or lecture jottings. After his stay in Paris the notebook was used for a very long time, at least a good way into the 1730s, for notes on a wide variety of subjects, everything from history to

---

[87] *Photolith.* II, 1–100; Rodhe, 42–48; Rodhe has arrived at this through mathematical analyses of the manuscript. In a forthcoming article he explains Palmqvist's importance for Swedenborg's mathematical education.

[88] Eneström (1890), 27f; Hyde, no. 143; Nordenmark (1933), 2.

astronomy. Among other things, he took notes from Wolff's *Ontologia* (1730) about Euclid, Descartes, Newton, and so on about the division of geometrical figures into genera and species.[89] From Pieter van Musschenbroek he also took descriptions of the strength and geometrical forms of different kinds of timber, such as cylindrical alder, square oak, cylindrical elm, etc.[90]

When weighing up *Geometrica et algebraica* it is necessary but not sufficient to look at the content. One must also ask why he wrote it and what function it filled. We then arrive at a different conclusion from those previously suggested. We know that Swedenborg intended to learn higher mathematics by seeking out the people who had already mastered it, such as Varignon, La Hire, and Palmqvist. We also know that he bought *Usage de l'analyse* in Paris in 1713, which later formed the basis for his manuscript. One possibility is that it happened as follows: in Paris, where he had an excellent opportunity to learn the subtleties of mathematics, he took the letter of recommendation he had received from Benzelius, visited Varignon, and asked him for advice on how he could proceed in his mathematical studies. He may have said, 'I want to study higher mathematics, how should I go about that?' 'Read this,' Varignon would no doubt have answered, showing him Reyneau's book, a book that he used himself, and gave him some suitable exercises. 'Do this and that, and then we can meet again in a while'. But it is very likely that he already became familiar with Reyneau's book in Utrecht while discussing mathematics with Palmqvist, who owned this book.

It is not inconceivable that Swedenborg received some private tuition from Varignon, or possibly attended his lectures in mathematics at the Collège Mazarin. A less likely alternative is La Hire, who was sceptical about infinitesimal calculus and Newton's laws of dynamics.[91] Since Swedenborg's knowledge of French was probably inadequate, there remained only one language they could communicate in—Latin. Varignon had to be able to read and correct Swedenborg's calculations. That is why *Geometrica et algebraica* is in Latin and not in French, which would perhaps have been more natural if it had been a mere collection of excerpts. That is also why it is not in Swedish either, as it ought to have been if it had been envisaged as a continuation of *Regel-konsten*. It would have been somewhat odd if he had used a French book which he translated into Latin so that he could then publish it in Swedish. *Geometrica et algebraica* may therefore have been written under the supervision of Varignon, which would also explain the unusually high—for Swedenborg—mathematical level, far above anything else he wrote, a work virtually free of the miscalculations and careless errors with which he otherwise strewed his works, generously and heedlessly. It therefore cannot be automatically assumed to be a totally independent work. The differences that go beyond Reyneau's book, as well as the corrections and changes in the manuscript, may have to do with the tutorials in mathematics that he received in Paris in 1713–1714.

---

[89] *Geometrica et algebraica*, 278; Wolff (1730), § 246; notes from the same section are also found in KVA, cod. 88, 286f.

[90] *Geometrica et algebraica*, 205–223.

[91] *Dictionary of Scientific Biography* VII, 577f.

Unrest Disturbs My Work

Education, that beautiful edifice of wisdom, rests on the twin foundation of language and mathematics, wrote Polhem, 'just as a large house, church, castle, palace, etc. necessarily requires two foundations, namely one below the ground and one above; whereas smaller and meaner houses require only one.' It must be a free-standing building, the wisdom of which has a greater pleasure if the foundation is 'pure truth' than if it is one based on other people's opinions and authorities. The first foundation of wisdom is the ability to read, write, speak, and understand one's own and other people's tongues, while the second is the ability to 'count, measure, compare, present, and execute all visible and palpable things to the benefit of oneself and others.'[92] An introduction to the latter wisdom is given by Polhem in the work edited by Swedenborg, *Wishetens andra grundwahl til vngdoms prydnad mandoms nytto och ålderdoms nöje* ('The Second Foundation of Wisdom, for the Embellishment of Youth, the Utility of Manhood, and the Pleasure of Age', 1716). This is a classical metaphor for theories aiming at sure and absolute knowledge. Theories are like buildings which require a sure, solid, and permanent foundation on which the superstructure can rest securely. In it Polhem explains elementary mathematics in a concrete way. Geometry or 'The art of measurement teaches how to measure lines and distances, arable fields, meadows, forests, a room, vessels and measures, corners, heights, and inaccessible distances.'[93] The explanation for a right angle is also given with examples, such as the corners that tables, chests, and books have.

Swedenborg himself would make an attempt at the genre of textbook with his *Regel-konsten författad i tijo böcker* ('Algebra Composed in Ten Books', 1718), a manual of algebra which some people think did not entirely fulfil its educational function. In January 1718 Swedenborg wrote to Benzelius that he used his spare moments at Brunsbo to write a work on algebra in Swedish without having any book or anything else to assist him.[94] Many people had started to read this subject in Lund and Stockholm and would need such a book, he said, asking whether Benzelius could help him to get it printed, and whether there was any 'algebraicus' in Uppsala who could check the proof. There may have been a purpose behind *Regel-konsten*, apart from the purely educational one. It could to some extent count as a qualification. Professor Elvius had died on 12 January 1718, thus leaving a potential opportunity open for Swedenborg. Benzelius encouraged him to apply for the vacant chair, but Swedenborg declined, even though he had entertained the idea of a post as professor in Uppsala for some time.[95] At any rate, *Regel-konsten* was

---

[92]Polhem (1716), [i]; cf. Polhem, 'Den menskeliga wijsshetenss fördelning och korta förklaring för des älskare ibland den ostuderade hopen'. KB, P 20:1, 125.

[93]Polhem (1716), § 3.

[94]Swedenborg to Benzelius, Brunsbo, *c.* 11–12 January 1718. *Opera* I, 276, cf. 283.

[95]Swedenborg to Benzelius, Brunsbo, 21 January 1718. *Opera* I, 278f, cf. 231, 245f.

sent in the same month to the printer Johan Henrik Werner in Uppsala, and Benzelius managed to find an Uppsala student, Nils Hasselbom, who could check the figures, a task that he evidently discharged carelessly. Hasselbom later became professor of mathematics in Åbo (Turku) in Finland.

Regel-konsten, despite the title, consists of only seven books. The other parts— about roots, differential and integral calculus, and all the figures—never appeared. Swedenborg wrote that he had three aims. Mathematics was to be explained so that it could be easily understood. He wanted to demonstrate the utility of mathematics for surveying and ballistics, and he wanted to coin new Swedish mathematical terms.[96] The use of the word *regelkonst*, literally 'rule-art', for *algebra* was presumably inspired by Georg Stiernhielm, who followed Simon Stevin's thoughts about mathematical terms in the national languages, and among other things created a series of new Swedish terms such as *runde* 'round' for circle and *fyrägg* 'four-edge' for square.[97] Besides algebra with Descartes' algebraic symbols and geometry, including analytical geometry, it contains quite a lot about applied mathematics, mechanics, screws, blocks, cogs, jacks, sloping surfaces, and other forces. In a manner similar to that of Polhem, under the heading 'Concerning some necessary terms' Swedenborg introduced a number of definitions of geometrical concepts which can be regarded as concrete examples rather than strict mathematical definitions.[98] Some samples may illustrate this: 'A circle or a round line that comes together', says nothing about the centre or radius; and 'A curve is any line that bends' simply refers to curves in general. A *kastlinie*, that is, a 'throw-line' or parabola, 'is a line made by bombs and everything else that is thrown.' A parabola is also described by the way ropes hang, water sprays, sails bulge in the wind, and by the shape of church vaults, bridges, dams, and ships. The hyperbola or *Öfwer-kast-linien* 'is likewise a curve that bends even more', but nothing is said about conic section or eccentricity. Such lines are bent by an external force, such as barrel hoops, fishbones, twigs, and can be used for megaphones, shawms, French horns, and lenses. He then continues with the area of a circle as 'A round, a circular board. also called a trundle, plate, disc', and the cylinder is a roller of the same thickness all the way along. It is obvious that his purpose here is not chiefly to provide unambiguous mathematical definitions, but rather to give examples and to introduce his own Swedish mathematical terminology, inspired by Stiernhielm. His new Swedish terms enjoyed some use. Among others, the surveyor Eric Agner employed Swedenborg's word for area, *omkast*.

Swedenborg's attempt to translate Latin mathematical terms into Swedish reveals an oral situation. The words were supposed to be spoken, not just seen. This difficulty in finding Swedish scientific words can be seen all the time in his letters and essays in Swedish, which not infrequently are formulated in a mongrel language. When Swedish words failed, one had to resort to Latin. Polhem, who

---

[96] *Regel-konsten*, preface, 2f.

[97] Nilsson (1974), 164f; Nilsson (1975–1976), 125; Nilsson (1992), 44f, 57.

[98] *Regel-konsten*, 4–6, cf. 84–86.

knew hardly any Latin, but also Roberg, von Hoorn, and Rydelius, tried to coin new Swedish words. The mother tongue had the advantage of being more concrete and transparent because it was closer to everyday thought. But Latin, with its scientific, more or less fixed terms, made certain kinds of learning and thinking possible. Written Latin, unlike the language used in everyday life, seemed less emotionally charged and could be associated with learning and objectivity. The capacity for thought had to be honed on the gleaming steel of quotations from Roman authors. Latin therefore functioned very well for abstract matters and science, not least for Swedenborg's later rational theology in which an eternal struggle was waged between reason and emotion. Swedenborg not only wrote but also thought in Latin. After his experiments in *Regel-konsten* and other essays from the 1710s he abandoned Swedish as the language of his writings and used Latin exclusively, partly because he was aiming at an international readership from the start of the 1720s. For this, only Latin would do.

In comparison with *Geometrica et algebraica*, his *Regel-konsten* is of an elementary character, besides which it has a rather astonishing wealth of misprints and miscalculations; no less than 81 have been counted.[99] For this reason, a modern-day critic's verdict on *Regel-konsten* was that it is not at all boring; on the contrary, it is the most entertaining thing Swedenborg ever wrote.[100] Many people have therefore concluded that Swedenborg could not have been a mathematician, or at least not a good one. Another conclusion that has been put forward in modern times is that an intradisciplinary mathematical analysis of *Geometrica et algebraica* reveals that Swedenborg was in fact a good mathematician.[101] We now have to decide: was Swedenborg a good or a bad mathematician? In reply to that one can answer that, as the question is posed, virtually all of the more detailed interpretations of Swedenborg's mathematical knowledge are historically misleading and not wholly convincing, irrespective of whether he is judged to have been a good or a bad mathematician. If one absolutely must answer the question, one first has to establish what a 'mathematician' *was* and then what a 'good' one *was*, not *is*. Being able to handle advanced mathematical operations or being guilty of many miscalculations does not provide an immediate answer to the question. The historically relevant conclusion is the one that views Swedenborg in relation to his own times, his context, what others thought about him, and his ability to apply his mathematical knowledge to the problems he set himself. Finally, one must ask whether an assessment of his mathematical ability is historically rewarding, and if so what it can say to us.

To begin with, the words 'mathematician' or 'mathematicus' had a somewhat different meaning in Swedenborg's time, a wider sense that did not necessarily require advanced knowledge of algebra and infinitesimal calculus; it referred more to a person working with natural science, chiefly from a mathematical and rational

---

[99]Eneström (1890), 21; Eneström (1889), 529–531.

[100]Kleen, I, 168.

[101]Rodhe, 48f.

point of departure. In that sense Swedenborg was undoubtedly a 'mathematicus', and this was what he was called by his contemporaries. It is harder to determine what is meant by a 'good' mathematician. We receive some guidance as to what people in Swedenborg's time thought should be the properties of a good mathematician from Harald Vallerius's dissertation, *Theorema de matheseos incrementis* (1694). In a corollary Vallerius explains what is required of a good mathematician: he should not only be familiar with what the ancients had discovered but should also know of more recent findings in the field and add something new himself.[102] Swedenborg certainly fell into this category by his contemporary standards. He knew his Euclid, he was well versed in the new calculus, and he also tried to create a new Swedish terminology.

When someone is called a good mathematician there are often rhetorical undertones. One such case, concerning King Charles XII, will be examined in the next chapter. For the moment it will suffice here to say that there may be personal advantages in calling a particular person a 'prominent mathematician' and exalting his knowledge of the subject, especially when one is dependent on him or in a generally inferior position. A statement to a third party can in certain cases give somewhat better guidance. Polhem, for example, in a letter to Benzelius calls Swedenborg 'a quick Mathematicum and a competent aspirant to mechanical knowledge'.[103] Rudbeck put in a good word for Swedenborg, 'whom I know to be skilled in Mathesi', when there was discussion of appointing a successor to Johan Vallerius as professor at a meeting of the board of Uppsala University in June 1719.[104] But since Swedenborg had not applied for the position, it went to the professor of Greek in Lund, Elof Steuch. In *Acta literaria Sveciæ* for 1721 there was a long review of *Regel-konsten* by Burman, who says that Swedenborg was born under a lucky star for the progress of learning, especially mathematics.[105] The following year, in the German journal *Neue Zeitungen von gelehrten Sachen*, Swedenborg is mentioned as a benefactor of mathematics and the first in Sweden to present the basics of algebra. He 'seems to have been born for the promotion of mathematics', the reviewer observes.[106]

In a translation and adaptation of a German textbook, *En klar och tydelig genstig eller anledning til geometrien och trigonometrien* ('A Clear and Distinct Pathway; or Guide to Geometry and Trigonometry', 1727), the Swedish publisher, Johan Mört, emphasizes in his foreword the significance of mathematics for education. It is essential to 'diligently remove ignorance from the young minds, to clear away all false concepts, and in their place let them discover sound and pure thoughts, and

---

[102] H. Vallerius and Dryander, 5; Eriksson (1967–1968), 165f.

[103] Polhem to Benzelius, Stjärnsund, 10 December 1715. *Polhems brev*, 114, cf. 112, 115, 117.

[104] UUB, Academic Consistory minutes, 10 June 1719, 147–150; *Green Books* II, no. 176. SLBA.

[105] Burman, reviewed in *Acta literaria Sveciæ* (1721), 126f; *Bokwetts Gillets protokoll*, 14, 36.

[106] *Neue Zeitungen* 14 May 1722, 378; translation in 'Swedenborg and his scientific reviewers', in *The New Philosophy*, January–June 2003, 463.

increasingly guide them to the truth'.[107] Alongside Polhem, Duhre, and Triewald, the publisher pays tribute to Swedenborg for having used the mother tongue to ensure that young people would warmly embrace the mathematical sciences. Later on *Regel-konsten* continued to be praised, most remarkably in a dissertation written under the supervision of the leading mathematician in Sweden in the eighteenth century, Samuel Klingenstierna. In *De usu algebræ* (1743), defended by Lars Julius Kullin, Swedenborg's book is held up as one of the pioneers in algebra, together with the works of Newton and Richard Sault.[108] Furthermore, Swedenborg was described by his friend Anders Johan von Höpken as a skilled and thoroughly learned mathematician, and the historian Sven Lagerbring stated that Swedenborg is probably 'among the first to have written about Differential and Integral Calculus in Sweden, although that chapter was not printed.'[109]

From this we may assume that, by the general standards of mathematical education in Sweden in the 1710s, Swedenborg's mathematical knowledge was regarded as being at a relatively high level, and he was considered a pioneer and introducer. But he was not alone in Sweden at the time in possessing a knowledge of the new calculus. Polhem had acquainted himself with the calculus and recommended algebra, infinitesimal and differential calculus as the best grindstones for sharpening the brain, and as good touchstones for ascertaining which brains were suited for higher studies.[110] In the mid 1710s Johan Vallerius lectured on 'the most subtle invention of our time', that is, differential calculus.[111] The same year that Swedenborg published his algebra book, he also acquired a rival in the textbook genre, an old student friend who was more successful. This was, as it says in a memorial sketch, the absent-minded and unpractical Anders Gabriel Duhre; although unsuited to an academic career, his private tuition was of great significance for Swedish mathematics.[112] His pupil, Georg Brandt, published his lectures which had been delivered in Swedish in 1717, *En grundelig anledning til mathesin universalem och algebram* ('An Exhaustive Guide to Universal Mathematics and Algebra', 1718).[113] It was addressed to the members of the Board of Mines, including Swedenborg. The preface emphasizes the utility of mathematics for manufacture, war, trade, and so on. Mathematics sharpens the mind, just as the

---

[107]Weidler, translation, [v, xf].

[108]Klingenstierna and Kullin, 6, 15f; refers to *Regel-konsten*, 61. The same Kullin later became a harsh critic of Swedenborg's spiritual doctrine in the Gothenburg Cathedral Chapter. Lenhammar (1966), 73, 78.

[109]Höpken to Christian Tuxen, Skänninge, 11 May 1772. Höpken, 461; Lagerbring, IV:3, 49f; cf. Prosperin, 11.

[110]Polhem, 'En lijten, dåck grundel. underre[te]lsse, om analysis infinitorum eller den sinrijka räkenkonsten, som tiena kan, icke allenast för en brynsteen att wässa hiernan uppå, utan och för en god probersten att huilka hiernor som till höga studier ähro bequema'. KB, X 705:1–2.

[111]Lindroth (1989), II, 474; see also *Catalogus prælectionum publicarum* 1715.

[112]Strömer, 12.

[113]Brandt (1718), foreword & preface; cf. Duhre (1721); Swedenborg owned Duhre (1722).

body is strengthened by physical exercise. It was not until the end of the 1720s that Swedenborg was seriously overtaken by younger mathematical talents, especially Klingenstierna. Swedenborg's heyday as a mathematician was over.

Merely committing an error in calculation does not necessarily make a person a bad mathematician who should be disqualified outright. Perhaps it could be objected that an experienced and good mathematician would immediately and intuitively have *seen* the mistakes in his calculations, as Swedenborg ought to have done in *Regel-konsten*; by this is not meant an inexplicable intuition, but the use of imprinted knowledge based on previous experience. But if one is not allowed to make mistakes, then Leibniz would not be numbered among the geniuses of mathematics, which no one doubts that he was. Leibniz miscalculated things in the epoch-making article where he put forward his method for studying curves and for the first time presented the new differential calculus.[114] For Swedenborg the important point was not whether he calculated correctly or not. For him the idea was almost always the primary thing, while the implementation of the theory was secondary. What mattered was the vision, the question, the thought process, not the practical execution or the finished product. There is scarcely a work by Swedenborg that is finished and complete. Virtually all his writings, both manuscripts and printed works, are unfinished torsos of enormous visions. They were drafts, sketches, the first test of an idea, or as he himself sometimes titled them, 'prodromus', a precursor of the total work that would be completed in the future. The prodromus is the north wind that forebodes the rise of Sirius, or as he perhaps regarded it, a stormy wind before the rise of his own star. But everything he undertook remained preliminary, half-finished, never completed projects. It was the vision, the initial idea that was crucial. That is why he could hand in *Regel-konsten* for printing without careful proofreading. The work was 'finished' inasmuch as the process of thinking it out and writing it down was over. The perfected work was already in his head. It was just an insignificant technical matter to transfer the idea to paper. Swedenborg was an impatient person. 'Literary occupations are my amusement every day,' he wrote in a letter to Benzelius in 1715. 'It is impatience alone that causes me some little unrest, and unrest somewhat disturbs my affairs here.'[115]

Another difficult test of 'the good mathematician' is the extent to which he can set up his own mathematical problems and draw on his mathematical knowledge to solve concrete problems, for example, in mechanics and physics. To answer the question it is essential to look at the purpose behind the mathematical writings and the problems he had to solve. When it comes to *Geometrica et algebraica* the chief aim was no doubt to learn the new calculus for himself. It may also have been used as his own compendium or summary of calculus for future needs. His notes on geometry show that Swedenborg was talented and understood the basics of the new calculus well, but not necessarily that he could have been an independent and creative mathematician. The real baptism by fire is instead if he could solve

---

[114]Leibniz (1684), 467–473.

[115]Swedenborg to Benzelius, Stockholm, December 1715. *Opera* I, 233; *Letters* I, 74.

new problems with the calculus, if he could apply it and transfer his theoretical mathematical knowledge to practical problems in mechanics. That is the difficult step, a step that those whom history calls 'great' mathematicians succeeded in taking, for example, Huygens, Newton, and Bernoulli. In Swedenborg, however, we find no serious attempts to use his advanced mathematical knowledge in physics. His mathematics and geometry can instead be said to have been part of a rhetoric for which he exploited their aura of certainty, rationality, and objectivity. He constantly claimed that natural philosophy builds on geometry, and he almost always uses geometrical concepts in his argumentation, but he rarely uses purely geometrical proof. The geometrical calculations that he inserted in his *Principia* do not seem to add very much or serve as proofs; they function more as rhetorical illustrations in figures of what he says in words and letters.

Finally, the question of the good mathematician is in many cases a matter of the time in which the historian of mathematics is living. A good mathematician belongs to the good, enlightened side, while a bad one belongs to the bad, dark side. With a negative verdict one can show how clever we are in our own time. We have come so much further! Distinguishing bad from good mathematicians is a question of deciding by the criteria of modern mathematics who belongs in history and who does not, of eliminating the losers from the winners' history. In a strict sense the historically relevant answer to the question whether Swedenborg was a good or bad mathematician is that he was regarded by his contemporaries as a prominent mathematician, that he had an unusually advanced mathematical education by Swedish standards, but that in the course of the 1720s mathematical 'development' in Sweden overtook him.

From Barbarism to Culture

Mathematics occupies a modest place in surveys of the history of ideas, even though most scholars acknowledge the crucial role of mathematics in the creation of modern science. The history of mathematics has not infrequently been a history viewed from within mathematics itself, a story of success and salvation within the discipline, the tale of those who calculated right, a history of the victors, thrusting forward with the arrow of advancement constantly pointing the way out of the scientific darkness, onwards and upwards to the enlightened present.[116] Those who counted wrong need not apply for a place in this history. This is partly due to the view of mathematics as a transcendent and abstract science. Mathematics is thus treated as a non-empirical regular world, not tied to earthly, bodily, human reality, but belonging to an eternal, immutable Platonic world of ideas. What can politics, economics, philosophy, art, and language be other than insignificant garnish, an appended ornament?

---

[116]There is, it is true, a genre of mathematical history which includes 'cultural history' in the title, but the term is virtually always used as an equivalent to 'popular history', giving the freedom to scatter anecdotes, illustrations, and timelines.

This was how Swedenborg regarded mathematics too. That was what attracted him to the promised land of geometry. Geometry functioned not only as a means to create order in spatial perception; it was not just a practically useful tool to describe nature and solve technical and scientific problems. There are some other factors that can explain why Swedenborg and other mechanists elevated geometry to the chief of sciences. One is that geometry was believed to discipline and civilize people; another is that it was objective, representing an ideal scientific method for acquiring knowledge and arguing proofs; moreover, it belonged to a transcendent reality beyond man, eternally true and rational. The perception of geometry as objective and transcendent is an underlying idea that is taken for granted, unquestioned.

But mathematics was created by man. A cognitive history of ideas would take a different view of the history of mathematics, bringing a new perception that may be felt as an iconoclastic sacrilege, making the divine human. Like other forms of culture, mathematics is a work of human beings; it proceeds from and is shaped by the human brain, by our conceptual system and our imagination, the experiences of the body and consciousness, our orientation and action in the world, and as such it also belongs to a cultural context. It arises from everyday experience of spatial relations, from human activities such as counting, measuring, building houses, playing games, from movement, change, the grouping and manipulation of symbols. Mathematics is grounded in human thought and is therefore—no more than rationality—transcendental, unique, or independent of man as a living being. From the nature of human rationality, in interaction with culture and history, it emerges as studies of structures that we use to understand and think about our experience—structures that are inherent in our preconceptual, bodily experience and that we make abstract with the aid of metaphors. Like literature and poetry, mathematics is a cultural phenomenon that concerns how we create terms for the world around us, using mental images and concepts to depict and express our perception of reality. Geometry is a way to try to understand our spatial experiences, and it concerns the human activity of making shapes and forms which are then used in order to understand other experiences of reality. Mathematics is not out there, beyond and independent of humans, eternally true and objective. Instead the mathematical descriptions of scientific laws are made by man in attempts to characterize the regularities that are experienced in the physical universe. The 'fit' between the world, that is to say, the regularities in the physical universe, and mathematics is something that exists in our thinking. The numbers and geometrical forms are in the head, not in a regular universe. It is the human ability to understand experiences with the aid of basic cognitive concepts that is the key to the progress of mathematics.[117]

It is essential to consider how people really think and interact with their environment and culture. Several historically important cultural ideas and world-views outside mathematics have managed to change the content of mathematics. There are rival schools in mathematics, ideas that are accepted by most people

[117]Lakoff, 354f, 366; Núñez and Lakoff, 109–124; Núñez, 333–353.

but not by everyone. Its theory formation is important, but it has turned out that a judgement of what is a good or bad mathematician can be misleading if one sticks to an internal historiography of the discipline. The historian of science finds it difficult to be an internalist and a Platonist, but tends to regard mathematics and science as an invention, not a discovery, by humans, produced by a culture and a social context, grounded in human cognitive capabilities. The historian is forced to have his hands tied behind his back, to put away his modern knowledge and notation, obliged to solve the mathematical problems with the tools of the time in order to try to see the period through the eyes and with the prior knowledge of the people who lived in it.

A cognitive history of mathematics considers the way people think, in metaphors, in circles, spheres, spirals, infinite series, or the way they draw incontrovertible conclusions from axioms and postulates. Mathematics has become a sign system through which people express themselves, a cultural phenomenon like art, music, literature, and poetry. The poetry of geometry expressed a dream of clarity, simplicity, order, and harmony. This was what made Swedenborg admire geometry. He realized the enormous strength of calculus, and for him infinitesimal calculus became the symbol of the boundless potential of advanced mathematics. The crucial thing here is not whether he could count or not, but his fascination with mathematics and his obvious 'geometrical thinking', or what could be called 'geometrism'. His writings, right from the early scientific efforts to the last theological works, are filled with geometrical analogies, signs, waves, spheres, points, spirals, infinitesimals, and other mathematical concepts and figures. The point here is the constant attraction of geometry to human thought and people's fascination with the glass-bead game of geometry. This is an appeal on behalf of all the bad mathematicians, all those who miscalculated, but who dreamed of the clear, simple, rational, non-corporeal, and eternal. Swedenborg was one of them. Those who were wrong and are condemned to eternal silence demand redress. It is a matter of freedom of speech, the right of dead books to make themselves heard.

A truly educated person during the baroque era had to know mathematics. In particular, he had to master the practical side of mathematics, fields such as fortification, surveying, and navigation. The theme comes up constantly in popular writings of the time, for example, in Polhem and Anders Celsius.[118] Mathematical knowledge could help people to arrive at the pure truth, find guidance for reason and the senses; it could quite simply equip young men with a clean and tidy head. Mathematics makes us disciplined, civilized, cultured beings, it was believed. '*Arithmetic* and *Geometry*', Roberg wrote, 'have transformed entire peoples from barbarism to *culture*.'[119] Euclid's *Elementa* could convert heathens, according to Christoph Clavius, in that the persuasive power of mathematics demonstrated an order that must have its origin in God.[120] When Aristippus was shipwrecked off

---

[118]Polhem (1745), 78; Celsius (1743).

[119]Roberg (1747), 27; cf. G. Polhem, 2f.

[120]Engelfriet, 30f.

Rhodes he found geometrical figures drawn in the sand, which convinced him that the island must be inhabited by civilized people.[121] Anyone who knows mathematics is civilized.

A famous Greek motto adorns the title page of the first scientific book in Swedish, Stiernhielm's *Archimedes reformatus* (1644), which illustrates how geometry can build walls. This is the motto that is said to have been inscribed over the entrance to Plato's Academy: 'Let no one enter without a knowledge of geometry.' Plato guarded the entrance to the holy inner room of philosophy in the work of the ballistics expert Niccolò Fontana (Tartaglia), and the motto was also used by Mikołaj Kopernik (Copernicus).[122] Geometry was the key to philosophical truth, dividing people into initiates and non-initiates, rational and irrational, civilized and uncivilized. The view of mathematics as objective and transcendent thus serves the purpose of holding the mathematical community together, as an élite writing solely for the initiated. This alienates outsiders, besides which the myth indirectly stratifies people socially and economically. Mathematics has a social function of excluding people from participation in the discussion of natural philosophy. There were thus social advantages for Swedenborg to occupy himself with mathematics.

Mathematics also has an important role to play in the disciplining of man in history. Disciplining has almost always meant social disciplining, learning to control one's feelings and behaviour, that the state, work, religion, and schools should foster, civilize and impose order. But the disciplining of geometry is an even more serious threat to the freethinker; it reaches to the depth of human thought. Thoughts have to be ordered according to logical principles. Man must be forced to think rationally, to draw specific conclusions from the dogmatically established axioms. With mathematics one becomes a civilized, disciplined, efficient, controlled, and rational being. Geometry is something that can be perceived as a creator of order, giving concentration and instilling confidence and security. For the chemist Robert Boyle mathematics, especially the laborious operations in algebra, is the best medicine for curing wandering wits.[123] It disciplines and forces the mind to pay attention, indicating a road away from the lax habits of romances.

The Immutable World

People want it to be true that mathematics is sure, that there is something certain and absolutely true in this world of uncertainty, that there is an order, that we can be rational, logical, and sure of our conclusions. Eulogies of mathematics were common at this time, in dissertations, in learned academies, in textbooks for the growing generation, in almanacs for merchants and farmers. Mathematics was

---

[121]Fontenelle, edition, 32f; cf. Edwards, 113–116.
[122]Tartaglia, translation, 18f, 63ff.
[123]Johns, 381.

hailed for being precise, consistent, stable in different ages and societies, under-standable across cultures, efficient as an instrument for description, explanation, and prediction. Geometry is the ideal objectivity.[124] The objectivist stance means, among other things, that concepts and thinking are transcendental and independent of the bodies of the thinking beings, that thinking is a mechanical manipulation of abstract symbols that are meaningless in themselves but are given meaning through correspondence with the things in the world. Mathematical concepts seem to be the same from all reference points, at all times and in all situations. This is stable and immutable knowledge. Three statements above all apply to mathematics.[125] First, it is objective, its truths are universal, absolutely certain, it is abstract and independent of man and the body, and it transcends or surpasses human beings. Second, its efficiency as a scientific tool led to the assumption that it exists in the structure of the physical universe; to learn mathematics was simultaneously to learn nature's language. Finally, mathematics is characteristic of logic and rational thought.

Geometry is of epistemological significance, as the very model of rational thought, of objectivity and the perfect method for arriving at sure knowledge. It was a matter of 'surety', finding the sure method for thought and science, a solid foundation on which everything could rest. The word of God was one solid foundation and geometry the other, God's Bible and Euclid's *Elementa*. The 'exactness' of the mathematical craft can be interpreted as a series of properties, such as correctness, legitimacy, rightness, precision, and certainty.[126] Mathematical proofs also had beauty. The 'most beautiful' or 'most elegant' solution can mean the 'simplest'; the most economical one requiring the smallest number of assumptions, or the one that is in some sense least 'alien' or most 'familiar'.

The geometrical or axiomatic-deductive ideal of science, the geometrical manner, 'more geometrico', goes back to Aristotle's *Posterior Analytics* (fourth century BC), but can also be associated with Parmenides of Elea and Plato, and it has its best-known application in Euclid's geometry. Briefly, this ideal requires that a science be built up with the aid of a hierarchical system of statements, either axiom or theorems. Axioms are evident, obvious truths, which are supposed to be general, necessary, universally valid, and incapable of being derived from or proved by other truths. Theorems are truths which can all be derived by deduction from the first type of truths, the axioms. From axioms and postulates it was possible to deduce indubitably true theorems, true for all people, everywhere and always, necessary and not contingent. Deductive thought was supposed to set the rules for scientific thinking, rules for deriving theorems from axioms.

An axiomatic-deductive system builds on a classical idea that categories are sets of objects with shared properties.[127] Axioms characterize precisely these essential properties which are independent of each other. With Euclid's *Elementa*, the idea of

[124]Lakoff, 154, 179.

[125]Lakoff and Núñez, xv; 339f.

[126]Bos, 412.

[127]Lakoff, 10, 220, 225f; Dunér (2006a), 65–85.

essences is led to mathematics. As a practical application of the axiomatic-deductive ideal, the *Elementa* served as a model for most sciences. The programme of Euclidean geometry sought to demonstrate how *all* truths in geometry can be shown to follow, with the aid of thought alone, from a small number of clear and intuitively obvious definitions, together with a small number of clearly comprehensible and obviously true propositions. All geometrical truths in plane geometry were to be derived from just five postulates. The clear, simple structure of Euclidean geometry is, in a way, a mathematical counterpart to the simplicity of Greek temples, which, like geometry, express order and harmony.

The geometrical ideal of science exerted a huge influence. In particular, it became closely associated with the mechanistic world-view that was Swedenborg's premise, according to which geometry expressed order and absolutely necessary validity, certainty, and clarity. The world was nothing but one large geometrical system. For Descartes, universal science would have mathematics as its pattern and method. Everything would go together in the same way as in the long, simple chains of geometry. I write about salt, snow, the rainbow, but 'the whole of my physics is nothing other than geometry.'[128] In his radical doubt in *Meditationes de prima philosophia* (1641) he ended up disbelieving all theorems, including analytical theorems such as 'a square has four sides', which seem to be true whether one is dreaming or awake. But a malevolent demon may have deceived him. The loss of geometry's certainty and surety fills him with horror: it felt 'as though I had suddenly fallen into very deep water, I am so taken unawares that I can neither put my feet firmly down on the bottom nor swim to keep myself on the surface.'[129] In his quest for the immobile Archimedean point he found rescue: as long as I think, I am something. From this incontrovertible 'axiom' he is able to systematically build up an entire philosophy. In one text Swedenborg travesties Descartes' famous statement, when he writes about geometrical truths that there is a kind of analysis in thinking, a sort of analogy, and with this analogy or analytical approach one can arrive at the conclusion 'I doubt, therefore I am rational.'[130] Geometry is in large measure about certainty, objectivity, rationality, and firm foundations. In the quest for certainty, as Descartes writes, his design had to be to cast 'aside the shifting earth and the sand in order to find the rock or the clay.'[131]

Blaise Pascal sums up the geometrical spirit of Cartesian natural philosophy in the words: 'What is beyond geometry is beyond us.'[132] In *De l'esprit géométrique* (*c.* 1658), Pascal explains that, thanks to natural light, geometrical science can penetrate nature and its wonderful properties. The true method consists 'in defining every term, and in proving every proposition.'[133] In Pascal and in the geometrical

[128]Descartes to Mersenne, 27 July 1638. *Oeuvres* II, 268.

[129]Descartes (1641), II; *Oeuvres* IX, 18; translation, 102.

[130]*Psychologica*, 6; cf. Jonsson (2000), 10.

[131]Descartes (1637), III; *Oeuvres* VI, 29; translation, 47.

[132]Cited in Cassirer (1970), 143; cf. Pascal (1986), commentary, 76.

[133]Pascal (1657/58); translation, 429.

ideal there is a quest for the simple, general, and everyday things, beyond esoteric and grandiose, puffed-up mannerism. At the same time, the geometrical
method is extremely assuming, with a self-assured belief in its ability to arrive
at the unreachable, at what is concealed by nature and reason. The axiomatic-
deductive tradition was taken to its extreme in Baruch de Spinoza's *Ethica ordine
geometrico demonstrata* (1677), where he proceeds from God, definitions, axioms,
and conclusions in an attempt to build up a system of ethics modelled on sure,
eternal geometry. Others rejected the geometrical method; as Henry More put
it, the manic search for exact demonstration, the mathematical sickness, 'morbus
mathematicus', was a fundamental error in Cartesian philosophy.[134] In a dialogue
by Marin Mersenne the Sceptic says that mathematics is a sheer dream, that calling
someone a mathematician is the same thing as calling him a fool.[135] But these were
exceptions in the otherwise unison praise of geometry.

The geometrical spirit would spread, according to Fontenelle, who wrote in 1699
that it would bring order, clarity and exactitude in everything it touched.[136] And
this had happened in natural law. For Huig de Groot (Grotius) the rules of natural
law were as valid as those of geometry. With Euclidean geometry as a pattern, the
Cartesian jurisprudent Samuel von Pufendorf sought to make law an exact science.
The state was a machine designed by humans. Bishop Swedberg had studied under
Professor Pufendorf in Lund, but he disliked scholastic logic: 'To brood too much
over scholastic fancies, and to busy oneself with nothing but teaching and waging
word-squabbles and pen-wars, and fastening a lot of dry questions, definitions,
distinctions, and limitations in the brain is something that I have never cared for.'[137]
Yet he found Pufendorf's natural law useful; it was just a shame that he made himself
mad, 'had his corn lanced, whereby his life was ended [by gangrene].' During his
studies in Uppsala Swedenborg took part in disputation exercises on Pufendorf's
De officio hominis et civis (1673).[138] Pufendorf's term 'quantitates morales' led
Klingenstierna into mathematics, and in Gestrinius's edition of Euclid he found a
science consisting 'solely of irrefutable truths'.[139]

Swedenborg and other natural scientists in Sweden exploited the rhetorical
potential of Euclidean geometry. The candidate defending the dissertation *Theorema
de matheseos incrementis* tried to prove that the new mathematics was superior to
the old, with the aid of axioms, definitions and *reductio ad absurdum*. The seventh
axiom runs: 'Clear and lucid knowledge and a discipline that is handled with a
sure method is more perfect than unclear and obscure knowledge and a discipline
presented without order.'[140] Another dissertation by Harald Vallerius about the line

---

[134]Cassirer (1970), 133.

[135]Mersenne (1625), book V.

[136]Heilbron, 1.

[137]Swedberg (1941), 30, 551f.

[138]*Constitutiones nationis Dalekarlo-Vestmannice Upsaliæ*, 20f.

[139]Hildebrandsson, 7.

[140]H. Vallerius and Dryander, 3ff.

emphasizes the clarity of geometrical proof, its usefulness, necessity, and foundation for the whole of mathematical science.[141] Wolff's rationalist philosophy was later to exert particular influence in Sweden; Swedenborg especially felt an affinity to its clear logic. The Wolff-inspired Anders Celsius proved the existence of the soul through the Euclidean ideal in axioms, theorems, definitions, observations, demonstrations, and scholiae.[142] The geometrical method was one of the pillars of Wolffianism. It was an ontological vision of the order of science, in which it would be possible to illuminate the deductively related concepts through simple definitions and axioms.

The geometrical method was also Swedenborg's, but more as an ideal or vision than as a consistent model for philosophical activity. To a large extent Swedenborg's method involved trying to perform analytical deductions, or rather generally rational conclusions, from certain experimental facts. In *Principia* he tried a priori to build up an entire world system based on the mathematical point. In *De infinito* the whole of reality can be derived from the infinite. He cherishes the dream of reality as a deductive system based on God as the axiom. Occasionally he used the magic formula of geometrical thought, *QED*, *quod erat demonstrandum*, 'which was to be demonstrated', to signal rationality, reason, consequential logic, and necessary conclusions. This need not mean that the reasoning actually follows strict logic; above all it has a rhetorical significance: Look here, I think rationally. Now you *must* believe me! The geometrical method and the mathematical examples are there to convince, to decide between true and false, and between what is possible or impossible. A quadrilateral triangle is an impossibility, Swedenborg noted from Rydelius.[143] Or as the inspector of mines Samuel Troili wrote in a letter to Swedenborg, these facts put before his eyes are as clear 'as that 2×2 is 4.'[144] Geometry and mathematics are the indisputable truth, QED.

An objectivist stance assumes that there is an objectively true rationality in the universe which transcends all beings and their experiences. Swedenborg, like other mathematicians of the time, tended to view geometry as Platonic, belonging to an eternal, unchanging world of ideas, as a pale reflection of the perfect forms in the world of ideas, a utopian dream of a transcendental world of timeless, perfect, absolute, and universally valid truths independent of the finite reason of earthy beings: a beautiful thought, full of the desire to escape from the earthly, material, and corporeal. Pure mathematics liberated thought from the unreliable senses in the Platonic tradition. Geometry draws the soul towards the truth, Plato writes in *The Republic* (fourth century BC), it makes it easier for us to see the idea of good.[145] Perfect geometrical forms exist only in the world of ideas, or, as Proclus writes following Plato, geometrical forms distance us from the things of the material world

---

[141]H. Vallerius and Ternerus, 1, 3.

[142]Hermansson and Celsius.

[143]*Note book*, 114; Rydelius (1737), 320.

[144]Troili to Swedenborg, Falun, 7 October 1742. *Letters* I, 495.

[145]Plato, *Politeia*, 7.526c–527c; cf. Spole and Alinus, 11, 28.

and stimulate us to turn towards Nous.[146] For the Pythagoreans, Proclus wrote, the theorems of geometry were a step upwards, drawing the soul towards the higher world. In the mechanistic world-view there is the same transcendental geometry, which means that the ideal geometrical forms were independent of the mind or reality. The form and nature of the triangle, says Descartes, are immutable and eternal, irrespective of whether it is in the mind or somewhere else in the world.[147] With manna as an example, Locke expresses the same idea: 'A Circle or Square are the same, whether in *Idea* or Existence; in the Mind, or in the *Manna*'.[148]

Mathematics was the way into a world of timeless truths independent of the senses, disconnected from life and time. In this way mathematics became a haven for eternal truth, free from the risk of being perceived as heretical, beyond the temptations of the impenetrable world. Reason has been regarded in western thought as something abstract, divorced from perception, the body, and culture, and also from the metaphors and mental images of the imagination. In the modern science that emerged in the seventeenth century, science was supposed to deal with an objective world outside man, a science without values or context, liberated from its surroundings. Science must strive to understand the world, Bacon wrote, 'ex analogia universi', not 'ex analogia hominis'.[149] Scientific thought seeks to distance itself from all personal and anthropocentric elements, and in its aspiration for exact scientificity it seeks to erase all anthropomorphic features in the world-view. The machine metaphor could be the solution.

For Swedenborg, geometry and mathematics must have meant the essential in the material, that is to say, matter is geometry, and the central metaphor thus became *the world is geometry*. The geometrical method was perfect objectivity and rationality, the ideal for science and thought, with the central metaphor *to think is to count, to calculate*. Thoughts and men are civilized and disciplined with this rationality, which distinguishes them from wild beasts and foolish people. Geometry shows the way to a transcendental world, eternally true and independent of humanity; it draws our gaze to perfection, beauty, divinity. These were values that Swedenborg admired. That is why he could not avoid mathematics. Geometry must have exerted an irresistible attraction on the young, go-ahead Protestant Swedenborg in his love of truth, justice, and the sublime.

---

[146]Proclus, 40, 50, 84.

[147]Descartes (1641), V; *Oeuvres* IX, 51.

[148]Locke, book II, ch. VIII, § 18; cf. Berkeley, intr. § 15.

[149]Bacon (1620), I, § 41; cf. Cassirer (1953), 286.

The Sign

IIIIIII is called *la le li lo lu lyl*. mnfsstf is called *ma ne fi so su tyf*. nmstvt is called *ne mi so tu vyt*. nmstt is called *ni mo su tyt*. nmst is called *no mu syt*. mts is called *mu tys*. ts is called *tys*. nmo is called *numy*.[1]

Swedenborg, *En ny räkenkonst* ('A new system of reckoning', 1718)

Everything Is Silent, No One Knows Yet the Destination

Arithmetic, the art of counting from 1 to 8, is what the cryptic exercises in pronunciation represent. They concern figures, digits, and numbers, with *la le li lo lu lyl* being another way to say two hundred and ninety-nine thousand five hundred and ninety-three. In October 1718, during the construction of the canal known as Karls Grav ('Charles's Ditch') outside Vänersborg, Swedenborg wrote his proposal for an octal numeral system, a new arithmetic with base eight, entitled *En ny räkenkonst*. The full title in English translation is 'A new system of reckoning which turns at 8 instead of the usual turning at the number 10 whereby everything respecting coinage, weights, dimensions, and measures, can be reckoned many times more easily than in the ordinary way' (1718). This was an attempt to reform our way of counting, a task that seems to have amused both himself and the man who assigned it to him, Charles XII. The work is dedicated to Charles XII, yet not to a king but to 'a profound Mathematicus'.[2] Whether it should be 64 or 8 instead of 10 was the question. Which number is best suited as a base for counting coins, weights, and measures, while simultaneously being geometrical? An advantage of the number 8 is that it is 'geometrical', unlike the number 10, in that 8 is equal to the cube of 2, and can be halved all the way down to 1 with no need for fractions. Moreover, octal

---

[1] *En ny räkenkonst*, § 6; edition, 258; translation, 18.

[2] *En ny räkenkonst*, dedication; edition, 255.

arithmetic would correspond better to the way of counting and dividing coinage, weight, volume, and dimensions, all of which—with a few exceptions—could be derived from base 8.

The fact that it was in 1718, the last year of absolutism, that *En ny räkenkonst* was written may be important for understanding Swedenborg's octal system. This was a year of upheaval, with a dramatic finale that soon gave way to a hope for peace. Sweden had been at war for 18 years. The Swedes had had a king who seemed to have forgotten his homeland and was wandering around Europe, plundering. One army after the other had been annihilated. The country was devastated. Farms were abandoned. The economic situation was miserable, with galloping inflation and a system of token coinage, known as emergency coins, which had few supporters. Taxes, compulsory loans, press gangs, and unpaid wages did not make matters better. Occasional rebellions against suffering and oppression under a despotic regime were crushed. Swedenborg himself witnessed how the peasantry of Vadsbo had gathered in March 1710, rioting, lynching, and stabbing a bailiff to death.[3] Discontent was widespread and the mood was one of nervous insecurity. 'All is silent, no one knows what will happen. The people are griping about the many changes to the coinage,' wrote Benzelius in his *Anecdota Benzeliana* in May 1718.[4] The country was sighing and writhing in torment. Epidemics of plague had struck the country, and many people had been 'hidden under ground', Swedberg remembered when he thought back on the time.[5] Crops had failed for three years in a row, leading to starvation. In June 1718, in the third year of crop failure, Benzelius noted: 'There is great lamentation and famine here. The people travel 20 or 30 leagues around, and cannot buy ½ a bushel of rye. God have mercy on us.'[6]

Based on Swedenborg's work about the octal system and the discussions about divisions of the coinage and counting and measuring systems during the years when the absolutism of the Caroline period collapsed, one can depict another side of the critical mood that was prevalent as regards the lack of political freedom, the despotism, and the economic situation. The absolute monarchy can be illuminated by means of arithmetic and Swedenborg's *En ny räkenkonst*. Yet his work can not only be construed in a geometrical-arithmetical sense, but also interpreted in terms of history of science, economics, politics, and rhetoric. Besides this, we have a number of sources which shed light from new angles on the history of the numbers 64 and 8 and the political context, particularly a scarcely noticed little pamphlet that Swedenborg published in 1719, the year after the king had fallen in a muddy ditch at Fredrikshald. The discussion of a new numeral system had its origin in some conversations about mathematics between Charles XII and Swedenborg in the years 1716–1718. These meetings are mentioned in most of the biographies of Charles

---

[3]*Opera* I, 203f; *Ludus Heliconius*, edition, 62f.
[4]Benzelius (1914), 58.
[5]Swedberg (1941), 259; cf. Swedberg (1711), 588ff.
[6]Benzelius (1914), 59.

XII, which simultaneously emphasize his genius for mathematics.[7] The evidence cited for this is the testimony of Swedenborg, who is thought to have painted a trustworthy picture of a royal mathematician and to have admired the mathematical genius on the throne. In the literature about Swedenborg one can also find the view that he sympathized with the king and that he advocated an octal system, but that he later changed his political stance. These assumptions are exaggerated or incorrect, or at least a careful scrutiny of the sources fails to corroborate them.

The story of Swedenborg and the art of counting to eight also concerns signs and the interest in the symbolic language of mathematics. Numeral systems, algebra, and universal mathematics reveal a linguistic side of mathematics, as a language that can express regularities in nature. Especially when it comes to systems of weights and measures, coinage, and universal language, there is a human aspiration to categorize reality, to order and divide it, to make it easier to handle and survey. The discussions between Swedenborg and Charles XII about numeral systems also show how numbers are associated with the body, with human action, and with our environment. Numbers are related to the fingers (we count on our ten fingers or digits), to writing (numbers are symbolized by written characters), to space (numbers can be imagined as being on a line, or in squares and cubes).

In the baroque period the world was a meaningful fabric of numbers and signs. The universe was a system of signs with an infinite number of links to other signs. The signs reflected out to reality, formed sympathies and antipathies between things, and joined the world together in a large chain. By reading signs a person could be led from a knowledge of nature to a knowledge of the divine. God's providence must have created something good and meaningful. The natural philosopher became a solver of riddles, searching for the true reference and meaning of signs, an interpreter of signs who read the visible symbols in order to find nature's hidden message. The baroque thinker lived in a symbolic world, he created and interpreted signs, searched for similarities and analogies so that he could discover a meaning and impose order on his experiences.[8] The baroque thinker had a particular fondness for expressing himself in metaphors, emblematic images, symbols, allegories, anagrams, parables, and pastoral elegies. He took an interest in hieroglyphs, Chinese characters, myths, proverbs, fables, dreams, astrology, and alchemical symbols.

The semiotic world-view of the baroque is visible in Swedenborg not just in his metaphorical baroque poetry but also in his way of thinking. He was constantly thinking in signs that stand for something else, of a macrocosm that is reflected in the microcosm, of semantic links between different phenomena in reality, which would later result in his deciphering of the different levels of meaning in the Bible and the symbolism of the spirit world. As an interpreter of signs he had long

---

[7]Siljestrand, 2f; O. Sjögren, 622–625; Quennerstedt (1912), 197f; Quennerstedt (1916), 140f; B. Hildebrand, 93; K.-G. Hildebrand; Bengtsson, 346–348; Hatton, 430; Jonasson, 10; Liljegren (1999), 65–69. After reading this chapter Liljegren has embraced a more critical judgement of the king's mathematical ability. Liljegren (2000), 317, 402; cf. Dunér (2001), 211–238.

[8]Eco (1986), 12f; Scholz, 3ff.

experience. Swedenborg the metallurgist and chemist looked for signs in nature and was especially familiar with the diversity of chemical sign systems. Books such as David Kellner's *Berg- und Saltzwercks-Buch* (1702), which stood on the shelf in his home, contain long lists of chemical symbols that were crucial knowledge. In his anatomical studies he could rely on a long tradition of interpreting signs, as in the doctrine of signatures, where it was assumed that God had given plants a special form to signify their medicinal properties. Or as in symptomatology, which had been a self-evident part of medical training since Galen's inductive analogy, where the physician read the signs on a sick person's body to draw conclusions about the unknown causes and then make a diagnosis. Roberg's idea for a hospital in Uppsala was good, in the opinion of Rudbeck the Elder in 1700, because students of medicine could learn 'through what Signa the disease is to be recognized.'[9] It was indeed as a scholar of medicine and semiotics that Rudbeck displayed himself in the title plate of *Atlantica*, cutting a section in the planet earth in search of signs to reveal the hidden reality. Even the Bible urges mankind to interpret signs, as when Christ in the Gospel of Matthew 16:3 talks of 'the signs of the times'.

A semiotic reading of nature was generally embraced during the period, with a long previous history of astrologers, alchemists, chiromancers, weather prophets, and oneirocritics. One of the most central metaphors of the time was of a semiotic kind, that of the Book of Nature as a counterpart to the Book of God. The Book of Nature is open to our eyes, but unless one knows its special vocabulary of triangles, circles, and other geometrical figures 'one cannot understand a word of it, but must wander about forever as in a maze', as Samuel Duræus cited from the famous passage in Galileo's *Il Saggiatore* (1623).[10] The world has a Euclidean grammar. The mechanistic natural philosophers, like Swedenborg, often proceed in their thinking from basic metaphors such as *the world is a text* and *to know is to see*. There is an element of semiotic interpretation in the mechanistic world-view, according to which deductions were made from visible signs to invisible causes. Why should nature not have established a special sign that causes our experience of light, Descartes wondered, even if the sign shows no similarity to the experience?[11] Has not nature created laughter and tears to teach us to read joy and sorrow in people's faces? The actual paragraph in Locke's *An Essay Concerning Human Understanding* (1689), where the science of signs, 'semiotikè', was first coined, attracted Swedenborg's attention.[12] Semiotics, Locke says, is a science that studies the signs that the mind uses in order to understand things, or to convey its knowledge to others. This chapter is about signs, chiefly the different mathematical sign systems in arithmetic, algebra, and measurements. The new arithmetic and the discussions about a new numeral system can reveal how mathematics is linked to semiotics and rhetoric, and also economics and politics.

[9]Cited in Å. Dintler, 75.

[10]Duræus, preface.

[11]Descartes (1664a), ch. I; *Oeuvres* XI, 4.

[12]*Oeconomia*, II, n. 212; Locke, book IV, ch. XXI, § 4; Jonsson (1969), 123, 343.

Learned Games with the Number Sixty-Four

During a few winter days in 1716 Swedenborg was visiting Polhem in Stjärnsund. He had finally been invited and his plans could be realized. During his visit he seems to have made an impression on Polhem, who wrote to Benzelius: 'Some time ago Mr Swedberg was here with me, when I became aware of his intelligence and good qualities and therefore was all the more willing to pass on to him my small knowledge in order to instil light, for which he is inclined and capable through his prior learning and knowledge of mathematical matters.'[13] Three years of close collaboration would follow between them, which meant a great deal for Swedenborg's continued development as a natural scientist. Polhem brought him along on technical assignments to Karlskrona, Vänersborg, and Bohuslän. Swedenborg read, fair-copied, edited, and published Polhem's essays on various mechanical and experimental topics in *Dædalus Hyperboreus*. But the secretarial task that Swedenborg had undertaken was not all that easy to bear, as we understand from what Polhem writes to him: 'as long as I live, I hope material for printing shall not be wanting, so long as Min Herr is pleased to take the trouble to calculate, draw, write up, and prepare all that pertains thereto; for such work wearies me.'[14]

One of the first assignments Swedenborg took on was to edit and publish Polhem's little textbook of counting and measuring for the young generation, *Wishetens andra grundwahl*, in the spring of 1716.[15] The 'art of counting' (*räkenkonsten*) or arithmetic, Polhem explains, involves writing and pronouncing figures (*ziphertahl*) and adding together different sums into one sum.[16] In the first book in Swedish to use Descartes' algebra and coordinate system, he gives his explanation of what algebra, or 'the art of rules' (*regelkonsten*), is: 'Algebra teaches how to seek out all the rules that all manner of questions require for their solution, and its foundation consists in addition, subtraction, multiplication, division, equality, and equation.'[17] The foundation of algebra is the equality of two quantities.

Algebra would play a not insignificant part in the forthcoming conversations about a new numeral system. Algebra was the model of a logical, unambiguous sign system and was thus universally useful, becoming a pattern not least for Swedenborg's and Polhem's ideas about a universal language or universal mathematics. In general it may be said that algebra is a system of rules, of structures and relations between concepts, with an abstraction that distinguishes it from the individual considerations and clarity of geometry. The algebraic letter designations, introduced by François Viète, could be studied by Swedenborg in a book that he himself owned, the second edition of the Cartesian Johan Bilberg's *Elementa geometriæ planæ*

---

[13]Polhem to Benzelius, Stjärnsund, 6 March 1716. *Polhem's Letters*, 117; cf. *Opera* I, 239.

[14]Polhem to Swedenborg, Stjärnsund, September 1716. *Opera* I, 260; *Letters* I, 118.

[15]*Polhem's Letters*, 117, 120f; *Opera* I, 240f, 250, 253; cf. *Stockholmiske kundgiörelser* (1717:14).

[16]Polhem (1716), § 2.

[17]Polhem (1716), § 5, cf. § 55.

ac solidæ (1691). Descartes' analytical geometry, which combined extension and numbers, geometry and algebra, rests on metaphorical thinking, with the central metaphor 'numbers are points on a line', which allowed him to conceptualize arithmetic and algebra in geometrical terms and visualize functions and algebraic equations in spatial terms.[18] This is a step in the direction of liberating mathematics from the physical, from a specific, concrete subject matter, towards something purely mental.

In the autumn of 1716 Swedenborg and Polhem set off south to Lund, where Charles XII had set up his headquarters. They travelled through a poor November landscape where the crops had not grown as they were supposed to. A few days ago, on 18 November, they left Brunsbo, Bishop Swedberg reported, and wrote anxiously: 'In this country there is already a dearth of grain and peas, which were spoiled by the long autumn rain.'[19] In Lund, Swedenborg was quartered in the same house as the king's secretary, Bernhard Cederholm, whom he had previously met in Greifswald.[20] They were soon granted an audience with Charles XII, where Swedenborg presented his *Dædalus Hyperboreus*. The book is signed 23 October 1715 with a dedication to the king; in 'deepest humility' he places some mathematical essays at the feet of His Royal Majesty. He emphasizes the utility of mathematical studies for manufacture, the art of navigation, artillery, and shooting, and ends with the declaration: 'I remain to the hour of my death, Your Majesty's my ever gracious king's most humble and faithful subject Eman. Swedberg.' In the preface to the first issue we also get a glimpse of the lamentable state of the sciences in these hostile times: 'Foreigners do oft-times claim that our cold and Northern lands are little suited to Scientias Mathematicas; but they judge blindly.' The reason is that they have better opportunity and encouragement for scientific activity. But there is hope that matters will change 'when the all-prevailing God grants our Incomparable Monarch Peace and quiet from his many and rancorous enemies.'[21] *Dædalus Hyperboreus* pleased the king. It lay on his table for three weeks and provided matter for many discussions and questions, Swedenborg states with pride. On Polhem's recommendation, Charles XII appointed Swedenborg as extraordinary assessor in the Royal Board of Mines on 10 December 1716, with the special task of assisting Polhem in his role as head of division. He had not expected this kind treatment, he wrote in a letter to Benzelius: 'What pleases me most is that he expressed an extremely kind and gracious judgment concerning me, and himself defended me before those who thought the worst of me, and afterwards assured me of further grace and consideration'.[22]

It was a rather strange meeting. The stammering Swedenborg, the cross-eyed Polhem, and the king with the receding hairline engaged in discussions on mechanics, geometry, arithmetic, algebra, and other mathematical sciences. They

[18]Lakoff and Núñez, 260.

[19]Swedberg to G. Benzelstierna, Brunsbo, 20 November 1716. UUB, G 20a, no. 75.

[20]*Letters* I, 60, 124.

[21]*Dædalus Hyperboreus*, dedication, preface; cf. *Polhem's Letters*, 112.

[22]Swedenborg to Benzelius, Karlskrona, December 1716. *Opera* I, 262; *Letters* I, 135.

touched on the subject of the freezing point of different liquids. Here Charles XII was able to contribute from his own experiences. One winter in Poland was so cold that even the Hungarian wine froze to ice, so he simply took out his rapier and cut up the wine for the soldiers.[23] But one subject in particular was discussed, a conversation topic that was kept alive until the autumn of 1718. It concerned the design of a new numeral system with base 64. Swedenborg has a cursory description of these conversations in his *Miscellanea observata circa res naturales* (1722), and in a letter published in *Konung Carl den XII:tes historia* (1740), a biography written by the former court preacher, Jöran Andersson Nordberg, who had accompanied the campaign to Poltava. It is from these two sources, and especially the latter, that all the stories of Charles XII as a mathematician derive.

In the register of loans from Lund University Library there are a few hitherto unknown records that shed light on the mathematical conversations between Swedenborg and Charles XII.[24] On 24 November 1716 Swedenborg wrote his name in the ledger; he was borrowing the first part of Sturm's *Mathesis juvenilis* (1699), which he kept until 2 February. The question is why. After his stay in Paris he cannot have had so much to gain from this rather elementary survey of arithmetic, algebra, geometry, and trigonometry, and the part about mechanics and statics could hardly have been new to him. The only thing he had not previously worked with was military and civil architecture. At any rate, *Mathesis juvenilis* has a part that is connected to the mathematical conversations with Charles XII. It is about squares and cubes, geometrical series, and Napier's bones, but also about how to calculate the sum of measurement and coins of different sorts.[25] Among other things, Sturm deals with the question of a decimal division of coins and measures, which would entail fewer difficulties with comparisons, and less risk of confusion in trade. This is precisely what a great deal of the discussion about numeral systems between Swedenborg and Charles XII concerned. A week later, on 1 December, Cederholm also visited the University Library and borrowed two books about algebra. They were James Hume's *Traicté de l'algèbre* (1635) and Jacob Brasser's *Regula cos of algebra* (1663), which were not returned until 22 June 1717.[26] As royal secretary it may have been on behalf of the king that Cederholm visited the library in the Lundagård building. In the same month Cederholm also borrowed a book about the geometrical principles of fortification by Sébastien le Prestre de Vauban, and presumably Vallemont's book about divining rods.[27] In both Hume's and Brasser's books Charles XII could read about logarithms, squares and cubes, and geometrical series, in French and Dutch respectively.[28]

---

[23] See K.-G. Hildebrand.

[24] LUBA, FIa 1, 103f, 107f.

[25] Sturm (1699), I, 30–54, 70–72, 75, 110.

[26] A book with an almost identical title under the pseudonym D. Henrion has similar content, but since this copy was in duodecimo format, it cannot be same as this, which is in octavo. Cf. Mangin, 2–5.

[27] Presumably Vauban; Jobert; Vallemont (1696).

[28] Hume, 1–5, 209; Brasser, 1.

At this time many mathematicians, historians, and linguists were speculating about the history of numeral systems, arguing over which was best, and comparing the words by which numbers and figures were known in different languages. Mathematical tuition now included—as astronomy already did—the Babylonian sexagesimal system, and of course the Latin and Indo-Arabian sign systems. The prior history of mathematics had examples of several other numeral systems, but none based on 64. With reference to Ovid and others, Benzelius speculated about our numerals. Our ancient ancestors counted to twelve, which was a sacred number, he states, and this explains why we do not say, by analogy with 'thirteen' and 'fourteen', the numbers 'oneteen' and 'twoteen'.[29] Benzelius would later correspond with the Orientalist Gottlieb Siegfried Bayer in Saint Petersburg about similarities and differences between counting systems and numbers in Chinese and Arabic, and those used in Europe and India.[30]

The number 64 was Charles XII's original idea. He wanted to establish, as Swedenborg wrote in his letters to Nordberg, a system 'better built and constructed on a geometrical ground'. Proceeding from his own name, he invented new designations and signs for numbers, with the 64 numbers divided into eight classes so that anyone who knew the first eight could easily work out the others. Swedenborg praised the king's mental ability, for a numeral system with base 64 was not the simplest thing to master. The reason the choice fell on this figure, according to Swedenborg, was not just that it was a geometrical number, but also that the king wanted to display his acumen and profound thought even more than his energy for great deeds. Swedenborg and Polhem, however, were at first sceptical about the system based on 64 since they realized the difficulties that would attend it: 'We did not, however, fail to warn that such a number would be too high, troublesome, and almost impossible to manage.' One would have to count to 4,096, or 64 times 64, before a third figure would be added. Calculations would be cumbersome and intricate: 'But the greater the difficulty was made, the more desirous the mind would be to make an attempt, and show the feasibility of a thing that we declared to require greater thought than can be hastily brought to order and completion.' The king charged Swedenborg with resolving the mathematical difficulties, by drawing up a table showing the difference between the different numeral systems, in both name and number. Perhaps it was all an innocent learned game, a 'Ludum literarium' as Swedenborg himself called it.[31]

The learned games about the number 64 are not only mentioned in the two works cited above. In Swedenborg's letters to Benzelius he paints a picture with greater contrasts. The idea of a new numeral system seems to have gained momentum at Swedenborg's meeting with the king in Lund in the early summer of 1717. On 26 June, little more than a week after the celebration of the king's 35th birthday in Lund, Swedenborg sent a letter to Benzelius in which he told him about the

---

[29] Benzelius (1762), 45. Ovid, *Fasti*, 3.119–134.

[30] G. S. Bayer to Benzelius, St Petersburg, 10 September 1730 & 30 April 1731. Benzelius (1979), II, 329, 337.

[31] Nordberg (1740), 599–601.

new numeral system that the king had devised and given new signs and names. Moreover, the king had given a draft of this new system to Swedenborg, but he had already realized how clumsy it would be for multiplication, although he could see the advantage that it contained the square, the cube, and the fourth power of a whole number. 'His Majesty has powerful penetration', Swedenborg noted.[32] Six months later, after a meeting with the king in Lund, his tone was different: 'I got to talk with his Majesty no more than two times, and that was all concerning fancies in *mathesis*, puzzles in algebra, etc.; for the sake of the Herr Councillor of Commerce, I have sought with all diligence not to get this grace more often.'[33] Swedenborg seemed palpably irritated by the king's brain teasers and zest for recreation. But the numerical games had a refreshing childishness that matched Charles XII's somewhat puerile streak and strange humour. He could have an insensitive and naïve way of joking.

Swedenborg nevertheless settled down at the bishop's residence of Brunsbo outside Skara and began to outline a new numeral system: 'His Majesty was right well pleased with a numbering of this kind.'[34] He now had a little time over between his work with the naval dock in Karlskrona and the digging of Karls Grav. Yet what he was working on was not a system based on 64 but a simplification of this method, an octal system, along with practical examples to show that it could be of general utility. He sent the manuscript to Benzelius on 7 January 1718. Swedenborg suggested that, before it went to the printer, it should first be read by the professor of mathematics Johan Vallerius. Benzelius, however, does not seem to have been too enthusiastic about this new way of counting, presumably because it was impractical. Moreover, it could entail a new ordinance introducing new units for coinage, weights, and measures, which would not be welcomed by the people. Yet another reason may have been that Swedenborg's octal system was contrary to the king's wishes. It would be tactless to put forward a different mathematical opinion from the king's. With a degree of pride over his 'inventa mathematica', Swedenborg replied:

> I wish I had as many *noviteter* [novelties], yea, in *re literaria*, a *novitet* for each day of the year; then the world would find pleasure therein. In *seculo* [a century] there are enough of those who go in the beaten track to be in accord with that which is old: but perhaps there are 6 or 10 in a whole *seculo* who bring forth *noviteter* which are founded on reason and on another foundation.

But Swedenborg understood how tricky it would be to introduce a new counting system, to apply not only to coinage and the economy, but also to ordinary people's deeds and lives: 'God grant that all projects were of the same kind; no subject will suffer therefrom, not even if one or two were to count in a different manner, which I know is not likely to be the case. Since the King has already *improberadt*

---

[32]Swedenborg to Benzelius, Lund, 26 June 1717. *Opera* I, 272; *Letters* I, 158. The number 64 actually has no fourth power as an integer.

[33]Swedenborg to Benzelius, Brunsbo, December 1717. *Opera* I, 274; *Letters* I, 164.

[34]Swedenborg to Benzelius, Brunsbo, 7 January 1718. *Opera* I, 275; *Letters* I, 165.

[disapproved] the 8 numbering.'[35] The reason the king did not approve of the octal system was probably not just to do with political and economic circumstances, but also a matter of opinion and taste. The octal system was far too simple, both conceptually and in practice. The number 64 was more difficult and therefore better. It was a number that contained both cube and square and could be halved down to one, but was also linked to the number 8. In yet another letter Swedenborg tried to convince Benzelius that the essay should be printed, but he was simultaneously painfully aware of the limits to freedom of speech:

> I had intended the *New Reckoning* for the learned. I hope that my Brother will order it to be printed. I take all responsibility on myself, and warrant that no such publication will be forbidden. In respect to laws, war, and taxes, the King has free disposition; but in respect to words, language, and reckoning, none at all. One has, indeed, cause to be weary at all the *noviteter* [novelties] which are going on. God grant that such had not been the case in the coinage, etc., but only in the reckoning in connection with the coinage; with this, the country would have been better off.[36]

Polhem, however, had written to Swedenborg to inform him that the king had conceived 'unfavourable thoughts' about Swedenborg's failure to publish *Dædalus Hyperboreus* at the same pace as before. 'I should much like to take down with me something which falls in with the King's liking,' Swedenborg wrote to Benzelius. 'Let nothing interfere with my new way of reckoning; it may be very useful for those who will use it. The responsibility I take upon myself.'[37]

Swedenborg met the king once again in the late summer in Strömstad. He had been ordered there by Polhem to lead the transport of warships to the front against Norway. Seven small galleys were dragged overland, across lakes and snow-free ground, through bogs and swamps on logs and cable bridges, along the 17-mile road from Skagerack to Idefjorden, in order to take the Danes and Norwegians by surprise at Fredrikshald.[38] With the aid of 800 soldiers, Swedenborg struggled to free the larger brigantine *Luren*. Swedenborg was surprised at the unexpectedly gracious way he was received by Charles XII—'a *bonum omen*', he wrote to Benzelius.[39] Every day he presented mathematical topics to the king, and he was given a special opportunity on 29 August when an eclipse of the moon was visible at nine in the evening. He and Charles XII went out to observe it and to discuss this celestial phenomenon. Once the ships, after enormous effort, had been launched in Idefjorden at the start of September, an anecdote records that Charles XII, the Duke of Holstein, and Swedenborg stood on the rock of Hällesmörk and watched as their fleet was swiftly annihilated by the Danes.[40]

---

[35]Swedenborg to Benzelius, Brunsbo, 21 January 1718. *Opera* I, 280; *Letters* I, 175f; Benzelius's letters to Swedenborg, sent around 19 January 1718, are lost; cf. *Letters* I, 173, 176.

[36]Swedenborg to Benzelius, Starbo, 30 January 1718. *Opera* I, 282; *Letters* I, 179.

[37]Swedenborg to Benzelius, Starbo, February 1718. *Opera* I, 283; *Letters* I, 182.

[38]Berggrén, 168–171.

[39]Swedenborg to Benzelius, Vänersborg, 14 September 1718. *Opera* I, 286; *Letters* I, 192.

[40]*Letters* I, 191f.

The plans for the printing of *En ny räkenkonst* had come to nothing. Instead Swedenborg resolved to present a handwritten version of the octal system to the king. In early October he took himself to Karls Grav, where he completed a copy while he was working with the canal lock. But it is uncertain whether the king ever saw it. Military movements commenced at the end of the same month, and on the last day of November the attack on Norway came to an abrupt end. At the fortress of Fredriksten on Sunday 30 November 1718, projectiles fell in parabolas over the battlefield. Constant grapeshot and musket fire, chains of salvoes in rapid succession, incessant cannonades and bombardment. Balls of burning pitch were fired into the air to illuminate the field. It 'made a beautiful sight', the young fortification lieutenant Bengt Wilhelm Carlberg noted. The French engineer officer Philippe Maigret said only: 'Behold, the spectacle is over, let us go and have supper.'[41]

In Brunsbo Swedenborg was able to breathe out: 'Praise God I have escaped the campaign in Norway, which had very nearly caught me, if I had not used plots to withdraw myself.'[42] But when the letter was written, on 8 December, the news of the king's death had not yet reached Skara. Between the lines one can suspect that Swedenborg may have fallen in disfavour shortly before the king's death. The reason is uncertain; perhaps he had refused to enrol as an officer, or might he have expressed opinions that were displeasing to Charles XII?[43] All we know is that their dealings on earth came to a definitive end on 30 November. One of those who was there at Fredriksten when the fatal shot was fired, the hereditary prince Frederick of Hessen-Kassel, was honoured soon afterwards by Swedenborg in a dedication in *Om jordenes och planeternas gång och stånd* ('On the motion and position of the earth and planets', 1719), signed 16 December 1718. Swedenborg laments 'the great general sorrow that has befallen us with the sudden death of our glorious *MONARCH*', but reckons it a good fortune for Sweden that Frederick has now come to the defence and rescue of the kingdom.[44]

The Geometrical Number Eight

Just over a month before the fatal shot was fired at Fredriksten, Swedenborg had finished the work on his octal system. The manuscript *En ny räkenkonst* is the only source describing the new numeral system in detail. Swedenborg had been charged with the task of solving the mathematical difficulties of the new system. 'By command the experiment was made with 8, which is a cube of 2', we read in the letter

---

[41]Voltaire (1731), II, 336; Maigret, 199; Carlberg, 225; Ahnlund, 77–95.

[42]Swedenborg to Benzelius, Brunsbo, 8 December 1718. *Opera* I, 288, cf. 287; *Letters* I, 202.

[43]Many years later the visionary Swedenborg wrote some enigmatic notes about Charles XII: 'unless the state had been changed from favourable into angry, with Charles XII, one [of us] would certainly have perished. This occurred with many circumstances, which it is not allowed to relate.' What made Swedenborg reticent to reveal this? Presumably it had to do with events in the spirit world; *Diarium spirituale*, n. 4704; translation IV, 95; cited in Bergquist (2005), 72.

[44]*Om jordenes och planeternas gång och stånd*; *Opera* III, 301.

to Nordberg.[45] Swedenborg was thus following a royal order when he tackled arithmetic. His octal system turned out to be much easier to handle than the system based on 64. With as many as 64 figures, one can really question the practical advantages of such a numeral system. It has the character of a game with figures, a pastime, rather than a serious mathematical reform with practical benefits. It was more a way of testing where the idea could lead. The fact that it was a pastime was not, however, an obstacle to the possibility that something of mathematical value could come out of it. An innocent game, of cards or dice, can very easily develop into something serious. Gamblers can arrive at Pascal's triangle or a probability theory, as Pascal himself did. Swedenborg's and Charles XII's game with numbers, however, did not result in anything permanent. The underlying logic was never elaborated.

Important reasons for the construction of both the base-64 system and the octal system are, on the one hand, a geometrical-arithmetical interest and, on the other, the possibility of practical utility. The geometrical premise meant that 64 and 8 were regarded as 'geometrical figures', that is, the number 64 contains both the cube of 4 and the square of 8, while the number 8 contains only the cube of 2. Their ideas about numbers as squares and cubes rest on metaphorical thinking, with the numbers considered as something spatial. In Greek mathematics, numbers were represented spatially with points, small stones or *calculi*, which in the Pythagorean triangular numbers were 1, 3, 6, 10, 15..., and the combination of two triangular numbers gives a quadratic number, 1, 4, 9, 16, 25... The perfect number was 10, which is the sum of its parts, $1 + 2 + 3 + 4 = 10$, the circular number that returns to unity. Charles XII and Swedenborg occupied themselves with a kind of profane numerology, unlike the Pythagorean, Biblical, and Kabbalistic numerology, but for them the number ten was obviously not sacred as it was for Pythagoras. In other contexts Swedenborg's interest in the meanings of numbers emerges; for instance, he mentions Pythagoras's numerology.[46] And in a poem composed for his father's 63rd birthday on 28 August 1716 he alludes to the special significance of the number seven. Every seventh year was a critical year in ancient times, and particularly in medicine and astrology the 63rd year was especially crucial. Jesper Swedberg had now seen the firmament circle nine times seven revolutions.[47]

Furthermore, 64 and 8 are divisible by halving all the way down to unity, 1, with no need for fractions. Unlike an arithmetic series, this halving forms a geometrical series with quotient (or base) 2. As Swedenborg explained in *Regel-konsten*, we have a geometrical progression when 'a linear number has itself as the first to the second, then the second to the third, then the third to the fourth, and so on. E.g. 1.2:4:8:16:32.64.128. is a linear number.' Arithmetic progression, on the other hand, is when 'A figure is numbers with equal differences between', as for example 1, 3, 5, 7, 9.[48] Both the base-64 and the octal system can therefore be said to be related

[45] Nordberg (1740), 600; *Letters* I, 461.

[46] *Oeconomia* I, n. 630.

[47] *Cantus Sapphicus*; *Ludus Heliconius*, edition, 68f, commentary 154f.

[48] *Regel-konsten*, 41f; *Photolith.* II, 32; cf. H. Vallerius and Erichsson.

to the binary system that Leibniz devised. Moreover, Leibniz was fascinated by the numerical series 1, 2, 4, 8, 16, 32, 64. These geometrical properties are not found in a decimal system. The number 10 can only be halved once before we are forced to use fractions, and its square root, cube root, and quartic root are not whole numbers.

A utilitarian argument that was put forward for a new arithmetic system is that the numbers 64 and 8 agree better with the customary way of dividing coins, weights, volumes, and linear measures. A new arithmetic with base 8 would therefore be of general benefit, Swedenborg argues, and not just for Sweden. With some exceptions, it was the rule in the early eighteenth century that all coins, weights, and measures, in Sweden, as in many other countries, could be doubled from 1 up to their highest amount, and then halved again down to 1. Why 10 once became the base in our arithmetic, Swedenborg guessed, was that people 'like our peasants at the present time, seem first to have counted on their fingers'.[49] But if there had been 'some fundamentally learned and deep-minded master-reckoner' in the beginning, perhaps a different number could have been chosen as base. If man had been created with eight fingers, we would have had an octal system instead. Swedenborg, in other words, is close to the idea of the physical origin of mathematics, that thinking proceeds from the body, or is dependent on it. We count on our fingers. Polhem is thinking along the same lines when he writes in *Wishetens andra grundwahl* that 'figures have ten signs, equal to our ten fingers', and gives the somewhat superficial instruction to students: 'The others in between can be learned from paginis [pages] in books.'[50] To sum up, Swedenborg cites three main advantages of his new octal system: it avoids fractions; with simple addition one can achieve the same as with complicated multiplication; and the different divisions in volume, weight, dimensions, and coins follow in a row.

The number 8 proves best suited for all coins, weights, and measures, for square and cube functions, and for halving. One could thus count to 8, where the number would turn, becoming two figures, and from there proceed to 16, 24, 32, and so on up to 64, which is the same as 8 times 8 and the first number written with three figures. To ensure that all this will not lead to 'confusion in its comprehension', the existing figures should not be used, in Swedenborg's opinion; new figures with new names should be constructed instead.[51] Best of all would be to invent 64 new signs or figures. Since the printing presses have nothing but the old familiar figures and letters, however, we have to make do with them.

Swedenborg's octal system therefore proceeds from the seven letters *l, s, n, m, t, f, ú,* corresponding to 1, 2, 3, 4, 5, 6, 7. Behind this seemingly arbitrary choice of letters one can discern that Swedenborg actually had a well thought-out plan. Five of them are pronounced with an initial *e*, like the ordinary letters: *ell, ess, en, em, ef.* The last two are the letters that come immediately after *s* in the alphabet, that is, say *t* and *ú.* They were to be pronounced *et* and *u.* The seventh figure, which no one has

---

[49] *En ny räkenkonst,* § 1–2; edition, 256; translation, 14; cf. Leupold, tab. I & III, 2–4, 17f.

[50] Polhem (1716), § 6.

[51] *En ny räkenkonst,* § 3; edition, 257; translation, 14.

hitherto been able to explain, is written both *ú* and *v* and correspondingly varies in pronunciation.[52] It is pronounced *v* before a vowel, as in Latin, and otherwise *u*. Swedberg's orthographical rules for Swedish in *Schibboleth* (1716), the title of which alludes to the catchword that some people found difficult to pronounce in Judges 12:6, gives some guidance as to the correct pronunciation. Swedberg's spelling reform was controversial, coming under harsh attack from archiater Urban Hiärne, who was more conservative in matters of spelling and orthography. A war of words is both pernicious and reprehensible, Swedberg declared in *Schibboleth*, but despite this he incurred criticism from Hiärne, who remarked that he had done away with the frequently used *h* and double vowels.[53] In Lund on 3 January 1718 Swedberg defended his opinions in a disputation with Humerus as opponent, with a splendid outcome. All the opposition was totally crushed, according to Swedberg himself, and 'if the chief heretic the Pope of Rome had come there, they would not have had anything more to say.'[54] But Charles XII did not bother to get out of bed that day, however that should be interpreted, and failed to turn up on account of 'a severe and rather grim winter and chill, as the King himself informed me, as if to excuse himself.'[55] Yet Charles XII was scarcely a man who feared bad weather. Later Swedenborg tried to mediate between Hiärne, who was vice-president of the Board of Mines, and his father Jesper to bring an end to the dispute. 'But this is entirely against my will,' Hiärne wrote to Benzelius, 'for it is better that the bishop vents his full rage and bitterest gall, so that I have occasion to attack his person as he has done against me.'[56] 'I shall therefore throw flying ants in his beard, so that he will be occupied for a while pulling his reverent beard.'

In *Schibboleth* one can find Swedberg's view of the use of *u*, *v*, and *w*, if we may assume that Swedenborg followed his dear father's orthographical rules. At any rate Swedenborg declaims his tribute in a poem, *In parentis mei Schibboleth* (1716), intended for his father's book: 'liber est tuus ille Magister', your book is our teacher.[57] There are three kinds of *u*, Swedberg says: both *u* and *v* are always vowels in Swedish, whereas *w* is only a consonant. The difference between *u* and *v*, according to Swedberg, is that

> U is always written after a consonantem: but never at the start of any syllable. Hustru-jungfru, nu, tu, stufwa, [et]c. In contrast v is always written at the start of the letter, and not after. As in; vtur, förvtan, vnder, vmgås, förvndra. I never find it printed thus; utur, förutan, under, umgås, förundra.[58]

---

[52]Neither Acton nor Oseen have managed to unravel Swedenborg's use of *ú/v* in their editions. Swedenborg in fact uses the designation *ú* in only three sections at the start and the finish (§ 4, 5, and 17); otherwise he has the sign *v*.

[53]Swedberg (1716), 13, 17ff, 78f; Hiärne (1716/1717); Swedberg (1719), 6.

[54]Swedberg (1941), 555.

[55]Swedberg (1941), 570.

[56]Hiärne to Benzelius, 26 November 1719; *Green Books* II, no. 189.

[57]*Ludus Heliconius*, edition, 136f; cf. *Opera* I, 275, 278.

[58]Swedberg (1716), 26f; cf. Swedberg (1719), 65.

The figure *ú* has an accent, partly to distinguish it from *v*, partly because Sweden-borg's idea was that the first four figures should resemble the second four, to which a bar is added. Thus, *l* (1) is like *t* (5), *s* (2) is like *f* (6), *n* (3) is like *ú* (7), and *m* (4) is like *lo* (8).[59] Moreover, if one arranges the first six figures in alphabetical order and puts them in pairs, then all the pairs add up to the number 7, just as *ú* in itself is 7. In this series of consonants it may seem strange that it is broken by the vowel *ú*. The idea behind this choice may be that, if one adds the three pairs, that is, 3 times 7, one obtains the sum of 21, and *ú* is the 21st letter of the alphabet. Nor is it unlikely that Swedenborg's designations for the figures are a cipher. Writing messages in a code based on conventions known only to a group of initiates was common in military contexts. Steganography, with its construction of ciphers and its search for the code, also had some mathematical, mechanical, and philosophical interest. Wallis the mathematician solved codes, Polhem designed a cipher machine, and cryptology for Leibniz was a model for universal mathematics, like a general symbolic mathematics that could solve nature's cryptic texts.[60] Charles XII, in his base-64 system, had constructed new figures based on the letters in his own name. Puns with the letters of his Latin name, CAROLUS, can be found elsewhere. The headmaster of the gymnasium at Anclam, Christopher Pyl (Pylius), immortalized the occasion when Charles XII returned to Swedish Pomerania in November 1714. At the celebration boys carried a letter each of his Latin name and changed places during the ceremony, so new letter combination was formed: SOL CURA, 'O Charles, our Sun, care for us!'[61] Unfortunately, if there is any cryptology behind *En ny räkenkonst*, it is not revealed in the book.

The vowels of the alphabet are used instead to express powers in Swedenborg's system. Corresponding to the empty quantity of the decimal system, *0*, Swedenborg has the vowel *o*, the pronunciation of which depends on the size of the number following the alphabetical order, from the greatest eighth power to the least, *a*, *e*, *i*, *o*, *u*, *y*.[62] When one pronounces a higher number, they come in the natural order. In the same way as 0 in the decimal system expresses the square of 10, these vowels are used as powers of 8. For example, the number *lo*, or 8, is pronounced *ly*; the number *loo*, or 8^2, is pronounced *lu*, and so on to *loooooo*, or 8^6 (=262,144), which is pronounced *la*. Swedenborg's octal system is thus a position system like the decimal and binary systems, where the meaning of the figure depends on its place in the number (Fig. 1). But when it comes to the pronunciation one could in principle have greater freedom as regards the sequence, although Swedenborg never exploits this. Swedenborg's system has surprising similarities to that of Leibniz, as proposed in the fragment *Lingua generalis* (1678). To what extent Swedenborg and Charles XII

---

[59] *En ny räkenkonst*, § 17; edition, 268; translation, 33f.

[60] Polhem, 'Några Mechaniska Inventioner, som fuller icke änu blifvit practicen wär[k]stälte och försökte; men likwäll på god grund byggde att de man tar sitt försök'. KB, X 267:1, fol. 58; Pesic, 688, 692.

[61] Helander (2004), 459; *Festivus applausus*, edition, commentary 136–138.

[62] The vowel *y* in Swedish is pronounced [y], like French *u* or German *ü*.

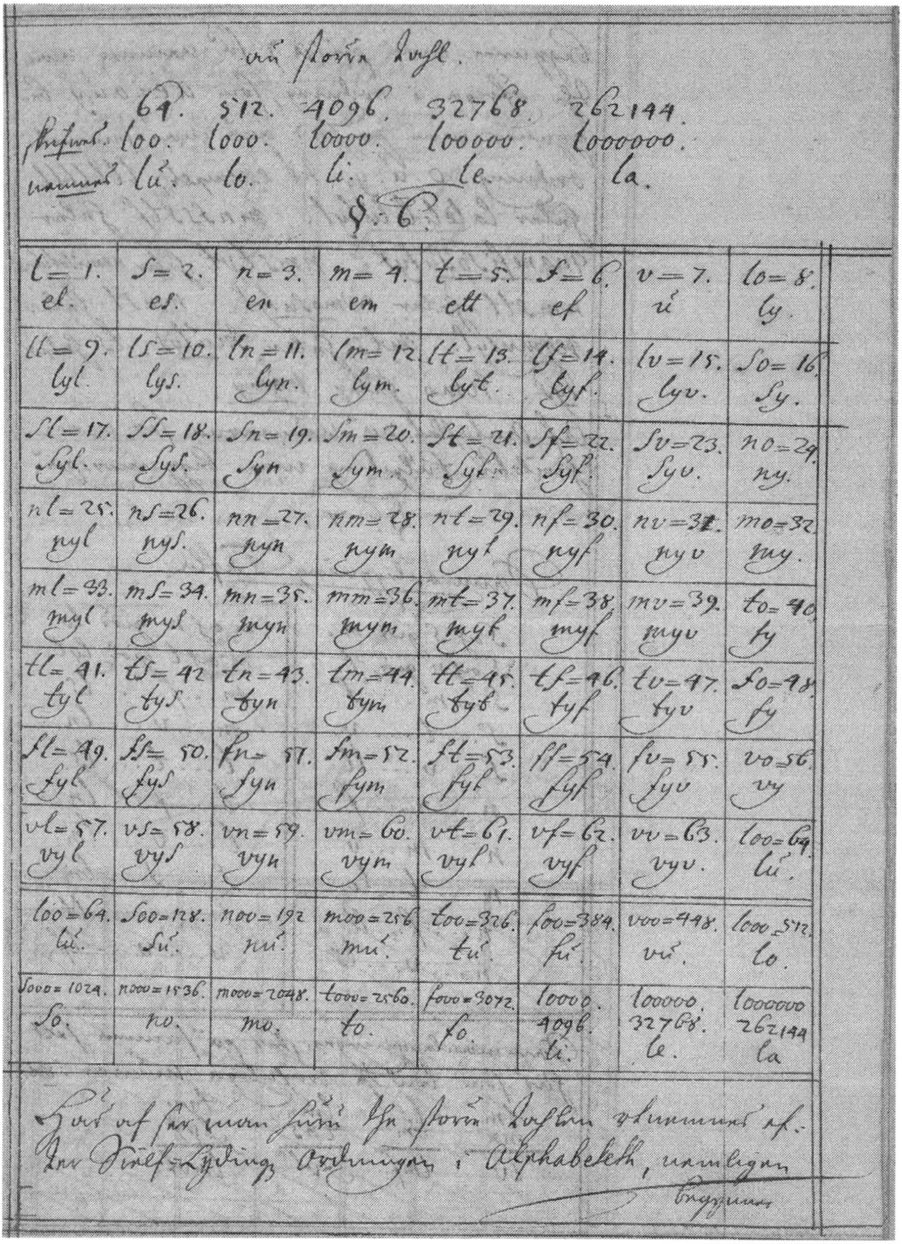

Fig. 1 Table of Swedenborg's octal system (Erratum: *too* is not 326 but 320. Swedenborg, *En ny räkenkonst* (1718))

knew about his experiment is highly uncertain, although Leibniz had had the honour, Voltaire says, of entertaining Charles XII for a quarter of an hour in Leipzig in 1707, but according to Leibniz's own letter to the British ambassador to Berlin he only saw the king eating.[63] To every single idea Leibniz assigned a number, and then replaced the figures 1 to 9 with *b, c, d, f, g, h, l, m, n*, allowing the five vowels to stand for the decimals 1, 10, 100 ... The number 81,374 would therefore be pronounced *Mubodilefa*, but the consonants are independent of their position, and the same sum could just as well be obtained from *Bodifalemu*.[64] This is an artificial phonetics not unlike Swedenborg's *lalelilolulyl*.

After presenting the basic rules on which the octal system rests, Swedenborg goes through the application of the system to the different arithmetic operations, addition, multiplication, and division, and provides the reader with a table for converting octal figures to decimal. Swedenborg also uses John Napier's method of division and multiplication. The Scottish mathematician Napier, who introduced the decimal point and invented logarithms, had speculated about alternative numeral systems and found that, if one has weights in the series 1, 2, 4, 8 ..., it is possible to weigh any amount in whole pounds. This idea led to logarithms. They were introduced to Sweden in Spole's dissertation, *De trigonometria* (1687), and the first logarithmic table was published by Elvius in 1698.[65] Harald Vallerius also considers logarithms in a dissertation, *De logarithmis*, from 1700. Swedenborg had no doubt learned about logarithms according to Napier and Briggs in Bilberg's *Elementa geometriæ*, and above all from Reyneau's *Usage de l'analyse*, in which he noted passages dealing with logarithms, and he brings up the subject himself in *Regel-konsten*.[66] In *Rabdologiæ* (1617), a work that Johan Vallerius lectured on early in the eighteenth century, Napier presented a mechanical multiplication method called 'Napier's bones'.[67] In a way Napier's bones can be regarded as a simple predecessor of the adding machine, like those of Pascal and Leibniz later in the century. With this method one could avoid the need for long multiplication tables. According to the instructions, you have to cut out each column of figures and paste them on square rods. You then put the two rods together and read off the right product. This is an early example of Swedenborg's fascination with the idea of the machine, the possibility that it can be used to calculate all truths, all knowledge. These thoughts about numeral systems point the way forward to a logical-philosophical vein, abstraction, the creation of a logical system independent of human anatomy, our ten fingers. It was to lead on to his speculations about universal mathematics.

---

[63] Voltaire (1731), II, 342; Kemble, 457.

[64] Rossi, 183.

[65] Elvius (1698).

[66] Swedenborg's register in Reyneau, *Usage de l'analyse*, 705, 797, 813, 820, 879, 880; *Regel-konsten*, 43.

[67] J. Vallerius, 'Joannis Neperi Baronis Merchistonii Scoti Rabdologia'. KB, X 219.

The Useful Number Eight

The practical significance of the new way of counting is underlined in *En ny räkenkonst*. The description of the arithmetic properties of the octal system is followed by the practical metrological application, that is to say, its use for counting coinage, weights, and measures. Swedenborg's system arose partly due to an asymmetry between the base 10 of arithmetic and the non-decimal base of coins, weights, dimensions, and volumes. Virtually all even numbers between 2 and 20 could be used as divisions of the units. The situation was chaotic because of the abundance of different measurement systems that existed. Each province, sometimes even certain manufactures and smaller regions, could have separate measures. For example, money was counted according to the metal content of the coins, whether it was silver or copper, and units of volume depended on whether it was wet or dry goods, in addition to which dry measures could be heaped. Although each system served its purpose on the local market, this confusion caused difficulties for trade, craft, and communications between the different parts of the country. The counting could occasion a great deal of trouble at the marketplace. For people who were interested in figures there were popular handbooks which they could use to calculate the weights, measures, and coins to which they were accustomed in their everyday life.

Mathematics in the service of society became increasingly prominent in Sweden later in the eighteenth century, in political arithmetic, statistics, and tables of population and land. Man and society were quantifiable in exact numbers, in relations and proportions. The expansion of trade inside and outside the country meant that the search for a new and more universal measurement system became increasingly urgent. Presumably the scientific aspect of the problem was particularly important for Swedenborg. The diversity of units was regarded as far too irrational and arbitrary for modern science. This was something for learned scholars to change, a task made more pressing by the need for a rational, scientific system for weights, volumes, and dimensions. This was particularly important for experiments in chemistry and physics, when weighing chemical substances or measuring heights of fall. Scientists wanted to be able to compare their results with each other. There was thus a need for a more uniform, systematic, and universal measurement system suitable for science and trade.

The first step in this direction in Sweden was Stiernhielm's standardization of weights and measures, which he had been commissioned to draw up in 1661. Yet even before this he had touched on metrological matters in his *Archimedes reformatus*. 'The knowledge of measurement', he wrote, alluding to The Wisdom of Solomon 11:20 (but referring to 2:22), 'consists in measure, number, and weight; and encloses in it, and extends out into, and over everything that God has created and sent.' The world was created according to mathematical laws. The knowledge of measurement, which is 'in the separation of light from darkness; in the vaulting of the firmament; the extension of the airs; the foundation of the earth; and the bounding of the oceans, has become like a right hand to its father, and a secret counsellor.' And the astute Pythagoras burst out in amazement in the words 'Deus est numerus; & Numerus Deus est.'[68] The result of Stiernhielm's work was *Linea Carolina*, a line engraved on a square brass rod. It was kept in Stockholm and was to serve as a standard for all linear measures, and it could also be used to convert measures to other systems. In 1657 Stiernhielm wrote to the king after whom it was named, Charles X Gustavus, declaring that with this rod he could measure astronomical degrees, the perimeter of the heavens, the distances of the stars, the size of the planets, and the circumference of the earth, with it he divided seas, areas, fields into leagues, stages, and acres, with it he measured and weighed metals, marbles, and stones.[69]

Stiernhielm's system, which gave the measurement system a more logical structure, was based on the international mass unit called the *ass* (approximately 48 mg). An *ass* was supposed to function as an unchangeable and absolute unit, or as Bengt Horn, councillor of the realm, put it: 'The ass is like number 1, and cannot be changed.'[70] Furthermore, with the aid of water Stiernhielm connected the definitions of mass, length, and volume, 'as a general element identical all over the world, uniform, and everywhere the same weight'.[71] Stiernhielm's special relationship to water can be seen in a detail of David Klöcker Ehrenstrahl's portrait of him from 1663.[72] In the bottom right-hand corner there are some objects alluding to his work as a geometrician and metrologist: the coat of arms with the star as a symbol of mining; the brass cube representing the cubic measuring vessel that he introduced in his role of metrologist. A black eight-pound cannonball in cast iron refers to his artillery standard for measuring the weight and diameter of balls made of different material, but it may also symbolize the perfect geometrical figure, the black sphere of which he speaks in his natural philosophy, without beginning or end. We also see his Carl-Staf and Linea Carolina, along with the compasses, the geometrician's symbol. All this in front of a broken column invoking the classical tradition as the symbol for *fortitudo*. But what has not been noticed before is that the icosahedron in the picture is precisely the Platonic body that is associated with the element of water.

This was the origin of the 1665 decree establishing the divisions of weight, length, and volume that were to be valid until the 1730s.[73] It was this decree that applied at the time of *En ny räkenkonst*. But now, in 1718, the situation was still not wholly satisfactory. This was partly because Stiernhielm's radical reorganization had not been implemented in all its parts, although there was now less confusion about weights and measures. It is almost certain that Swedenborg

---

[68]Stiernhielm (1644), dedication.

[69]Stiernhielm to Charles X Gustavus, 1657. *Samlade skrifter* III:3, 197f.

[70]Cited in Ohlon (2000), 181.

[71]Cited in Ohlon (2000), 191; Stiernhielm thus did not reckon that the density of water was dependent on the temperature.

[72]Cf. Bruzelli and Carlestam, 21; Pipping (2000), 174f.

[73]*Kongl. May:ts förnyade förordning*, 27 May 1737, & 29 May 1739; Pipping (1968), 54f, 64f.

knew of and proceeded from Stiernhielm's metric system and the decree of 1665.
What Swedenborg regarded as the greatest advantage of the new arithmetic was its
adaptation to the division of Swedish coins, weights, and measures. No fractions,
no conversions were required. When it came to coins, each sort could be divided
by eight; in other words, 1 pure mark (*lödig mark*) is made up of 8 riksdaler,
1 riksdaler of 8 marks, 1 mark of 8 silver öre (*öre silvermynt* or *öre s.m.*), and 1
öre s.m. of 8 half-öre. 'That is to say, in our coinage, we have a perfect octonary
system,' Swedenborg observed.[74] The coins are wholly in symmetry with the octal
arithmetic. As a result, one can immediately see how much one has, no matter what
kind of coins one is counting. There is thus no need for fractions of conversions, all
the sorts stand as if in a row, so that in a single sum one can see both the smallest
and the largest coin, divided into sorts or in just one sort. There is never any need to
recount the coins in another unit; they are always regarded as just one number. The
difficulty that existed at this time in converting coin sorts for a merchant or farmer
is evident from all the ready reckoners that were published.[75]

But there were exceptions to the octal system in the counting of coins,
Swedenborg acknowledged, which forced him to modify his system. The problem
was the carolin, because there were *nl* or 25 öre s.m. in 1 carolin. The ratio of 25
to 1 had applied since 1716, and in that year Swedenborg had already examined the
calculation of the carolin in the second issue of *Dædalus Hyperboreus*.[76] After an
article by Polhem about a geometrical method for calculating compound interest
through a triangle, Swedenborg had introduced a table for converting the carolin
to other types of coinage and had written an article about carolin calculation, with
a title meaning 'Another handy manner to convert Carolin numbers (the Carolin
calculated at 25 styver) into any sort of money one pleases, through halving'. In it
Swedenborg proposed that if one has no conversion table one could use an 'easier'
halving method instead of calculating with numerical relations. Soon, however,
Swedenborg had fears that someone had stolen his method. He expressed his
worries to Benzelius: 'In *avisen* [the gazette] there was something concerning a
new method of reckoning from Carolin dalers into dalers, etc.; I hope that Werner
in Stockholm has not copied and published mine.'[77] He may have seen a notice
about *Vträkning, som wijsar hwad kopparm:t giör uthi caroliner, efter den nya
valvationen* ('Computation showing what copper coins amount to in carolins, after
the new valuation' 1716) by an anonymous author, with a conversion table from
daler and copper öre to daler, öre, and carolin pennies.

The bearing idea in *En ny räkenkonst* was that all counting with coins, both multiplication and division, fitted the octal system, in the same way that Stiernhielm's
decimal yardstick fitted the decimal system. The solution was to convert carolins

[74]*En ny räkenkonst*, § 12; edition, 261; translation, 23.
[75]Lagerholm; anon. (1710), 36–56.
[76]*Dædalus Hyperboreus* II, 31–39; cf. *Opera* I, 239f; *Regel-konsten*, 61–66; Åmark, 3.
[77]Swedenborg to Benzelius, Brunsbo, April 1716. *Opera* I, 250; *Letters* I, 102.

into öre s.m., which were a part of the octal system. If one had a sum in carolins and
multiplied it by 25, the result was a value in öre s.m. Similar conversion to öre s.m.
could be applied to coins other than carolins, such as *daler karoliner, femstycken*
(pieces of five), and *kronor* (crowns). The octal system applied not only to silver
coins but also copper ones (*kopparmynt*, abbreviated k.m.), to *daler k.m., halv
marker k.m.* (half-marks), and *halvöre* (half-öre). There was no need to introduce
a decimal system for coins, in Swedenborg's opinion. Instead people just had to
change their way of counting:

> From this it follows that it is not necessary that our coinage be brought to the decimal or ten
> reckoning, according to the assertion of some of the learned; the use to be expected from
> this would be for one country alone. But if the reckoning itself were changed to an octonary
> system, all lands which should use it would then have the same advantage, since most of
> the coins of the world are divided into 1, 2, 4, 8, 16, 32, 64.[78]

One may wonder which would be simpler: to change the division of the coinage or
the way of counting. At any rate, Swedenborg's proposal for a changed arithmetic
indicates how firmly rooted the traditional divisions of coins, weights, and measures
were, and that it was not as easy to abandon them as one might think.

Weights could also be brought into an octal system, if 8 oz as used by
goldsmiths equals 1 lb (*skålpund*). The octal order of weights should therefore
be, according to Swedenborg: mark, ounce, quintine (*kvintin*), which agrees well
with the 1665 decree on weights and measures. Exact determination of weight was
essential for scientists. A rational division of weights, along with the calibration and
improvement of weighing instruments, was therefore an urgent matter, as is evident
from Swedenborg's description of Polhem's division of the steelyard ('Betsmans-
vtdelning') in the third issue of *Dædalus Hyperboreus*. In this question, about the
point of equilibrium on the balance, the mechanic turns to geometry for assistance,
he says.[79] Likewise, the octal system should apply to yardsticks and measures of
length, with the order being the rod (*stång*, 16 ft), cubit (*aln*, 2 ft), half-quarter (*halv
kvarter*, a quarter-foot). For Swedenborg 8 cubits make 1 rod, but this division was
not wholly self-evident at this time. The rod, which was an implement for surveying
land, could be divided into everything from 5 to 10 cubits depending on the region
and the context.

Swedenborg's scientific interest makes itself felt when he combines his octal
system for linear measure with the fall of balls, weights, stones, and water. We
know from experience, he declares, that weights fall 8 cubits on the first pulse
or second, then 3 rods on the second pulse, 5 on the third. In this connection
he also mentions the geometrical inch, one of the few decimal measures at this
time (Fig. 2).[80] Volumes follow the octal order: barrel (*tunna*), firkin (*fjärding*),
jug (*kanna*), quart (*kvart*), and 'half-virgin' (*halvjumfru*). This division corresponds

---

[78] *En ny räkenkonst*, § 12; edition, 264; translation, 26.

[79] Polhem, *Dædalus Hyperboreus* III, 41f, cf. 50.

[80] *En ny räkenkonst*, § 13; edition, 266; translation, 30.

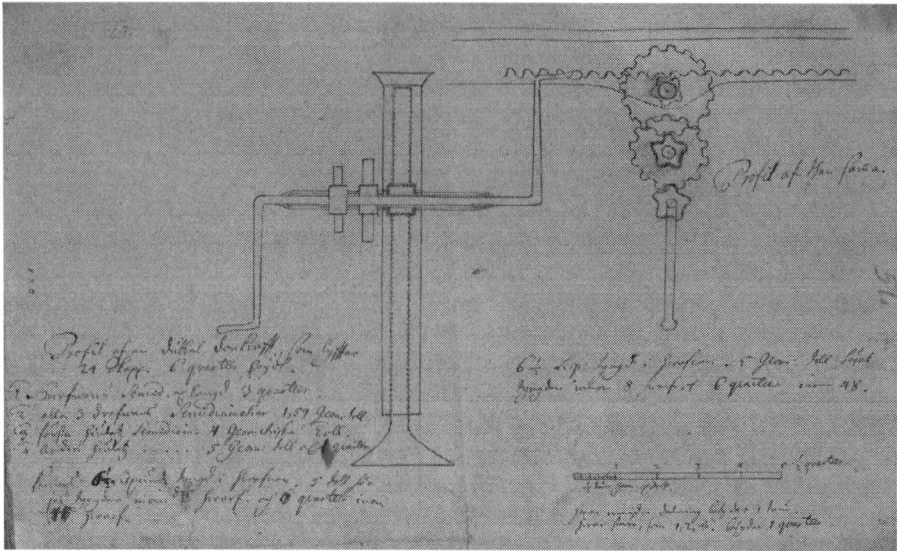

Fig. 2 The value of the geometrical inch in relation to the older Swedish inch (*verktum*) can be read off on a scale on one of Swedenborg's technical drawings. 1 half quarter = 2.5 geometrical inches, 1 geometrical inch = 1.2 old inches (Swedenborg, *Projekt och uträkning på domkrafter* (1716))

to the rules for Swedish dry measure. The proportions refer to the volume of the measuring vessel. In practice, however, one had to reckon with the heaped part, the cone above the brim of the vessel, which meant that the measured amount of goods was much larger. In addition, other measures of volume applied to wet goods. The mercantile problems of a farming country, peasants who shovelled wheat, rye, and oats in barrels and jugs, was also a mathematical problem in Swedenborg's *Regelkonsten*. Here he also deals with the fact that everything weighs less in water than in air, and even less in salt water and least of all in mercury. A person 'weighs 200 marks in air, 30 in water; so he consists of 120 marks of flesh and bone, 50 marks of blood or water, 30 of less or half that weight.'[81]

Since all counting with coins, weights, and measures is octal, one can easily see, for example, what a commodity costs by weight or volume. If you are told the price of a barrel you immediately know what a firkin or a jug will cost. Coins, volumes, and arithmetic are in symmetry, all with base 8. It is not just on the local market that the octal system can be beneficial. It is of astounding utility for the entire learned world, explained Swedenborg:

> For the rest, advantages could be found and shown with respect to many numbers; but since brevity gives more pleasure in what is new and unusual, I will postpone this to an occasion

---

[81] *Regel-konsten*, 81f, cf. 75, 78f.

when it will be acceptable. We can then submit examples of new advantages which are now left unmentioned. Meanwhile, should the practice of the use, and the use of the practice give its approval, I suppose that the learned world will gain incredible benefits from this octonary reckoning.[82]

The Lord Is Wrathful

There was widespread discontent in the country, in all classes of society, during the years when these numerical speculations were being formulated. In Swedenborg's immediate surroundings there were several examples of indignation, but also suppressed opposition. Benzelius, for instance, Swedenborg's close friend, has been assumed to have been one of the leading oppositional figures within learned circles, although it is difficult to find sure evidence of this. Benzelius is said to have belonged to a group headed by the county governor Per Ribbing, who was preparing a new form of government and wanted to remedy the damage caused by absolutism.[83] In *Anecdota Benzeliana* we read that

> King Charles XII was a great hero, but brought his kingdom into serious insolvency through his thirst for war; otherwise he had a number of incomparable virtues, such as generosity, restraint, temperance, and one can say of him as of M. Cato about Julio Cæsare: ad evertendam Rempublicam sobrius accessit.[84]

The cultivated humanist Benzelius interpreted Charles XII's temperance metaphorically through Suetonius' description of Julius Caesar as 'the only sober man who ever tried to wreck the constitution'.

Bishop Swedberg, who asserted the divine sanction of theocratic absolutism, fulminated from the pulpit about these godless times. In his *Betenckiande om Sweriges olycko* ('Meditation on Sweden's misfortune' 1710) it was obvious to Swedberg, with reference to the Psalter, that the country's military and political calamities were a punishment visited by God on a people living in sin:

> We have been as drunkards, fallen deep in a hard sleep of sin: and have let God's thunder rumble upon us in vain. We have not wished to be inconvenienced. Therefore God has now brought enemies upon us: and has now let another thunder rumble from cannon and suchlike; so that the earth shakes and gives way, and the foundations of the mountains shift and tremble: for the Lord is wrathful.[85]

The decisive battle was not fought on the fields of the Ukraine, but between evil and good within people themselves. And he who evades the sword is taken by the plague,

---

[82] *En ny räkenkonst*, § 17; edition, 269; translation, 34.

[83] Forssell, 382–384; Nyström, 75f, 93; Thanner, 36, 368; Ryman (1978), 4f; *Festivus applausus*, edition, commentary 22.

[84] Benzelius (1914), 64f; Suetonius, 1.53.

[85] Swedberg (1710), 3f; cf. Ps. 18:8; cf. Swedberg (1709a), [2]; Swedberg (1941), 259, 444; cf. Jer. 17:21–27.

he warns in *Gudelige dödstanckar, them en christen altid, helst i thessa dödeliga krigs- och pestilens tider, bör hafwa* ('Divine thoughts of death that a Christian should always have, especially in these times of war and pestilence' 1711).[86] War, plague, and famine as punishments for sinful behaviour had historical parallels in the history of Israel. This is a case of metaphorical thinking, in which Swedberg understands his time through the Old Testament. The pattern could already be found in the Bible, in the Psalms, in Jeremiah, and in other books. The fact that Charles XII was far away in the land of the Turk was a consequence of our grave sins and the neglect of the sanctity of the Sabbath in Sweden. Despite the critique of the king's war, the bishop could, at least in principle, give the institution of the monarchy his approval, as for example in his collection of sermons, *Gudz barnas heliga sabbatsro* ('God's children's holy Sabbath rest' 1710–1712), where the whole world is regarded as being tied together through 'command and obedience'.[87] The king is to command and the subjects to obey. The Table of Duties in Luther's Small Catechism, printed in the Swedish hymnbook, emphasized the need to obey one's superiors. Yet the choleric bishop could not remain silent when God's law was not followed, nor did he conceal his discontent with the devastating war which had brought misfortune on the country and brought it to the brink of disaster.

In January 1718 Swedenborg informed Benzelius about his father's forthright but unpalatable truths: 'Father has come home from his journey. He has a lot of things to tell about, and also a lot concerning the straight truth he there spoke out to the King.'[88] In Lund, on the second Sunday in Advent 1717, he had delivered a sermon to Charles XII, speaking of the signs in sun and moon, of earthquakes in our time and the terrible sign that had been seen in the heavens on 6 March 1716, and about the starving maid Ester Jönsdotter. But when he compared Charles XII to Rehoboam, King of Judah, he was incautious. Rehoboam's policies, as described in First Kings 12, had been unsuccessful and disastrous. This led to civil war and a humiliating defeat at the hands of Pharaoh Shishak of Egypt. 'But may the Lord preserve him from Rehoboam's counsellors!' he warned from the pulpit, referring to Charles XII. 'But this was displeasing to many, as I soon learned and heard. A servant of the Lord may not spare the truth; however badly it may be received and interpreted, and he must suffer for it: as I too now suffer these days, and will suffer.'[89] Pharaoh had listened to the wrong counsellors, 'and this happened here too. Sweden's plague *Gördz* [Georg Heinrich von Görtz] had too much to say. But he also got his well-earned punishment, the gallows. And rightly so. Dixi, salvavi animam meam.'[90] I have spoken and saved my soul.

Charles XII also expressed his disappointment over the bishop's sermons. Swedberg was nevertheless allowed to sit at his table and dine on the tin-plated

---

[86]Swedberg (1711), introductory intercessional hymn.

[87]Swedberg (1710–1712), 697.

[88]*Opera* I, 281, cf. 274, 278; *Letters* I, 177, cf. 164, 172.

[89]Swedberg (1941), 559.

[90]Swedberg (1941), 561.

iron dishes that Polhem had given to the king, a whole service with vinegar jars, sugar boxes, and the like. 'Not a word was spoken there: all was silent. The king ate rather quickly,' Swedberg remembered.[91] Polhem's tinned plates were the subject of a little piece written by Swedenborg, which can be viewed as a kind of advertising brochure, *Underrättelse, om thet förtenta Stiernesunds arbete, thess bruk, och förtening* ('Information about tinned Stjärnsund work, its use, and tin-plating' 1717). It concludes with the words: 'N.B. It has been noted that if children who are plagued by worms eat food that has stood on untinned iron vessels overnight, exceedingly sour, then the worms are consumed hereby, which is left to others to investigate further.'[92] Charles XII, however, preferred the man who had shared a room with Swedenborg in his Uppsala days, Andreas Rhyzelius, whom he appointed as First Court and Bodyguard Preacher and his father confessor in June 1717. Rhyzelius belonged to the Lutheran orthodoxy and distrusted modern science, but he too displayed a slight irritation over Charles XII's uncompromising refusal to commence peace talks. There was evidently some rivalry between Rhyzelius and Swedberg. They did not esteem each other very highly. Rhyzelius wrote that Swedberg in Lund had 'got to do all manner of displeasing things'.[93] Charles XII was not satisfied with Swedberg's sermon on the second Sunday in Advent about Rehoboam, and said to Rhyzelius: 'He does not preach as well as for the late king; for he is old now.' Rumour had it that Swedberg left Lund in shame, which he himself claimed was an outright lie in his autobiography.

Swedenborg was probably also discontented with the way the war and the national economy were being managed. He was well aware of all that was wrong. In London in 1712 he had bought a pamphlet prohibited in Sweden because it was critical of the autocracy of Charles XI and Charles XII; he then sent the pamphlet to Benzelius.[94] In *Camena Borea* (1715), a poetic allegory of the events in the Great Nordic War, one can discern Swedenborg's critical attitude to the political situation. Leo, that is to say Charles XII, is struck in the poem by Jupiter's thunderbolt and falls down from Olympus, leaves the abode of the Muses and ends up amidst the din of war drums and trumpets.[95] The allegorical name Leo alludes to the Paracelsian prophecy about the lion, which Hiärne had identified with Charles XII and others before him had interpreted above all as Gustavus Adolphus. The lion from the North would defend the true faith and defeat the eagle, that is, the Russians and the Poles. In the same work there is also a mystical allegorical figure named Dejodes, who symbolizes peace and a return to the old constitution.

---

[91]Swedberg (1941), 556; Bring (1911b), 236; J. Hultman, 120f.

[92]*Underrättelse, om thet förtenta Stiernesunds arbete*, [4]; C. Sahlin (1923); advertisement in *Stockholmiske kundgiörelser* (1717:15).

[93]Rhyzelius (1901), 81, cf. 76; Swedenborg asked Benzelius to send Sturm's *Mathesis juvenilis* I to Rhyzelius. *Opera* I, 253; cf. Liljegren (2000), 72.

[94]Benson (1711); *Opera* I, 220; *Letters* I, 41.

[95]*Camena Borea*, edition, 80–83, 120f, commentary 16, 21.

Swedenborg's pessimism about the prevailing situation is also reflected in a letter to Benzelius from June 1716. The country would soon be making its last kick. He saw no possibility of change as long as the king lived: 'Probably many desire that the torment may be short and we be delivered; yet we have hardly anything better to expect *si Spiritus Illum maneat.*'[96] The obscure Latin words can be interpreted as 'if life remains in him'. Two years later, in June 1718, Sweden was still at war. Swedenborg was disconsolate: 'when a country in general leans toward a state of barbarism, it is likely to be vain for one or two persons to hold it up.'[97] With peace, circumstances would improve. And rumours of peace came and went. It is bruited about in Lund that peace will come soon, Swedenborg reported. But it was essential to remain silent. Not everything could be said out loud with the limited freedom of expression under absolutism, as the county governor Germund Cederhielm put it in a letter about the political situation in the summer of 1718: 'But best be quiet, for watching and listening, silence and smiles, rarely go wrong.'[98]

A Peripeteia on the Decimal

In the light of all this, how should we now assess *En ny räkenkonst*? Was it written primarily out of an interest in arithmetic and geometry, as a way to benefit society, or could it have been a necessary tactical device, a means to gain advantages and obtain the king's grace? Did Swedenborg himself believe in the mathematical content, did he believe in his octal system? Probably not. One of the major reasons for this assumption is a small work printed by Swedenborg in November 1719, almost exactly a year after the king's death. It was sold by the bookbinder Dalbeck's widow in Nygatan in Stockholm for 4 öre s.m. and had the title *Förslag til wårt mynts och måls indelning, så at rekningen kan lettas och alt bråk afskaffas* ('Proposal for the division of our coins and measures, so that counting can be facilitated and all fractions eliminated' 1719). To be written by Swedenborg this is unusually clear, brief, and concise. Yet in this work Swedenborg advocates the decimal system!

> But if the division of coins, and also of measures, were according to the decimal, which is based on tens, then the stupidest person could count like the wisest; a peasant as well as a tax accountant. In addition, all trade and dealings, and the general and special economy of the realm, would derive incredible benefit and pleasure therefrom.[99]

The arguments for the decimal system resembled those applied to the octal system. If the systems for counting, coinage and measures were all decimal, calculations would be easier and one could avoid fractions. Each species of coin would thus

[96]*Opera* I, 252; *Letters* I, 105; *Festivus applausus,* edition, commentary 21f, 111f.

[97]Swedenborg to Benzelius, Vänersborg, June 1718. *Opera* I, 285; *Letters* I, 187; cf. Benzelius (1914), 53.

[98]Cited in Liljegren (2000), 328.

[99]*Förslag til wårt mynts och måls indelning*, [2].

be divided into tenths, from 1 mark of pure silver, via the riksdaler, pieces of five, wittens, down to pennies. And the same decimal system is also suggested for volumes, from 1 load (*läst*) which is then divided into barrels, measures, jugs, and glasses. Coins and measures would thus be in symmetry with arithmetic. By adding or removing a zero one could obtain the amount in the desired sort. If one had different units they could easily be placed in a row to form one sum. In just one number it would be immediately possible to see the larger and smaller parts of the amount. There would be no need for conversion, fractions could be avoided, and people would be spared from mental toil. Knowing one unit, a person would instantly know all the others. If you knew, for example, what a load cost, you immediately knew the price of the other volumes into which it could be divided. The introduction of a decimal system would thus mean that

> no more fractions or circumstantiality would be needed in any calculation: trade and dealings and the country's economy, as regards income and expenditure, would thereby find good arithmetic, order, and correctness; and an advantage over other nations in the world in all its counting and calculating.[100]

These eight pages 'about the decimal in our coinage and measures', Swedenborg writes to Benzelius in November 1719, 'will be my last word, since I notice that only Pluto and *Invidiae* [the Envies] occupy *hyperboreos*, and one secures greater fortune if one acts a fool rather than as a rational man etc.'[101] Swedenborg is resigned. In Sweden only envy rules. Better then to be an idiot. But the learned games with the numbers 64 and 8 had led up to the number 10. With *Förslag til wårt mynts och måls indelning* it is clear that Swedenborg advocated a decimal system for counting and for the division of coins, weights, and measures. The question is whether Swedenborg had now changed his opinion as to the most suitable base for a numeral system, or if he perhaps had really preferred a decimal system all the time. The king's death was probably not without significance in this context. It now became possible to recommend a decimal system, probably because the dependence on Caroline absolutism was over. He no longer needed to show any consideration for Charles XII's numerical speculations and was no longer required to await his approval. It may very well have been the case that Swedenborg as a scientist had always—both before and after 30 November—regarded the decimal system as the most suitable, although it cannot be entirely ruled out that he made an about-turn, in the space of a few months, concerning the base for a numeral system.

What he did know about, and what certainly served as a foundation for *Förslag til wårt mynts och måls indelning* and its decimal system, was that philosophers of nature wanted a logical and scientific system of measurement, and there were also Stiernhielm's metrological ideas. The linear standard of the Linea Carolina is based on a decimal division, and in several manuscripts Stiernhielm describes

---

[100] *Förslag til wårt mynts och måls indelning*, [7].

[101] Swedenborg to Benzelius, Stockholm, 26 November 1719. *Opera* I, 295; *Letters* I, 221.

decimal arithmetic.[102] He had originally had the intention to accomplish a complete decimalization of the measurement system and had advocated a decimal scale of length based on the smallest thing in the material world—the point. Ten points make a line. But the suggestion was ignored; it was still far too radical. It was not until the decree of 1733 that a decimal division was introduced for linear measurement, to be used alongside the traditional duodecimal system with 12 in. to a foot. In the circle around *Dædalus Hyperboreus* Stiernhielm was admired for his Linea Carolina and his Swedish poetry. In 1716–1717 Benzelius, Polhem, and Swedenborg discussed including a biography of Stiernhielm, '*Stiernhielms vita*', in the journal together with the Linea Carolina.[103] Swedenborg wanted to borrow Elvius's copy of the Linea Carolina and also asked him what he knew about it. Polhem returned several times to Stiernhielm's experiments, studied and discussed his determinations of weight and measure.[104] In the spring of 1717 Maja and Mensa, as Polhem called his daughters Maria and Emerentia, brought a letter to Swedenborg in Stockholm. Even Sweden has geniuses, writes Polhem, referring to Stiernhielm:

> Although the Sun gives Sweden short and cold days in the winter, they are so much the longer and more delightful in the summer, so that southerners have nothing to boast about in this respect when the year is ended; in the same way, and although Sweden engenders the stupidest people that other nations rightly disdain, there are on the other hand such clever ingenia that other nations cannot surpass or teach, although these two extremes together do not do more than intermedia in other places or vice versa.[105]

A decimal division of weights and measures had been advocated by many scientists and mathematicians, including, in Sweden, Anders Bure and Mathias Björk, but above all by the Dutchman Stevin, who put forward the decimal system at the end of the sixteenth century, which was later introduced to Sweden by Gestrinius.[106] Polhem too was a supporter of the decimal system. In *Wishetens andra grundwahl* Polhem writes about linear measure, that craftsmen divide the Swedish foot into 12 in., but surveyors 'otherwise more correctly into 10 *tolls*; one toll into 10 lines, and a line into 10 points.'[107] He also mentions that 'a measuring rod is 10 foot; a measuring tape 10 rods or 100 foot', and the decimal cubic foot, a jug equivalent to 100 cubic tolls.[108] In another context he referred to Swedenborg's *Förslag til wårt mynts och måls indelning* and wrote that a decimal division is more natural than the artificial and unnecessary fractions, 'I, like all wise Mathemati[ci], regard this division as the most perfect.'[109]

---

[102]Stiernhielm, 'De numeris geometricis siue quantitatibus algebraicis'. KB, Fd 15, fol. 7; KB, Stiernhielm, X 727.

[103]*Opera* I, 256, cf. 267–269.

[104]*Polhem's Letters*, 46f, 49, 52–54, 62; *Polhems skrifter* III, 115, 124.

[105]Polhem to Swedenborg, Stjärnsund, 27 March 1717. *Polhem's Letters*, 127.

[106]F. Hultman, 7f; Falkman, II, 15f; Jansson, 60.

[107]Polhem (1716), § 21; cf. *Polhem's Letters*, 96.

[108]Polhem (1716), § 22, 29, 31.

[109]Polhem, 'Om Sveriges lösa ägendom', *Polhems skrifter* II, 166; cf. Polhem 'Arithmetica eller Reknekonst'. KB, X 705:1, fol. 70f; Polhem, 'De mensura comuni'. KB, X 706, fol. 37f.

In *En ny räkenkonst* Swedenborg wrote that it was not necessary to introduce a decimal system as the learned recommended. The same advantages could be obtained through an octal system. But the time was not ripe to introduce a decimal system across the board, neither in Stiernhielm's nor Swedenborg's time. Geometrical measure, that is, a decimal linear measure, was nevertheless used by land surveyors. The only examples of decimal division to be introduced were the subdivision of the foot and the definition of the volume of a jug.[110] Nor was 1718–1719 the best time to carry out a reform, in view of the economic difficulties and the widespread annoyance about the emergency coins. But Swedenborg's *Förslag til wårt mynts och måls indelning* did not pass unnoticed. It was discussed in Bokwettsgillet and reviewed in *Acta literaria Sveciæ* for 1720, and it appeared in a new edition as late as 1795.[111] The Russian historian and mine owner Vasily Nikitich Tatishchev, visiting Sweden in the mid-1720s, discussed the advantages of the decimal system with Swedenborg.[112] He himself tried, without success, to introduce the decimal system for weights and measures in Russia. Swedenborg then gave a copy of his work to Tatishchev, who immediately had it translated and sent to the imperial cabinet in Saint Petersburg just after New Year 1725.[113] This manuscript is the very first translation of Swedenborg into Russian, and it still exists, discovered by me in the archives of the Russian Academy of Sciences.[114] In the accompanying letter to Ivan Cherkasov, Tatishchev gives an account of his meeting with Swedenborg, whom he describes as a prominent physicist, mathematician, and mechanic. Tatishchev also acquired a complete collection of *Dædalus Hyperboreus*.

In December 1741 Swedenborg donated to the Royal Academy of Sciences a copy of his *Förslag til wårt mynts och måls indelning*, which the academy then delegated to the president of the Board of Trade, Anders von Drake, to peruse.[115] Swedenborg's colleague in the academy, the surveyor Jacob Faggot, published an essay in the proceedings of the academy the following year entitled 'Om Tijotälning, eller Decimalers häfd i Bokhålleri och Räkning, som rörer Mått, Mål, Wigt och Mynt, utan rubning i de wanlige inrättningar' ('On counting in tens, or the decimal tradition in accounting and arithmetic, concerning dimensions, measures, weights, and coins, without disturbance to the accustomed institutions). Among other things, Faggot proposed emulating the Chinese, the only people who used the decimal system, and dividing the coinage decimally from rundstycken, slantar, and daler to purses, bags, and sacks of money for the really wealthy. He also mentions Swedenborg's and Charles XII's discussions of a new numeral system:

---

[110]Ohlon (1989), 124. No decimal system came into existence until the 1855 statute on weights and measures. Nystedt, 74.

[111]*Bokwetts Gillets protokoll* 22 January 1720, 11; *Acta literaria Sveciæ* (1720), 22; *Neue Zeitungen* 2 June 1721, 352.

[112]Küttner, 121, 159; Jukht (1985), 201f.

[113]Tatishchev to Cherkasov, Stockholm, 2 January 1725. Jukht (1990), 105.

[114]Tatishchev, *Tetradi Tatishcheva k ego radotash po geometrii*. RAN, Razrjad II, Opis 1, No 211, pages 8–12.

[115]*Svenska Vetenskapsakademiens protokoll* I, 364, & II, 155.

Indeed I have been recently informed that the late King Charles XII intended to implement
such a system, following the method suggested by a learned man, had not death unfortunately intervened. For the rest, I have also, since this was composed, been presented by a
good friend with a well-wrought publication, under the name of a proposal for the division
of our coinage and measures, printed in 1719: both the manner and the utility of the decimal
system are clearly revealed therein. But the method adopted for this in the present work is
so vastly different from the aforementioned fine proposal, that the aim here is to retain the
established institutions.[116]

It was Swedenborg's *Förslag til wårt mynts och måls indelning* that Faggot's good
friend (perhaps Swedenborg himself?) had presented to him. Perhaps Faggot had
also come across Nordberg's biography. But no extant sources say anything about
Charles XII and Swedenborg having discussed the introduction of a decimal system
as Faggot states.

By 1719 times had changed in several respects. King Charles was no more. Not
only had Caroline absolutism been buried; the Görtzian gods had also been toppled
from their Olympus. Widespread discontent and distrust had been provoked by the
emergency coinage minted from the spring of 1716 onwards by the king's first
minister, Georg Heinrich von Görtz, with tokens bearing mottoes such as 'Wett
och Wapen' (Sense and Arms) and 'Flink och Färdig' (Quick and Ready). They
were introduced to finance the increasingly costly war, although the idea cannot
be blamed on Görtz, but possibly on the department head Casten Feif.[117] The
emergency coins, with a nominal value of one silver daler but a substantial value far
below this, were minted until 1719. The idea was that they should function roughly
like banknotes, or like the tokens used at the mines, but they were made of metal
and could be cashed in by the state. A Turkish creditor was amazed that Charles XII
had such power that he could make his subjects believe that a copper coin had the
same value as a silver one.[118] Therein lay the problem. Hardly anyone believed in
them, and the result was rampant inflation.

This tampering with the value of the coinage and the metal content was criticized
by Swedenborg and many like him. Swedberg may also have been referring to coins
of this kind, admittedly not struck so badly but with a metal content lower than the
nominal value. In *Schibboleth* he likened old hymns

to old current riksdalers, which have good silver and good bronze in them, and are of
much worth in weight and value, having the price within them, although they are outwardly
shabby, rough, and unsightly. But many new hymns sound good to the ears, give a pleasant
sheen to the eyes, and run well; as new coins with a fair stamp and well ornamented; but
their content and material are weak. They do not always possess such good quality and
reverence.[119]

[116]Faggot (1742), 56; cf. Faggot (1739).
[117]Tingström (1995), 195; Tingström (1997), 104.
[118]Liljegren (2000), 264.
[119]Swedberg (1716), 104.

Fig. 3 Coining machine in Avesta. 'But it is not our business to explain here the instrument and machinery used in this process,' Swedenborg writes. 'It is enough to subjoin some drawings thereof.' The picture shows a coining machine at Avesta copper works in 1715, designed by the master builder Magnus Lundström. At *G* there are two cylinders between which the copper is pressed into thin plates. To the *right* a water wheel and in the background a window looking out on a lone, slender spruce tree (Swedenborg, *De cupro* (1734))

Benzelius noted in May 1718 that 'the Russians bring fine goods, but no one can trade because of the coinage. The country sighs, and execrates *Giörtz*.'[120] Nor did Polhem have any understanding for the idea behind the tokens, although he had taken part in their production (Fig. 3).[121] In a memorandum to the Board of Trade

---

[120]Benzelius (1914), 59.

[121]Polhem (1716), 125; *Polhem's Letters*, 122; Chydenius, § 4; Bring (1911a), 55.

in 1715, Polhem protested against the changes in the value of the coinage, which he termed 'childish madnesses or mere rogueries'. And in an essay from 1716 he remarks: improving the state's finances by debasing the coinage, letting 1,000 dalers become 1,500 dalers, was like 'turning an ass into a horse with a gold-embroidered saddle and caparison, and using a high title to turn a peasant into a Doctor Philosophiæ'.[122]

After the death of the king in the trenches and that of Görtz on the gallows, the problem arose of how to redeem the coin tokens. Swedenborg wrote a piece on the matter which was read to the Secret Committee in February 1719.[123] If these coins were not abolished in good time, inflation would grow like an incurable disease and ruin the country. His stance was that the value of the coins should not be reduced. Swedenborg also criticized the ideas about 'the valuation of the coinage', raising or lowering the value, and thus opposed token coinage in a pamphlet from 1722 entitled *Oförgripelige tanckar om swenska myntetz förnedring och förhögning* ('Unassailable ideas about the lowering and raising of Swedish coinage'). This caused 'quite a stir' in Stockholm, wrote Swedenborg.[124] The censor Johan Rosenadler, who had previously gone under the name Upmarck, had to defend himself for having let it pass. Later, however, Swedenborg's pamphlet was praised by the leading political economist in Sweden, Anders Chydenius, and it was reissued as late as 1771.[125] Linnaeus also wrote about the valuation of coins in a work about assaying, *Vulcanus Docimasticus* (1734), expressing ideas in line with those of Swedenborg: 'Nothing can exhaust a country as much as incorrect and false coinage; for if the coin is *too good*, it is taken out of the country, and therefore is less plenty; *too poor* coin is minted abroad and brought in. E.g. coin tokens.'[126] Swedenborg concludes *Oförgripelige tanckar*, his unassailable thoughts, by declaring

> that a raising is extremely damaging to a country; so too is a lowering, [...] to stir and disturb a coin in a situation where it is not necessary would be to stir everyone's chief business, and to disturb the noblest thing in one's keeping, and to make a change to the commerce of the entire kingdom, and as a consequence thereof, to submit oneself to be answerable to each and every one of those who will ultimately suffer from it.

Peace had finally come when this was written, as a result of the Treaty of Nystad in August 1721. When the peace was celebrated in The Hague on 8 December—with fireworks, wheels of swirling flame, fire-spouting mountains, and military music—Swedenborg wrote a poem about the Russian eagle which had defeated Mars. Blood

---

[122]Polhem, 'En discurs emellan Oeconomien och Commercien uti Sverige af assessoren H. Chr. Pollheimer författat uti 7bris och octobris månader åhr 1716', *Polhems skrifter* II, 36f; cf. *Polhem's Letters*, 139f.

[123]Swedenborg, 'Förslag till myntetecknens och sedlarnas inlösen'; translation, *Letters* I, 205–211.

[124]Swedenborg to Zacharias Strömberg, Stockholm, 7 November 1722. *Opera* I, 311; *Letters* I, 280; Benzelius to G. Benzelstierna, Linköping, 2 July 1737. LiSB, Bf 11, no. 175; Benzelius (1791), 266.

[125]Chydenius, § 17.

[126]Linnaeus (1925), 73.

had flowed, but now rivers of nectar would flow.[127] But was absolutism gone, could people now express their opinions freely, was Charles XII really dead? A couple of years later, rumours were circulating that Charles XII was still alive. But it turned out to be a Finn, whose real name was Benjamin Dünster, who had pretended to be the king and claimed that he had never left Turkey, but had sent a bodyguard to Sweden in his place. That was the man who had died at Fredrikshald, not Charles XII. Dünster was arrested in December 1724. In a letter from his inquisitive brother Jesper in February 1725, Swedenborg was asked what sentence was passed on the false Charles XII.[128] He was pronounced insane and placed in a lunatic asylum, where he died in 1730.

A Million Million

Swedenborg returned on two later occasions to discussions about a new numeral system. The first time was in 1722. With his presentation of the king's base-64 system in Latin in *Miscellanea observata*, Swedenborg wanted to make it known to an international audience. Here he paints a portrait of a storied hero, whose powers of thought were most acute in everything to do with mathematical calculations, and he emphasizes how thoroughly the king had penetrated the deepest secrets of arithmetic. A slight degree of scepticism can nevertheless be detected. He had expressed some doubts about the king's insistence on the number 64. The series of numbers would be far too long and unmanageable, not just in subtraction and addition, but especially in division and multiplication. But the more he had pointed out the difficulties, the more eagerly the king had persisted in his experiment with the number 64. To Swedenborg's surprise the king, after one or two days, had devised new signs and figures resembling the letters in his own name. This great monarch, with a brain of heroic power, was not only my rival, but also my better in my own field, Swedenborg exclaimed.[129] This presentation of the king's base-64 system made some impact.[130] The essay in *Miscellanea observata* was reviewed in 1723 in both *Acta eruditorum* and *Neue Zeitungen von gelehrten Sachen*. Christian Wolff also mentions this discussion in the Geneva edition of his *Elementa matheseos universæ* (1732).

Being cited by the famous Wolff probably filled Swedenborg with pride. At any rate, he was not slow to mention it in his letter to Nordberg: 'Herr Court Councillor and Professor Christian Wolf has referred to it in his *Geometrie*, and made it

---

[127] *Ludus Heliconius*, edition, 70f, commentary 156–158; *Letters* I, 258f.

[128] J. Swedenborg to Swedenborg, Brunsbo, 26 February 1725. *Letters* I, 371f.

[129] *Miscellanea*, IV, 5; translation, 116.

[130] *Acta eruditorum*, March 1723, 96f; *Neue Zeitungen* 29 March 1723, 235; Wolff (1732a), § 46; Swedenborg, 'A curious memoir of M. Emanuel Swedenborg, concerning Charles XII. of Sweden', *The Gentlemen's Magazine*, September 1754, 423f.

somewhat known to the learned world.' This letter, possibly written after his return from Germany in summer 1734, was to be printed in 1740. The letter is probably based on Swedenborg's own notes in *Miscellanea observata*. Now he conveys his recollections of the discussions about the number 64 to this Nordberg, who had been given a parliamentary commission in 1731 to write a history of Charles XII. Swedenborg had been present himself at this session of the diet and was well aware that his letter to Nordberg could be published and be included in his biography. This conclusion can also be drawn on stylistic grounds. Unlike his other letters, the one to Nordberg is more correct in language and better organized. We see once again the picture of a king with 'depth of mind and penetration'. He says of Charles XII that it emerged during conversations on mathematics that he 'must have had a greater intelligence than he allowed to appear outwardly'.[131] For the king one was only half a man if one had no insight into mathematics. In the letter Swedenborg once again states that he owns an outline of the king's system '*in originali*'.[132] The two hand-written sheets have never been found. In *Miscellanea observata* he had put forward plans to have them published. Perhaps this never happened because of the typographical difficulties of procuring 64 new types.

If the king had been allowed to live, he might have caused the sciences to flourish, Swedenborg conjectured: 'And if the gracious Providence of the Most High God had been that he should have ruled his kingdom in peace and tranquillity, he would probably have brought studies and the *scienter* to a higher *grad* and a greater *flor* [flower] than had ever before been in the kingdom of Sweden.'[133] The same tribute had been expressed by Swedenborg in the dedication to Halley in *Försök at finna östra och westra lengden* ('Attempt to find eastern and western longitude'), but Charles XII was still alive then. At that time he had declared that, 'if the supreme God grants our Great Monarch a long life', Sweden would be more encouraged in bookish arts, 'since His Majesty not only shows great interest and care for it, but also has a great understanding and profound judgement in such matters; so great that many of the great Mathematici cannot match him'.[134] Swedenborg, however, ends his letter to Nordberg by mentioning two technical projects which had been started during Charles XII's reign: the dock in Karlskrona and the locks at Trollhättan. In both these projects Swedenborg himself had played a very active part. Was he perhaps applauding himself?

Swedenborg portrays a royal mathematician in these texts. This picture has made an indelible impression in historiography and has become a romantic tale of two colleagues engaged in learned conversations about mathematics in Lund, an anecdote about a mathematician on the throne. The picture is greatly exaggerated. One is tempted to interject that not even his 'applied ballistics' was particularly successful. When people speak of Charles XII as a mathematician, they are largely

---

[131] Nordberg (1740), 599, cf. 601; *Letters* I, 459, 464.

[132] Nordberg (1740), 601; *Letters* I, 463; cf. *Opera* I, 272, 275; *Miscellanea* IV, 5; translation, 116.

[133] Nordberg (1740), 601f; *Letters* I, 464.

[134] *Försök at finna östra och westra lengden*, dedication.

referring, directly or indirectly, to Swedenborg's letter to Nordberg. Yet a close reading of all the other existing documents actually reveals nothing more than that Charles XII was interested in mathematics. This is scarcely surprising. As crown prince he would have been taught through Bilberg's *Elementa geometriæ*.[135] And mathematics, geometry, and trigonometry were part of the training of an officer. A knowledge of fortification and ballistics was useful in times of war. Swedenborg's account can in fact speak against the king as a mathematician rather than in favour. If we look at the purely mathematical matters in the texts, we find that they are rather non-committal and imprecise on that point. There are no details to provide any concrete information about the level of the king's mathematical education, but rather what can be interpreted as hints about his lack of expert knowledge, for example, his failure to use algebra in the solution of a problem. 'And I can give assurance', writes Swedenborg, 'that to me it was incomprehensible how this could be found out by mere thinking and without an application of the method of algebraic calculations commonly in use.'[136] Here Swedenborg is laconic, but he is all the more lavish with his general assessments, for instance about the king's profundity and penetration. There is not a single convincing and objective argument for the king's mathematical talent. What he says about the king's mathematics in fact concerns the simplest task of all when it comes to creating a new numeral system: devising new names and figures. This requires no mathematical knowledge whatsoever, just a little imagination, patience, and a delight in such mind games. Anything that might be more complicated—the work of compiling tables and arithmetic rules—was a task the king ordered Swedenborg to perform.

Of course, Swedenborg may really have been impressed by the king's mental ability, and this was not an unusual sentiment. During a conversation about numeral systems with the king and Swedenborg in Lund, Polhem happened to mention the figure of 'a million million', declaring that so many seconds had not passed since the creation. The most common calculations, for instance by Eric Benzelius the Elder and Peringskiöld, placed the creation around 4000 BC.[137] Polhem's statement was immediately checked by the king. He had 'an uncommon talent for multiplying large numbers', Polhem noted in a mathematical fragment.[138] Immediately after this in the same fragment, he wrote, but then crossed out, the words: 'like Lasse på Jorden', which is high praise indeed for Charles XII. Such a gift for mental arithmetic as that displayed by Charles and Lasse was also found in the learned doctor John Wallis, whom Polhem himself had got to know in Oxford.

Lasse på Jorden (Lasse on Earth), or Lars Bengtson Granberg as he was really named, was one of the many memory artists and arithmetical geniuses who travelled round showing off their skills. In the correspondence between Leibniz and Magnus Gabriel Block there is a discussion of a Spanish monk and other mnemotechnicians.

---

[135] Nordenmark (1959), 92.

[136] Nordberg (1740), 599; *Letters* I, 460.

[137] Benzelius the Elder, 7; Peringskiöld, 93; Linton, 276.

[138] Polhem, KB X 705:1–2, fol. 107.

Block tells of an unusually gifted Swedish 12-year-old, 'un ragazzo di anni 12', who, despite being illiterate and unable to write the figures with which he counted, was amazingly quick at mental arithmetic.[139] The philosopher and mathematician Leibniz was fascinated by this person who not only shared his mathematical talent but also had an almost identical physiognomy. Lasse, who was now 20 years old in 1698 and a great spectacle at the Danish court, had previously displayed his skills to Charles XI and Bishop Swedberg. It may also be Swedberg who was behind the publication at this time of Lasse's *Curieuse och snälle uthräkningar* ('Curious and swift calculations'). Among his virtuoso arithmetical performances was the calculation of the number of peppercorns in the cargo of 1,800 warships:

> When 1 ship carries 5,000 Skieppund, 1 Skieppund is 20 Lispund, 1 Lispund is 20 Skålpund, 1 Skålpund is 32 Lod, 1 Lod is 4 Quintin. When there are 18 peppercorns to 1 Quintin, when I split 1 peppercorn into 19 parts. Now I want to know how many million thousand peppercorns 18 hundred of the biggest warships in the sea could be loaded with. Total 157: 1000: 1000: 1000 times a thousand, 593: 1000: 1000 times a thousand, 200: 1000: times 1000:[140]

An admiration for the king's intellectual talents need not necessarily mean that Swedenborg also liked his policies. We should bear in mind that Swedenborg was actually heavily dependent on the king.

Attempts have been made to show Charles XII not only as a mathematician but also a philosopher. In this case there really is a source that can be attributed to Charles XII. This is his 14 theses, 'Anthropologia physica', which were printed in Nordberg's *Anmärkningar, wid hög'stsalig i åminnelse konung Carl den XII:tes historia* ('Notes on the history of the late King Charles XII', 1767).[141] These brief notes, quickly jotted down after philosophical conversations in Lund with the Hessian court councillor David von Hein, are not particularly convincing either. The fragment begins with an almost hedonistic thesis: 'The natural drive of all living things is what is called passion, or the desire to enjoy pleasure.' Otherwise the theses have clear points in common with the Cartesian distinction between mind and body, the idea of life-spirits in the blood, and the belief that the five senses could be reduced to one, feeling. The philosophy of Rydelius can also be glimpsed, along with what can supplement previous interpretations of the notes, links to Paracelsian alchemy when he talks about salt and sulphur as two of the basic principles of matter. A certain deterministic belief in fate pervades the theses as a whole. Feif advised Charles XII not to send the theses to Christian Thomasius and thereby challenge this professor of law in Halle who was highly critical of Descartes' philosophy: 'It would thus be better for your Majesty, in truth, to give up this war of pens and stick to the sword.'

---

[139] Block to Leibniz, Stralsund, 30 October 1698. Block, 207, cf. 209, commentary 248f.
[140] Granberg; Platen, 49f.
[141] Nordberg (1767), 55–60; cf. KB, Nordberg, D 809, D 812 & D 814.

The only person who saw through the 'evidence' for Charles XII as a mathe-matician was François Marie Arouet de Voltaire, with his expertise in science and mathematics. He was not terribly impressed:

> Some people would describe Charles as a good mathematician; he possessed, no doubt, a great degree of penetration, but the arguments they make use of to prove his knowledge in mathematics, are by no means conclusive: he wanted to alter the method of counting by tens, and proposed to substitute in its place the number 64, because that number contains both a cube and a square, and being divided by two is reducible to a unit. This only proves that he delighted in every thing extraordinary and difficult.[142]

These lines about Charles XII's base-64 system appear in the revised editions of Voltaire's *Histoire de Charles XII Roi de Suède* from 1749. The earlier editions, the first of which appeared in 1731, do not have this passage. This suggests that Voltaire had taken the information from Carl Gustaf Warmholtz's French edition from 1748 of Nordberg's biography of Charles XII, despite the fact that Nordberg was Voltaire's rival in this field.[143] Voltaire's statement can also be viewed as a critique of Nordberg. Indirectly, then, Voltaire had received the information from Swedenborg himself. Voltaire, who had no need to show any regard for the Swedish political situation and national considerations, could thus paint a less embellished, more brutal picture of the time just before Charles XII's death, in contrast to many Swedish historians well into the twentieth century. Voltaire describes a country mostly consisting of old men, women, and children, a land where the clergy were taxed, which made Charles XII hated by the whole nation, and where the priests openly called him an atheist since he demanded money of them. His generosity degenerated into wastefulness and ruined Sweden, and his power bordered on tyranny.[144] A similar opinion had been voiced by Voltaire's host, Frederick the Great of Prussia, in 1757, expressed in figures: 'In all the books that talk about Charles XII, I find splendid praise of his moderation and chastity. But 20 French cooks and 1,000 mistresses in his retinue and 10 troupes of actors in the army would never have caused a 100th of the damage to his kingdom that his fiery vindictiveness and boundless ambition produced.'[145]

Rhetorical Arithmetic

Both in the account in *Miscellanea observata* and in the letter to Nordberg, the main topic is *Charles XII's* idea about 64 and *his* 'penetration', and not really about mathematics itself. Swedenborg's own efforts are placed in the background.

---

[142]Voltaire (1749), 277.

[143]Nordberg (1748), III, 278f & IV, 304–306; Voltaire to Nordberg 1744. Voltaire (1880), 278–283; Brulin, 25; Holm, 51.

[144]Voltaire (1731), II, 324; Voltaire (1749), 253, 270, 276.

[145]Cited in Oredsson, 278.

Although he mentions the octal system, there is little that can be connected to *En ny räkenkonst*. One reason may be that, as a scientist, he did not actually believe in the practical value of the octal system. He preferred the decimal system. Would it not be simpler to introduce decimal divisions for coins, weights, and measures, instead of inventing a whole new way of counting? Swedenborg the scientist also understood that the system based on 64 was in the nature of a royal diversion of limited scientific benefit. Could the author of *En ny räkenkonst* have preferred a different numeral system? To obtain an answer to that question we must first ascertain the purpose of these texts, his underlying intentions. The aim of the two accounts of the number 64 was nothing other than scientific and historical. The essay in *Miscellanea observata* and the letter to Nordberg can be viewed against the background of rhetorical and social rules. The main purpose was to follow the convention of erecting an honourable memorial to the late warrior king. These texts tell us nothing about Swedenborg's true personal opinions. We cannot know from them what he really thought, only that it is perfectly possible that in private he could have had a completely different opinion from the one expressed in the words. They cannot be accepted as wholly credible and truthful sources. In this rhetorical genre a literal reading instantly leads one in the wrong direction.

When it comes to the interpretation of the text in *Miscellanea observata*, we should be attentive to the fact that it follows immediately after the dedication to Ludwig Rudolf, Duke of Brunswick and Lüneburg, another high-born man with whom it was beneficial to be on good terms. It would therefore have felt appropriate to begin with a text about royal arithmetic. It soon emerges that this is indeed a panegyric, which is to be interpreted rhetorically. One should therefore read Swedenborg's assessment of the king critically. The same applies, perhaps even more so, to the letter to Nordberg. His biography of Charles XII, which was supported by the diet, can of course be viewed as a nationalistic project with the warrior king as a revanchist symbol intended to whip up enthusiasm for war. The work was subjected to critical scrutiny, amounting to censorship, and parts of it were deleted so that it would paint as good a picture of the king as possible.[146] The Hat Party was thirsting for revenge, and war with Russia was imminent. The student Anders Odel composed a heroic poem about Sinclair to fire the senses, depicting how Major Malcolm Sinclair, who had been murdered by Russian soldiers in 1739, is received by Charles XII in the afterlife.[147] The year after the publication of *Konung Carl den XII:tes historia*, Sweden attacked Russia. Once again it ended in disaster. These texts do not give plausible, objective verdicts. The aim was to present an encomium, a biographical speech lauding a person's virtues and good qualities.[148] The task of the epideictic author, in this case Swedenborg, was not to be objective; on the contrary. The idea was to enhance and exaggerate a person's merits,

---

[146]Nyström, 231; cf. Holm, 42–44; Sandstedt, 144–149, 275f; Sandstedt does not mention Swedenborg.

[147]See Odel.

[148]Ridderstad, 91f, 261.

to paint as favourable a portrait as possible. In this genre, for example, a foolhardy man became courageous. In Charles XII's case the need for diversion became profundity, an interest in arithmetic became mathematical genius, and Swedenborg's partner in conversation became his mathematical better. These epideictic memorials were part of a pattern of social behaviour. It was not a matter of telling the truth, but of saying the right thing. It is rhetoric. Swedenborg was expected to write a panegyric to a great king, and this can also be seen as a gesture of submission to the ruling king at the time, Frederick I.

It is perfectly clear that Swedenborg was subordinate to Charles XII, but also to Polhem and Benzelius. In that sense one can detect a patron-client relationship. Swedenborg makes constant declarations of loyalty to Charles XII. He seeks to show his humility and obedience, which is a strategy, indeed an obvious code of behaviour, for a client aspiring to rise. Having a powerful and influential patron was essential for anyone who wished to move up in the hierarchy. Charles XII, like Polhem, functioned as hoists for the young Swedenborg, dragging him up from the dark depths of unestablished scientists to the light. Swedenborg severed his ties with both of them. In his letters to Benzelius he consistently addresses him as 'highly honoured' and rounds off with turns of phrase like 'humblest brother and servant', 'most obedient and faithful brother', and even sometimes 'most faithful brother even to my death'.[149] This signals that he adopted a role as willing servant to his patron. It was also important to acquire a reputation for being competent and enterprising, an aspiration that certainly shines through in his letters from the 1710s. Benzelius was patron to many humble priests.[150] But for Swedenborg it was rather the case that Benzelius, as his brother-in-law, was his patron more through family ties than in a traditional client relationship. They were also very close to each other, and a genuine friendship is expressed in the letters between them. Swedenborg functioned in turn as patron to the penniless astronomer and future lecturer in mathematics, Birger Wassenius.[151] His dissertation on the planet Venus, *De planeta Venere* (1717), probably written by himself under Elvius's presidium, is dedicated to his patrons Swedenborg and Polhem. Swedenborg likewise displayed great benevolence to 'Little brother Eric', the son of Eric Benzelius the Younger. Swedenborg led him into the study of science. From Rostock Swedenborg wrote in 1714 with concern about the situation in Uppsala now that the Russians were approaching. He longed to see his nephew once again and wondered: 'He can now perhaps make a triangle or draw for me, if I get him a little ruler.'[152] Little brother Eric later became a member of the Board of Mines.

Swedenborg's thorough familiarity with classical rhetoric is shown by his attempts at poetry and his dissertation on some maxims of Seneca and Publilius Syrus. Dissertations and disputations were in fact a rhetorical genre in which the

---

[149]Swedenborg to Benzelius, Stockholm, 21 November 1715. *Opera* I, 232, cf. 265; *Letters* I, 70.

[150]Ryman (1988), 42–50.

[151]*Opera* I, 274f, 284f.

[152]*Opera* I, 228, cf. 267, 316; *Letters* I, 59, cf. 146, 382.

essential thing was to convince, not necessarily to present new truths. The rhetorical interpretation of Swedenborg's panegyric to Charles XII and his mathematics can also be compared to the poetic tribute that Swedenborg wrote to Charles XII on his return to his kingdom, more specifically to Stralsund in Swedish Pomerania on the night of 11 November 1714. Swedenborg was then in nearby Greifswald, where he was enrolled at the university nine days later as the last Swedish student before the Danish occupation.[153] But Greifswald, he says, is 'a very scurvy Academy' and the professor of mathematics, Jeremias Papke, is fit for any subject but that.[154] This place would not do much for his mathematical improvement. Instead he devoted his time to publishing *Festivus applausus in Caroli XII in Pomeraniam suam adventum* (1714/1715) and other poetic works.

What emerges from this is that one cannot, despite the praise, immediately draw the conclusion that Swedenborg was an admirer of Charles XII.[155] It turns out that Swedenborg was not sincere in this tribute; on the contrary, it is highly likely that he did not embrace the sentiments expressed in it. Swedenborg's private political opinions were different. The language is hyperbolic and inflated. This was a part of the way of writing eulogy, a genre that required panegyric judgements that were not expected to be an expression of the author's personal views, but rather of well-adjusted and socially accepted roles. Swedenborg, who was at this time on his way home to Sweden, lacked employment and a secure economic and political position. During these years he was simultaneously driven by a zeal, an almost overambitious craving, to make a career as a scholar and gain the recognition of the scientific world. What could favour a career like this more than good relations with the absolute monarch? That was virtually a *sine qua non*. Being on good terms with the sovereign could hasten his career considerably. Even before they met, he knew from the widely spread rumour that the king was interested in mathematics, as is evident from the dedication in *Dædalus Hyperboreus*. The mathematician Agner passed the reports on to Benzelius in 1711, informing him that he had heard that the king appreciated mathematical sciences and was 'tolerably well versed in them'.[156] Moreover, the king had asked about his book on logarithms. Swedenborg therefore called on the king in Lund, presented his projects for the improvement and edification of the country, and showed his journal. It paid off. He was appointed extraordinary assessor and given a number of engineering assignments. *En ny räkenkonst* should be interpreted in this context. It was a pawn in a tactical game, consciously or unconsciously.

It therefore seems natural that Swedenborg exaggerates Charles XII's intelligence in *En ny räkenkonst*. It was after all dedicated to and intended for the king. The underlying motive was perhaps not so much scientific, an expression of an interest

---

[153]Seth, 21; Önnerfors, 27.

[154]Swedenborg to Benzelius, Greifswald, 4 April 1715. *Opera* I, 229; *Letters* I, 62.

[155]*Festivus applausus*, edition, see especially Helander's commentary 13, 18f.

[156]Agner to Benzelius, Bönesta, 23 June 1711. Agner (1940), 52; Agner (1710); cf. *Polhem's Letters*, 88.

in geometry or public utility, as a tactical move, self-assumed or enforced, to acquire advantages or avoid falling from grace. My hypothesis is that *En ny räkenkonst* is an example of a 'rhetorical arithmetic'. Swedenborg's intention in writing it was not to present the unshakable, eternal mathematical truth, nor his own personal opinion about numeral systems. Instead he wrote about the octal system, partly because it was at the direct behest of the king, partly because he could see that it would benefit him politically and further his career. His personal views about numeral systems could be set aside. Now it was a matter of pleasing the king. This means that a scientist's opinions do not always have to agree with what he explicitly defends in his writings. There can be a difference between the scientist's private scientific opinion and his publicly aired one. A previously known example in the history of Swedish science showing how official scientific opinions need not correspond to what a person really thought is the case of Anders Spole. In his printed works he defended the Ptolemaic system and criticized the Copernican system for being in conflict with the Bible. In his unpublished notes, however, he is a firm Copernican and Cartesian.[157] What Swedenborg expresses so clearly and in agreement with the template, that is to say, following the hyperbolical language of the rhetorical genre, may thus be of less value as a source than the things that deviate from the norm, such as the unfavourable statements or negative judgements that go against the rhetorical rules. What Swedenborg says between the lines and in the more or less private letters to his intimates is presumably closer to the truth. Even in sober science, in seemingly objective mathematics, there is thus a rhetorical element with the specific purpose of convincing and persuading, and of obtaining personal or other advantages.

Swedenborg was aware of and discontented with the precarious situation in the country. He was also critical of the economic policy pursued by Görtz. When he wrote *En ny räkenkonst* he could very well have been an opponent of absolutism and Charles XII. Voicing criticism of absolutism was of course risky and not particularly tactical as long as the autocratic ruler was alive. Immediately after the king's death, however, the general aversion to Charles XII came up to the surface, not least in the diet. In 1722 Polhem looked back at the time around 1710, and observed that things had not become better:

> at that time the Kingdom was for the sake of the king, now the king is for the sake of the Kingdom. Then the Kingdom was at risk of the people being devastated by plague and war, now it is in danger of more population growth than book, pen, and sword can support, and its foundations of plough and miner's sledge can maintain. Then the economy and culture of the Kingdom were sick, now they are on the verge of death etc.[158]

Later we also find Swedenborg expressing overtly critical opinions about both the autocratic ruler and absolutism. It is not necessarily the case that Swedenborg changed his mind about Charles XII. The fact is that it was now politically possible, no longer risky, to be critical of absolutism. It even became politically correct to distance oneself from despotism. Swedenborg's rejection of absolutism goes

---

[157]Lindroth (1989), II, 503.
[158]Polhem to Benzelius, Stjärnsund, 2 September 1722. *Polhem's Letters*, 146.

together with his cautious policy in the diet, where he was against waging war. In a paper presented to the diet in 1734, opposing war with Russia, he declares that a country's prosperity does not depend on its area and the number of its provinces, but on flourishing trade.[159] It was the economy and freedom of expression that mattered. Impressed by the prosperity in Holland, Swedenborg noted in his travel journal in August 1736 that the reason for this was 'that it is a *Republique*, in which Our Lord seems to take greater pleasure than in *sovereign* realms'. In a republic the citizens cannot 'lose their courage or their free rational thoughts through fear, timidity, and reticence, but can with full freedom, without repressed minds, place their soul or *elevate* it to the honour of the highest'. He seems to be referring to his own experiences of absolutism. He remembers what it was like to be forced to live in a country under a sovereign ruler: there 'such minds repressed under *sovereignties* are fostered to *flatteries* and duplicities, to speak and act differently from how they think'.[160] This perhaps applied to his own actions as well.

Swedenborg, like the author Olof Dalin, belonged to the circles within the Hat Party who opposed the war. The conclusion of Dalin's *Sagan om hästen* ('The Story of the Horse' 1740) can be read as a statement to the diet against the Russian war, with opinions similar to Swedenborg's. It is a story that also shows how allegory and metaphorical thought could be used to veil a message. Härkuller on his horse Grollen chanced upon 'Pelle of Holmgård, who had done damage to his forest', but Grollen got stuck in the quagmire of Holmgård with its bottomless mud, and was forced to spend the night there in the moonlight. This was a reference to Charles XII and Sweden, who encountered Peter the Great in the Ukraine and were forced to stop under the Turkish crescent. But when Grollen later lost his master, 'Poor creature, he was so fond of his Härkuller, that when he first noticed that he was gone, incredibly, he trembled like an aspen leaf.' In the conclusion Dalin observes that it is better to stay at home and tend one's farm and take care of one's implements than to 'rove in search of adventure, which seldom gives full barns and storehouses'.[161]

Swedenborg would later, like Sinclair, see Charles XII again, first in his dreams during the dream crisis and then as a visionary in person in the hell of hells. In the diet Swedenborg openly defended political freedom and supported the constitution of the Age of Liberty, warning against the dangers of absolutism.[162] In a paper presented to the diet in 1761 he declares that one must above all be careful not to engender 'discontent with our good established Government', lest one 'fall into Scyllam when one wishes to avoid Charybdin, and so that an egg that is intended to bring forth a bird of paradise does not produce a basilisk'.[163] At the same time he wrote several memoranda of his meetings with Charles XII in the spirit world.

---

[159] Swedenborg, *Projekt* 1734. KVA, cod. 56; *Letters* I, 470.

[160] *Resebeskrifningar*, 70.

[161] Dalin (1740), 21, 23, 25.

[162] *Festivus applausus*, edition, commentary 26; Bergquist (2005), 347–350.

[163] Swedenborg, *Ödmjukt memorial* (12 January 1761). KVA, cod. 56, 30; edition, II, 55; cf. Swedenborg, LUB; *Dela Gardiska archivet*, 191–195; *Documents* I, 545–549.

The spiritual diary gives an undisguised, violent, almost sarcastic picture of a ruthless, power-hungry desperado. Swedenborg's thoughts in his spiritual notes are freer, less tied to earthly life, and not so entrenched in rhetorical convention as his explanations of the base-64 system. Since these were his own private notes, he did not need to show any consideration for what was politically correct or opportune. It could be said that the dreamer spoke more truthfully than the waking man on the subject of where he thought that Charles XII belonged.

Trees, Boxes, and Universal Mathematics

Thus far the purpose here has been to demonstrate that Swedenborg's thought, like that of all other people, is dependent on the surrounding culture, society, politics, and economy. This applies just as much to abstract thought, including mathematics. In his thinking he was forced to follow rhetorical conventions, heedful of political, social, and other considerations. But the picture of Swedenborg's thought is not complete with this. In line with the cognitive history of ideas that was outlined in the introduction, one must go on to ask how he thought on the basis of his cognitive abilities. It is this interaction between, on one hand, the personal thinking inside the head, and the outside environment on the other, that makes the history of ideas extremely complex and elusive. As regards the personal level, in the story of Swedenborg and the numeral systems we find some fundamental metaphors that steered his thought in a particular direction; these were more or less unconscious, and he rarely or never reflected on them. These metaphors also recur throughout his thinking, not just about the number eight, cans, inches, and carolins. One of the metaphors is that *categories are containers* or that *the world is a construction kit*; another is that *thinking is counting* or that *thought is a machine*. These also capture two central problems in baroque thought: the classification and atomization of things and concepts, and the question of the correct method for communicating and acquiring knowledge or creating new learning.

The mathematical conversations between Swedenborg and Charles XII about numeral systems, measures, and coin sorts reveal a fundamental cognitive ability in human thought and perception. It is the categorization of reality. Numbers were to be grouped in 64, 8, or 10; coins in marks and half-öre; weights in pounds and quintines; length in rods and half-quarters; volumes in barrels and half-virgins; and many other categories. It was essential to establish order, to master the world with concepts, categories, names, and classes.[164] Division was also an exercise of power, as definitions served to exclude, to disqualify things and people from belonging to a particular category. Dividing and linking categories with each other achieves better order in the chaos of reality. That was precisely what Swedenborg's octal system and later his decimal system sought to do, to link the measurement system and the

---

[164]Cf. Foucault, 71; Yeo, 241–266.

numeral system, so that greater order would be imposed on the categorized reality in that all the categories would be based on eight or ten. In Swedenborg there are constant attempts of this kind to master nature and the world through the imposition of order, natural laws, and fixed systems. Many of his most central ideas later in time rest on this categorization of reality: the universal mathematics and its extension, the doctrine of correspondences.

Categorization is about man's eternal quest for order in chaos. It is easier to live in an ordered world than in a chaotic one. Man faces something greater than himself, something that cannot be grasped with his limited understanding. One way to make the world comprehensible is to try to make it logical, mathematical, and geometrical. This implicit desire is the foundation of Swedenborg's geometrical world-view. By dividing the world into unambiguous categories and clear geometrical forms, he simultaneously made the world apprehensible and turned it into an ordered and perfect world machine worthy of the creator and designer of the universe—God. But the systematization, the categories, the regularities, the laws, the mathematical description of nature does not really exist out there; it is chiefly in the human brain. The categories are necessary for man's interpretation of the world, which is why they often say more about man than about the world itself. That is also the reason why Swedenborg's natural philosophy, with its categories and geometrical figures, says such a lot about how he thought and observed the world, but less about the 'true' world and nature.

When approaching an understanding of how Swedenborg thought, the question of how people categorize is of crucial significance.[165] The view that Swedenborg embraced was of the classical kind, regarding categories as defined by the objectively given properties shared by members of the same category. Categorization thus depends on knowledge of a particular category's essential properties. In his mechanistic-geometrical natural philosophy, Swedenborg searched for the essential characteristics of objects, of matter, and of physical phenomena, the properties that make a thing the kind of thing it is and allow us to predict its behaviour. In the geometrical world-view the essence of things can be determined through their geometrical form. Particles of a certain shape, round or square, give rise to different properties that are dependent on that shape. What Swedenborg follows is the informal 'theory' in human thought, that of essences; in other words, man regards every thing as a sort of thing, belonging to a specific category; he thinks that all things have a collection of essential properties that make the different things the kinds of things they are, and that this essence is an inherent part of the thing. The model for this way of thinking came from Aristotle's definition of 'definition' as a list of properties that are both necessary and sufficient for something to belong to its kind of thing, and from which all the properties of the thing derive.[166]

This Aristotelian logic, this idea of essences that is implicit in Swedenborg's thought, builds on the metaphor that *categories are containers*. Concepts can be

[165] Rosch; Taylor; Gärdenfors (2008c); Dunbar.

[166] Aristotle, *Analytikōn ysterōn*, 2.3.90b30–31; Lakoff and Núñez, 107.

placed in different, clearly distinguishable compartments, like different sorts of coins. By understanding one's experiences with the aid of objects and substances, one can categorize and group them, quantify and reason about them. This has obvious consequences for the universal mathematics and the doctrine of correspondences elaborated by Swedenborg, in which every word, every concept, is assumed to represent certain essential properties. It was natural for him that categories exist outside human consciousness, which simultaneously means that he also assumes a universal, transcendental logic that goes beyond man. God has created the logic of the world. But these thoughts about categories and essences were not actually something he reflected on; they were more or less unconsciously assumed, taken for granted. An understanding of Swedenborg's thinking is in large measure a matter of finding such underlying cognitive assumptions, ascertaining the 'containers' or categories into which he divided the world. For these are dependent on his experiences, notions, perceptions, movements, and the culture around him, but also on metaphors, metonymies, and mental images. Swedenborg's division of the world into categories tells us something about himself. The categories are not found 'out there' in the world as ready boxes, but 'in there' in his thoughts.

A typical example of a categorization of reality in which every thing and concept was supposed to be placed in its predetermined box, the only proper one, is the universal language or universal mathematics that was much sought after at this time. If the former emphasizes the linguistic, communicative aspect of the universal idea, then the latter stresses the algebraic, deductive side. Swedenborg too nourished this dream. The idea was to create a kind of formal language or a calculus that would be unambiguous and independent of the various human languages and could be used to ascertain all that can be known. What should be added here to the current image of universal mathematics is its origin in human cognitive capacities. It combines the cognitive ability to categorize reality with the baroque interpretation of signs.

Concepts were viewed as being distinct, separated in a vacuum. Thoughts, words, can be divided, analysed into atoms, split into parts like small particles of information. *Döden molmar i Mull*—'Death moulders in soil', wrote Stiernhielm the philosopher of language in his *Hercules* (1668), where the letters and phonemes had special meanings; for instance *mo* and *mu* stand for ageing and darkness.[167] In the mechanistic world-view there is a far-reaching 'atomization' of reality, in almost every sphere, from the theory of particles to universal mathematics. People thought with the aid of the metaphor that *the world is a construction kit*. Everything was envisaged as consisting of components, blocks that were put together to make a world machine. Thoughts consisted of simple ideas, words of letters, music of notes, nature of numbers. Even machines and mechanical movements had their smallest single parts. For his teaching of mechanics, Polhem constructed a 'mechanical alphabet' consisting of a large number of simple, instructive wooden models demonstrating the basic laws of mechanics, the building blocks of all engineering. Each one corresponded to a 'letter' in the mechanical alphabet. Just as

---

[167]Stiernhielm (1668a), *Samlade skrifter* I:1, 11; Sellberg (2000), 152f.

a poet can write the most beautiful poetry with the aid of the ordinary alphabet, an engineer could learn the mechanical alphabet and form 'sentences' from the mechanical letters, that is to say, construct complicated machines which could perform useful work.[168]

The idea of a universal language was based on the premise that every concept should have a designation, that there is a limited number of concepts related to each other, in a hierarchy in which they are subordinate or superordinate to each other. Substances were divided into classes, into a hierarchy from the highest to the lowest. In Swedenborg's library there is a very clear example of this ramification of classes that was used in Aristotelian philosophy, particularly in the dichotomous tables of Ramism, known as the 'Porphyrian tree' after Porphyry of Tyre with his *Isagoge* (third century AD).[169] In Johann Coster's *Affectuum totius corporis humani* (1663) mental ailments, states of mind, and their treatment were logically divided into widely branching trees.[170] The Porphyrian tree was an attempt to reduce the labyrinth of reality to a two-dimensional tree, a way to tame the world labyrinth. The wild primeval forest has a hidden order.

The categorization and classification of things and concepts acquired particular significance. Kircher and Gaspar Schott categorized; John Wilkins and George Dalgarno tried to compile a total encyclopaedia which involved devising a new alphabet in which each 'letter' would designate a single concept.[171] By placing things in their proper categories one could apply a syllogistic logic to create new knowledge. As assistant to Polhem, Swedenborg almost certainly came into contact with his encyclopaedic universal mathematics, most closely inspired by Wilkins. Around 1710–1711 Polhem wrote *Orda teckn på naturens materialer och dess egenskaper* ('Words on Nature's Materials and Their Properties'), presumably intended for the Collegium Curiosorum, which constitutes a general physics in a single system where physical principles could be ordered with the aid of a deductive method.[172] During his visit to London in 1711, Swedenborg came across a copy of Thomas Baker's widely read *Reflections upon Learning* (1699), the fourth edition of which appeared in 1708. He wrote to Benzelius that he had read it through 'three times, finding in him my greatest delight; but I wonder that he *approberar* [approves of] nothing, but makes all that has been *inventerad* [discovered] and written, incomplete and unworthy of his *esteem*'.[173] In a chapter about language Baker polemicizes against 'a Real Character and Philosophical Language', that is, Wilkins's attempt at a universal language. Baker finds it as lofty and impossible as Wilkins's flying chariot and journey to the moon. Swedenborg was probably not

---

[168]Cronstedt, TM 7405; Polhem (1729), 75–77; Lindgren, 363.

[169]Aristotle, *Analytikōn ysterōn*, 2.13.96b25–97b14; Eco (1986), 80, 84; Sellberg (1979), 65f.

[170]Coster, passim.

[171]Wilkins; Schott (1664), VII, 483ff; Eco (1997), 203–207.

[172]KB, Polhem, X 519; transcript J. Troilius, KB, X 521; Polhem, 'Korta ordatecken på naturens materialer och dess egenskaper'. KB, N 60; Dunér (2007), 133–164.

[173]*Opera* I, 209; *Letters* I, 20; Baker (1714), 19f; idem. 4th ed. 1708, 21f; Jonsson (1969), 114, 342.

equally convinced that Wilkins's universal language was impossible. Much later, in 1739, Swedenborg wrote that, without a mathematical universal philosophy, it would be easier to reach the moon than the human soul.[174] Reaching the moon is virtually impossible, but with a universal mathematics the secrets of the soul are within reach.

It was not until now, in the 1730s, that Swedenborg seriously started working on his own attempt at a universal mathematics, largely inspired by Wolff. He also touches on one of the models for universal language, the Egyptian hieroglyphs that fascinated many through their enigmatic ambiguity; it was assumed that they built on signs standing for a whole concept, that they were ideographic. The hieroglyphic ideograms were symbols, graphic representations of ideas and concepts. When reading Wolff's *Psychologia empirica* (1732) some time around 1733–1734, Swedenborg made a note about hieroglyphs: a hieroglyphic sign is a designation whereby one thing is transferred or allowed to represent another, for example a triangle denoting the divine trinity.[175] Fantasies or dream images, he continues, are perfect examples of hieroglyphs, a kind of hieroglyphic sign with similarities to other given things. The ancients represented dogmas and historical subjects with hieroglyphic figures, as was customary among the Egyptians, but as some also say among the Chinese. One's thoughts here go to Leibniz, but also to Kircher, who saw the assumed ideographic character of Chinese signs as clues to a universal language. It can also be compared to the vivid interest that the circle surrounding the Bokwettsgillet, headed by Benzelius and Rudbeck, showed in the Chinese language, the Norse runes, and the supposed link between Sámi and Hebrew.[176]

When he wrote about hieroglyphs, Swedenborg was living in an age that was interested in the interpretation of signs. He went on to combine hieroglyphic signs with his doctrine about the tremulations of the body, and added that it should be possible to design a kind of writing that would contain almost everything, with which one could write more in a single line than can now be done on several pages. In *Psychologia empirica* Wolff puts forward his ideas about a universal language of this kind, an 'ars characteristica combinatoria'. His conception involves the use of different signs to indicate things and sensations. With these signs one could then work out new knowledge, create new learning, and discover hidden truths with the aid of a sort of calculus. His predecessor in this universal mathematics was his teacher, Leibniz, who envisaged a similar 'characteristica universalis', which would be simultaneously a new language, a new logic, method, mnemotechnique, combinatorics, Kabbalah, and encyclopaedia in which all sciences were collected as in an ocean.[177] It would contain all knowledge and even all future knowledge.

---

[174]Swedenborg, *De via ad cognitionem animæ* 1739. KVA, cod. 65; *Psychological Transactions*, 10.

[175]*Psychologica*, 90–93; Wolff (1732b), § 151, 153.

[176]*Bokwetts Gillets protokoll*, 72f, passim.

[177]Yates, 370.

No more controversies, no more insoluble problems. People would only need to sit down and say to each other: let us calculate! Universal mathematics is the dream of an all-embracing method.

Swedenborg likewise dreamed of a science of sciences, which he developed into concrete attempts during his anatomic-physiological period. In the manuscript *Philosophia corpuscularis in compendio* (1740) Swedenborg speaks about globules, bullulae, and intermediate tetrahedra and cubes, a particle theory that is closer to the theories from the start of the 1720s than that in *Principia* from 1734. The manuscript can be interpreted as a summary of his corpuscular theory after a reading of Leeuwenhoek's *Arcana naturæ detecta* (1695) where the author examines square salt particles and small animals in ordinary water, with a critique of Descartes' theory of snake-like water particles.[178] Swedenborg rounds off with the enigmatic words 'These things are true because I have the sign.' Swedenborg makes an attempt at a universal language in another manuscript, *Philosophia universalium character-istica et mathematica* (1740). He tries to construct a philosophical language with letters or signs for general concepts. S stands for blood, A for artery, M for muscle, and N for nerve. In addition there are accompanying signs: a, or for continuous combination, nc, for cohesive substances such as fibres, muscles, membranes, or adjacent combinations, nf, through contact as in liquids, water, oil, blood, and air. The more perfect this union is, the more perfect must the form of the parts be, this is to say, circular. To obtain correct knowledge of these adjacent substances, one must become acquainted with their form and construction. Quantity is of two kinds: size (continuous quantity) Qc, and number (discreet quantity) Qd. As for the minimum and maximum of the quantities, the smallest or unity is designated 1, medium 2, and the greatest 3. Finally, Swedenborg gives an example: $AAAQc3$ denotes the aorta or the powerful heart.[179]

The dream was to elevate oneself from actual reality to the formal, regular world. The idea was that universal language and universal mathematics would apply to the whole of humanity, independent of languages, cultures, and human cognitive capabilities. The somewhat ironic thing in this context is that the attempts by Swedenborg and others to achieve a universal mathematics were in large measure dependent on how their own culture organized the world. Categorization of the world and concepts is often bound to a particular culture and does not really concern the 'true' division of real things. But that was not at all how Swedenborg and other natural philosophers viewed it. Division into classes and concepts was not arbitrary. The concepts of signs of the universal language were actually supposed to correspond to things in reality, in the same way as the hands on a clock correspond to the movements of the universe. It was thus assumed that there was a similarity between the structure of the universe and human thought, an analogy between the order of the world and the grammatical order governing the symbols in language.

---

[178]*Opera* III, 267f; *Photolith.* VI, 318; *Treatises*, 162–164; Jonsson (1969), 56f; Jonsson (1999), 65–67.

[179]*Photolith.* VI, 265–269; *Treatises*, 165–171.

Concepts were a reflection of the universe, and the ordered classification reflected the cosmic harmony. The designations and relations of universal mathematics correspond to, are isomorphic with, the inherent properties and relations of things. By learning universal language one would simultaneously acquire knowledge of nature and where the lacunae in our knowledge could be found.

At this time mathematics was transformed, as it increasingly shifted from geometrical to algebraic thinking. And it was precisely the success of symbolic mathematics, algebra, and arithmetic in manipulating symbols in order to obtain new knowledge of reality that made it a model for universal mathematics. With inspiration from differential and integral calculus and probability calculus, scholars developed the idea of a universal calculus which could work out all that could be known, even what was outside the domains of mathematics. Calculus as a system of variable and constant signs or symbols, with set rules for combining signs and with rules for transformation or derivation, became the perfect method for a science. The rationalist tradition of *more geometrico* cultivated universal mathematics as a purely mechanical processing of signs, with the different signs representing thoughts, concepts, ideas, things, properties, which could be combined according to set rules, thus making it possible to solve all problems and arrive at absolutely certain conclusions. Universal language would eliminate everything that makes thoughts obscure; it would improve and build up a sure science.

To Think Is to Count

Swedenborg's idea of a universal mathematics is based on the underlying metaphor that thinking is mathematic calculation, *to think is to count*. When Swedenborg and others speculated about mental processes, they imagined them as a mathematical activity. In the same way as numbers can be represented with symbols, it is possible to represent ideas and thoughts with written symbols. Just as mathematics counts with figures, the mind calculates with ideas. As composite numbers can be broken down into the ten digits of which they consist, composite concepts and ideas can be broken down into simple ideas. As with mathematical calculation, thinking is done according to systematic, universal principles proceeding step by step. Reason and thought are universal, exactly like mathematics and numbers. 'Therefore the intellect is represented more by words than by the eye', says Swedenborg in his commentary on Wolff, and the intellect can also combine things with each other, forming sentences like $2 + 3 = 5$ or like the notes in music. Thinking is thus for him a kind of manipulation of symbols, demonstration in a syllogism, a proof leading from premises to a conclusion that is harmonious and can be represented with letters and figures.[180] In this way one can reach the unknown.

---

[180] *Psychologica*, 118–121.

The machine metaphor also has consequences for the outlook on rational thought in the mechanistic period. In both mathematics and thought, counting is a mechanical activity, a game played with signs. Correct thinking follows set rules. In other words, here yet another metaphor falls into line with Swedenborg's mechanistic world-view: *thinking is a machine*, the mind is a machine, ideas are its raw materials, and conclusions are its products. Step by step, the thinking machine assembles its thoughts into a finished product, and if everything has worked as it should, it then spits out a well-fashioned, incontrovertible truth. The knowledge machine had been a dream ever since the concentric circles of the Catalan Franciscan friar Ramón Llull over 400 years previously, and in the seventeenth century it was developed into the mechanical calculators of Pascal and Leibniz, which tried in their way to emulate the human calculator. Thinking, wrote the philosopher Thomas Hobbes in *Leviathan* (1651), is a kind of mathematical calculation. Logicians think with words as mathematicians do with lines, figures, angles, proportions, and so on. 'For REASON, in this sense, is nothing but *Reckoning* (that is, Adding and Subtracting)'.[181] Rydelius, who combined Cartesian rationalism with Locke's empiricism, defined 'Ratio' in the second edition of his *Nödiga förnufts-öfningar* (1737), which Swedenborg often read and excerpted from, 'as a rough calculation, and since reason with its ideas like counting tokens figures out the truth, it has therefore acquired such a name.'[182] Basic ideas are the small change of reason, which are added together in 'rational sums'. Besides the counting metaphors, Rydelius also describes thinking with the aid of images borrowed from machines and craft work. The simple-minded masses, that is to say, ordinary people, have thoughts that are mostly driven by symbols in the mind and by the senses, 'not entirely unlike a machine that is propelled by its wheels, and the wheels by the wind, water, weight, or springs.'[183] And it is not without good reason that our ingenuity on the building site of thought can be called our internal implements or tools.

The interpretation of Swedenborg's thoughts about different numeral systems, the rationalization of systems of weights and measures and coinage policies had as their primary goal to demonstrate the dependence of thought on the interplay between man's cognitive capacities and the world around him. The metaphors that govern Swedenborg's thought are fundamental, such as that categories are containers, that the world is a construction kit, that thinking is mathematical calculation, a machine. It has also been possible to interpret the history of signs, numbers, and measures in several complementary ways. First it can, of course, be interpreted mathematically and logically. We see a geometrical interest, in how the new method of counting is related to logarithms and numerical series, and where numbers, as in ancient Greek mathematics, were regarded geometrically. Instead of powers he spoke of the cube, the quadrate, the biquadrate, and so forth. At the same time, Swedenborg's octal system contained pre-logical tendencies, a desire

---

[181] Hobbes, part I, ch. V, 111; cf. Lakoff and Johnson (1999), 406; Sawday, 237–241.

[182] Rydelius (1737), 112, cf. 15; cf. Almquist, 65.

[183] Rydelius (1737), 162, cf. 75.

for abstraction, a quest for the exact language through which to capture concepts in abstract symbols, showing the way towards universal mathematics and the perfect mechanical calculator. An interpretation in terms of the history of science shows how the construction of the perfect numerical system was intended to rationalize the systems for weights and measures. A more logically ordered and scientific division was what contemporary scientists wanted in order to perform increasingly exact experiments and measurements.

An economic perspective makes itself felt in the coin divisions, the emergency coinage, inflation, and valuation. In political terms we are brought into issues of autocracy, despotism, republicanism, opportunism, freedom and repression of opinion. The political interpretation shows that abstract scientific ideas, which may seem to be far removed from the ideological battlefield, can in some cases be dependent on political situations, that a scientific idea is not created solely out of inquisitiveness about the world and scientific motives, but is also influenced by political stances. Science was in the service of war and autocracy at the end of Sweden's period as a great power. A linguistic-rhetorical interpretation reveals a picture that differs from the traditional one. The king was not 'a profound mathematician'. Swedenborg himself did not uphold the octal system, nor was he an admirer of Charles XII. Instead he may have been a member of the stifled opposition who was heavily dependent on the king. It is not necessarily the case that Swedenborg changed his opinion of the king, nor that he abandoned an octal system for a decimal one. Swedenborg may actually have preferred the decimal system, following Stiernhielm, when he wrote *En ny räkenkonst*.

On the psychological plane we find the ambivalent man seeking protection, perhaps an opportunist and a careerist, in pursuit of a future, a title, a name. But it was not just an advantage to have won the king's distinctions. Accused of having been a royal favourite, in the years following the death of the king Swedenborg was forced to consolidate his position in the Board of Mines and to prove his competence in the management of mines and mineralogy. His ideas about systems of weights and measures, finally, reveal a social image of Sweden. We see an age of social hierarchies, a country where the rural population sold their wares in pounds and jugs in return for copper coins and carolins, as war and epidemics raged. The war claimed countless victims, and freedom of opinion was curbed. Some people mused about the art of arithmetic, others counted on their fingers.

The Wave

It also frequently happens that a person falls into the thought of another person, that he perceives what another is doing and thinking, that is, that his membrane trembles from the tremulation of the other person's cerebral membranes, just as one string is affected by another, if they are tuned in the same key.[1]

Swedenborg, 'Bewis at wårt lefwande wesende består merendels i små Darringar thet är Tremulationer' ('Proof that our living being generally consists of small quivers, that is, tremulations', 1718)

The Water Waves in Leiden

To live is to tremble. The body shakes and twitches in constant convulsions. A vibrating, oscillating wave movement runs through the nerve threads, causing the whole living being to tremble with it. The slightest touch, the smallest movement of thought, can spread in seconds over tissues and ligaments, like the strings to the sounding board in a clavichord. But death is petrifaction, rest, as tremulations abate and waves level out. Swedenborg's idea that tremulations constitute life itself proceeds from geometrical-mechanical forms of movement—waves. The wave metaphor is extended from the visually perceived arching of water waves, spreads through the air waves of sound and the swift ether waves of light, finally ending up in the wave impulses of the nervous fluids. Water waves serve as a model and a source of metaphors through which he tries to understand sound, light, and nerve movements. As a geometrical figure, waves describe a rhythmic change, a recurrent movement up and down, back and forth.

---

[1] *Dædalus Hyperboreus* VI, 13; translation, *On tremulation*, 6; Dunér (2003a), 177–201; Dunér (2006b), 27–48.

D. Dunér, *The Natural Philosophy of Emanuel Swedenborg*, Studies in the History of Philosophy of Mind 11, DOI 10.1007/978-94-007-4560-5_4,
© Springer Science+Business Media Dordrecht 2013

Swedenborg's idea of tremulations, put forward in the latter part of the 1710s, is a piece of iatromechanics or iatromathematics, that is to say, an attempt to explain human anatomy and physiology in mechanical, mathematical, and geometrical terms. This trend developed especially in Italy, but by the start of the eighteenth century it had many followers all over Europe. For the iatromechanics, the human body was a machine. The body's movements and sensory perceptions were nothing more than the result of a subtle particle mechanics beyond the limits of the microscope, where geometrical forms determined how the perceptions were expressed. It is a mechanistic, Cartesian world-view, according to which everything in nature, all scientific truths, could ultimately be explained by two concepts: matter and motion. Physics could be reduced to mathematics. All sensory impressions could be boiled down to fundamental, primary qualities such as extent and motion. From this it follows that vision acquired a supreme position among the five senses. Vision with its abstract interpretation of the world in terms of extent was broadened at the expense of the other senses. The object of vision could explain the experiences of the other senses. It is the detached gaze of the scientist, registering the true properties of things from a distance. He does not need to taste, feel, listen, or smell, merely to observe from a distance. Swedenborg interprets the world, the senses, and life as tremulations, as wave movements, undulations, and tremulations. Everything shakes and quakes. Everything quivers like an aspen leaf in the wind, vibrates like the strings of a lute, shudders like a church organ, billows back and forth like waves beating against a shore.

Water is where the wave metaphor originates. On a day in the early spring of 1713, Swedenborg was walking alone by the edge of the canals through the Leiden of the ancient Batavians. The water flowed gently. The flow twisted and turned in changing forms past the Hortus Botanicus and the university famous for its medical education. He followed its winding course wherever it led, and was carried along in a poetic mood. 'And the water takes me with it in a circle [...]. It is like some Divine force, which now, living in the waves, constantly follows my path.' This poem, which is dedicated to a famous man on the arrival of his wife and newborn child in Utrecht in 1713, constantly returns to the movement of water, the winding and circling movements. The Latin words are rhythmically repeated and duplicated, surging forth like waves: aqvam—undis—aqvis—unda . . . water—waves. The rhythm of the hexameter and pentameter lines of the distich, Swedenborg counted on his fingers.[2] He is holding a sheet of paper in his hand:

as the sheet is held open above the surface of the water,

the sheet is seen to be reflected in the surface of the waves,

it hangs as if it had been written in the middle of the waters.

And that which I write to you and which seems to be written in the waves,

that which I wish is almost wished in the water itself.

---

[2]Swedenborg, 'Lusus extemporalis ad amicum qvendam Oxoniæ 1712', *Ludus Heliconius,* edition, 96f, commentary 194.

And the vows which I double are again doubled by the waves.

So what I wish, the Naiad wishes that, too, in the surface of the water.[3]

The 'famous man' is the mathematically interested Swedish diplomat Johan Palmqvist, who had been sent to monitor the peace negotiations in Utrecht after the War of the Spanish Succession. At Christmas 1712 Swedenborg had arrived in Holland, where he visited The Hague, Amsterdam, Leiden, and other cities on his way to Paris. From Paris Swedenborg wrote in August 1713 to tell how, during his visit to Utrecht each day, he had gone to Palmqvist's to discuss algebra: 'He is a great *mathematicus* and a great *algebraiste*.'[4] The poem about the waves alludes not just to mathematics and astronomy but also plays with the words in Palmqvist's name, meaning 'palm branch'. A dove comes flying with a palm branch, which with the baroque fondness for analogy and puns expresses the theme of peace.

The erosive property of circulating water recurs in the poem *Fabula de Perillæ et Nerei amore* (1714/1715).[5] Inspired by Ovid's *Metamorphoses*, Swedenborg fashions a polyvalent baroque symbolism. God himself runs around Nereus in a heavy current. 'Nereus' is a mythological sea god, but can also be interpreted as the Russian tsar or the erosive power of the sea that shaped Scandinavia. 'Perilla' or 'Ariadne' is the elusive nymph of myth, but also Sweden and Scandinavia which are formed by the sea. In no less than eight ways, in accordance with the baroque aesthetic, he varies the words for water and sea.[6] The poem also contains a word often used by Swedenborg, the Neo-Latin *tremulo*, 'tremble', which Polhem had previously used in Swedish in a letter from 1711 where he talks of 'the tremulations of strings and bells'.[7]

Fish moved under the calm surface of the oceans, causing trembling curves, and the sun quivered in tremulations over the waves.[8] Swedenborg seems to be combining in his mind water waves with light waves, no doubt referring to the same image that Descartes had used to explain the whirl theory. Whirls in the universe resemble the way fish move under water, how their fins push aside the water which then flows back in a circle to the place the fish has left behind.[9] This shows that the circular movement is easy and natural for things in the universe. Harald Vallerius also writes in *De aquarum motu per circulum in globo terraqueo* (1704) about how fish propel themselves through the water.[10] The Cartesian fish was intended to show

---

[3]Swedenborg, 'Ad virum illustrem in adventum uxoris ejus comitante illam nova prole, Ultrajectum ad Rhenum 1713', *Ludus Heliconius*, edition, 94f, cf. 78f, 90f, commentary 187; translation, 95.

[4]Swedenborg to Benzelius, Paris 9/19 August 1713. *Opera* I, 223f, cf. 229; *Letters* I, 51, cf. 61f.

[5]Swedenborg, 'Fabula de Perillæ et Nerei amore', *Ludus Heliconius*, edition, 108f, commentary 198f; cf. *Camena Borea*, edition, 84f.

[6]*Ludus Heliconius*, edition, commentary 37; cf. *Camena Borea*, edition, 120f, commentary 28.

[7]Polhem to Benzelius, 3 November 1711. *Polhems brev*, 77; Nilsson (1975–1976), 128.

[8]*Ludus Heliconius*, edition, 102f.

[9]Descartes (1664a), ch. IV; *Oeuvres* XI, 19f; cf. Descartes (1644); *Oeuvres* VIII:1, 58f.

[10]H. Vallerius and Ramzelius, 8.

that no vacuum is needed to explain movement; everything returns in a circle. It was precisely the opposite, the existence of vacuum, that Lucretius sought to defend with his fish in *De rerum natura* (*c*. 50 BC), to which Descartes was almost certainly alluding. There Lucretius objects to those who say that, as water gives way to scaly fish and opens wet paths that leave room behind them where the displaced masses of water can flow together, other things can also move and change place even though everything is massive.[11]

God flows around you in a gigantic stream, Swedenborg continues in the poem about Nereus and Perilla, and nourishes you, like a heart. He rolls around in circles, flows in and out, in towards the inner parts of the country. In poetry, not least in the baroque aesthetic, waves are particularly expressive. Ships tossed back and forth on angry waves are an image of life exposed to the elements; the unprotected person among turbulent waves, in the midst of the violent storm that is the world. Water possessed a poetic symbolism, served as a source of metaphors, could express purity and a life-giving force. Water, fountains, and rivers stood for return, as in Swedenborg's *Camena Borea* where the river Meander flows back to itself.[12] And peace is like a still, calm sea, as Stiernhielm the assayer put it: 'It lieth, like a gleaming mirror / Like molten silver in a crucible.'[13] This is the abode of Pax and Musa.

Water represents change and movement. Alternating metamorphoses, waves shaping the landscape and washing up shells high above sea level, stand unchanged in Swedenborg's thought. Few classical works flow through Swedenborg's texts as profusely as Ovid's *Metamorphoses*. In this book of changes Ovid speaks of billowing, foaming waves, the crash of breakers, the river Meander winding in playful curves, rivers pouring forth with waves of milk and nectar in the eternal spring. The Ovidian language and the metamorphoses in Swedenborg's texts should not be regarded, as is sometimes claimed, as a mere mythological appendage and stylistic adornment not to be taken literally.[14] In fact, I would claim, the mythological element in Swedenborg's reasoning is a part of his metaphorical thought. Myth conveys knowledge about causal relationships, from myth to reality. The narrative form of myth also makes the sequence of thoughts easier to remember. Mythological and metaphorical thought is especially noticeable when Swedenborg turns to the primeval history of the earth to explain the forms of today's landscape and the falling water level at our coasts. When Hiärne took up the question of the sinking water in *Den beswarade och förklarade anledningens andra flock, om jorden och landskap i gemen* ('The reason answered and explained, part two, on the earth and landscape in general', 1706) he cited book 15 of Ovid's 'Book of Change', about countries which have lain under a great river and waves, plains bestrewn

---

[11]Lucretius, 1.372–376.

[12]*Camena Borea*, edition, 56f.

[13]Stiernhielm (1668b), *Samlade skrifter* I:1, 106.

[14]Cf. Högbom, 16.

with seashells, and ancient anchors found on mountain tops, far from sea and wave.[15] For Hiärne everything was in metamorphosis, 'nothing else is permanent, but impermanence itself.'[16]

There is an existential element in these changes. Mersenne spoke of the uncertainty of the world as the constant ebb and flow of things, like running water.[17] Wave and storms are a common theme in baroque poetry, like the antitheses of shipwreck in life and death, heat and cold. Samuel Columbus lets the calm, gentle sea rest on soft *l*-sounds and then, in the raging fury of the storm, causes the frothy surf to roll on *r*-sounds: 'How grim its waves resound, its dreadful rumbling roars / A ship is torn and thrown, soon to be led off course.'[18] Swedenborg wrote in a poem that, 'when you have ended up, shipwrecked, in so much water, / there will not be for you a *stream* any longer, nor a *vein*.'[19]

The Surging of the Sea

Water is not just a fundamental element; it also carries a large mechanical force. It is a high-technology source of energy, something for man to capture, to hold out, to bridge. A great many of Swedenborg's inventions are connected with water. He designed a submarine, a steam engine, a water clock, siphons and pumps to extract water from mines, but also water machines for sheer enjoyment. In the technical draft *Machina siphonica* (1716) he explains how water machines can be constructed on the basis of geometrical forms, such as a cylinder or a globe with many holes spouting water of different colours, white, red, yellow, and green, in the shape of lilies, columns, and houses.[20]

As assistant to Polhem, Swedenborg worked on a couple of major water projects, which he would later describe in *Underrettelse om docken, slysswercken och salt-wercket* ('Information about the dock, the locks, and the salt works' 1719). At Lindholmen in Karlskrona, work began in 1716 on digging the 'Polhem Dock' in the primeval rock.[21] The project was directed by Polhem and the master shipbuilder Charles Sheldon, while Swedenborg was responsible for the geometrical

---

[15] Hiärne (1706), 268; Ovid, *Metamorphoseon*, 15.262–269.

[16] Hiärne (1706), 384.

[17] Mersenne (1625), book I, preface & ch. III.

[18] Columbus, 21; Jonsson (1969), 214; cf. ships encountering tornadoes. Lagerlööf and Högwall.

[19] *Ludus Heliconius*, edition, 134f, cf. 132f.

[20] Swedenborg, *Machina siphonica* (1716). LiSB, N 14a, no. 24; *Photolith.* I, 20; *The Mechanical Inventions*, 27.

[21] Swedenborg, *Underrettelse om docken*, [1–3]; Swedenborg, 'Artificia nova mechanica receptacula navalia et aggeres aquaticos construendi', *Opera* III, 215; *Principles of Chemistry*, 232; Swedenborg, *Memorial på de förbättringar som wid Carlzcrona stå att practisera* (1717). LiSB, N 14a, no. 49; *Photolith.* I, 127–129; Polhem (1729), 31–37; rev. in *Acta eruditorum*, May 1722, 269; KrA, *Journaler öfwer arbetet på dockan*.

calculations. Another project, as we read in a speech in memory of Benzelius, was 'to make our extensive peninsula navigable'.[22] The archival researcher Benzelius had raised the idea after having read a letter by Bishop Hans Brask from 1523 about a sailing route between Stockholm and Gothenburg, over the lake Vänern and the river Göta Älv. Benzelius sent a copy of what he had found in the archive to Swedenborg, who then brought it to the king in Lund in 1716. Polhem penned a memorandum about canal locks at Trollhättan in June 1717 and Swedenborg wrote a fair copy.[23] Work commenced the following year. Swedenborg's task was to build a caisson lock with a height of 5 m and to clean Karls Grav at Vänersborg, while Polhem worked on high-lift locks blasted through the rock at the Trollhättan Falls.

The design of hydraulic machines, docks, dams, locks, canals, pumps, mills, water wheels, and other devices confronted the engineer with problems concerning the properties, physics, and movements of water and other liquids. This called for advanced research in hydrostatics and hydrodynamics. We discover in Swedenborg a theoretical interest in the physical properties of water, its flow, mechanical force, and internal structure. In *Regel-konsten* he takes mathematical examples from the dock in Karlskrona, from dams, water wheels, and pumps.[24] Swedenborg found Polhem's models of great value, and he tried to save them from decay. In 1725 they were kept in a room at the office of the Board of Mines at Mynttorget in Stockholm, and Swedenborg was worried that the windows were in a poor state, letting in snow in the winter and rain in the summer.[25] Polhem performed hydrodynamic and aerodynamic experiments in his Laboratorium Mechanicum; among other things he designed a hydrodynamic experimental machine for testing the effect of different kinds of waterwheels and the inclination of chutes. This was a rather early and unusual attempt to test scientific, systematic, and geometrical methods in engineering. But after about 25,000 experiments which did not lead to any result because of errors in the design, Polhem confided to Göran Vallerius in November 1710: 'To sum up, just between ourselves, this work is as useful as the fifth wheel on a wagon.'[26] Polhem later wrote to Swedenborg about experiments with water wheels and the resistance of matter, that they are of use for 'calculations for water sprayers, garden fountains, as well as bombs and cannonballs etc., which all have their mathematical rules in agreement with the actual practice'.[27]

In *Dædalus Hyperboreus* Swedenborg speculated about various experiments for finding out how thick ice can become, and experiments on the water resistance

---

[22]Dalin (1744), 18; Benzelius (1791), xxiii; Lagerbring, IV:3, 22f; Almqvist, 143.

[23]*Underrettelse om docken*, [3–5]; RA, Polhem; KrA, Kungsboken 16:1. The pictures, one of them with Swedenborg's name, are probably by Rudbeck the Elder; Dahl, 110f; Widegren, 94.

[24]*Regel-konsten*, 99–103.

[25]RA, Bergskollegii arkiv, EIV:169, fol. 272f; RA, Bergskollegii arkiv, AI:71, 136–140; *Letters* I, 367, cf. 75, 445; *Opera* I, 233; Dunér (2009b), 14–33.

[26]Polhem to G. Vallerius, Stjärnsund, 12 November 1710. *Polhems brev*, 38; Neumeyer, 10; Lindqvist (1984), 71f.

[27]Polhem to Swedenborg, Stjärnsund, September 1716. *Polhems brev*, 125, cf. 115.

of different types of ship's models, in order to find the ideal design for yachts, warships, brigantines, and frigates.[28] The basic idea is that the mechanical arts can be assisted by geometry and its experiments and trials, as here, in trying to ascertain the unknown 'bow-lines' or curves of ships. Swedenborg proposed building small models of the same weight, but of different shape, to test the properties of ships, 'one with a pointed, wide, flat, hyperbolic, or parabolic bilge; the other with a high, low, pointed, or wide breast'.[29] Underlying this are ideas that there is a correspondence between the small and the big world. Small ship's models in a bath behave in largely the same way as big ships on the sea. From this experiment one could draw conclusions as to the form a ship should have. Swedenborg's trials with models found a successor in Gilbert Sheldon. Yet Fredric Henric af Chapman also described tests with models in a water trough as a way to ascertain water resistance, and arrived at the conclusion that parabolic hulls were preferable.

Swedenborg was interested in the processes that shape the surface of the earth, the forces that carve out coasts and continents, how water sinks and creates new coastlines. One day in May 1715 he visited Kinnekulle in Västergötland when the hill was in its spring greenery. 'In regard to pleasantness and fertility it is a little Eden. Fruit trees grow there as if they formed a wild wood: walnuts in the orchards; and everything that is there planted is so favored by the air and the soil that it flourishes better than in other places.'[30] From the top of the hill 'the sight is lost in Lake Venner, nothing but the sky and the water being seen.' It would be an ideal place to build the most peerless observatory in the world. In *Om watnens högd och förra werldens starcka ebb och flod: Bewjs vtur Swergie* ('On the Height of Water and the Strong Ebb and Flow in Former Times: Evidence from Sweden', 1719) Swedenborg presents some evidence indicating that the water once reached higher up. He found one proof of the terrible height of the water in the stratigraphy of Kinnekulle, formed by the forces of the water, which he perceived as 'horizontal layers' of different kinds of stone forming the cut-off tip 'of a pyramid'. When he looked at different hills, musing about the processes that created them, he saw parallel lines, central points, balanced waves, oblong figures: 'it has often been a pleasure to me to notice with my own eyes, how this has happened, according to

---

[28]Swedenborg, 'Experimenter som kunna werkstellas i wintertiden, förmedelst wår swenska köld, gifne af åtskilliga wid handen', *Dædalus Hyperboreus* II, 30f; cf. D. Tiselius (1730), 25, 50, 89.

[29]Swedenborg, 'Ett experiment eller prof hwar med skepsbygnaden kan befodras', *Dædalus Hyperboreus* VI, 8; cf. Swedenborg, *Nya sett at segla emot strömmen ther wind är med, fast streken är emot* (1716). LiSB, N 14a, no. 44; *Photolith.* I, 86f; cf. Swedenborg, 'Modus mechanice explorandi virtutes et qualitates diversi generis et constructionis navigiorum', *Opera* III, 222–224;*Principles of Chemistry*, 239–241; rev. in *Acta eruditorum*, May 1722, 270; on geometry concerning ships, Swedenborg owned Bernoulli (1714), which he read on 22 June 1733, and Poleni (1717), see e.g. 3, 7; *Documents* II:1, 23, 34; Lindroth (1989), III, 118, 120.

[30]*Om watnens högd*; *Opera* I, 7, cf. 231, 292, 299; *Treatises*, 20; *Letters* I, 65, 216, 230f; *Bokwetts Gillets protokoll*, 6, 8f; Burman, rev. in *Acta literaria Sveciæ* (1720), 5–11; *Opera* I, 43–48; rev. in *Neue Zeitungen* 31 March 1721, 202–206; D. Tiselius (1723), 13; Linnaeus (1747), 54–57; Rohr, IX, 229f; Nathorst, 357.

a straight geometrical line as guide, and assumed a shape different from others.'[31]
The ridges once formed on the bottom of the sea follow the direction of the compass
towards the north. He looks for signs and geometrical forms in the landscape. The
world consists of horizontal planes and round stones.

Water has the power to carry stones back and forth as boats and ships are driven
by the waves. A deep sea that swells, rises, and tumbles has much greater force to
remove and roll away things that get in its way than a shallow sea. The reasoning is
based on an analogy with air. At the bottom of the sea, virtually the same phenomena
take place as at the bottom of our atmosphere, where the wind can overturn things
that are much heavier than itself, and correspondingly the currents and tempestuous
waves at the depths of the ocean can shift large stones. The hilltops of Billingen and
Hunneberg, his geometrical vision continues, are completely flat and horizontal,
'drawn as it were with a level'.[32] The slopes are vertical, steep as walls, with filed-
off edges, sometimes pressed together into points—all created at the pleasure of
waves and winds. In other places he finds round holes in the rock where the water
whirled around, abraded, drilled, and shaped pits. If a stone came into the whirl, it
wore down the sides and extended the hole more and more. Kettle holes like this
could be found in Strömstad, but also elsewhere according to Rudbeck the Younger,
Swedenborg tells us (Fig. 1).[33] As he wandered through the landscape he found
further traces of the sea, such as mussels, shells, and other marine creatures thrown
up by the waves. He had visited the shell banks at Uddevalla, which he took as
obvious proof that an ancient sea had covered Sweden (Fig. 2). And in some notes
about mussels and shells in limestone, he says that at Billingen beside Öglunda
church there is a flat rock where some petrified shells are embedded in the clay.
'Such snails are found in abundance round about Billingen, especially during the
spring.'[34]

Another proof that the water had formerly been higher was the giant bones found
at Norra Vånga in the Skara district in 1705.[35] It was first assumed, perhaps with the
inspiration of Kircher's giant in *Mundus subterraneus* (1664/1665), that they were
the bones of a giant, but when they had been examined by Moræus and brought to
Uppsala, it was discovered, presumably by Roberg, that they were actually the bones
of a whale. While waiting to travel abroad in March 1710, Swedenborg wrote to his
friend Andreas Unge on the eve of his disputation in theology, honouring him with

---

[31]*Opera* I, 9; *Treatises*, 23; cf. *Miscellanea*, 12; translation, 9.

[32]*Opera* I, 17; *Treatises*, 34; In Swedenborg's *Om jordenes och planeternas gång och stånd* at the
Linnean Society, London, between pages 32 and 33, the same is noted by hand, taken from *Om
watnens högd*. See *Green Books* II, no. 171.11.

[33]*Opera* I, 18f; *Treatises*, 36f; *Opera* III, 315; Rudbeck the Younger (1701), 18–21.

[34]Swedenborg, *Anmärckningar om musslor, sneckor etc. i kalcksten och om skifwer* (1716). LiSB,
N 14a, no. 18; *Opera* I, 31, cf. 19 & III, 315; *Treatises*, 4; cf. *Opera* II, 124; *The minor
Principia*, 149.

[35]*Opera* I, 21; *Prodromus principiorum*, edition, 5; translation, 4; SLBS, Carlmark 34:18; Kircher
(1665), II, 56; *Polhems brev*, 99; Aurivillius; Frängsmyr (1969), 162.

Fig. 1 A kettle hole at Strömstad. In *Om watnens högd och förra werldens starcka ebb och flod* (1719) Swedenborg writes about holes created by water: 'Several at Strömstad and one right beside the town, where along a rocky promontory there is a large crack inside which a large number of round and polished stones have been found. These holes are called giant pots' (Photo by the author, 2001)

Fig. 2 The shell banks at Uddevalla. Swedenborg took this to be an ancient beach (Photo by the author, 2001)

a playful poem about the giant bones.[36] The gigantic skeleton lacks a brain, unlike you, Andreas, who are a genius! How could the whale have stranded so far inland if the sea had not once covered the plains of Västergötland? That theory gained general support.[37] Swedenborg's theory of sinking waters could be perceived as a defence of the Truth and the Word of God. There were still visible traces in the landscape of the great flood that God had visited on man. The reason the sea level is now lower, Swedenborg explains, is that the speed of the earth's rotation has decreased. In the past the vortex of the moon pushed the water at the equator towards the poles. Now the sea has withdrawn to the south and lost its original, more oval form and become increasingly round, that is to say, flatter at the poles and rounder at the equator. There is no doubt that Sweden was once surrounded by water. Rudbeck the Elder had demonstrated that the country was an island in pagan times, the real Atlantis, the island of the blessed, an Ultima Thule. The water had then receded and 'Neptune, by turning up his rugged back' gave the land on which we have been able to build our homes.[38]

In the spring of 1716, in a beautiful grove a short distance from the vicarage in the parish of Ryda in Västergötland, Swedenborg discovered three springs with different chemical properties which burst forth in a row. One of them had ordinary water, the second moderately mineral water, and the third highly mineral water. He examined the reddish or bright orange stones over which the springs flowed, and found a coal-black mud that could be used for dyeing fine cloth 'as permanent and beautiful as the Parisian'.[39] The fine white sand underneath could be used in hourglasses. Swimming in the pond were elusive, blackish fish. There was a fine and wholesome air around the three spas, so those 'who dwell in the vicinity insist that for the last 50 years they have not known of any cattle having died from sickness, nor of any ailment in man or beast.' In April he wrote to Benzelius to tell of his secret discovery in Västergötland of a clay that is used in Holland and England for making stoneware vessels and tobacco pipes: 'Here in Westergyln [Västergötland] is found a white clay which I *subsonerar* [suspect] to be of the same kind. Should that be the case, it would be worth many 1,000 Riksdalers; but silence with regard to this. N.B.'[40]

In Noah's time the water spouted up from the earth through the heat of the sun, broke the crust and burst out in a huge flood, creating the irregularities that we now see here. In *Om jordenes och planeternas gång och stånd* ('On the Motion and Position of the Earth and the Planets') he demonstrates the movements of water as an important formative force:

---

[36]Palmroot and Unge, 14; *Ludus Heliconius*, edition, 62–65; Andreas's younger brother Jonas later married Swedenborg's sister Catharina; cf. *Opera* I, 204; *Letters* I, 9.

[37]Benzelius's note is from *c.* 1719/20. Benzelius (1762), 10, 14; Frondin and Forssenius, 8f; Dalin (1747), I, 4, 6; Ferner, 10f, 25.

[38]*Opera* I, 26; cf. *Miscellanea*, 46f; translation, 30f; Wallin, 98.

[39]Swedenborg, *Om åtskillig slaggz jordmohner och gyttior* (1716). LiSB, N 14a, no. 48, fol. 126; *Photolith.* I, 94; *Treatises*, 2; cf. *Miscellanea*, 50f; translation, 33; *De ferro*, 381f.

[40]*Opera* I, 251; *Letters* I, 102.

The land itself shows that it has been a sea bottom, that the water flowed back and forth, that protracted surging has dug out hills and dales; that is to say, the sea overflowed Sweden and the Nordic countries in the past; there has been a terrible ebb and flow, streaming back and forth; dreadful falls down cliffs, all of which the land bears the marks of.[41]

It can be shown geometrically that ebb and flow come from the orbit of the earth around the sun. The round whirl of the earth is pressed into an oval on one side by the movement 'like a round ball travelling in the air or against water.' In the past, when the earth moved faster, the sea was a 100 cubits higher than now. The water then cut through mountains and valleys, leaving mussels and shells behind it.

The marvellous properties and behaviour of water bewildered Swedenborg and his contemporaries. Questions about mystic phenomena in Swedish lakes were also a frequent item on the Bokwettsgillet agenda. The source of knowledge, the wellspring that pours forth life-giving water in a constant cyclic movement, became the very emblem for Bokwettsgillet (Fig. 3).[42] Spring water, water mirrors, and fountains are the symbol of eternal life—the source of life. Through central perspective one can see in the printed version in *Acta literaria Sveciæ* a garden with geometrical flowerbeds and a straight row of poplars or cypresses, underlining the geometrical order of the world and enabling us to survey it. Floods and stagnant rivers were the topics of dissertations by Harald Vallerius, and Hiärne wrote about the miraculous power of water, introducing hydrotherapy to Sweden.[43] Block, who was described as a genius 'more bubbling than a rushing torrent', tried to explain why Motala River suddenly stopped in its course one winter's day in 1706.[44] The reason, he assumed with inspiration from Boyle, was that the cold air contained pointed particles of salt which mixed with the water and turned it into ice. Swedberg, firm in his belief in supernatural signs, was frightened by Lake Barken in Dalarna when the water was transmuted into blood one summer's evening in 1697. Thick, coagulated blood drifted out on the lake—a divine miracle, Swedberg observed.[45] Linnaeus later explained this phenomenon as being due to many millions of small water-fleas colouring the water red.[46] In 1720 Swedenborg sent Bokwettsgillet an account he had heard about a strange phenomenon at Lake Väsman near Ludvika, not far from Barken and Starbo where his sister Hedvig lived with her husband, the inspector of mines Lars Benzelstierna. From Larsberget they had seen the sun over the horizon and reflecting in the lake for a whole night. If this were true, according to Swedenborg, it must be due to the lake itself.[47] At another meeting Benzelius told

---

[41]*Opera* III, 315f.

[42]Benzelius, *En vignett föreställande en vattenkonst, med öfverskrift 'Collecta refundit' och underskrift 'Societas Litterœria Upsalensis, instituta Anno 1719. d. 18 Nov.', ex inventione E Benzelii.* LiSB, N 14a, no. 1, fol. 5; *Bokwetts Gillets protokoll,* 46f; Dal (1995), 19–25.

[43]H. Vallerius and J. Vallerius; H. Vallerius and Wennerwall; Mansén (2001), 192ff.

[44]*Acta literaria Sveciæ* (1722), 331; Lindroth (1973), 164–167, 222.

[45]Hiärne (1702), 32–35; Swedberg (1710–1712), I, 77f; Grönwall and Resenberg, 65.

[46]Linnaeus (1746), 25–27, cf. (1739), 26, (1749), 29.

[47]*Bokwetts Gillets protokoll* 28 April 1720, 23; cf. *Opera* I, 302f; *Letters* I, 235.

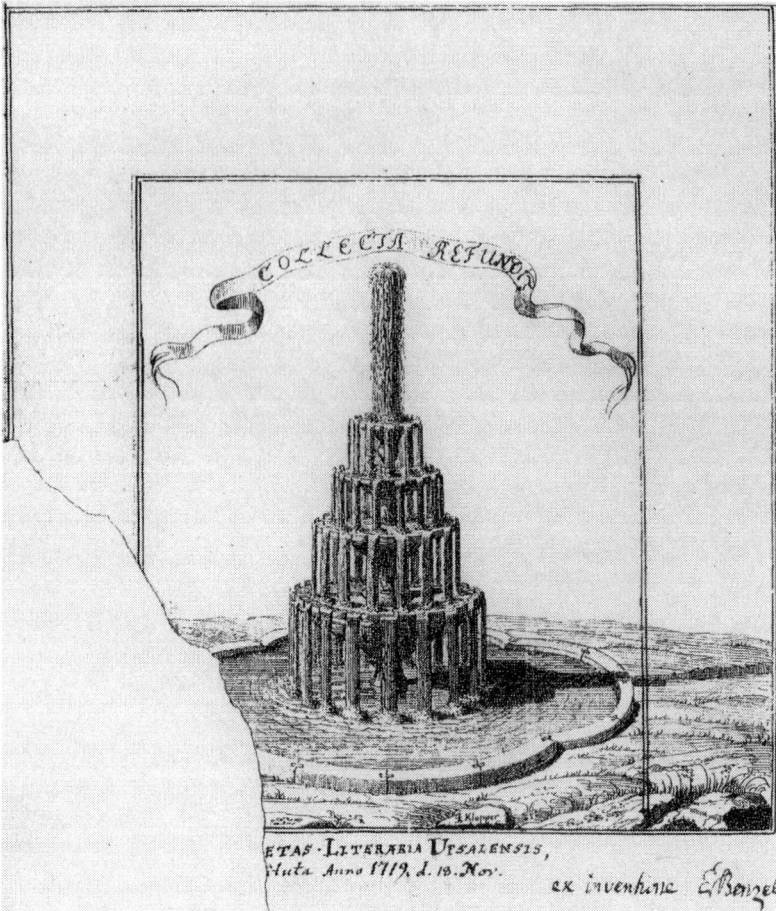

Fig. 3 What she has collected she pours out again (The emblem of Bokwettsgillet, following the proposal by Eric Benzelius, drawn by Jan Klopper *c*. 1719)

of a phenomenon that is sometimes observed at Lake Vättern, and that he himself had experienced. He had heard a loud noise that lasted for three or four hours as in a storm, but the water nevertheless lay perfectly still, and the noise was followed by rain within 24 hours.[48]

A horizontal line was explained by Swedenborg in *Regel-konsten* as 'A water-line follows the edge of the water, like a level'.[49] This is the topic of his manuscript about the varying water levels of Vänern, *Om Wennerns fallande och stigande* ('The Falling and Rising of Vänern', 1720), in which the geometrical vision and the concepts of geometry are prominent. He talks about areas, horizons, central or vertical lines, figures, volumes, proportions, lengths, widths, oblong forms, balances,

---

[48] *Bokwetts Gillets protokoll* 22 July 1720, 27.

[49] *Regel-konsten*, 4.

vertical slopes, undulations, and so on. It was an unsolved riddle how the water level in this lake could vary so much, and it could not be due to the tributaries and out-flows, Swedenborg noted. What he concluded, among other things, was that, since salt water is heavier than fresh water, it levels out more evenly along the horizon. Fresh water is lighter, more viscous, and does not as easily lie horizontally. When water is mixed with salt, the form of the particles is better suited to transparency. The cause of these properties can be found in the geometry of the particles:

> Although we are groping in darkness as to all that which concerns the finer constitution of nature, still it may be that we will gradually be enlightened concerning it, by leading ourselves forwards by means of experiments and by supporting our thoughts by geometry and mechanics; even as to water and its nature we are indeed very foolish and ignorant.

But as long as we lack evidence, says Swedenborg, we should not defend different principles and hypotheses, 'as they would then rather deserve to be called figments than principles.'[50]

For the Latin version of Swedenborg's essay on Vänern, produced by Roberg and published in *Acta literaria Sveciæ* (1720), some additional details about the water level in Vättern were obtained from the scientifically interested vicar Daniel Tiselius of Hammar in Närke.[51] Among other things, there is a report of an event on Vättern on 27 September 1712, when a sudden loud roar arose, like cannon, which could be heard far afield, with no lightning or earthquakes. Moreover, one can often see a peculiar mist under the water, which fishermen say looks like a naked woman with flying hair. Tiselius in turn referred in *Uthförlig beskrifning öfwer den stora Swea och Giötha siön, Wätter* ('Detailed Description of the Great Swedish and Gothic Lake, Vättern' 1723) to Swedenborg's idea that fresh water has larger, more open pores and is thus lighter and more viscous than salt water, which is heavier and levels out better along the horizon.[52] Tiselius measured variations in water level in Vättern, studied standing waves, calculated the size and location of the lake mathematically. On these matters Tiselius eagerly tried to make contact with Swedenborg, but met him only once, at the Medevi spa in July 1722, when he did not even have the opportunity to show him what he had written about Vättern.[53] However, through Benzelius he did gain access to a handwritten copy of Swedenborg's *Prodromus principiorum*.

---

[50]Swedenborg, *Om Wennerns fallande och stigande och huru wida thet härröra kan af wattnets tillopp eller aflopp igenom strömmar* (1720). LiSB, N 14a, no. 96; *Photolith.* I, 120–126; *Opera* I, 40, cf. 10; *Treatises*, 72f; For his studies of the rising water of Vänern, Swedenborg used a letter from 21 March 1715, written by Jonas Unge and read aloud in Bokwettsgillet 15 January 1720. *Bokwetts Gillets protokoll*, 10; Hag (1983), 11.

[51]*Bokwetts Gillets protokoll*, 28, 33; review by Roberg, 'De celeribus utriusque Gothiae lacubus, Vennero Vetterque, . . .', *Acta literaria Sveciæ* (1720), 113; *Opera* I, 50f; *Neue Zeitungen* 7 August 1721, 502f; cf. *Acta literaria Sveciæ* 1723, 471–476; D. Tiselius (1730), 50, 55; Wassenius, 16–22; H. Richter, 258f.

[52]D. Tiselius (1723), 6, cf. 95, 105.

[53]Tiselius to Benzelius, 28 July 1722. UUB, G 19:7 a; 11 July 1721. UUB, G 19:6 b; 8 July 1723. UUB, G 19:7 b; 21 October 1723 & 28 September 1724. UUB, G 19:8 b; Hag (1983), 14, 23f; E. Tiselius (1951), 110f; *Letters* I, 263–266.

Fig. 4 Tiselius explains how changes in water level are due to the air above ground pressing out the underground air in spiral and vortical passages, *a*, to the lake, *A*. Under the lake is its subterranean source or water reservoir, *C*. The water whirls up and the water level in the lake rises, 'causing loud noise and thunder, so that it roars and bellows' (Tiselius, *Ytterligare försök och siö-profwer uthi Wättern* (1730))

Relying on Kircher and Hiärne's geology, Tiselius believed that there was an underground circulation of water, an idea that was not alien to Swedenborg either (Fig. 4). In an article in *Philosophical Transactions* (1704–1705), Hiärne had just put forward the idea of subterranean currents under Vättern.[54] The geological foundation for Hiärne was Kircher's *Mundus subterraneus* and Johann Becher's *Physica subterranea* (1669), two works that were also to be significant for Swedenborg. Hiärne imagined water circulation caused by a mechanical pressure through the earth instead of through the atmosphere. The Maelstrom at Lofoten that Olaus Magnus wrote about and Kircher drew in his *Mundus subterraneus*, or the eddy current Charybdis that swallowed ships off the coast of Sicily was interpreted by Hiärne as the water's secret whirling entrance into the underground, as a stage in the constant circulation of water through the earth, up into the mountains and down the rivers, out into the sea and back.[55]

With the subterranean water Swedenborg was able to explain more phenomena, such as earthquakes and the Flood. Hiärne, who had previously tried to explain geological forces, wrote that a sound is heard underground during earthquakes 'like many wagons rolling past, with a great din, noise, and racket, and then an outburst of howling, roaring, and bellowing.' A smell of sulphur breaks forth, the water

---

[54]Hiärne (1704–1705), 1940–1942.

[55]Kircher (1665), I, 101; Frängsmyr (1969), 28, 35.

rumbles, 'seethes and bubbles as in a pot.'[56] Swedenborg in turn argued in *En ny theorie om jordens afstannande* ('A New Theory about the Retardation of the Earth' 1717), that when the subterranean water sinks it leaves behind cavities which cannot withstand the weight. The earth is rent asunder and collapses, 'so that lands and towns are swallowed up.'[57] God's words about a terrible darkness and earthquakes at the end of the world are confirmed by this. Fascinated by the force of the earth, he noted in his travel journal from 1733 data on vibrations during earthquakes, how clocks and lamps began to swing, how waves arose on otherwise calm canals.[58]

Flattered by the way his theory of water reduction in *Om watnens högd och förra werldens starcka ebb och flod* had attracted attention abroad from Jacob à Melle in Lübeck in his book *De lapidibus figuratis* (1720), Swedenborg responded to him in a friendly letter.[59] Swedenborg's reply was read aloud in Bokwettsgillet in May 1721 and included in the proceedings. He once again considered the attempt to show on hydrostatic grounds that large stone blocks can be moved on the bottom of the ocean by the swelling sea and transported by currents over the world. The irregularities of the earth's surface were caused by fluctuations and inundations, but not necessarily by Noah's Flood, which lasted only a year. It seems to have taken a very long time for the sea to transport all the shells, all the mud, sand, and stones that shaped the eskers and polished the stones to perfect roundness, as if in a lathe, by a constant rolling movement.[60]

A reviewer in *Historie der Gelehrsamkeit unserer Zeiten* criticized Swedenborg's idea that the water at the bottom of the sea has a greater force than the surface water, and can thereby move large boulders.[61] This would be in conflict with the laws of hydrostatics, according to the anonymous reviewer. But perhaps the critic himself was misled by his own assumption that water particles cannot be compressed. A cubic foot of water deep down does not contain more water particles than a cubic foot at the surface, he says. This attack led Swedenborg to publish an article about the laws of hydrostatics in *Acta literaria Sveciæ* in 1722.[62] Once again he defended the idea that stones and fragments of blocks on the bottom of the ocean can be moved hither and thither by a large flood (Fig. 5). Similar phenomena can also be observed in the air, at the bottom of the atmosphere. The debate did not end with this article.

---

[56]Hiärne (1706), 151f.

[57]Swedenborg, *En ny theorie om jordens afstannande* (1717). LiSB, N 14a, no. 34; *Photolith.* I, 28–65; *Opera* III, 278; translation, 51.

[58]*Resebeskrifningar* 13 July 1733, 26.

[59]Melle, 4, note c; *Epistola ad Melle*, 192–196; *Bokwetts Gillets protokoll*, 50; rev. in *Neue Zeitungen* 14 September 1722, 723–726; English translation, Swedenborg's first, in *Literary Memoirs of Germany* (1742), 66–68.

[60]*Opera* I, 56; translation *Miscellaneous*, 151f.

[61]*Historie der Gelehrsamkeit* (1722), 315–327.

[62]*Expositio legis hydrostaticæ*, 353–356; translation, 156–159; *Bokwetts Gillets protokoll*, 75; *Opera* I, 311; *Letters* I, 288f.

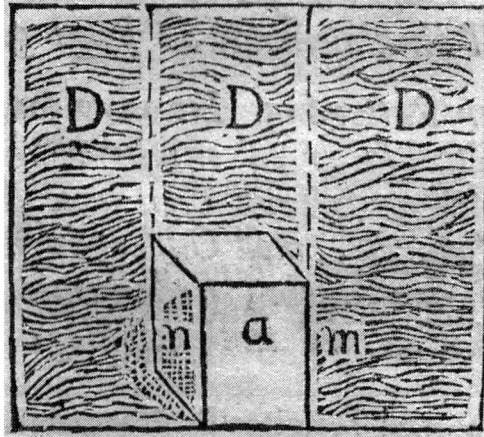

Fig. 5 The power of water. A square stone block, *a*, lies at the *bottom* of the sea and is pressed from all sides by the water. A column of water, *D*, *above* the stone exerts a certain pressure depending on its height. The water flows from *m* to *n* and exerts pressure on the stone. Swedenborg explains the ability of water to move large boulders: 'We have ocular demonstration of this fact in every body of water in violent agitation; for if it strike against any quiescent object, as from *m* to *n*, a sort of sinking of the water appears around *n*, or rather, it is like an *empty space*, which resembles either a hyperbolic or triangular cone' (Swedenborg, 'Expositio legis hydrostaticæ', *Acta literaria Sveciæ* (1722))

In 1724 he received an anonymous reply in *Neue Zeitungen von gelehrten Sachen*.[63] Swedenborg wrote to Benzelius saying that he wished to respond to the new critique with a 'clear proposition on the subject, *allegare* [adduce] *data, experimenta* drawn from *hydrostaticis* and *hydraulicis*, set forth the geometry applying to them, and so will establish a clear conclusion without mentioning the *canaille* who seeks *gloire* of dragging one into a dispute in an *ignobelt* fashion.'[64] He would then send this proof to Polhem and, after his verdict, send a copy to Wolff in Halle.

Swedenborg's geological ideas tend towards a form of Neptunism, in which water is the first matter, a primeval force that once shaped the landscape. In *Miscellanea observata* he states that 'the original matter of the earth may have been water.'[65] We read in the creation story that the Spirit of God moved upon the face of the water and divided the waters which were above the firmament from those under the firmament. This explains why the earth is round. If it had consisted of harder material it would have been more irregular and angular. He begins his *Miscellanea* with a number of geological essays, taking a particular interest in one of the enigmas of the time, 'figure-stones', petrifacts in interesting

---

[63]*Neue Zeitungen* 23 March 1724, 230f, cf. 28 February 1724, 167f.

[64]Swedenborg to Benzelius, Stockholm, 26 May 1724. *Opera* I, 314; *Letters* I, 335; *Bokwetts Gillets protokoll*, 108.

[65]*Miscellanea*, 43f; translation, 28f; cf. Becke (1674), title page with the solar rays of the trinity against a sea with animals, 'Ex aqua et seminibus', from water and seeds.

shapes. Steensen had previously investigated fossil shark teeth, and Hiärne had explained these objects in terms of petrifaction by water. As his predecessors in palaeontology, Swedenborg mentions Roberg who explained trilobites not as some amusing petrifacts but once living crustaceans; Bromell, whose large collection of marine insects turned into stone he had seen for himself; and also his cousin, the physician Johan Hesselius.[66]

It was with Hesselius as his travelling companion that Swedenborg set off for Holland at the end of May 1721. Hesselius's purpose was to obtain a degree as doctor of medicine in Harderwijk. The two men celebrated Christmas in Aachen and Liège, when they took the opportunity to explore the geology of the district.[67] Bending down towards the ground, they looked for petrified plants and animals. At Aachen Swedenborg found a large number of shells, reflecting how the sea had been higher in the past (Fig. 6). He found stones with leaves and stripes running diagonally upwards, or horizontally or in parallel, and noticed the orientation of the layers of rock, and he tried to convey what he saw in geometrical terms such as circular, convex and concave curves, elliptical and parabolic shapes, horizontal positions, and serpentines caused by an ocean that covered the whole world long ago. At the monastery of Chartreux near Liège he performed a visual inspection, through a geometrician's eyes, of the circular fractures of different sandstones.[68] If the blocks of stone were split they were seen to contain circles with colours differing towards the middle (Fig. 7:1–9). He drew the conclusion that some kind of liquid surrounded the surface of the stone, penetrating into it and shaping the circles. Boyle has found a liquid that penetrates stones and colours them, says Swedenborg. It was with a geometrical gaze that Swedenborg regarded the landscape and the world. He saw geometrical forms all around him. In a now lost essay by Swedenborg from 1725, intended for publication in *Acta literaria Sveciæ*, he made geometrical calculations of oval burial mounds.[69] This was very likely the barrows at Gamla Uppsala. The geometrical world can be described with geometrical methods. An observant eye can see the geometrical forms in nature.

Sound in the Mountains of Lapland

When a stone hits a smooth water surface, the waves spread from the centre as troughs and peaks in concentric circles. It is in the waves of water that the wave metaphor has its origin. Swedenborg, and also Polhem, envisaged sound as

---

[66]*Acta literaria Sveciæ* (1721), 192; *Opera* I, 54; *Miscellaneous*, 149; *Opera* III, 315; *De ferro*, 287.

[67]*Miscellanea*, 13, 22–24, 26; translation, 9f, 15f, 18; *Letters* I, 254.

[68]*Miscellanea*, 34, 40, 69f, 76f; translation, 23, cf. 26, 44f, 48f.

[69]*Opera* I, 316f; *Letters* I, 382.

Fig. 6 Fossilized shells found by Swedenborg at Aachen. Shells with oblique lines or furrows, *T*, or with lines without granules, *G*. Others have fine lines, *FZH*. One shell, *E*, filled with smaller shells and containing a turbine-shaped shell. *S, X, C,* and *W* are spiral 'turbinites'. *Y*, a large heap of worms. *K*, a shaft of reed. *R*, small shells (Swedenborg, *Miscellanea observata* (1722))

an analogous undulatory phenomenon, although through air and not water. The ancient Pythagorean Archytas of Tarentum knew that sound is produced through bodies striking each other. It was a blow or motion that was propagated through the air, through the ears to the brain, the blood, and finally to the mind, as Plato

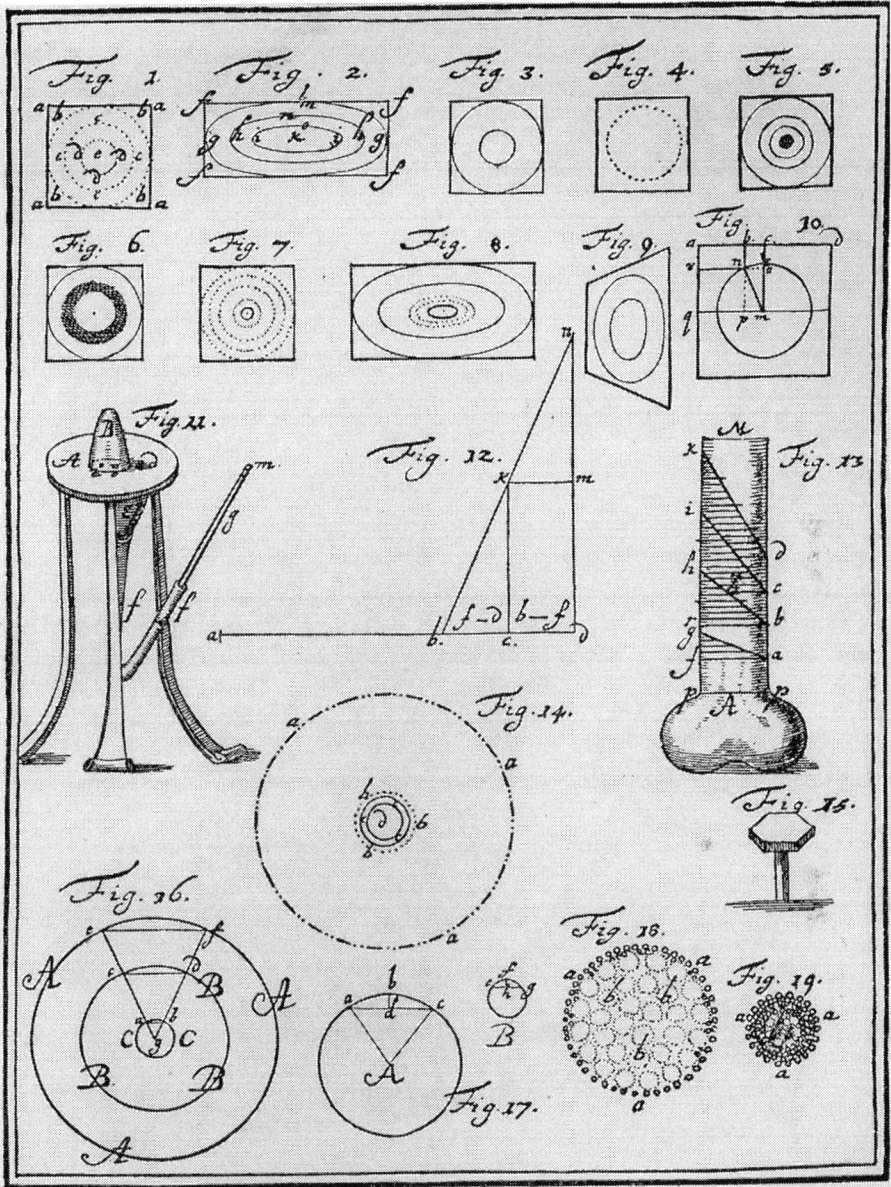

Fig. 7 The circular fractures of different sandstones. Increasingly round circles from *bbbb* to an exact circle, *ddd* (*1*), or to a perfect ellipse, *ii* (*2*). When the stones were exactly cubic the core was exactly round (*3–7*), but when they were of a different shape the core was neither oval nor round (*9*). Instrument to ascertain proportions between metals, which Swedenborg calls 'Archimedes' glass' (*13*). Water crystal shooting up from ice in an exactly hexagonal form (*15*). The air particles are bubbles with extremely small fire particles, *aaa*, on the surface (*18–19*) (Swedenborg, *Miscellanea observata* (1722))

explained in *Timaeus* in the fourth century BC, and Aristotle described sound in a similar way.[70] Through the Pythagorean tradition of a connection between sound and mathematics, the mechanistic world-view found it rather easy to incorporate sound in geometry. For Mersenne, sound was tremors in the air, while the atomist Isaac Beeckman propounded a theory of sound particles in 1616, according to which a vibrating string cut the air into globular particles of air which were propagated in all directions by the vibrations. Descartes believed instead that sound consists of waves resembling those on water, and Kircher in turn described sound as undulatory tremulations in the air.[71] Sound was a regular consequence of impulses, it was thought towards the end of the seventeenth century. It was a shock wave of alternately condensed and attenuated air. But many questions arose, for instance, about how sound spread, whether it was affected by altitude above sea level, velocity, and how this geometrical motion could give rise to sensory impressions. These questions fascinated Swedenborg, Polhem, and other members of the Collegium Curiosorum.

In October 1710 Polhem wrote a letter to Benzelius intended for the Collegium, in which he put forward some ideas about the motion of air and ether as a cause of hearing and vision. The motion of water waves is the starting point for Polhem's reasoning about the nature of sound. It is a wave phenomenon, like the waves on the water, as when 'one stirs up calm and stagnant water, then small waves spread in round rings, which gradually diminish and disappear'. If one hits the water in a gully, the waves go forward, but when they encounter an obstacle they go back at the same speed. The motion of waves can also be observed in the 100-fathom ropes in mine shafts. If the rope is hit with a stick at one end, 'then by means of the oscillation of the rope waves run like the wriggles of a snake all the way back and forth along the rope', a wave motion that also turns back.[72] A coarser, heavier rope gives slower waves. If you have a pile of fish bladders and flick one of them with a feather, the whole pile will quiver. Spinning an object in the air causes a sound; the faster the spinning, the higher and louder the noise. The fact that sound takes time to travel through air could be easily observed by watching a cannon being fired at a distance; the fire can be seen long before the bang is heard. Using Pierre Gassendi's experiments with cannons, Marin Mersenne calculated the speed of sound through the difference in time between the bang and the flash from the muzzle of a cannon at a specific distance.[73] A person crying upwards has to shout loud, whereas someone shouting downwards can be heard better. 'Those who have been on the mountains of Lapland can report that if one shoots, or if there is thunder above the mountains, those who live in the valleys hear it very shrill and sharp, but those who are on top of

---

[70]Plato, *Timaeus*, 67B; Aristotle, *Peri psychēs*, 2.8.419b9ff.

[71]Ullmann, 3; Buhl, 47, 49, 54f.

[72]Polhem to Benzelius, 30 October 1710. *Polhems brev*, 24, cf. 76; cf. *Dædalus Hyperboreus* I, 6f; *Polhems skrifter* I, 161; *Regel-konsten*, 21.

[73]Ullman, 6; *Polhems brev*, commentary 236.

the mountain hear little.'[74] Shots and thunder high up in the mountains of Lapland have 'a much duller and softer sound, as trustworthy people have told me'.[75]

Polhem may have got the idea for the cannon experiment from Harald Vallerius. Under his chairmanship Daniel Touscher defended a dissertation about the mountains in Jämtland, *Adumbrationem Alpium, quas habet Jemtia, perbrevem* (1694). He mentions an experiment performed in the Carpathians by the Hungarian mathematician David Frölich; this information may have been derived from Sturm.[76] Frölich had found that when a gun is shot at different heights, the sound of the bang on a mountain top is no louder than the breaking of a stick, but down in the valley the gun makes a dreadful din like a cannon of the largest calibre.[77] The noise continued for 7–8 min, penetrating distant mountain crevices, rebounding against rock faces which further amplified the air waves. Fröhlich's experiment is also mentioned in another dissertation presented under Vallerius, *De vallibus* (1708), on valleys, defended by one Eric Dahl, whose surname appropriately means 'valley'.[78] Hiärne too describes the same experiment in *Den beswarade och förklarade anledningens andra flock* ('The cause answered and explained, second part'). On mountain peaks the air is 'more subtle', whereas valleys are filled with coarse, earthy and thick, watery particles. All 'din, bangs, and sounds are larger and more vehement in the thick air', but in the thin air on mountain tops the boom of a gun makes little more sound 'than if one strikes a pair of Russian mittens against each other.' Further down the air is thicker because of the ascending vapour. There the gun makes a louder bang that is doubled against the many bends and angles of the mountains, 'so that it sounds like peals of thunder.'[79]

The relationship between volume and pitch was a subject of investigation. Aristotle believed that higher notes spread faster than lower ones, and that sound was heard better at night than in the daytime. Kircher likewise thought that the speed of sound was dependent on the volume, that a louder noise goes faster than a weaker one. The reason for the phenomenon in the mountains of Lapland, Polhem says, is instead the air pressure, the atmosphere compressing the air. The sound occurs because the air is packed and tense like a pile of bulging air bubbles. If a bubble is touched, the motion spreads like waves in every direction. In the mountains the air is thinner and the sound coarser, whereas in the valleys the density of the air is greater, which gives rise to a finer sound.

Neither Swedenborg nor Polhem had been in the Swedish mountains, so what they knew about the acoustics of mountains must have come from reading or hearsay. In the popular standard work by Johannes Schefferus, *Lapponia* (1673),

---

[74] *Polhems brev*, 25; cf. *Dædalus Hyperboreus* I, 8; *Polhems skrifter* III, 211, 275.

[75] Polhem, 'Discurs om eldens och vattnetz strijdigheter' (1711 at the latest), *Polhems skrifter* III, 78.

[76] Frölich, II:1; Sturm (1685), II, 158f.

[77] H. Vallerius and Touscher, 17f; translation, 32; cf. Rudbeck the Younger (1968–1969), commentary, 96f.

[78] H. Vallerius and Dahl, 30f.

[79] Hiärne (1706), 197f.

about the nature and inhabitants of Lapland, nothing much is said about sound phenomena in the mountains other than, in an addition to the first Latin edition, what was reported by the Lapland priest Nicolaus Lundius. As a Sámi himself he could draw on his personal experience. To the second English edition from 1704 he added 'that the Inhabitants of the Valleys among these Mountains relate, that if they happen to make any noise in the Evening, or the Dogs fall a barking, there appears frightful Specters to them, and they hear doleful Voices.'[80] Nor does the Lapland traveller Jean François Regnard seem to say anything about special acoustic conditions in the mountains, but he does write that the Sámi remind him of machines with springs: 'their legs are thin, their arms long, and the whole of this little machine seems to move on springs.'[81]

Swedenborg's uncle, the future inspector of mines Daniel Swedberg, took part in Samuel Otto's expedition in 1675 up along the Dalälven river. Among other things they surveyed Mount Ida in the parish of Idre, which turned out to have a height of 9,135 cubits, or 5,426 m above sea level. Rudbeck the Elder was not slow to quote this datum in his *Atlantis* (1679), proudly claiming that the Swedish mountain was identical with the Mount Ida where Aeneas in Homer's *Iliad* was conceived.[82] His son Rudbeck the Younger, one of the members of the Collegium Curiosorum, was among the few who could report from his own experience about sounds in the mountain world. Together with the draughtsman Anders Holtzbom and two young counts from the Gyllenborg family, he had been on a natural-history expedition that reached as far north as Torne Träsk, from where he had turned south towards Kvikkjokk and on through the Sámi lands of Lule Lappmark. In his diary of the journey through Lapland in 1695 he wrote about the bewildering mountain regions where one could easily get lost and be separated from the other members of the expedition. The sound seemed to disappear above the tree line: 'And no matter how loudly one shouts, often no sound arrives unless the weather is particularly favourable and there is a following wind.'[83] On the same expedition, dedicated to the Swedish sun-king Charles XI, Spole and Bilberg were also sent to the district of Torneå and the Kengis copper- and ironworks at Pajala, to study the midnight sun. Yet they do not seem to have reached any high altitudes. Swedenborg mentions the expedition in his proposal about the utility of an observatory.[84] In *Ludus Heliconius* (1714/1715) he also wrote a poem about the Sámi in the distant land of the midnight sun, where Phoebus buries his sunset in the middle of his rising.[85]

The Swedish natural philosophers seem to have longed for the mountains, away from the pestilent houses of Uppsala. The interest in acoustic phenomena, and especially the sounds in the mountains of Lapland, influenced the planning

[80]Schefferus (1673); French translation, 407; English translation, 371.

[81]Regnard, 129; translation, 163.

[82]Rudbeck the Elder (1679), I; edition, I, 502f; Dahl, 195.

[83]Rudbeck the Younger (1987), II, 54; cf. *Bokwetts Gillets protokoll*, 66.

[84]*Photolith.* I, 4; Nordenmark (1933), 41; Bilberg (1695).

[85]Swedenborg, 'Ultimorum Lapponum descriptio', *Ludus Heliconius*, edition, 108f.

of the Collegium Curiosorum for Henric Benzelius's expedition to Lapland in 1711. This was to be the society's largest experimental enterprise, as a follow-up to Spole and Bilberg's investigations of the midnight sun. The aim was to perform an experiment on the speed of sound with the aid of cannon, to find out how sound changes depending on differences in altitude and the refraction of the sun's rays at the summer solstice. Polhem submitted proposals for experiments that could be performed in the mountains of Lapland. His *Förtechning på några experimenter som på LappFiällen och i des Dahlar wore nödige att werkställas* ('List of experiments which it would be necessary to perform in the mountains and valleys of Lapland', 1711) was sent to Upmarck so that he could pass it on to Eric Benzelius's brother Henric. In the accompanying letter Polhem complains of his poor eyesight, the cost of postage, and his private economy: it is not easy to be 'curious when worries about bread-winning spoil everything'.[86]

In 20 points the practically minded Polhem displays a wealth of ideas and a keen curiosity about basic research. Trigonometry was to be used to ascertain the true figure of the earth, a question raised by the Académie des Sciences in connection with Spole and Bilberg's observations, which was later answered after the expedition to measure a degree of the meridian, mounted by the astronomers Pierre Louis de Maupertuis and Anders Celsius in 1736–1737. Is the earth oblate or prolate? In other words, is it flattened at the poles or the equator? The latter—Cartesian—view was embraced by the members of the Collegium Curiosorum, but it was the Newtonian view that was confirmed by Maupertuis's expedition. The general idea of Polhem's experiments was to compare different phenomena on mountain peaks and in valleys. He suggested investigating air pressure with barometers, measuring air resistance, and studying the gravity and density of air with a pendulum or a falling cannonball. The inspiration came from Pascal, who had demonstrated on the summit of Puy-de-Dôme in 1647 how changes in the mercury barometer varied with altitude above sea level. Sound waves and the speed of sound were also to be investigated. It is important, Polhem says, to find out 'the tremulation of the air, and the speed of its waves', which could also lead, by analogy, to a better knowledge of ether waves in light and the time they take to travel between the sun and the earth. Furthermore, experiments should compare the sound of two bells or pipes, one high up and the other down in the valley. According to the theory, pitch ought to be dependent on air pressure, with the sound being brighter or finer when it goes downwards and coarser when it goes upwards. It should also be ascertained whether a wax candle burns with the same flame on the mountain as in the valley, whether air pressure had any effect on the force of gunpowder, and whether a cannonball can be fired further in the mountains than in the valleys. The expedition was also recommended to study the refraction of sunlight, glacier ice, plants, and animals.

The speed of sound was to be calculated by measuring the time between the flash and the boom of a cannon, 'both in the deepest valley and on the top of

---

[86]Polhem to Upmarck, Stjärnsund, 15 April 1711. *Polhems brev*, 71.

the highest mountain'.[87] But should a cannon really be pulled up above the tree line? Those experiments with cannon are not feasible, Roberg objected at one of the society's meetings.[88] It would work equally well with a musket. The society discussed purchases and preparatory experiments. It was no doubt Roberg and Rudbeck, with their interest in natural history, who also suggesting asking hunters the names they gave to birds, fishes, and other animals, and collecting seeds of the flower known as 'King Charles's sceptre' (*Pedicularis sceptrum-carolinum*, moor-king lousewort in English). Henric Benzelius set off the same spring, after having defended a dissertation in which he touched on Diophantus and the mathematical sciences in ancient Alexandria.[89] In the course of his travels he sent letters to Eric reporting on the expedition. He managed to get as far north as Jukkasjärvi and Torne Träsk, and remained in Norrland into the new year. But the experiments were not entirely successful. He had problems with the apparatus and its use. The results were meagre. Soon, however, the orientalist Henric Benzelius set off on the next major journey, this time to the near east, to Charles XII in Bender and on to Constantinople, Damascus, Jerusalem, and Egypt. In the late autumn of 1711 Polhem wrote to report on a remarkable similarity shown by waves. They are the same in sound, light, and water. The waves or tremulations also have their mechanical rules and proportions 'as for example when the string of the octave is as 1 to 2, then their tremulations in the same time are 1 to 4. Likewise, the fifth as 3 to 2 has a tremulation of 9 to 4 and so on.'[90] Polhem the mechanic compares this with the fact that the gearing of cogs requires special proportions. The gearing of cogs is of the same proportion as the waves of water, sound, and light. Ouch! 'Forgive me for writing badly,' Polhem excuses himself, 'for I have cut my finger and besides I am in haste.' The letter ends there.

The spring of 1716 saw the appearance of the first issue of *Dædalus Hyperboreus*. It contained no less than six articles about sound. They were about experiments with the properties of sound, about waves, resonance, acoustics, echoes, and various instruments such as ear-trumpets and a 'boom-tube' or megaphone. This issue includes extracts from Polhem's letters to Benzelius about experiments on the nature of sound, and his proposals for experiments in the mountains of Lapland. There are still experiments to perform, according to *Dædalus Hyperboreus*, which can reveal the hidden properties of sound and 'show that our country is more bounteous for performing experiments than others; which could then serve for enlightenment in our machines, which would also yield something remarkable, useful both for other societies and for our own country.'[91] To Polhem's list of

[87] *Polhems brev*, 72.

[88] *Collegium Curiosorums protokoll*, 63–65; Rudbeck the Younger (1720).

[89] H. Benzelius, 154–161; Törner and H. Benzelius, 48–55; cf. *Opera* I, 224, 230; *Letters* I, 52, 63.

[90] Polhem to Benzelius, Stjärnsund, 3 November 1711. *Polhems brev*, 77; cf. *Dædalus Hyperboreus* I, 9.

[91] Polhem, 'Experimenter som ännu återstå i wårt land at giöra om liudet', *Dædalus Hyperboreus* I, 10.

acoustic experiments Swedenborg has added a few other pieces of data, probably from his own experiences. The third experiment seems to be based on what emerged from the discussions in the Collegium Curiosorum about Polhem's suggested sound experiments. In the minutes for 8 May 1711 we read: 'NB. questions asked about echo up in Lapland.'[92] In *Dædalus Hyperboreus* it is suggested that echoes should be studied where there are peaks and valleys. It should be ascertained how many times the sound repeats, with simultaneous records of readings on a barometer and thermometer. In Italy, Swedenborg interjects, there is said to be a building where the echo is reproduced as many as 18 times. The echo phenomenon reminds him of the reflections in a mirror:

> Otherwise the reverberations resemble two mirrors set opposite each other, with a pyramid or chandelier between them, so that it is seen in a mirror multiplied 10 to 12 times, set one after the other in beautiful order, until the furthest one appears as in an obscure darkness.[93]

This is an example of Swedenborg's metaphorical thought, how he pictures one phenomenon through another one, that it is possible to 'translate' an auditory impression into a visual one.

The transmission of sound is another source of questions and experiments. By putting one's ear to bare ice, one could find out the distance from which one could hear another person talking on the ice. It had been discovered that a cannon or mortar can be heard between 60 and 120 miles away at ground level. 'And if one talks by the wall in a vaulted chamber, a person at the other side hears the other's speech' as in the Whispering Gallery in Saint Paul's Cathedral in London, 'up in the round iron railings or gallery'. Swedenborg himself had visited the cathedral, designed by Christopher Wren, just a few days after it had been completed. But when Polhem was in London at the end of the seventeenth century it had not yet been built. He also gained experience of sounds in the world around him from the galleries and underground vaults in the Falun mine, where it was possible to study how voices oscillate ten fathoms up and down the shaft. Other questions deal with the discordant sounds that arise from inclement weather and how organ notes vary depending on the weather. Swedenborg the constant traveller has added: 'In strong weather and preferably in a contrary wind a lovely melody is heard from a wagon as a noisy racket; and there are more dissonances than on a clavichord.'[94] The music of travel.

Polhem's questions seem to have remained unanswered when Polhem's list of experiments was published in Latin in *Acta literaria Sveciæ* in 1722.[95] However, the acoustic experiment of shooting from a particular height and comparing the volume and pitch was later listed by Linnaeus as one the questions to consider on his trip

---

[92] *Collegium Curiosorum protokoll*, 64.

[93] *Dædalus Hyperboreus* I, 11.

[94] *Dædalus Hyperboreus* I, 11; cf. *Opera* I, 207; *Letters* I, 13.

[95] Polhem (1722), 285–289; translation P. Martin. *Bokwetts Gillets protokoll*, 65; cf. *Polhems brev* 144.

to Lapland in 1732. He would 'shoot on the highest mountain when the first is not heard but others long after'.[96] It is uncertain whether he really tried the experiment as a lone traveller, but during his journey in Dalarna he performed it with his fellow travellers on 9 July 1734. On the top of Gopshusberget, a mountain between Mora and Älvdalen, he writes, there was a marvellous prospect over mountains, rivers, wild forests, lakes, and churches. They fired their pistols, shooting up into the air and down into the depths. 'From above the shot sounded very briefly, as if it had been quickly silenced, but for a long time afterwards there was a thundering in the hills opposite like a thunderstorm.'[97] They went down the mountain to hear the bang from there, yet it was not heard 'louder than on top, but continued for quite some time without interruption.'

In the Baroque Echo Temple

Swedenborg's observation of the mysterious mirror reflections of sound in *Dædalus Hyperboreus* touch on the mythological and metaphorical meanings of the echo phenomenon. The very Greek word *ēkhō* can be traced back to Ovid's mythological account in *Metamorphoses*. The constantly replying Echo was the nymph of the voice, who distracted Juno with her long stories, giving Jupiter a chance to amuse himself with the mountain nymphs. But Juno discovered this and punished Echo by making her unable to speak for herself, only able to repeat what others said. '"Is anyone here?" and "Here!" cried Echo back.'[98] Then, when her feelings for Narcissus were not reciprocated, she vanished into the woods, pining away until all that remained of her body was the voice. The echo is often repeated in emblematics and poetry. Swedenborg himself used the echo symbol in *Camena Borea*, where a muse leaves the land of the Sarmatians when she fails to find an echo to answer her.[99] He was, of course, not unfamiliar with the emblematic tradition. Among other things, he used a standard work in the genre, Filippo Picinelli's *Mundus symbolicus* (1687).[100] In emblematics there is a metaphorical dialogue in which the image is a silent poem and the poem a speaking image which mutually support each other. Emblems expressed the intellectual through the sensual, and the abstract was understood in terms of palpable things, sounds, pictures, and words.

Spegel uses the echo in his posthumous *Emblemata*, from the late seventeenth century, when he describes how the cry of the God-fearing person is answered by God.[101] Spegel refers, for example, to the passage in the First Book of Kings,

---

[96]Linnaeus (2003), II, 333.

[97]Linnaeus (1953), 37; translation, 89.

[98]Ovid, *Metamorphoseon*, 3.380.

[99]*Camena Borea*, edition, 118f, cf. 102f, commentary 193f.

[100]Picinelli; Jonsson (1961), 57ff; translation, 46; Jonsson (1969), 153.

[101]Spegel (1966), 21; Luke 11:9, Ps. 33 & 144, James 5:[1]6.

chapter 18, when Elijah prayed for rain. God heeded the plea and the rain fell. Spegel's successor as bishop of Skara, Jesper Swedberg, likewise used the echo for edificatory purposes. In *Schibboleth* he explained 'echo' as 'reverberation' or 'response', and referred to the Wisdom of Solomon 17:19, where 'a rebounding echo from the hollow mountains' spread terror.[102] The echo also functioned as a resounding poetic effect, as in the melancholy lamentation of the shepherdess Floris in a poem by Hiärne:

Let me for my company choose

Desolated wilderness,

So that naught may me oppress

But for death, which cannot lose.

Out in this deserted wood, where the echo does complain;

It alone to me is true, and with me will e'er remain.[103]

Echoes are about calling and receiving a response, making contact and finding connections in reality, not being alone. Nature is a state of mind. Jacob Frese looks out through the window from his sickbed, towards the outer world that can be glimpsed between the curtains. He sees the effect of the sun in May:

When song of lark and waterfall,

Cause Echo to resound her call

    Against the mountain side:

And fishes blow their bubble-foam,

A-swimming in their watery home

    Along their racecourse wide.[104]

A figurative expression for the lingering echo preaching its message over the world is found in Frese's tribute to Charles XII on his name-day, *Echo, å Sweriges allmänne frögde-qwäden* ('Echo to Sweden's general songs of joy' 1715). The sun is shining, bassoons are blasting, bells are ringing, the earth is rejoicing, and heaven replies in an echo on Charles's name-day.[105]

Baroque people entered nature's hall of mirrors, which reflected the light from other parts of the universe; they listened to the sounds that were thrown between the walls of reality in the echo temple of life. Baroque man observed the countless connections of the reflections, listened to the harmony of things and Echo's reverberation in the universe. Meanings were cast back and bound to each other.

---

[102]Swedberg (1716), 267; Ps. 38:15.

[103]Hiärne, '[Floris hwar om suckar du]', in Hiärne (1995), 176, cf. 60–62.

[104]Frese, 'Solens wärkan i maii month, uti en brud-skrift öfwer handelsmannen, ähreborne och högwälaktad herr Michael Hysing och ähreborna och dygd-edla jfr. Margaretha Frodbohm, då åhr 1717 den 30. maii deras äktenskap beslöts', in Frese (1726), II, 55f; cf. Lamm (1918), I, 144–146.

[105]Frese (1715), [9].

The world consists of counterparts. It is on this way of imagining the world as an echo temple that Swedenborg's doctrine of correspondences partly rests. Properly posed questions receive answers in nature's echo temple. For Bacon, Echo the nymph becomes the symbol of scientific language. He compares human reason to a mirror that is not adapted to the rays of objects but mixes in a part of itself and distorts and pollutes the image.[106] The individual is in his own cave which refracts and distorts nature's light. In natural philosophy there was also a more concrete interest in echo effects and reflections. Kircher, Schott and many others describe in detail acoustic effects and experiments with mirrors (Fig. 8). Guiseppe Biancani wrote an 'Echometria' and Bacon dreamed that the New Atlantis would have a special sound-house with various strange and artificial echoes, tossing the voice back and forth.[107]

Acoustics and optics constituted a major part of scientific study in Sweden too. Optical illusions and reflections in water fascinated people, as when peasants in Skåne at the outbreak of war saw troops marching across the sky, or when Spole, standing behind Mount Omberg, saw the island fortress of Visingsborg reflected via the clouds in Lake Vättern.[108] The future archbishop with the appropriate name of Spegel ('Mirror') mentions in *Guds werk och hwila* the burning-glasses of Archimedes, whose reflections set fire to the enemy ships during the siege of Syracuse, so that 'burning rays flew like arrows'.[109] And Swedenborg tells in his work about iron, with a reference to an article from 1697 in *Acta eruditorum* by the German scientist and inventor Ehrenfried Walther von Tschirnhaus, how a certain kind of burning-glass could melt iron nails in 30 seconds.[110] In Bokwettsgillet a letter from Master Jonas Unge was read aloud, about an unusual echo beside Lake Vättern. At Undnäs there are 'five headlands, each of which makes an echo, one after the other, for as many as five or six stanzas, which, (now the whole world rests,) and not before the verse is over'.[111] In Spegel's hymn, the echo wanes out over the lake in the dark of night.

Swedenborg's manuscript *Experiment om echo* (c. 1716) is an expression of the contemporary acoustic interest in sound, hearing, music, and the numerical proportions of harmonies.[112] More specifically, it concerns an experiment in acoustics exploring the nature of resonance, which is one of the very few examples of Swedenborg as an experimenter. He tries to ascertain the nature of echoes by determining the mathematical proportions and wave movements, which is a characteristic theme in his thinking, the idea of nature's harmonious proportions and

---

[106]Bacon (1620), I, § 41f; Eriksson (2002), 558f.

[107]Bacon (1627).

[108]Spencer, 28.

[109]Spegel (1685), edition, I:1, 194, cf. 328f.

[110]*De ferro*, 360; Tschirnhaus, 414–419.

[111]*Bokwetts Gillets protokoll* 16 December 1720, 39; *Then swenska psalm boken* (1697), Ps. [3]75.

[112]Swedenborg, *Experiment om echo* (1716). LiSB, N 14a, no. 154, fol. 407; *Photolith.* I, 205f; translation, 835–839; cf. Hyde, no. 97, cf. 85.

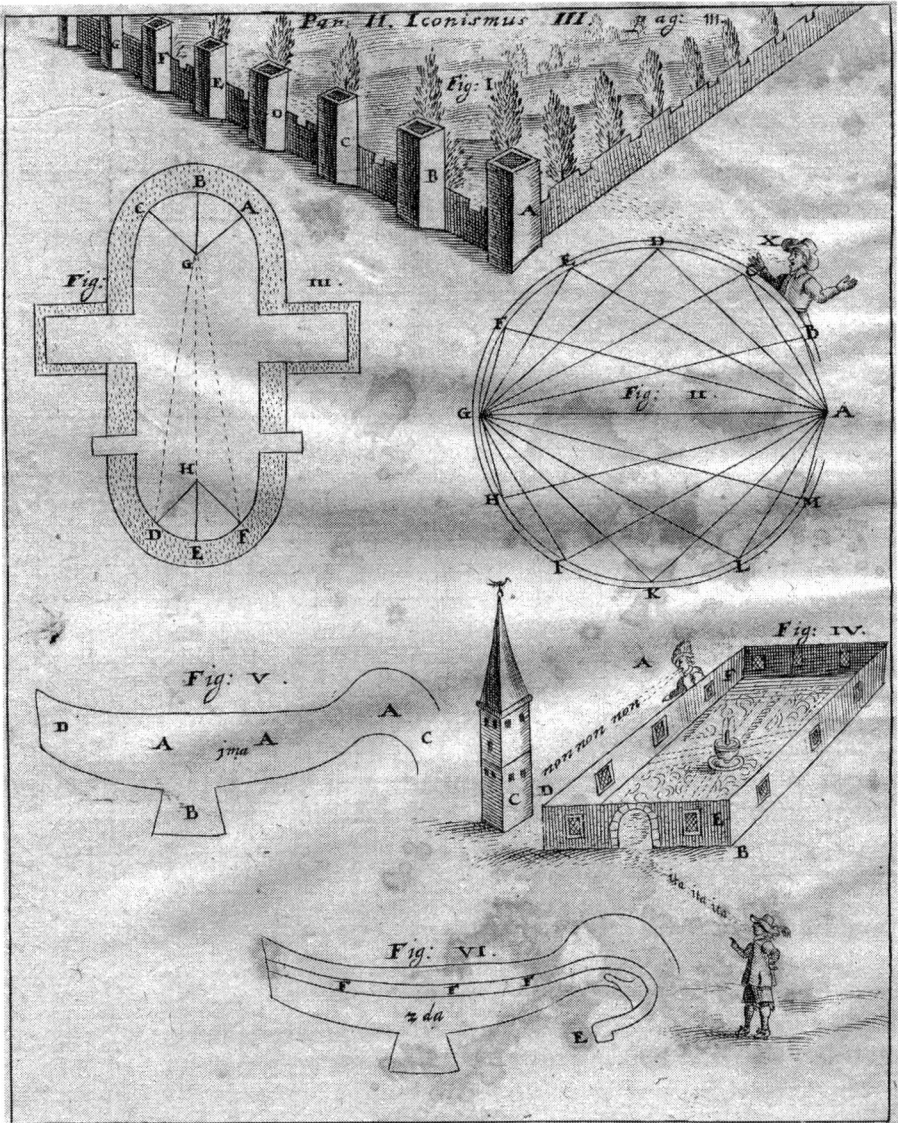

Fig. 8 The resounding echo; 'non non non … ita ita ita' ('no, no, no … yes, yes, yes') is the sound of the echo in Schott's *Magia universalis naturæ et artis* (1657)

the numerical proportions of geometrical bodies, waves, and sounds, the concept of nature's arithmetical and geometrical structure. The manuscript was probably written in connection with the first issue of *Dædalus Hyperboreus* concerning sound, and was most likely intended for publication in that journal. Swedenborg's *Experiment om echo* has points in contact with Polhem's suggested experiments in the mountains of Lapland and the echo experiments in *Dædalus Hyperboreus*. This means that it could have been written by the end of 1715, and in other words could be up to a year older than previously assumed.

Swedenborg, however, is not between two mountain peaks, but between two walls in an unknown building in southern Sweden. He strikes the walls with a stick and counts the sound as it bounces. The echo was heard four times, and a fifth time when it was so faint that it could scarcely be heard. The explanation why the echo did not rebound more times than this was that the timber in the building muffled the sound. Where he performed the experiment is not stated, but from the calculations we can work out that the distance between the inner walls of the building was 30 cubits, that is to say, just under 18 m. The echo was heard four times within half a second, meaning that the sound travelled 400 cubits in 1 s according to his calculations. What he seems to be aiming at is to measure the speed of sound. Mersenne was one of those who had previously tried to calculate the speed of sound with the aid of echoes.[113] But the result obtained by Swedenborg corresponds to 238 m/s, which is a good way under the true speed of sound, 330 m/s. There are some features of the experiment that do not seem right. He claims that his result has 'often been observed and found to be so'. In a letter of 7 December 1715, however, in which Polhem comments on the papers in the first issue of *Dædalus Hyperboreus*, he points out that Christiaan Huygens had found that sound had a speed of 180 fathoms a second, or 321 m/s.[114] Swedenborg's unsuccessful calculation of the speed of sound may be one of the reasons why *Experiment om echo* was never published.

Sound travels at 24,000 cubits per minute, Swedenborg goes on to say in his experiment with echoes, and thus far he is correct based on his results, but he thinks that 24,000 cubits corresponds to half a Swedish *mil* plus 6,000 cubits. Since 1 *mil* equals 18,000 cubits (just over 10 km) according to the decree of 1665, it ought to be 1 *mil* and 6,000 cubits. This means that he is also wrong to state that sound travels one whole *mil* in one and a half minutes; it should be 1 *mil* in half that time, three quarters of a minute. If the decipherment of the manuscript is correct, then Swedenborg either reckoned with 36,000 in a *mil* or in his haste he confused the cubit (the *aln* or 2 ft) with the foot. But the nice thing about his calculation is that, thanks to the 'mistake', he gets the speed of sound to be in perfect proportion to the spread of water waves. Water waves move 24 cubits a minute or 36 cubits in one and a half minutes. And perhaps this is what was important to him: not the accuracy but the harmony. Furthermore, he discovered that, if one stood half-way between the walls one could hear a quavering sound for eight beats. He also came to the conclusion that loud noises have a greater angle of reflection and that echoes are not propagated in a circular or oval motion but in a straight line. The experiment on echoes ends with Swedenborg leaving the building: 'where the opening was, I walked out through it.' Outside the echo faded.

---

[113]Ullmann, 6; see also Musgrave, 433–438.

[114]Polhem to Swedenborg, Stjärnsund, 7 December 1715. *Polhems brev*, 113; *Dædalus Hyperboreus* I, 8, 10; KB, Polhem, X 267:1, fol. 104r; Huygens (1690), 9.

Thunder and Organ Peals

The first issue of *Dædalus Hyperboreus* contains four articles about different instruments for amplifying sound. An article entitled 'Assessor Polhammars Instrument at hielpa Hörslen' ('Assessor Polhammar's instrument for helping hearing') describes ear-trumpets for people whose hearing is impaired by age or 'slaps and booms', causing the eardrum to burst or become laxer and thus no longer able to perform 'the usual tremulation' that transmits the movement in the air and leads to an auditory impression.[115] It is basically a spherical funnel with a hemisphere where the sound rebounds into the hole, which can amplify sound 20–30 times (Fig. 9). An 'ear tube' of this kind had been made of silver for 'Her Late Majesty the Dowager Queen, now with God'. The ear-trumpet is based on acoustic laws of resonance which are calculated and explained geometrically, with the angle of incidence of the sound being the same as that of the outgoing sound. The way sound bounces back and forth in the conic section follows the rules of Apollonius. Polhem also describes another ear-trumpet, this time with a parabolic and elliptical shape rather than spherical. However, it has the disadvantage that it 'drives the sound back and forth with a double echo in the ear [and] is mentioned here only as something to choose if one

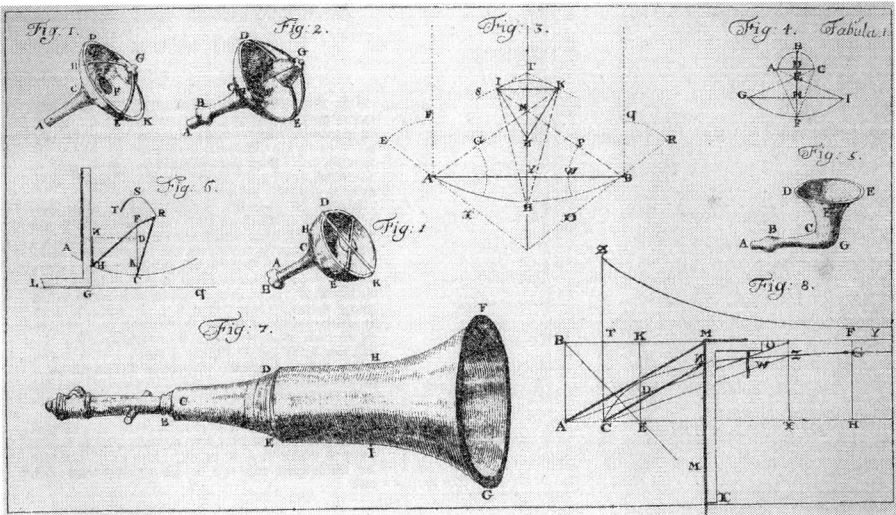

Fig. 9 Ear trumpets and a megaphone. Polhem's ear trumpets of *spherical shape* (*1*), of *parabolic* and *elliptical shape* (*2*), and an English 'ear tube' for the hard of hearing, with a hyperbolic opening (*5*). He also explains how to draw a parabola and an ellipse with the aid of a thread and a *set-square* (*6*). Polhem's megaphone (*7*) (*Dædalus Hyperboreus* I (1716))

---

[115]Polhem, *Dædalus Hyperboreus* I, 1.

finds it pleasing.'[116] It was no doubt Sturm's *Collegium experimentale* that was the chief inspiration for Polhem when he set out to design these aids for the hard of hearing.[117]

In the same issue there is a description of yet another acoustic invention, a 'boom tube' (*dåntub*), which is a kind of hyperbolic megaphone or amplifier of metal, built into a wall. The opening is from eight to ten cubits, that is, five or six metres in diameter. If this is placed on a high mountain and a cannon or mortar is placed at the mouth, the result is a remarkable sound effect, a bang which sounds at a distance of two *mil* as if it were only a quarter of *mil* away, fooling people into believing that there is a whole battery of ordinance nearby. The loud boom from two such tubes requires one to plug one's ears with wax. It is unlikely that Charles XII used this instrument in battle, but in the bishop's residence of Brunsbo outside Skara Swedenborg seems instead to have installed a communication system consisting of a speaking-tube that led down to the kitchen so that he could order his favourite drink, coffee.[118] The acoustic instruments may have been inspired by Kircher's *Mvsvrgia vniversalis* (1650), which is about acoustic chambers and spiral speaking-tubes. Through the differing ability of the echo to reflect against different walls, trees, and so on, Kircher regarded it as an analogy to optics.[119] The smoother the wall, the stronger the echo, as in a mirror. Moreover, echoes are heard better at night than in the daytime. Schott describes the same spiral tubes. Speaking-tubes, ear-trumpets, and hearing instruments belong to the genre of the curious, as with Valentini, or in Andreas Riddermarck's dissertation about acoustic tubes designed to cause air to undulate out on the sea.[120]

The magnificence of baroque music makes the body tremble. Horns and trumpets blast, drums and timpani thunder. There are cannon salutes and the earth quakes, as in Stiernhielm's ballet *Freds-Afl* ('The birth of peace', 1649), where the war god Mars enters the stage and roars: 'I shall thunder so that heaven and earth will quiver like aspen leaves.'[121] The palpable corporeality of sound, its ability to make a person tremble, is particularly evident in Swedenborg's early anatomical-physiological thought. The body is a musical instrument. Sound and music have to do with the body, with mathematics, acoustics, and technology.[122] In the circle around the Collegium Curiosorum, baroque music can be heard in the background, like the accompaniment of table music, played allegro. Music was not just about the pleasure bestowed by the notes. It was also a mathematical, scientific, and technical problem. Science and music had long since been combined in mathematics in the quadrivium

---

[116]Polhem, 'Ett annat dylikt Instrument', *Dædalus Hyperboreus* I, 4.

[117]*Polhems brev*, 113; Sturm (1685), II, 142–164.

[118]Kirven and Larsen, 11.

[119]Kircher (1650), book IX, 237–244, 302, cf. Kircher (1646), 131f; Schott (1657), II, 155; cf. Wald.

[120]Valentini, III, ch. XIV, 55–57; Riddermarck and Lychovius, 8f.

[121]Stiernhielm, 'Freds-Afl', *Samlade skrifter* I:1, 83.

[122]Cf. Wardhaugh.

of classical education, the four roads of the liberal arts, in which music was taught alongside arithmetic, astronomy, and geometry. In the Pythagorean tradition music was a matter of numbers as proportions and relations, while arithmetic considered numbers as units. Similarities were discerned between sounds, the elements, the motion of heavenly bodies, and the proportions of music. Music is often described in water metaphors. The very term 'music', according to Hugh of Saint Victor, comes from the word *aqua*, 'water', because no euphony is possible without moisture.[123] Music flows into the room.

There was something arithmetical and code-like, an element of universal mathematics, about baroque music that must have appealed to natural philosophers like Swedenborg, Vallerius, and others. Music, Leibniz wrote, is a secret mathematical exercise.[124] But Leibniz also praised Benzelius for his seriousness when he said he was ashamed of having wasted time once by going to the opera.[125] Even music contained metaphorical thought. The composer thought in analogies, as in the musical symbols of the theory of figures, such as the figure *kyklosis* or *circulatio*, that is to say, a melodic movement that circulates around a tone that stands for enclosure, chain, serpent, wheel, etc. At the St Thomas School in Leipzig, in the city where Swedenborg stopped on several occasions, the conductor of the choir was a man with a special taste for the sublime, Johann Sebastian Bach, who incidentally had a brother, Johann Jakob, who had been an oboist in Charles XII's army. Bach's scores conceal numerology, esoterics, secret signs, with symbolic figures using the notes B-A-C-H.[126] With Bach the polyphonic style reaches its zenith, with the fleeting, chasing voices of the fugue. There is something mechanical in Bach's rhythmically pulsating flow. Sometimes the German term *Fortspinnung* is used to describe a typical feature of baroque music, an asymmetrical principle whereby the melody, the musical idea, is spun out along unlimited falling or rising lines. A little of this constant, boundless forward motion can also be found in Swedenborg's work. He does not pause as he writes, but hastens forward in infinite variation on the same theme, in a composition without end. There is a synthetic endeavour that finds expression in the construction of a system, a holistic perspective, the doctrine of correspondences, and in the quest for associations and similarities between outer and inner, between macrocosm and microcosm, a structural system not unlike the fugues of baroque music, where a number of independent voices are interwoven, resting on a theme that shifts from voice to voice according to set principles.

Musical instruments became scientific instruments which could reveal nature's properties, its proportions, and the nature of sound. The most advanced sound instrument at this time was the organ with its pumps, controls, mechanical transmissions, and acoustically designed pipes. The organ combined subjects such as acoustics and engineering with music. In pace with technical development, compositions changed.

---

[123]Hugh of St Victor, 56.

[124]Leibniz to Christian Goldbach, Hanover, 17 April 1712. Leibniz (1734), I, 241.

[125]Forssell, 143f, 139ff.

[126]In continental musical notation, H represent B while B stands for B flat.

Tremulations and echoes were not uncommon features in baroque music. Baroque organs could have tremulants or tremolo stops, that is, devices which together with different stops make the notes vibrate to give a special drama, like the trembling flute stop known as *unda maris*—'wave of the sea'. The Dutch composer Jan Sweelinck, who was of some significance for Swedish musical history as the teacher of Andreas Düben the Elder, wrote echo fantasies for organ, and Antonio Vivaldi in Venice used echo effects such as weaker repetitions. The only extant composition by Rudbeck the Elder, *Sorg- och klage-sång till Axel Oxenstiernas begravning* ('Dirge and lament for the funeral of Axel Oxenstierna' 1654), allows echo effects.[127]

The organ is an advanced machine created by man. The world is an organ created by the most skilful organ builder, the most musically gifted organist of all—God. In Kircher's *Mvsvrgia vniversalis* the organ alludes to cosmological visions. This book in which music is regarded as a part of mathematics was purchased by Uppsala University and used for studies in higher musical theory.[128] Kircher compares the world and creation to an organ with pipes, stops, manuals, and pedals. It is God who plays the organ. In Mersenne's *Harmonie universelle* (1636–1637) it is God's hand that tunes the whole divine monochord of the universe. The harp is the symbol of harmony. A similar metaphor about the world as a musical instrument was put forward by the astronomer Sigfrid Forsius. The discs of heaven and the stars run against each other and drive each other in harmony: 'For as in an artificial organ or stringed instrument, one pipe or string harmonizes in different voices with the other, and makes a delightful melody and consensus, of thirds, fifths, octaves, etc.' as a 'sweet and delightful sound', like the spherical harmony of which Pythagoras and Plato speak.[129] The planets revolve in circular orbits around the earth at set distances in simple numerical proportions which bring out the harmony of the spheres. The crystalline music in the harmony of the spheres is in fact a metaphor adopted from the sound of a stone thrown on the ground. It is a metaphorical idea in which the harmony of music expresses the cosmic harmony. The ensemble of the cosmos could be understood through the concordance of a musical orchestra.

Coaxing sounds out of musical instruments was one of Swedenborg's great interests. During his years as a student in Uppsala he learned how to play the harpsichord, perhaps from Harald Vallerius, and in the spring of 1710, while waiting to set off on his travels in Europe, he deputized as organist in Skara.[130] The cathedral organist in Skara, Johan Ross the Younger, often neglected his duties, leading a life of drunkenness and debauchery, so Swedenborg quickly had to step in to lead the music in the cathedral. He played a 24-stop Cahman organ. While abroad, moreover, he made sketches of a universal instrument that could allow even the

---

[127] Moberg, 204.

[128] Kircher (1650), book X, 366; Kjellberg and Ling, 60.

[129] Forsius, 88.

[130] *Opera* I, 203, cf. 226; *Letters* I, 7f, cf. 57; cf. Hallengren (1997b), 10; the organ, designed by Hans Heinrich Cahman, was completed in 1704 by the organ builder Johan Åhrman and inspected by Johan Niclas Cahman. Räf, 20, 25, 27.

most musically unschooled person to play any tune. Automatic musical machines were a popular theme in books, as he could find, for example, in Kircher and Schott, and they also attracted Polhem.[131] On his travels he listened to church music and attended operas. He particularly admired the ability of Italian music to affect the senses and emotions.[132] His notebooks contain a significant number of lines about harmonies in music. Back home in Stockholm, he went to concerts given by his fellow academician Johann Helmich Roman and played Lutheran chorales on his simple chamber organ in the summerhouse.[133] Carl Christoffer Gjörwell saw Swedenborg at a concert in 1767: he 'now looked rather old, and mostly turned his eyes towards the ceiling. We all thought that he was talking to Roman', about how the master found the performance of his work.[134] Swedenborg shared his interest in music with many of the people in his immediate surroundings. There was a vibrant musical life in Uppsala at the start of the eighteenth century. His father the bishop wrote hymns, and his lodger Johan Hesselius played godly hymns every evening on the double bass. Polhem contemplated the mechanism of the organ, while Lars Roberg scraped at the viola da gamba and Olof Rudbeck the Younger plied the clavichord and wrote some hymns.[135] Harald Vallerius was a professional musician, speculated about the theory and practice of music, and performed concerts with his sons in his home.

Jesper Swedberg found the greatest of pleasure in music:

> For God has placed a very strong force in delightful and well-arranged music. And John the Evangelist tells in several paces in his Book of Revelations that there is beautiful music in heaven. He also speaks of a voice as of many waters, which also arouse and increase joy and happiness. And I confess that I feel the greatest invigoration of the mind and delight of the heart when I hear how rivers rush and cascade, and how mills run.[136]

In God's honour he composed a great many hymns for the Swedish hymnal, *Then swenska psalm-boken* (1694). With Swedberg's new hymnbook God hears our prayers and organ chorales (Fig. 10). But his hymnal was harshly criticized and was soon withdrawn for theological reasons. It was because the angels in Swedberg's hymnal cry aloud on the day of judgement: 'Up from the earth, up from the earth.'[137] This was evidence of false doctrine, ran the charge, since according to the New Testament only God can revive the dead. The angels may only sound their trumpets. But Swedberg himself claimed that the real reason for the confiscation of his hymnal was 'the mercilessness, unreasonableness, the wrath and the envy, and its origin

---

[131] Kircher (1650), book IX, 342ff.

[132] *De sensibus*, n. 487f; translation; Jonsson (1969), 221; *Note book*, 469–473, with notes from Brossard (1740).

[133] Swedenborg owned Sohren (1683); Erici and Unnerbäck, 468f.

[134] Gjörwell to Lidén, Stockholm, 1 June 1767. UUB, G 151:b, 61.

[135] Swedberg (1941), 189, 525; Dintler, 16.

[136] Swedberg (1941), 525; Up 5:8, 14:2.

[137] *Then swenska psalm-boken* (1694), Ps. 474:3; cf. 2 Cor 4:14, John 6:39, Rom. 8:11; Kleen, I, 18f.

Fig. 10 'Praised be the Lord for he hath heard the voice of my prayer.' When Swedberg preached, choirs of angels were heard singing hymns to the sounds of an invisible organ (Frontispiece of Swedberg's *Then swenska psalm-boken* (1694))

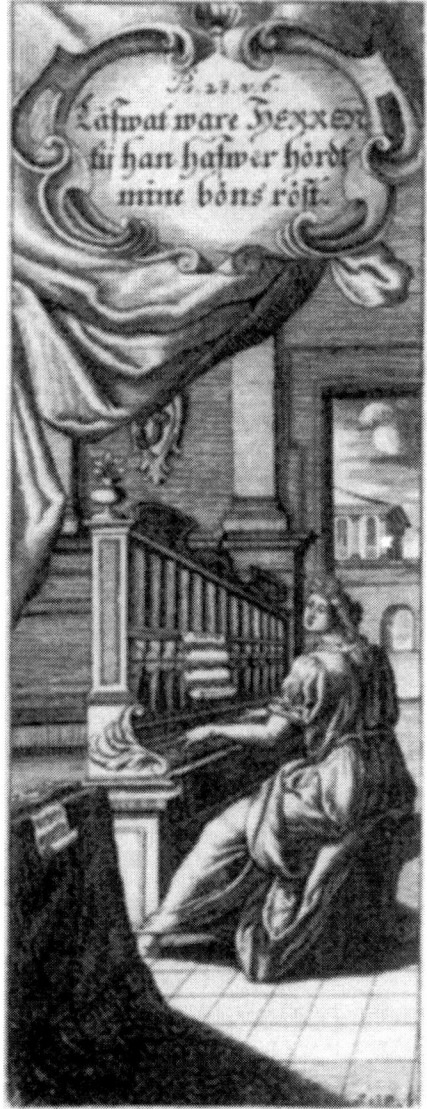

is Satan.'[138] Swedberg's book contains many hymns that make the body tremble from the roar of the organ and the eternal punishment of the last judgement: 'Ah be frightened, shudder, quake in awe, / For a long eternity!'[139] Think on the word,

---

[138]Swedberg (1941), 183.

[139]*Then swenska psalm-boken* (1694), Ps. 474:14.

eternal! eternal! The same shocking terror and fear in the human body vibrates in Lasse Lucidor's hymn about hell:

Quake, tremble, shake, my hair does stand on end,

As limbs and body quail, my thoughts descend

To the abyss, the torment there to see,

That gnaws without consuming, endlessly.[140]

Organ roar and trembling terror. Say a prayer when it thunders. In Swedberg's hymnbook there are also prayers for nature's thunder and lightning. These are not harmless natural phenomena; they have an inner meaning and are signs of something. In *Brontologia theologico-historica, thet är enfaldig lära och sanferdig berettelse, om åske-dunder, blixt och skott* ('Brontologia theologico-historica, that is, a simple doctrine and true account of thunder, lightning, and shots' 1721), Rhyzelius explains that thunder and lightning are not the work of the devil and his henchmen, nor products of mere nature; it is God who thunders, fires bolts and flashes in the sky. Based on the thundering God of the Old Testament, he warns sinful man of 'the terrible cannon shots, dreadful bangs and roars of the almighty God.'[141] It is not difficult to imagine that Swedberg's hymns vibrated and resounded in Swedenborg throughout his life. In his dreams he heard the hymn 'Jesus is my best of friends'.[142] Swedberg's Lutheran power coupled with an openness to the reality of spirits and angels and a pietist emotionality are far from alien to him. Death, the body, and music belong together. In his autobiography, *Lefwernes beskrifning*, Swedberg tells of Swedenborg's brother, Albrecht, who died just after their mother in 1696, at the age of 11. Albrecht was asked on his deathbed by his father Jesper: 'What will you do when you get to heaven?' 'I shall play all manner of stringed instruments,' Albrecht replied. 'But you have not learned how to play any,' Jesper wondered. 'I can learn there.' 'What more will you do?' '*I shall pray for my father and mother.*'[143]

If we look at Swedenborg's circle of closest acquaintances we find a musical orientation of a more secular, scientific, and mathematical character. It was chiefly the technical aspects of musical instruments and the rules of acoustics that interested Polhem. When he was repairing the astronomical clock in Uppsala Cathedral at the end of the 1680s he may simultaneously have been able to observe the work in progress on the country's biggest organ hitherto.[144] A driving force behind both the organ building and Polhem's work and studies in Uppsala was Rudbeck the Elder. Among other things, Rudbeck wanted some of the pipes concealed in the floor of

---

[140]*Then swenska psalm-boken* (1694), Ps. 479:11.

[141]Rhyzelius (1721), 14, 33.

[142]*Swedenborgs drömmar*, 13; translation, 17; *Then swenska psalm boken* (1697), Ps. 245; Bergquist (1988), 103f; translation, 136.

[143]Swedberg (1941), 230.

[144]*Polhems skrifter* IV, 392, 399; Polhem (1729), 5–10; Benzelius (1762), 243.

the gallery to create a sound resembling a distant echo.[145] Polhem wrote about the geometrical form and figure of church bells, about the proportions of organ pipes, about the undulations of the air and how they affect sound and notes. He invented a kind of metronome, 'A machine that beats and pauses the tempo of music faster or slower, as desired.'[146] In an essay on the barometer Polhem explains with musical examples, on the subject of thunder, how sound reaches our ears.[147] If you have two kettledrums and strike one, the other skin will be seen to vibrate if there is a pea on it. It is remarkable, Polhem notes, how a screaming person can make an empty glass crack, how the whole body feels the movement of a powerful organ stop, such as principal or mixture, or smaller stops such as kettledrums, trumpets or shawms. In connection with organ-building and bell-ringing he also wrote an essay on musical theory, *Om linea musica* (*c.* 1711), which deals with the harmonious mathematical proportions of music in octaves, fifth, fourths, thirds, and seconds, and about the numerical proportions of different keys with the aid of a monochord.[148] It was Pythagoras he had in mind, the philosopher who used the hammer blows of the smith and the monochord, the single-stringed instrument, to find the harmonious mathematical proportions between pitches. With the aid of 'self-moving playing-clocks' or automatic musical machines one could compensate for the shortage of organists, he wrote to Harald Vallerius, who had commented on his essay.[149] The letter is about music, although he apologizes for being so bold as to judge on a matter that he does not understand. Swedenborg's and Polhem's colleagues at the mine in Falun also engaged in music. The mine surveyor (*markscheider*) Samuel Buschenfelt designed musical instruments and is said to have been the first in Sweden to make 'hautbois and flauto dolce'.[150] His successor in the post of surveyor, Johan Tobias Geisler, was also a musician.

The Rudbecks were a music-making family. Olof Rudbeck the Elder 'played the fiddle and blew the trumpet' and sang with a powerful voice that, according to a song, made the organist Christopher Zellinger the Younger despair, since Rudbeck's powerful voice caused such an echo in his organ.[151] Rudbeck the Elder stimulated musical life in Uppsala; he composed pieces for special occasions and wrote thundering baroque musical condolences for the death of Charles XI, which made the windows rattle, as the earth quaked and was pelted with heavy rain. His son Olof struck up a mighty, thunderous fanfare at the coronation of Charles XII:

---

[145]Moberg, 192f.

[146]Polhem (1761), 124; cf. *Polhems skrifter* I, 133–135, 158–163; *Polhems brev* 69f.

[147]Polhem, 'Christopher Pohlhems anmärckningar om barometerns stigande och fallande på följande luftens förändringar', *Polhems skrifter* III, 209ff, 275, cf. IV, 264.

[148]Polhem, 'Om linea musica', *Polhems skrifter* IV, 355–367; *Collegium Curiosorums protokoll*, 62.

[149]Polhem to H. Vallerius, 1 February? 1711. *Polhems brev*, 63, cf. 65, 68f.

[150]Lindroth (1955), 683, 686.

[151]Moberg, 196.

Blast away, bugles and horns, trumpets and shrill-sounding pipes,

Blow hard, ye cornets and flutes, play on those jingling bells!

Timpanis sound out here too, with war-drums and thundering guns,

Loud violins, everything that thunders and makes the ground shake,

All play in concert with us, to honour the bold Charles the Twelfth.[152]

The black Nordic clouds of mourning had now been dispersed.

It was Rudbeck the Elder who trained Harald Vallerius in the art of music. In 1674 Vallerius defended a dissertation about the physico-musical properties of sound, and for a long time he occupied not only the chair of mathematics but also the post of cathedral organist and the university's director of music by having married the widow of the previous occupant.[153] In addition to this he held concerts in his home twice a week. In consultation with Rudbeck the Elder he adapted the melodies for the 1697 choral hymnbook, particularly the rhythmic arrangements.[154] Rudbeck is said to have assisted in the proofreading by singing through all the hymns. Harald Vallerius chaired numerous defences of dissertations concerned with musical theory. Music was examined from mathematical, physical, and psychophysical angles. Olof Retzelius defended a dissertation in music, *De tactu* (1698), which analysed the special rhythm of the *polska*, a peasant dance. Schooled musicians and village fiddlers marked the beat differently. The physical premise of the dissertation was that sound is nothing but the tremulations of bodies and the motion of air, which in Cartesian spirit is believed to consist of feather-like or branch-like particles.[155] Air waves come at intervals, affecting the eardrum and the nerves. With reference to Mersenne, rhythm is explained as a period of time like the pulse or the second hand on a clock. In the rhythm these musically interested natural philosophers could feel a regular pulse, a sense of time, as with the alchemist Oswald Croll, for whom the pulse of the blood corresponded to the motion of the heavenly bodies.[156]

In 1706 Harald Vallerius's son Göran defended a dissertation about ancient and medieval music, and another son, Johan, also a musician, succeeded Harald as professor of mathematics.[157] Johan likewise chaired a couple of public doctoral examinations on the theory of music, one of which concerned a dissertation about musical instruments including the organ.[158] Eric Burman himself wrote two dissertations on Pythagorean proportions and musical harmonies, *De proportione harmonica* (1715–1716), under the supervision of Elvius and Johan Vallerius. Music

---

[152]Rudbeck the Younger (1869), 28; Eriksson (2002), 594, cf. 578f.

[153]Norcopensis and H. Vallerius; Norlind, 126f.

[154]Norlind, 112, 127; Moberg, 207.

[155]H. Vallerius and Retzelius, 2f, cf. 14f.

[156]Hannaway, 66.

[157]Bellman and G. Vallerius.

[158]J. Vallerius and Bergrot, 12–20.

is mathematics, and vision could extend its claim even to music, as in a poetic tribute
to Burman from his friend Andreas Örström:

> Once 'twas just with naked ear
>
> We'd distinguish and compare.
>
> But now the eye can also see
>
> The ratio in harmony.[159]

Being able to *see* music, that the reality is not in the *listening*, that its core does
not lie in the sounds but in the numerical proportions, is in line with the Cartesian
tradition that reduces auditory impressions to primary properties, to mathematics.
In the notes one could see the music, in an instant one could grasp a chorale or
a fugue that in reality lasts for several minutes. Burman later became a central
figure in the musical life of Uppsala in the 1720s. Alongside his professorship of
astronomy he was organist in the cathedral and *director musices* at the university,
and in 1726 he started a Collegium Musicum together with some musical professors.
Perhaps they all found in the interwoven compositions of baroque music, like the
fugue, the mathematical beauty, the incentive to investigate sound, the metaphysical
visions of the world machine, a system of metaphors and emblems, an acoustic echo
of the divine music. The theory of harmony and acoustics was a link connecting
mathematics, music, and theology.

Fire and Colours

Water moves in waves, poetry moves in waves, sound moves in waves through
the air, and by analogy with this so too does light. For Swedenborg light is a
wave movement in which the interstellar medium consists of ether. In *Miscellanea
observata* he explains the light as nothing but the waves of its rays, that is to say,
vibrations in the ether.[160] The light in our eyes cannot be an occult phenomenon,
he says, for we find a mechanism to receive the rays. He also sees a connection
between the vibrations of air and those of ether. When the air vibrates, the ether can
start to undulate, which in turn can cause even smaller particles to fluctuate, and
so on. Air would then be an undulation in the ether caused by the vibrations of the
air particles.[161] That would explain why sound can pass through hard bodies such
as stone, iron, and glass. But where there is no air, there is no sound either. The
first impulse for Swedenborg's theory of light waves can be found in metaphorical
thought. Light can be imagined on the basis of water waves and sound waves.

Light as waves, or more exactly a mechanical pressure, can be found in Cartesian
philosophy, for instance as it is presented in Hoffwenius's *Synopsis Physica* and

---

[159]Elvius and Burman, 82; J. Vallerius and Burman.

[160]*Miscellanea*, 167f, cf. 84; translation, 104f, cf. 53.

[161]*Miscellanea*, 145, 168; translation, 91, 105.

other Swedish dissertations, such as Spole's work on light rays.[162] Descartes conceived of light as spreading out in straight lines from the sun when the ether particles, that is to say, the globes of the second element, bump into each other due to a centrifugal pressure in the solar vortex. The argumentation in Descartes' *Principia philosophiæ* constantly invokes analogies alluding to the reader's intuitive experiences of palpable reality. This was part of the explanation for the success of his natural philosophy. It can be intuitively perceived as true. He compares the rectilinear propagation, as pressure through a medium, to the way that balls tend to fall through to the lower part of a container like the sand in an hourglass. This pressure motion takes place at an instantaneous speed, in the same way as when one knocks a stick against the ground.[163] Scientists increasingly questioned whether light spreads at an instantaneous, infinite speed. In 1675 the Danish astronomer Ole Rømer was able to demonstrate that light had a finite speed by observing the changes in the phases of Io, Jupiter's moon, in relation to the position of the earth.

Swedenborg's and Polhem's theories of waves are partly a continuation of the Cartesian theory of pressure, with elements of later theories of light. Light is a wave movement or vibration with very small oscillations. Hooke had tried to explain gravitation by assuming that vibrations in the sundry parts of the earth are transmitted to other bodies through a vibrating ether.[164] Benzelius had purchased Francesco Maria Grimaldi's *De lumine* (1665) for the university library in 1703, and Swedenborg had read Milliet Dechales during his years in Uppsala.[165] The most famous name when it comes to the wave theory of light is Huygens. In his *Traité de la lumière* (1690) light was propagated through the ether and spread, like sound, in a sphere in all directions, reminiscent of the waves on the surface of water.[166] It was not a movement up and down, but back and forth, a longitudinal wave movement, a pressure wave in the direction of the wave's motion. But Huygens lacked a theory to explain colours. The main opposition to Huygens's wave theory came from Newton's particle theory, presented in 1675 and later in *Opticks* (1704). An argument against the wave theory was that light does not curve around an object in the way that water waves do. Light seems to move in a straight line. Light is thus of material substance, according to Newton, consisting of particles travelling in straight lines, and light rays are streams of particles. Swedish natural philosophers were aware that there were two conflicting theories: the particle theory versus the wave theory. They sided with the wave theory, more specifically the Cartesian pressure theory. Roberg simply dismissed Newton's doctrine as incomprehensible.[167]

Fire and colours were the subject of a remarkable exchange of opinions between Polhem and Swedenborg. In August 1722 Polhem wrote to Benzelius to thank

---

[162] Spole and Petriin, [11]f.

[163] Descartes (1644), III, § 64; *Oeuvres* VIII:1, 115, cf. VI, 114ff; Hoffwenius (1698), VIII:7f, 32f.

[164] Taton and Wilson, 227f.

[165] Liljencrantz (1968), 141.

[166] Huygens (1690), 4; Andriesse, 283–316; Dijksterhuis, 213–254.

[167] Berg, 75f.

him for publishing his essay on sound experiments in *Acta literaria Sveciæ*. But
he complained that his findings had been ignored, 'since I must bear the shame
of being like a useless rotten egg in my old age.'[168] Here in Sweden he found a
widespread scorn for both theory and practice in mathematical, mechanical, and
physical sciences. This was to the irreparable detriment and ruin of the nation. For
his own peace of mind and to pass the time, however, he had thought out nature's
causes, which could be demonstrated by purely mathematical and mechanical
means, without recourse to guesses as was the customary practice. The following
month Polhem sent Benzelius a 'Speculation' on physics, the first part of *Om
Elementernas jemvicht* ('On the equilibrium of the elements'). He wanted to hear
what the members of Bokwettsgillet thought of it, to see whether it could be a basis
for continued research.[169]

After Benzelius had read Polhem's essay to Bokwettsgillet he must have
delegated the task of reviewing it to Swedenborg. It was thus in fact Swedenborg
who was responsible for the harsh critique of Polhem. This has escaped the attention
of historians, and even Polhem himself seems to have been unaware of who he was
actually debating with. There is an essay by Swedenborg that has been erroneously
assigned to a completely different context and dated, on uncertain grounds, to
1717.[170] Swedenborg's *Om eldens och fergornas natur* ('On the nature of fire and
colours') is among the documents left by Bokwettsgillet, and in my view it can
be demonstrated without doubt to be a point-by-point response to Polhem's *Om
Elementernas jemvicht*. Nor is there any doubt that it was written by Swedenborg. It
is in his hand and it also contains typical Swedenborgian expositions of the concepts
of tremulation and undulation. It is thus obvious that Swedenborg's text about fire
and colours was not written in 1717 at all, but some time after 7 September and
before 5 November 1722, when Polhem posted his reply to the critique. A copy
of Swedenborg's response was sent by Benzelius to Polhem, presumably in the
latter half of October. But Swedenborg's involvement is not mentioned at all in the
minutes of Bokwettsgillet. This hidden polemic gives good insight into the differing
opinions about fire and colours espoused by Swedenborg and Polhem.

Swedenborg is extremely harsh in his critique. His fierce attack gives an
impression of not being entirely balanced. Their previous cooperation in matters
of engineering and science had probably not been without friction. There is ample
testimony to Polhem's unruly temper. The work on the lock at Vänersborg left
Swedenborg in a bad mood, wishing to start on a lock of his own where he himself
could be in command.[171] After the death of Charles XII their communication broke

---

[168]Polhem to Benzelius, Stjärnsund, 9 August 1722. *Polhems brev*, 145; *Bokwetts Gillets protokoll*
17 August 1722, 69f.

[169]Polhem to Benzelius, Stjärnsund, 2 September 1722. *Polhems brev*, 147; *Bokwetts Gillets
protokoll* 7 September 1722, 71.

[170]Swedenborg, *Om elden och fergornas natur* 1722. LiSB, N 14a, no. 43; *Photolith.* I, 80–85;
Opera III, 237–241; *Treatises*, 9–16; cf. Hyde, no. 114; Woofenden (2002), 30.

[171]*Opera* I, 286; *Letters* I, 193.

down completely. Polhem was worried when he wrote to Benzelius in April 1719 wondering what was wrong; three letters had been returned to him unopened, but he still hoped to get an answer from the man who was loved 'like our own dear son.'[172] Swedenborg was evidently furious—or perhaps disconsolate. The reason for this vehement reaction should probably be sought in personal and emotional rather than scientific or professional matters. An unconfirmed and perhaps not entirely reliable story claims that Swedenborg had conceived 'an irrevocable liking' for Polhem's youngest daughter, Emerentia, who was then only 13–14 years old. A contract was written between Polhem and Swedenborg, promising that she would be his in the future. But Emerentia 'was so consumed by such daily sorrow over this' that her brother Gabriel took pity on her and stole the contract from Swedenborg, 'who had no other satisfaction in his amour than the daily reading of this.' Swedenborg then saw her sorrow and relinquished his right, and 'departed from the house with a solemn oath never more to think about any woman, much less enter any liaison, whereupon he started on his foreign travels.'[173] He definitely could not keep his first oath, but he succeeded better with the second, at least in his earthly life. It is reliably recorded, however, that Polhem's older daughter, Maria, was in an 'engagement' to Swedenborg. But she took a liking for another and was betrothed to the court steward Martin Ludvig Manderström. Swedenborg comforted himself with the statement that Emerentia 'is, in my opinion, much prettier.'[174] A few years later she married the district governor Reinhold Rücker. Perhaps Polhem was referring to Swedenborg in one of his maxims: 'An early proposal of marriage, without regard for his own state and her circumstances, indicates governance by nature rather than intellect.'[175]

Whatever the psychological background to their more or less completely severed contact, Swedenborg's critique is rather merciless. Swedenborg's objections demolish the entire basis of Polhem's argumentation. Polhem works by thinking in analogies, believing that with the aid of known reality one can draw conclusions about the unknown. But Swedenborg argues that, if the aim is to arrive at certainty about the properties of fire and colours, one cannot make comparisons with the properties of water, since many things can have superficial similarities yet be essentially different. It would be better to draw on our experience of the properties of things:

> For, if we proceed by means of other elements, which are more visible to our sight and knowledge, we may soon be deceived into accepting certain notions which in themselves are contrary to the truth—just as if one were to make conclusions respecting the shape of angels from the shape of men, simply because both are living and have the general senses, such as sight, hearing, etc.[176]

---

[172]Polhem to Benzelius, Karls Grav, 18 April 1719. *Polhems brev*, 141.

[173]'Tillägg ur Exegetiska och Philantropiska Sällskapets Handlingar', in Robsahm, 75f.

[174]Swedenborg to Benzelius, Vänersborg, 14 September 1718. *Opera* I, 286; *Letters* I, 193; Christian Tuxen to August Nordenskiöld, Helsingør, 4 May 1790. *Documents* II:2, 437.

[175]Polhem, '[Maximer]', *Polhems skrifter* IV, 277.

[176]*Opera* III, 237; *Treatises*, 9f.

The remarkable thing is that this is precisely how he thinks himself, with analogies and metaphors. The only difference is that Polhem uses more mundane and concrete analogies and examples.

After the initial objections to the analogical method, Swedenborg presents point-by-point arguments against Polhem's ideas about what fire, light, and colours actually are. Fire, light, and heat are regarded by both Polhem and Swedenborg as closely related phenomena. Fire, Polhem assumes, has to do with the elasticity of air and its liberation 'from its prison', as when dammed-up water is released.[177] It is true, to be sure, that fire 'is nourished' by air, Swedenborg replies, but one cannot assume that the actual expansion or 'bulging' causes fire.[178] They seem to be on the track of the nature of combustion, guessing that there is something in the air that causes the burning. Several experimental facts argue against Polhem's opinion on fire, Swedenborg continues. If one places a candle in a pump from which the air is removed, the flame dwindles in proportion to the amount of air that is pumped out. On high mountains where one can scarcely breathe, the flame of a candle is extinguished or dimmed because the air is thinner; one can also feel a penetrating cold there, in contrast to the greater heat at sea level.

The non-existence of the vacuum was virtually taken for granted by most Swedish natural philosophers. The Cartesian world-view with its *horror vacui* required the presence of matter, more specifically ether, in the air pump. Harald Vallerius, for example, defended a dissertation on the vacuum in which he followed Descartes in his opinion that no vacuum can exist.[179] It is inconceivable and impossible to produce a vacuum in an experiment precisely because space is identical with matter. What the air pump shows instead is that there is an association between respiration, life, and fire. It is found that animals die, candles go out, and sounds disappear in it. They all need something from the air. The air pump became a symbol of the new natural philosophy and a standard feature of the apparatus used for scientific demonstrations and public lectures. With the air pump one could show that sound needed air, as in the experiment mentioned in *Dædalus Hyperboreus* when Kircher tried to ring a bell in a vacuum.[180] As for the question why light remains and can still travel through the pump, the general answer was that it was due to other kinds of particles, namely, ether.

In the third issue of *Dædalus Hyperboreus* Swedenborg published two articles about the air pump. He acknowledged Boyle and Guericke as the creators of the air pump and Hauksbee as its improver. Polhem frequently cited the experiments with copper hemispheres from 1654 by Otto von Guericke, mayor of Magdeburg, as an example to explain the cohesive force of matter.[181] In a transcript that Swedenborg made of Polhem's *Discours mellan Mechaniquen och Chymien om Naturens*

[177]Polhem, 'Om elementernas jemvicht [1]', *Polhems brev*, 152.

[178]*Opera* III, 237.

[179]Steuchius and H. Vallerius; Eriksson (1974), 112.

[180]Kircher (1650), book I, 11–13; cf. *Dædalus Hyperboreus* I, 10; *Polhems brev*, 113.

[181]*Polhems skrifter* III, 82, 168; Schott (1664), 39.

wäsende ('Discourse between Mechanics and Chemistry on the Essence of Nature', 1718), Miss Mechanica tries in Polhem's typical way to reason on the basis of 'the coarse things that we can grasp.' A lucid example is the Magdeburg experiment, in which not even several pairs of horses could separate two hemispheres from which the air had been pumped out. Mechanica sums up by saying that whatever has water outside and air inside, or air outside and ether inside, or ether outside and fire inside, 'holds together as if it were pitched and glued.'[182] On 23 March 1711 Swedenborg had visited the instrument maker Francis Hauksbee in Fleet Street in London and bought a copy of his *Physico-mechanical Experiments* (1709).[183] Swedenborg says that he was also given a description of the accessories for the air pump, 'of which I have [the] *author's original*.'[184] In Hauksbee's trials with the air pump, Swedenborg wrote in 1720, there 'are many fine experiments in respect to fire, the magnet, etc., in *vacuo vel in moto*.'[185] It is above all on Hauksbee that Swedenborg bases his articles about the air pump in *Dædalus Hyperboreus*.

'Om en Wäderpump. Eller Antlea Pneumatica' ('On an air pump, or Antlea Pneumatica') describes the principles of the barometer and the air pump and gives instructions for building one of these coveted but expensive air pumps. The principles of the barometer, it says, rest on Galilei's statement that water cannot reach higher than 30–31 ft, or about 9 m. Both Swedenborg and Polhem know from their own experience in draining mines that it is not possible to raise water further than this with a pump. In 1711 Göran Vallerius had performed trials with barometers in the Fleming Shaft and at Gruvrisberget in Falun to investigate air pressure and measure the depth of the mines.[186] Swedenborg writes that if a pipe full of water was turned upside-down in a bath of water, the column reached 30–31 ft and 'left all the other space empty'.[187] But in a barometer consisting of heavier matter such as mercury, the column reached only one fourteenth of the height of the water. Balance is maintained by the air pressure. The barometer, the instrument for measuring atmospheric pressure, showed that air has weight and presses us down. For Evangelista Torricelli, who constructed the first mercury barometer in 1644, we live on the bottom of an ocean of the element air.

The second article, 'Utrekni[n]g och afmätning för watnets och wädrets rymd och högd i sådana antlior' ('Calculation and measurement of the volume and height of water and air in such antlia'), likewise fails to take any clear stance on the true existence of the vacuum. In talking of a vacuum he means only the space above the water which contains neither air nor water: 'By the empty space is

---

[182]*Opera* III, 255.

[183]Hauksbee (1709). UUB; Beckman, 189; *Opera* I, 220; *Letters* I, 41.

[184]Swedenborg to Benzelius, London, August 1711. *Opera* I, 214; *Letters* I, 28f.

[185]Swedenborg to Benzelius, Brunsbo, 2 May 1720. *Opera* I, 304; *Letters* I, 237; cf. *Miscellanea*, 49, 101f; translation, 32, 64f; Swedenborg undertook to order an air pump from Hauksbee for the Board of Mines in 1725. *Letters* I, 367f.

[186]UUB, G. Vallerius; Lindroth (1951), 90, 192.

[187]*Dædalus Hyperboreus* III, 51.

understood here all the space that is above the water; both that of the receptacle and the antliae themselves. If the water stopped at height EE [...] then E P A B F was called the empty space: being the figure which is even or distorted, conical or cylindrical'[188] (Fig. 11). The experiments with the air pump also testify to Swedenborg's theoretical stance on science, in a quest for general rules that proceeds from an interplay between induction and deduction. First adduce the general rules from the individual cases, and then, conversely, draw conclusions about the individual cases through the general rule.

As regards colours, according to Polhem they arise from motion. Stagnant water and air lack colour, but when they start to move and are mixed, a foam of white colour arises. The expansion of air, its rapid movement through the ether, creates colours and fire. Swedenborg questions this idea by referring to experimental facts.[189] If one pumps air out of an air pump, so that only ether remains, and then lets the air flow back into the ether, no light or heat arises at all. No, colours have quite a different cause, in Swedenborg's opinion. They are not a phenomenon of motion but of refraction. Swedenborg thus rejects not just Polhem's theory of colour but also the tennis-playing Descartes' explanation in *La dioptrique* (1637), which used bouncing tennis balls to show how a prism caused the balls to spin and thereby acquire a colour they had not previously had. It was a mechanistic explanation based on the assumption that the speed with which the ether particles rotated caused the colours. It was thus possible to explain colours as measurable primary properties. Newton's *Opticks*, or the Latin edition *Optice* (1706) that Swedenborg learned of during his stay in London, if not before, explains colours instead as a refraction phenomenon.[190] Partly with the assistance of the law of refraction discovered by Willebrord Snellius, Newton was able to demonstrate that ordinary white light could be broken into different colours through a prism.

The year before the critique of Polhem's colour theory, Swedenborg had touched on colours in *Prodromus principiorum*, where he seems to have rejected the wave theory and approached Newton's particle theory of light.[191] All particles in the sublunar world, both hard and soft, are by nature transparent, such as ether, fire, air, water, salt, and so on, even the hardest metals ordered in a regular position are transparent, as when gold is dissolved in aqua regia. Air particles, he says, can pass through all bodies in nature. This theory can be demonstrated geometrically. Whiteness is caused by the reflection of the rays in many different angles, and redness arises when the bubbles in matter have two different kinds of particles on the surface. Yellow is a mixture of white and red. Finally, he promises a theory of light and colours. In his collection of excerpts about magnets he likewise refers to a manuscript on the subject of light, *De lumine*.[192] Apart from some scattered notes

---

[188] *Dædalus Hyperboreus* III, 58, cf. 60.

[189] *Opera* III, 238; cf. *Polhems brev*, 152.

[190] *Photolith.* II, 278f.

[191] *Prodromus principiorum*, edition, 99f; translation, 127f.

[192] *De magnete*, 264.

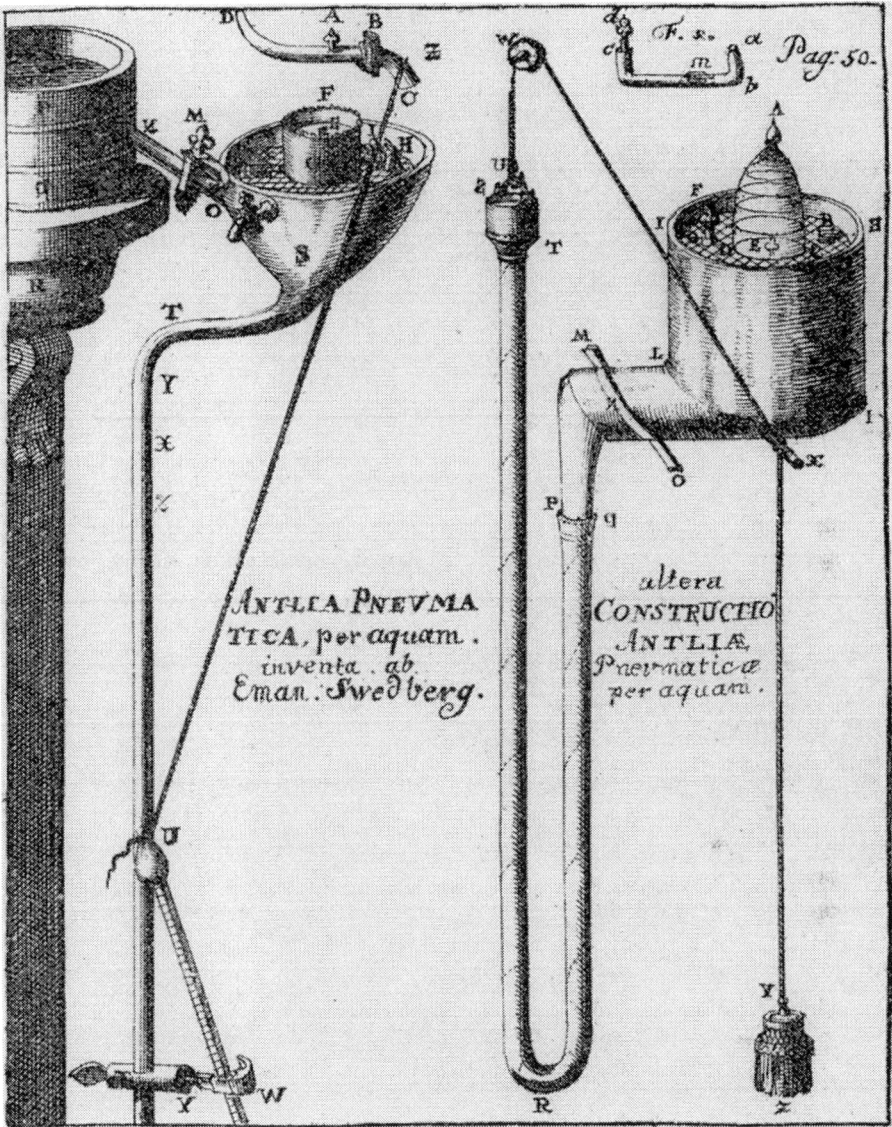

Fig. 11 The air pump to the *left* contains water and is made of copper or tin. The *narrow pipe* is 33–34 ft long, from T to y 31 ft. The *lower part* is placed in a bath of water. The principle is the same for an air pump with mercury, although it is smaller. To pump out air, place a glass in front of E, fill the bath Pq with water, open key M and let the water come down through pipe NO to the actual air pump. The air is then forced out through valve K. When all the air is out, close M again and draw VW upwards or open the lower pool. The water then runs down into the pool until it stops at a height that is in proportion to the air pressure. The procedure is repeated until all the air has been drawn out. The air pump to the *right* is based on the same principle (*Dædalus Hyperboreus* III (1716))

about optics, there is no such manuscript among the papers he left behind; this hints
at the existence of a vanished manuscript which has escaped the attention of his
bibliographers. Colours, as Swedenborg claims in response to Polhem, are the same
in both cold and warm spaces, in motion and at rest alike. They are caused by angles
of refraction and actually have nothing to do with the motion of waves. For example,
a soap bubble, an example also used by Newton, can display many colours if one
simply varies the vantage point or the refraction, as can be seen in experiments with
prisms. The white foam of water can be explained differently. If one looks more
closely at the foam in a microscope it is seen to consist of a great many uneven
bubbles. It is the unevenness of these bubbles, that is to say, the irregularity of the
particles, that causes the white colour. All other colours, on the other hand, require
regular particles. Polhem's reply is that it is the roundness that causes the whiteness:
'That coloured liquor gives white foam, and ice shavings white snow, comes from
the fact that everything around *in minimo* gives a white colour.'[193]

For Polhem each element corresponds to a particular sense: ether to our finest
sense, vision; air to hearing; and earth, mixed with water and air respectively,
corresponds to taste and smell. Fire, which is assumed to be of a more subtle
nature than ether, nevertheless gives rise to a coarser sense than vision: touch.
In the same way as the rushing of water gives rise to the movement of air and
produces a sound for our ears, so the pouring of air gives rise to the motion of
ether and produces a light for our eyes. There is no other cause of hearing than the
motion or tremulations of air. Nor is there any cause of vision than the motion and
undulations of the ether. Water can demonstrate this as a coarse model, according
to Polhem, 'when one considers its rings and small waves when it is gently
touched'.[194] Swedenborg admits that it is difficult to refute Polhem's statement that
the undulations of ether give rise to visual impressions, as the waves of air give rise
to auditory impressions, which he considers to be an 'ingenious speculation'.[195]
But Swedenborg has several objections to the idea that light is caused by 'the
streaming of the air' setting ether in motion. Light can shine where there is no
air, as for example in the ether between the sun and the earth's atmosphere, in
the vacuum pump, or in water. Swedenborg also finds it appropriate to make a
distinction between the terms 'undulation' and 'tremulation'.[196] Water waves and
the air waves of sound are of the former kind, that is, when a whole mass of
particles vibrates up and down, or back and forth. A 'tremulation', in contrast, is
the movement that takes place, for example in the air, when each particle vibrates
separately and not as a mass, or when one particle hits another. If the tremulation
occurs in the air it is not a sound that arises but an undulation in the smaller particles
of ether. 'But this, like the former, is an hypothesis which needs to be proved by

[193]Polhem, 'Förklaring på anmerkningarna mot Eldenss warelsse, och det i samma ordning punctwijss', *Polhems brev*, 167.

[194]*Polhems brev*, 153.

[195]*Opera* III, 239; *Treatises*, 12.

[196]Cf. *Miscellanea*, 143f; translation, 90.

many hundreds of experiments.'[197] The method is first to present a hypothesis, then perform experiments that can corroborate or disprove the hypothesis. Swedenborg's method is thus the hypothetical-deductive one, a method that is not chiefly geared to how we discover scientific knowledge, but rather to how we justify it. In *Prodromus principiorum* one can see precisely how he thinks in a hypothetical-deductive manner, when he more or less consistently makes a clear distinction between a priori and a posteriori propositions, in that order. First rational propositions that are independent of experience, then propositions proceeding from experience.

Water can spread a mist or vapour, which when viewed against the sun gives multicoloured rainbows. In the same way, Polhem explains, fire spreads heat around in the air. The comparison between heat and steam, Swedenborg replies, is a mere simile and nothing upon which to build a theory. A theory is better founded on experiment than on analogies: 'for by analogies one may catch at many ideas and theories which nature herself disproves in the experiments.'[198] He also points out with reference to the rainbow—indicating his familiarity with Descartes' explanation in *Les météores* (1637)—that the colours in the rainbow arise when ether or light rays are refracted through the round drops of water. The angles of refraction cause different colours. Polhem's comparison of the fire in fat and saltpetre to water rushing through sand, stones, and clay seems to reveal as many differences as similarities, Swedenborg says. Fire not only penetrates into the spaces between the particles in fat, oil, saltpetre, and other substances, but also derives nourishment from them. There is in them 'a matter which gives birth to fire'.[199] One simply cannot use a comparison with water running through sand, stones, and clay. But perhaps a different simile could be found? The matter that feeds fire which Swedenborg has in mind derives from two chemical authors that were well represented in his library. The allusion is to the phlogiston theory of Johann Joachim Becher, and above all Georg Ernst Stahl from the start of the eighteenth century, according to which all combustible substances contain phlogiston, the principle of combustibility, which Stahl also claimed to be the cause of the vortical motion of fire.

Swedenborg bypasses points six to nine and goes directly to point ten. Here Polhem claims, with data taken from Boyle's experiments, that ether particles are 556,000 times smaller than those of air, and air particles 1,000 times smaller than those of water.[200] Their speed also differs in the same proportions. There are experiments, Swedenborg replies, which indicate that ether is 55,600 times lighter than water, but this need not mean that ether is faster or more subtle in the same proportion. Swedenborg then suggests performing experiments with hollow balls of iron and wood. The particles of the elements behave in the same way as these balls. This shows that the particles of two different elements can be the same size yet

[197]*Opera* III, 239; *Treatises*, 13.

[198]*Opera* III, 240; *Treatises*, 13; cf. *Polhems brev*, 153.

[199]*Opera* III, 240; *Treatises*, 14; cf. *Polhems brev*, 168.

[200]Cf. *Polhems brev*, 12, 22, 168; *Polhems skrifter* III, 217f.

differ in weight and motion. Two different particles can also have the same weight but differ in velocity. This means that fire particles can have a much greater speed than air particles while simultaneously being just a fraction smaller in size. Finally, the speed of motion, around a centre or in a tremulation, depends on figure and size. The rounder or smaller the particle is, the faster is the motion. The reason is geometrical. A perfect little sphere has a smaller contact surface and thus encounters less resistance.

Swedenborg then tires of Polhem's analogies. Polhem's line of thought and his method of combining rational calculations and dizzying speculations is reminiscent of the grandiosity that pervades Rudbeck's visions of Atlantis. The subsequent analogy is the most challenging of all: 'When one compares the size of the sun with a thick root of dry wood that burns best, and the time it takes for such a root to burn up, one obtains a sum of years for the sun which consists of 28 figures, a length that is scarcely inferior to eternity.'[201] Scarcely any cosmic temporal perspective before or since has been as vertiginous as this. It must be the highest estimate by any scientist of the total length of the sun's life: at least a billion billion billion years. A few decades later Georges Louis Leclerc de Buffon, who is given the honour of having discovered the dizzying extent of geological time, managed in a similar experiment with cooling cannonballs to arrive at an age of the earth that is extremely modest in comparison, 74,832 years.[202]

Polhem sent a continuation of *Om Elementernas jemnvicht* to Benzelius on 18 September, but he had not yet received Swedenborg's comments. Polhem says that he has now proved that fire is not matter in itself but a 'transformation' with the aid of the motion between air and ether.[203] From this it can be concluded that ether is the first element, since we have no knowledge, whether from experience or reason, of the existence of any finer matter apart from the essence of 'God's angels and of souls', which cannot really be of any finite matter. Ether is the universal matter that fills everything. It was not until 5 November that Polhem responded to Swedenborg's critique in a letter to Benzelius. The first objection was the one that hurt most, that comparisons and analogies do not suffice as proof. 'If I am not allowed to use analogies and comparisons then I do not wish to reply to anything,' he writes, since it is 'for my part the only support and foundation on which everything else should be grounded.'[204] This reminded him of a commander who was reproached for not having fired a salute for a superior and defended himself with many excuses, one of them being that he had not had any gunpowder. That single reason was sufficient. Polhem claims that he has been totally misunderstood. In the appended manuscript *Utan jemförelsser och liknelsser komer man inte långt i Physicen* ('Without comparisons and analogies one will not get far in physics') he writes that if what one sense perceives may not be compared with what two senses

---

[201] *Polhems brev*, 154.

[202] Lachièze-Rey and Luminet, 150.

[203] Polhem, 'Om elementernas jemnvicht [2.]', *Polhems brev*, 157.

[204] *Polhems brev*, 161; *Bokwetts Gillets protokoll* 16 November 1722, 76.

testify to, such as vision and touch together, then the whole of our natural philosophy would consist of mere words and nothing else. How could one find the undulations of air that cause sound if one were not permitted to compare them with the rings on water? How could one otherwise find the difference between the colours? How could one find the refraction of light? Analogies teach us to understand, they have a didactic task:

> One could scarcely find out how habit can make pleasant the things that Nature has made unpleasant, if one were not allowed to compare the property of what our coarsest sense, touch, can display, namely, as a wire brush on the bare body itself is not very pleasant but through gentle stroking with it arouses an itch that tolerates the same brush better and better, just like bitter tobacco in the nose and mouth, a loud sound in music that livestock cannot bear to hear, and the sight of shiny gold that at first is painful to the eyes but then all too dear.

Without analogies and comparisons we remain as wise as we were before. If I am allowed to use mechanical analogies and comparisons, says Polhem, no question in physics will be too difficult for me. I find it hard to express my opinion, he concludes, 'since new things require new words, which I do not yet know myself.'[205]

Swedenborg's critique was painful to Polhem. On 16 November he wrote a second response, *Förklaring på anmerkningarna mot Eldenss warelsse, och det i samma ordning punctwijss* ('Explanation for remarks against the nature of fire, and in the same order point by point'). Polhem tries to counter all the objections about fire, colours, and light. He stresses that without eyes there are no colours. Their transformations are dependent on the angles of refraction. Colours and the different sizes of things appear differently in different eyes. In small pupils, things appear larger. Polhem has a rather interesting response to Swedenborg's objection to mixing the weight of elements with their size and velocity. He refers once again to Boyle, who had concluded that the ratio of the largest to the smallest form of air is 556,000 to 1. With the aid of the air pump and the barometer it had been found that air at sea level is almost a thousand times lighter than water. Since it had also been ascertained that sound has a speed of 1,000 ft/s, while the rings on water spread at roughly one foot per second, one could draw the conclusion from the mechanical theory that the undulations of the ether, or vision, must display the same proportion, 556,000,000 ft/s. Polhem thus estimated the speed of light at 165,076,400 m/s. This is an amazing speed, yet still lower than the results obtained by both Huygens and Rømer. He concludes the debate by writing that, based on the capacity of air for compression and elasticity, one can demonstrate anything at all mechanically, all chemical effects and operations. The notion that fire could be a substance 'is wholly absurd for those who understand the matter differently.'[206]

---

[205] Polhem, 'Utan jemförelsser och liknelsser komer man inte långt i physicen', *Polhems brev*, 163f.
[206] *Polhems brev*, 169; *Bokwetts Gillets protokoll* 23 November 1722, 76; Polhem (1731), 23–35; Lagerbring, IV:3, 53.

One Membrane Trembles from the Other's Trembling

The waves of water, the air waves of hearing, and the ether waves of vision have something to do with the very essence of life. Wave flow through all of Swedenborg's anatomical-physiological writings from this time. His is a mechanistic, Cartesian explanation of mankind. In an article in *Dædalus Hyperboreus*, 'Bewis at wårt lefwande wesende består merendels i små Darringar thet är Tremulationer' ('Proof that our living being generally consists of small quivers, that is, tremulations' 1718), he puts forward the idea that all organic life consists of tremulations of some kind. These tremulations or vibrations apply not just to the circulation of the blood, the nervous system, and the senses, but also to all the activity of the human mind and spirit. He starts by postulating nine rules of tremulation proceeding from different physical experiences. It is close to the deductive structure of the geometrical method using axioms: 'Before what is unusual and unknown can be made credible, it is necessary to establish some fixed and indubitable rules, according to which the theory may be proved.'[207] Yet Swedenborg is not very consistent in his deductions. It is not wholly evident that his conclusions must follow from the nine rules of tremulation.

The first rule states that the slightest blow can cause great tremulations in hard materials, just as a wagon can make big castles and mountains shake, or as a gentle tap on a walking beam can be heard at a distance of several miles, or when the strings on an instrument resound. A shot can set entire houses and towns aquiver. He then postulated that hard membranes tremble best, that hard objects tremble better than soft ones. The fourth rule states that trembling in one object can cause another object to tremble, like two lute strings with the same note, or an organ that can make a pulpit and a whole gallery vibrate with it. Even a glass can be cracked by a sound.[208] Fifthly, sound spreads in all directions, as when one throws a stone into water and rings spread from the centre. Water rings spread slowly, air waves faster, and ether even faster, so that light from the sun, as in Descartes, arrives here in an instant. A more problematic rule is the seventh one, which states that one tremulation does not prevent another. Two water waves or sound waves do not affect each other, he says at a time when the interference of light had not yet been explained. The tremulations nevertheless follow the rule that the angle of excidence is the same as the angle of incidence. It comes back in the same direction, like a rope in a mine shaft which moves up and down in 'serpentine coils', or like the rebounds of an echo. Finally, he declares that there are millions of different tremulations. A well-tuned clavichord can produce many different sounds. The sounds can be 'more flowing, broader, duller, or harsher. The sound and pronunciation of men differ like their faces.'[209]

---

[207] *Dædalus Hyperboreus* VI, 10; translation, *On tremulation*, 1.

[208] Cf. Morhof; Linder, chapt. III.

[209] *Dædalus Hyperboreus* VI, 12; translation, *On tremulation*, 4.

From these rules Swedenborg is able to arrive at astounding conclusions. Life consists of small tremulations. From the first rule we find that a single tiny particle in the body can set membranes, strings, blood, and the vital spirits in motion. Human speech is a tremulation, and hearing consists of tremulations running in through cochleae and membranes to the hammer and the anvil and on to the brain. Smell and taste are nothing but the movement of round, angular, and sharp particles which cause fibres and nerves to be moved in sympathy. Sight, 'which is the most delicate of our senses', is likewise a tremulation conveyed by the visual nerve. The sense of touch also consists of tremulations, just like motions of temper resulting from 'a stinging or biting of the bile'. If this trembling membrane becomes slack, it loses 'its heat, its blood, or its animal spirits, then the whole man becomes dull, heavy, and dead.'[210] When the nerves that have been tensed during the day relax, people fall asleep. This was also how Descartes conceived it. Tense fibres indicate wakefulness, while lax fibres mean a sleeping brain.[211] If there is damage to the membrane of the brain, where all the senses converge, Swedenborg says, then feeling is lost. A person faints, loses his capacity for thought and reason. If there is an overabundance of vital spirits, as in drunkenness, wild tremulations take the place of the normal ones.

Thinking is a kind of tremulation too. There are thought vibrations. Even 'the thoughts of the unreasoning animals are tremulations,' Swedenborg writes, thus distancing himself from the Cartesian view of animals as unthinking automata. Thoughts can move through space in the form of tremulations, like the wave movements of sound and light. It often happens, he says, that one falls into another person's thoughts, that one anticipates what another person will do and think. This is because one membrane trembles from the other's trembling, as when a string is made to vibrate by another when they are tuned to the same note. The possibility of working out other people's feelings and thoughts had occurred to him previously. In a letter to Benzelius from 1714 he says that he has invented a new analytical method for guessing people's wishes and moods.[212] He may have been on the track of a universal mathematics for thoughts. The science of the time did not find it wholly impossible that thoughts could fly through space. At his disputation Erik Odhelius, a doctor of medicine and uncle of the librarian Benzelius, defended a large number of arguments that special mental movements can pass between the bodies of friends even though they are in quite different places.[213] Polhem shared this opinion. Vital spirits could move through space, like sound. Solid matter moves our sense of touch, water with its salt gives taste, air with its volatile salt causes smell, the tremulations of air allow hearing, the undulations of ether bring vision, Polhem speculates by analogy with this in his *Tankar om andarnas varelse* ('Thoughts about the essence of spirits'), so 'why can some more subtle matter than ether not touch our thoughts or its organs in the brain?' It would otherwise be difficult to explain sympathies and

---

[210]*Dædalus Hyperboreus* VI, 13; translation, *On tremulation*, 5f.

[211]Descartes, *Oeuvres* XI, 174f, 197–199; Gaukroger, 23f.

[212]*Opera* I, 226; *Letters* I, 58; Jonsson (1969), 59.

[213]Bilberg and E. Odhelius; Spole and L. Odhelius, 9f; translation, 481.

antipathies and other things in nature if one were not permitted to assume that even thoughts have their matter. Two 'good friends can sense each other at a distance of many miles, especially children and parents, wife and husband, so that when some sorrow, mortal dread, or even great joy befalls one,' the other can sense it without knowing whence it comes.[214] Ghosts and shades of the dead can be explained in this way as the movement and effect of the finest mental matter.

The Beautiful Geometry of Tremulation

Swedenborg wrote a longer version of these tremulation theories which was given the title *Anatomi af vår aldrafinaste natur, wisande att wårt rörande och lefwande wäsende består af contremiscentier* ('Anatomy of our very finest nature, showing that our moving and living being consists of contremiscences' 1720). During the summer of 1719 he had the peace he needed to write some essays, including an '*Anatomie af wåra lifsrörelser*' ('Anatomy of our life-movements'). But 'I think,' he wrote to Benzelius, that this is 'likely to be my last word, since all such speculations and arts are unprofitable in Sweden and are esteemed by a lot of political blockheads as a *scholasticum* which must stand far in the background while their supposed finesse and intrigues push to the front.'[215] Through detailed studies of the anatomy of the nerves and membranes he had arrived at the conclusion that life consists of contremiscences. He wanted to demonstrate harmony with 'the beautiful geometry of tremulations'. His reasoning would accord with Giorgio Baglivi's theories. The medical scholar Baglivi from Ragusa (Dubrovnik) was a radical iatromechanic with a particular interest in the vibrations of fibres. The body machine consisted of a large number of smaller machines interacting. He compared the teeth to scissors and the blood system to a piped water system. To William Harvey's discovery of the circulation of the blood Baglivi wanted to add his own theory of a circulation of fluid in the nerves caused by contractions of the dura mater, the hard membrane of the brain. It was especially Baglivi's *De fibra motrice et morbosa* (1700) that caught Swedenborg's attention, a work to which he would frequently refer in later neurological studies.[216]

Swedenborg sent his geometry of tremulations to the influential medical society Collegium Medicum in Stockholm, where it was considered at a meeting in the late autumn of 1719.[217] The College decided to pass a verdict on it and began to circulate Swedenborg's essay among the members. Since then it has not been found, nor are there any further references to it in the documents of the College. Presumably it got

---

[214]*Polhems skrifter* III, 313f.

[215]Swedenborg to Benzelius, Stockholm, 3 November 1719. *Opera* I, 291f; *Letters* I, 215.

[216]*Addenda ad Transactiones de cerebro*. KVA, cod. 57, 42a–44a; *De cerebro*, translation II, n. 218; Baglivi, 237–394; Sturm (1685), II, 192–198.

[217]RA, Collegium medicum, A1A:2, 684f, no. 294. Minutes 30 October 1719; RA, Collegium medicum, A1B:1, no. 294.

lost while it was being circulated. However, Swedenborg wrote yet another version that he began to send in batches to Benzelius at the start of 1720. The extant version, which contains the first six chapters, and the thirteenth, should thus be properly dated to 1720 and not, as hitherto, 1719.[218]

Swedenborg wrote to Benzelius about his new theory in February 1720: 'It is true that Baglivi did indeed first open up the idea; Descartes likewise touched on it somewhat; later Borelli.'[219] But no one has proved and accomplished the matter as he himself had done, he asserts. The proof is new and his own, but opinions change. The medical authors he mentions clearly show where he is headed. They are all typical representatives of the iatromechanical school. The sequence of Baglivi–Descartes–Borelli does not conform to the historical development of iatromechanical thought, but is more concerned with his own inner line of thought, how Baglivi's ideas about the function of the meninges had first stimulated his imagination, which he then blended with classical Cartesianism and supplemented with Borelli. The three authors are combined in an explicit mechanistic description of the human body. Descartes regarded the body as a machine, an automaton manufactured by God's hand, but of much better design and more admirable than those made by man.[220] This iatromechanical thesis, incidentally, opened a dissertation defended in 1728 by the Swedish medical student Nils Rosén under the supervision of Roberg, about the mechanical method in medicine. Descartes compares the body to clocks, mills, and cogs, and above all to hydraulic and pneumatic machines, a metaphor in which liquids are pressed through pipes. In *De homine* (1662) he compares the circulations of the vital spirits in the brain and the nerves, that is to say, the neuromotor system, with a church organ and the hydraulics of fountains, as the body's own water organ.[221] The transmission of sensory impressions and the movements of muscles take place with ropes, levers, and pulleys. The heart is a fire machine, as when heat arises from rotting hay. The professor of mathematics Giovanni Alfonso Borelli is perhaps the most radical iatromechanic of them all. In *De motu animalium* (1680–1681), which was discussed in Collegium Curiosorum, he applied mechanical explanations not just to the external and visible muscular movements but also to the body's inner movements such as those of the heart the lungs, and the intestines.[222] Borelli had no firm opinion about 'nerve juice' or the mechanics of nervous impulses. What was transmitted through the nerves could be an immaterial force, steam, wind, liquid, movements, or impulses. Here Swedenborg instead presented a firm opinion: tremulations.

---

[218]Cf. Hyde, no. 130; Woofenden (2002), 36.

[219]Swedenborg to Benzelius, Stockholm, around 11 February 1720. *Opera* I, 297; *Letters* I, 227; because the younger Baglivi is mentioned before Descartes, Jonsson has suggested that 'Cartesius' could be read as Cortesius, thus referring to the Italian physician Giovanni Battista Cortesi. Jonsson (1969), 63f, 328f; Jonsson (1997), 99.

[220]Descartes (1637), V; *Oeuvres* VI, 55f; cf. Duchesneau, 122f; Des Chene; Roberg and Rosén, ii.

[221]Descartes, *Oeuvres* XI, 165f; Gaukroger, 189, 198f.

[222]*Collegium Curiosorums protokoll* 10 April 1711, 63.

Swedenborg reported that the doctors in Stockholm had expressed a favourable view of his ideas about tremulations, but he was still waiting for the return of the manuscript after the assessor Bromell had read it too. Benzelius took the newly written pages about tremulations to meetings of Bokwettsgillet, where they were also read aloud. At its meetings there were often readings from some topical book or newly written manuscript. This oral presentation of science played no insignificant part in the society. Reading aloud forces the person doing the reading to be accurate and fix on text. At the same time, the listeners lose some of the freedom they would have with silent reading in private. The accentuation, the rhythm, the stress, the melody are different, and there is no opportunity to reread a passage. In this way the assembled people are united in one reading, and are partly driven towards the same interpretation, which gives the text a unity in time and an existence in space. There is very little information about how the members of Bokwettsgillet, such as Benzelius, Bilmarck, Burman, Martin, and Roberg, received Swedenborg's ideas about tremulations. But united in a style of thought with shared metaphors, and with communal reading, they could understand him and arrive at similar interpretations of the text.

The first chapter in Swedenborg's *Anatomie af wår aldrafinaste natur* was read aloud on 12 February 1720, and the reading continued during the spring as Swedenborg submitted new contributions.[223] In one of the letters to Benzelius he makes it clear that theories based on vital spirits and chemical explanations, but without geometry, are insufficient. Instead he posits a theory in which the tremulations begin in the fluids of the membranes and then spread to the bones, so that almost the whole body starts to quiver gently in sympathy, resulting in a sensory impression.[224] At the end of February he had still not finished the passage about the mechanics of the passions and emotions, which he believed could be derived from the structure of nerves and membranes.

At the following meeting on 4 March, Roberg spoke about the illustrations in the German edition of Raymond Vieussens' *Nevrographia universalis* (1690) and Govard Bidloo's *Anatomia hvmani corporis* (1685) (Fig. 12). These two works provided Swedenborg with much of his sensory experiences of the interior of the human body. It was through anatomical pictures, which were oriented towards the optical and realistic—as the organs were to be seen with the eyes, not heard through texts—that Swedenborg saw mankind. His autopsies were done in the tidy plates illustrating books, not in the blood and stench of a half-rotten reality. At the same meeting the reading of Swedenborg's ideas about the tremulations of the nerves continued. Roberg commented that Swedenborg should be asked the tricky question how the trembling movement in one and the same nerve can go from the brain to the outer parts of the body and simultaneously go in the opposite direction.[225] After Swedenborg had gone to Brunsbo in March-April he had not had time to write

---

[223]*Bokwetts Gillets protokoll*, 15f; *Opera* I, 299f; *Letters* I, 230.

[224]Swedenborg to Benzelius, Stockholm, 24 February 1720. *Opera* I, 298; *Letters* I, 228.

[225]*Bokwetts Gillets protokoll* 4 March 1720, 18.

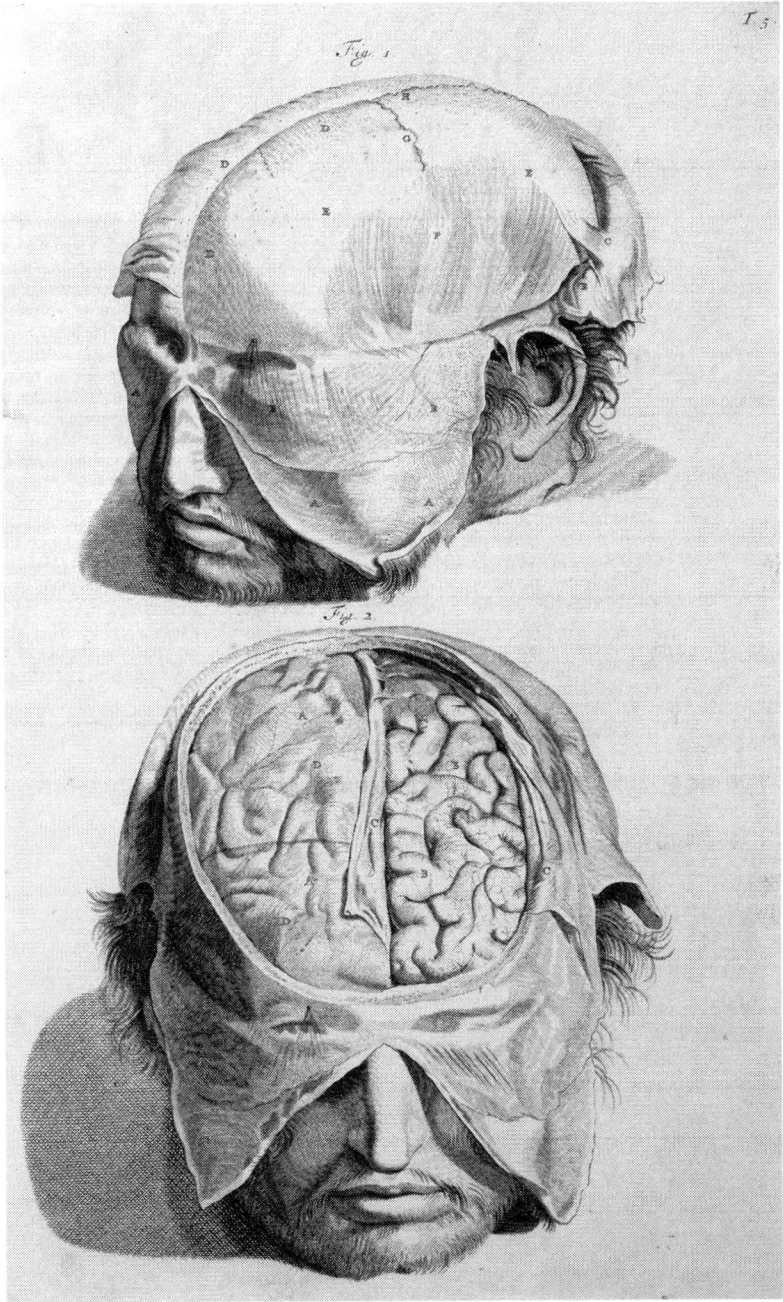

any more of his anatomy. The last time the tremulation manuscript is mentioned is 2 May. He had inadvertently left the draft with his sister Hedvig in Starbo and could not continue working on it since 'it would be some trouble for me to open up the *vestigia* [traces] which are already deeply *obducta alius generis cogitationibus* [covered over by thoughts of another kind].'[226] It is difficult to find the traces that are concealed deep under thoughts of another kind. His mind was now instead turned towards chemistry.

Swedenborg was at a stage where medical scholars had abandoned metaphors of craftsmanship in their description of the human body and had wholly embraced mechanical metaphors. In the Bible God is a potter and in Plato's *Timaeus* the demiurge is a baker or a smith when he makes human bones: 'Having sifted earth till it was pure and smooth, He kneaded it and moistened it with marrow; then He placed it in fire, and after that dipped it in water, and from this back to fire, and once again in water,' repeating the process until it was complete.[227] Rudbeck the Elder, who was at the transition to the breakthrough of Cartesian philosophy in Sweden, used a metaphor from cookery to describe blood: 'in the same way as with ordinary boiling, not only fire but also pots, spoons, whisks, and so on are needed to purify and skim what is boiled, so is the heart with its vessels in the preparation of blood.'[228] Hiärne made a comparison between the temple in Jerusalem and the edifice of the human body.[229] Polhem the mechanic was of course particularly inclined to imagine the body as a machine. An entire machine could be built of wood to resemble man, he speculated, 'moving all its limbs and members so perfectly, like a living person, only that instead of the soul there must be a keyboard to be operated by another person as an organ is by a musician.'[230] Swedenborg's manuscript on tremulations belongs to an iatromechanical or iatromathematical tradition, what could be called iatrogeometry. He focuses his gaze not only on the arithmetic proportions or the automatic mechanism, but particularly on the geometrical forms, the curves, and the structures.

Fig. 12 The convolutions of the brain. At a meeting of Bokwettsgillet on 4 March 1720, when Swedenborg's theory of tremulations was ventilated, Roberg also spoke about the figures in works on anatomy. Perhaps they were shown this picture of the meninx in Bidloo's *Anatomia humani corporis* (1685). In the *top picture* the scalp, A and B, has been peeled away. One can see fibres at C, and the periosteum, D and E, muscles, F, and sutures, G and H. The *bottom picture*: A, the meninx. B, the convolutions of the brain that are faintly visible under D, after parts of the meninx, C, have been peeled back

---

[226] Swedenborg to Benzelius, Brunsbo, 2 May 1720. *Opera* I, 303; *Letters* I, 236f.

[227] Plato, *Timaeus*, 73E.

[228] Rudbeck the Elder (1653), ch. III.

[229] Djurberg, 113.

[230] Polhem, 'Om naturens wärkan i gemen', *Polhems brev*, 35.

Iatromechanics was an attempt to explain the human body with mechanical metaphors. The human body is a machine. After Harvey's theory of the circulation of blood and Descartes' formulation of the iatromechanical programme, this school was above all developed in Italy. But iatromechanics did not really arise out of a need within biology; it was rather the application of mechanistic philosophy to organic life. A new mechanistic metaphor opened the door to a new way of looking and new discoveries. Closely allied to the iatromechanics were the iatromathematicians, who emphasized even more the mathematical description of the body. One example is the Dane Niels Steensen, whose iatromathematics proceeded from Euclid's axiomatic system and gave a purely geometrical description of the muscles.[231] The geometrical method was also used by Lorenzo Bellini in a chain of argumentation structured according to postulates, theorems, and corollaries. Bellini believed that blood was a fluid that flowed through conical vessels in accordance with Borelli's hydraulic laws, and envisaged that 'contractile villi' (which explain the body's reaction to stimuli) expanded, contracted, and vibrated like ordinary strings on a musical instrument.[232] Yet another example is a book that was in Swedenborg's library, Domenico Guglielmini's *De sanguinis natura et constitutione* (1701), which underlines the necessity of mathematics in medicine.[233] Or another book in the library, *Die höchst-nöthige Erkenntniß des Menschen* (1722), in which the author Christian Friedrich Richter talks of the body as a machine with mills, water pumps, bellows, musical organs, water jugs, trucks, spiders' webs, and so on.[234] On the other side of the board were the iatrochemists, who instead used chemical metaphors and consequently regarded the body as a chemical process based on Paracelsian theories and van Helmont's vitalistic teaching. For the iatrochemist François de le Boë (Sylvius), digestion was not a mechanical kneading but a chemical fermentation.

At the start of the eighteenth century iatromechanics was a common mode of thought at medical faculties in Europe. At the most important seat of medical learning, the University of Leiden, which was also very popular among Swedish students, Archibald Pitcairne occupied the chair of medicine and developed a physiological theory called 'hydraulic iatromechanics', in which the body was regarded—appropriately for Holland—as a canal system filled with fluids, mathematically and mechanically calculable.[235] But the most influential iatromechanic in Leiden was the professor of medicine and chemistry Hermann Boerhaave, a significant name in Swedenborg's library and in his thought. Boerhaave described the human body in geometrical and mechanical terms, as a hydraulic machine, combining the iatromechanical theories of Willis, Baglivi, Borelli, and Bellini. In *Institutiones medicæ* (1708) he explained the parts of the body as mechanical tools: levers, pillars, ropes, bellows, presses, sloping surfaces, block and tackle, filters, strainers, canals,

---

[231] Pedersen, 28.

[232] *Dictionary of Scientific Biography* I, 593.

[233] Guglielmini, 114f.

[234] C.F. Richter, part I, ch. VI, 173–195.

[235] Roos, 434.

and storage vessels.[236] Ultimately the human body, he assumed, consists of small particles which could explain different ailments. Knife-sharp particles could cut open small wounds, while amalgamated particles could block flows. The body has the same nature as the whole universe, and it differs from other machines merely by combining several machines operated by bodily fluids or humours. In Amsterdam there was another famous anatomist, Frederik Ruysch, whose *Observationum anatomico-chirugicarum* (1691), a book that Swedenborg owned and used, has a detailed examination of gynaecology and various deformities. One of those who attended Boerhaave's lectures in Leiden and became a faithful devotee of his medical theory and practice was Swedenborg's cousin, Johan Moræus, who, after having taken his doctorate about the chemical substance vitriol in 1703 under Roberg, had been given money by Jesper Swedberg to finance his medical studies abroad. In Amsterdam Moræus received training in practical medicine, for 'our learned Moræus knew very well that a theory without practice is like a foundation that never gets a house and roof on it,' as an anonymous obituarist wrote in the proceedings of the Academy of Sciences; it turns out to have been written by none less than his son-in-law Linnaeus.[237] Bromell, the man who was busy reading Swedenborg's tremulation manuscript in the spring of 1720, had also studied for both Boerhaave and Ruysch, in addition to which he had met Leeuwenhoek and defended a dissertation under Bidloo.[238]

Swedenborg's iatrogeometry came at exactly the right time. In Sweden there was a powerful mechanistic tendency in this period since the Cartesian controversy had waned at the end of the previous century. The defender of Cartesianism in the second battle, the professor of medicine Andreas Drossander, used his knowledge of experimental physics and regarded the human body as an automaton, 'an immensely beautiful and artful clockwork', in which sensory perceptions could be explained as the mechanical motion of the vital spirits through the nerves to the brain.[239] Much of Swedenborg's tremulation theory, however, is concerned with turning away from this Cartesian idea of vital spirits. He wanted to take yet another step in the direction of mechanics and sought instead for geometrical explanations of communication through the nerves. The volatile nature of the vital spirits could be perceived as running counter to the mechanistic explanatory model. Generally speaking, the vital spirits were something in the blood or in the nerves that represented life itself, transmitted it and sustained it. Inhalation brought in a living principle, the vital spirit, which made the blood healthy and alive. The Latin term for this active substance, *spiritus*, was ambiguous and could apply to explanations of both physiological and chemical phenomena. At bottom there is an obviously metaphorical mode of thought whereby the concept of vital spirit is formed through a metaphor from thin air to a fine, volatile spirit. That 'vital spirit' is a metaphor was

---

[236]Boerhaave (1708), 9, § 28; cf. Boerhaave (1709); Pehrsson, 41f.

[237]Linnaeus (1742), 299; cf. Linnaeus (1962), 103–106; Roberg and Moræus.

[238]Hult, 92.

[239]Cited in Lindborg (1965), 195.

something Rydelius had considered. To exemplify the way people find analogies he brought up 'thinking spirit', which he compared to a fine, volatile air. Even though spirit and air are completely different in kind, they share certain properties, such as that both are 'invisible, quick, and fleeting', but each in its own way.[240]

The concept of vital spirit can otherwise be traced back in history to Galenian medicine, whose figurehead, Galen of Pergamum, believed that nutrients engendered a *pneuma physikon* (*spiritus naturalis*) in the blood, which could be refined into *pneuma zotikon* (*spiritus vitalis*) through respiration, and in turn could be transformed into *pneuma psykikon* or *spiritus animalis* in the brain and then carried out through the hollow tubes of the nervous system. In the hands of Descartes the concept of *spiritus animalis* was clad in a mechanistic philosophy. He explained it as something similar to 'a very subtle wind' or a pure, lively vapour rising from the heart to the brain, then out through the nerves to set the muscles in motion.[241] According to Descartes, the tubular and hollow nerves contained a nervous fluid or fine particles of *spiritus animalis* which, when stimulated by a sensory impression, conveyed tremulations from the outside world and triggered similar tremulations in the nervous system, which finally set the body machine in motion. For Thomas Willis and Friedrich Hoffmann the boundary between vital spirits and nervous fluid was erased. The movement of the vital spirits gives way to a flow of stimulation, which in Willis is expressed naturalistically as the movements of chemical particles.[242] In this circulation of vital spirits in the nervous system, yet another metaphorical idea is evident. The circulation of vital spirits is understood by analogy with the circulation of the blood, as in Descartes, but also in Sylvius for whom the vital spirits moved from the blood to the brain, out into the nerve threads, to the lymphatic vessels and back to the blood. It is in this metaphorical thinking of spirits and circles that Swedenborg is positioned.

Polhem often returned to the vital spirits, regarding them as an infinite quantity of small living particles inside all living things, whether plants and animals or humans. During his years in Uppsala, Polhem may very well have picked up the idea of vital spirits from Rudbeck the Elder, who clung to the *spiritus animalis* of the brain. Yet the vital spirits are needed not only for movements of the body but also for thinking, says Polhem:

> when one thinks too much, the brain become tender and sore, and when one thinks of things one has never had before, the whole body becomes fatigued and powerless, which is a sign that thoughts require not only their spirits and their operations but the external movements of the body.[243]

Ether and prime matter are responsible for mental movements. The ether in the air is stretched out in the brain in specific shapes and figures, so that everything we think

---

[240]Rydelius (1737), 275.

[241]Descartes (1637), V; *Oeuvres* VI, 54; translation, 77; Hatfield, 346.

[242]Frank, 248f; cf. Thomson, 79–86.

[243]*Polhems brev*, 36, cf. 22, 100, 236; *Polhems skrifter* III, 9.

is nothing more than a depiction of what we have seen externally with our eyes. Polhem found it difficult to imagine anything that is not material.

One of the reviewers of Swedenborg's manuscript for the Collegium Medicum was the physician Johan Linder. He was an iatromechanic who had studied in both Harderwijk and Leiden. In *De venenis* (1707) he explained the effect of poisons as a result of their containing 'particles that are pointed and sharp or barbed. The particles sting, irritate, cut, abrade, burn, scrape, and tear.'[244] They thus cause small wounds to the internal tissues. These pointed particles penetrate the body's elastic fibres, which are then set in an oscillating motion. It works in the same way as a hammer driving in a nail, or like a builder's crowbar. The stronger the oscillations, the stronger the effect of the poison. People whose tissues vibrate more vigorously and have greater vitality therefore more easily fall victim to poisons than weak, lethargic, listless people. Iatromechanics acquired a brilliant representative in Sweden with the Cartesian Roberg, who proceeded from Boerhaave in his lectures in medicine. Roberg describes man as a pneumatic-hydraulic machine with horns, vacuum pumps, hoists, fire pumps, bellows, and kneading troughs.[245] People are living machines like two clocks which can make a third. For his disciple Rosén, mechanics was sure knowledge through which a doctor can mend a body with as much success as a mechanic can repair any machine. Our body, he says, has 'a lot of pillars, beams, wedges, mills, tongs, shears, levers, cords, and the like; we also find in it a lot of mechanical tools, such as: sieves, suction tubes, trap doors, tie-cords, pumps, hooks, and so on.'[246] Based on this mechanics, we can discuss anatomy with some certainty.

The need for nationalistic self-assertion in the 1710s brought a desire to find Swedish equivalents for Latin terms. Swedenborg attempted this in mathematics, while Roberg tackled the medical nomenclature. In the first anatomical textbook in Swedish, *Lijkrevnings tavlor* (literally 'Plates of corpse-ripping' 1718), Roberg writes about the remarkable creature that is man who, with his cheerful prattle, his fragile limbs, and his anxious disposition, is 'tardy and sluggish in wit and reason, quick to die, struggling with the soil and never content.'[247] The brain reminded Roberg of small intestines, and embedded among these intestines is the pineal gland which he assumes, following Descartes, to be the mediator between body and soul. The 'silver rope' or 'stringed instrument' (the spinal marrow, medulla oblongata) has a vital function that can be understood through his reference to Ecclesiastes chapter 12: remember thy Creator, indeed, 'Or ever the silver cord be loosed, or the golden bowl be broken.'[248] The musical metaphors and the tremulations that are so typical of Swedenborg can also be found in Roberg. The stringed instrument consists

---

[244]Linder, chap. I.

[245]Roberg (1747), 38f; Roberg (1748), 71; concerning iatromechanics in Sweden, Dunér (2009a), 111–149.

[246]Rosén von Rosenstein, introduction, 3; Pehrsson, 53, cf. 41.

[247]Roberg (1718), preface.

[248]Roberg (1718), 35; Eccles. 12:6.

of nerves such as olfactory strings, optic strings, eye-pulling strings, taste strings, and hearing strings. The body is a machine with tractive and elevating forces. Each muscle is a *drag* or pulling device; for example, the chest muscle (pectoralis major) is called the *bröst-draget* ('breast-drawer') or *famn-tage-draget* ('embracing-drawer')'. The mechanics of the face operate with the aid of eyebrow lifts, wrinkle drawers, humility drawers, etc. For breathing there is the 'cross-drawer', to help 'as much as they can in bellows-pushing.'[249]

Swedenborg may very well have peeped at Roberg's *Lijkrevnings tavlor* while he was writing about his tremulations. But above all, apart from the previously mentioned authors, he used anatomical descriptions by Willis and Vieussens. The advantage of Willis's work was that it not only gave anatomical descriptions but also showed the practical knack of performing a dissection. Swedenborg owned Willis's *Opera omnia* (1676), and in a notebook from the time he has written excerpts from his anatomical descriptions of the nerves and the folds and spiralling twists of the brain.[250] There was therefore, one might think, no need for him to perform any dissections of his own. Actually, there is nothing in his manuscript about tremulations that necessarily reveals him to have ever seen a dissected human body with his own eyes, although he did have a few opportunities to attend public anatomical demonstrations around 1716–1718 by Roberg in Uppsala and Bromell in Stockholm.[251] Rudbeck's anatomical theatre was in itself a rare observation instrument, with a telescope pointed at the illuminated cadaver. Swedenborg had never seen the trembling organs and tissues. They were found chiefly in his metaphorical thinking.

To Live Is to Tremble

What makes us live? If we proceed further and further into thought, Swedenborg writes in his *Anatomi af vår aldrafinaste natur*, we find that life is nothing but movement. To live and exist is to be in motion. The external senses consists of a wave movement that penetrates into the body and causes the fibres in the body to tremble, like small twitches, and gives rise to internal sensory experiences. Together all movements make 'a symbolum of life'.[252] The internal senses likewise consist of wave movements. The division between the external senses, such as vision, hearing, and so on, and the internal senses, the perceptions of the mind, was a part of the Cartesian philosophy of the mind, in which the vital spirits functioned as the link between outer and inner. According to Rydelius's 'pneumatics' there was a third sense, the innermost one, the mind's perception of itself, independent of the body.

---

[249]Roberg (1718), 61, cf. 54f, 63.

[250]*Photolith.* II, 261ff.

[251]Fredbärj, 12–14; Djurberg, 288; Weimarck, 36.

[252]*Photolith.* I, 132; *Om darrningar*, 69; translation, 9.

What is a thought, a living memory, if not a movement? Sound reason tells us that rest cannot play any part in life. Rest and life are contrary things, in the same way as death and life. Life as movement and death as rest is an example of metaphorical thinking that Swedenborg shared with Boerhaave and Hoffmann. It is the circular movement of the blood that constitutes life, Hoffmann explained.[253] The constant movement of the fibres, expanding and contracting, prevented the body from putrefaction. Sylvius, who proceeded from a chemical, not a mechanical, metaphor, concluded instead that life consists of the heat of the heart. Motion is of crucial importance for Swedenborg's iatromechanics. Life depends on the motion of particles, while death depends on their rest. If the motion is impeded, the spark of life is extinguished. But if motion is stimulated, life returns, as one can see among insects and other 'small beasts' in tiny drops of water. When the sun's rays touch their fibres, membranes, and arteries, they quickly revive. Life is ignited and the senses begin to live. Chill, however, blocks movement and life, and it can be geometrically proved that cold does not reach as far in large bodies as in small ones. It is movements that allow us to live, to use our senses, our thoughts, granting us perfect harmony or communication between them. Swedenborg is about to take the first step on the staircase of an unknown geometry. This subtle geometry of vital movements 'is closed to us and to our coarser senses.' A thousand steps remain before we reach complete knowledge of it. The movements of which life consists are the very subtlest of motions, 'such as cannot be seen or comprehended by any comparison with the grosser forms of motion.'[254]

The finest motion of all in nature is tremulation. If we investigate this tremulation we find that it is more like a motion around an axis or a centre, and not actually a local movement from one point to another. In hard things the tremulation is a motion up and down in the particles, like a search for balance in a ball bouncing against a floor in increasingly small bounces until it stops in equilibrium and rest. Tremulations are a quest for the balance that is in danger of being lost. The smaller and lighter the particle is, the faster the tremulation spreads. Water trembles slowly, air faster, ether even faster, and fire so quickly that the tremulation spreads from the sun to us virtually instantaneously. A thing can consist of both a tremulation and a local movement simultaneously, just as a bomb travelling through the air can also vibrate in itself. Tremulations can also enter each other. On a water wave there can be a smaller undulation, and on it a tremulation, and then yet another contremiscence and then an even smaller one which can almost be called something living. In a person on board a ship that is undulating up and down along with the waves, the heart and the brain can have their undulations, and these can undulate from a shiver, and the shiver itself from a tremulation, and the tremulation from even smaller tremulations of sound or something else.

What lives in us is the tremulations in the nerves, the membranes, the bones—the entire body system, Swedenborg continues. For example, an auditory impression

[253] Hoffmann (1748), I, ch. VIII, 25, § 12; cf. Hoffmann (1722).

[254] *Photolith.* I, 133; *Om darrningar*, 70f; translation, 10f.

is first a movement in the air, then in the membranes in the ear, then from one membrane to the other, from one nerve to the other, from one bone to the other. In other words, they are all connected to each other in a single system. A tremulation can set the whole body trembling in sympathy and give rise to sensory experiences. If one puts the tremulations of all the senses together, a person has a life: 'qvod erit demonstrandum'—which is to be demonstrated.[255] It is this theory of tremulations that distinguishes Swedenborg's iatromechanics. Yet Newton was following a similar track. A couple of questions in *Opticks* deal with vibrations, 'tremors', in different media transmitted through the optic and auditory nerves.[256] The body's movements take place with the aid of ether vibrations triggered in the brain through will power and conveyed through the nerves to the muscles.

The Circles of the Body

Large undulations cause smaller ones. The wave movement of the lungs gives rise to smaller tremulations. When the lung inhales air, it is blown up and expands, after which it contracts again, setting other parts in motion. The heart also has its own movement. It wrings itself in and out and drives the blood through arteries and veins. It is a circular movement, yet it can be called an undulation in the sense that it is a movement back and forth, a vibration as in a pendulum, although in spirals. Swedenborg is referring to Harvey's discovery of the circulation of the blood, but he also seems to combine it with what Huygens demonstrated in *Horologivm oscillatorivm* (1673), that an isochrone pendular movement can be geometrically described as cycloids. The regular motion of the pendulum in cycloids was naturally noticed by the watchmaker Polhem.[257] But Swedenborg's cardiac pendulations instead describe spirals. The brain has undulatory movements that are associated with the motions of the heart, and 'it has been discovered in our own age that the medullas, both the oblongata and the spinalis, vibrate and respire and rise and fall as if in fermentation.'[258] Swedenborg assumes there is a fluid in the nerves that is propelled in undulations, like the blood in the veins, from the body's outer parts, through the nerves to the membranes of the brain and out again. The brain is the source and origin whence this fluid flows through the nerves out to the membranes and keeps them tensed. Nature seeks to order everything in circulations, and above all in undulations or tremulations.

---

[255] *Photolith.* I, 136; *Om darrningar*, 73; translation, 14.

[256] Newton (1706), book III, qu. 14, 298; Newton (1721), book III, part I, qu. 14, 23f, 320, 328; Swedenborg took notes from Newton's *Optice*, qu. 19–23, 306–322. KVA, cod. 88, 363–365; cf. Park, 213f, 233.

[257] *Polhems skrifter* III, 440.

[258] *Photolith.* I, 136; *Om darrningar*, 74; translation, 16.

Swedenborg is using the metaphor of the circle, envisaging a process as a movement in a closed circle. It returns to the same point on the circle, a return that can continue infinitely. The circulation of the blood is a typical example of this circular metaphor. With the aid of this metaphor Harvey, Swedenborg, and others were able to show connections, discern a structure, reveal a mechanical movement in the chaotic, organic human body. By imagining the body's processes as part of a circle, it was easier to create mental images of them. It made thought concrete, easier to grasp and to communicate to others who could thus assimilate it. Harvey's idea of the circulation of blood can be understood as being grounded in and guided by a number of metaphors. In *De motu cordis et sanguinis in animalibus* (1628) he describes the movement of the pulse as an undulation, as when a horse drinks water and makes a sound in the throat at each swallow, and he says later: 'I began to think whether there might not be a motion, AS IT WERE, IN A CIRCLE. Now, this I afterwards found to be true.'[259] Blood follows a circular movement, he continues, in the same way as when Aristotle talks about moist earth being warmed by the sun, from which rises the vapour that is condensed and falls as rain to moisten the earth again. Many living things are produced in this form. 'The heart, consequently, is the beginning of life; the sun of the microcosm, even as the sun in his turn might well be designated the heart of the world.' For Harvey, the circulation of blood and procreation are a microcosmic movement in a macrocosmic circular pattern. Envisaging the motion of the blood as a circle means that other circular metaphors and associations are linked to the blood, such as the process of distillation, the circulation of water and plant fluids. It is also this in part that makes it so successful, although it was not until 1661 that the circle was closed by Marcello Malpighi when his vivisections of frogs revealed the capillaries that connect the arteries with the veins. With the circle metaphor it was easier to understand and accept the theory of the circulation of the blood.

The circle or circulation is also a concept in alchemy, which has the serpent 'Ouroboros' swallowing its own tail, a symbol of infinite time, the cyclic course of the seasons and the eternal cycle. It is a closed process that repeats itself. Among the alchemical literature owned by Swedenborg we find Johann Conrad Barchusen's *Elementa chemiæ* (1718), in which *circulatio* is explained.[260] It is generally used as another expression for distillation through an alchemical apparatus, involving a flow and a return flow, the heating of a liquid followed by evaporation, then cooling and condensation. It has been assumed that the circle in Harvey is linked to the *circulatio* in alchemy, which would then mean that the circle represents not just the cyclic return of the blood but also a chemical process for the perfection or purification of the blood.[261] This link between the circulation of the blood and alchemical purification is also found in an unnoticed Swedish example, in Rudbeck the Elder. In his dissertation on the circulation of blood, *De circulatione sanguinis* (1652), he

---

[259] Harvey, ch. VIII; translation, 51, cf. 32, 36.

[260] Barchusen, 73.

[261] Pagel (1981), 359–361.

considers Harvey's idea. In a similar way to Harvey, he makes a comparison in his dedication to Queen Christina, calling her the life-giving heart of the realm, after which he likens the circulation of the blood to the cycle of water in the big world. But he adds something extremely interesting which also seems to corroborate the thesis of the alchemical origin of the circulation metaphor. He writes that the circuit of the blood through the arteries resembles alchemical circulations in which a liquid is enclosed in a 'pelican'—that is, a retort with a 'beak' bent down towards the belly of the vessel—and is conveyed upwards by the heat, is condensed and falls again, as in a circle.[262] The following year, in his dissertation on the lymphatic system, he also put forward the idea that chyle circulates.[263] And later in *Atlantica* he compares the migrations in history that flowed from the Babylonian plain out into the world, and back again to the heart at the Euphrates and the Tigris, to the flow of the blood from the centre out to the periphery, from the heart to the body and back.[264] In other words, having once begun to think of processes in terms of circle metaphors, one can also find new circles in the body, the world, and history. Natural philosophers began to search for circulation movements not just in the blood and the gall, but also in saliva, in the fluids of the stomach and the pancreas, lymph, the vital spirits, and semen.

The circulation of blood through the arteries was read into a cosmological and metaphysical circular mode of thought which not infrequently proceeds from Aristotle. For Aristotle circular movement was the perfect movement, in that its continuity ensures that it remains the same and contains contrary things such as movement and rest, concavity and convexity, start and finish.[265] This movement can be seen in ageing, in the cycles of the seasons, and in the circular motion of celestial bodies. As Marcus Aurelius renders the words of the pre-Socratic philosopher Heraclitus: 'death for earth is to become water, and death for water to become air, and for air to become fire, and so on in backward sequence.'[266]

The cyclic transformation of the elements became a part of Stoic philosophy. Seneca, and after him Helmont, drew parallels between microcosm and macrocosm; like the human body, the earth had its veins and arteries, such as subterranean flows and sources.[267] The earth is a human body with veins. This parallel is common in the literature on geology and mining with which Swedenborg was familiar. He could find how Agricola likened veins of ore to blood vessels making their way through the globe, or how Kircher pictured water circulating through the earth like blood through the body.[268] This parallel between microcosmic and

---

[262]Rudbeck the Elder (1652), thesis II.

[263]Rudbeck the Elder (1653), ch. VIII.

[264]Rudbeck the Elder (1679); edition, I, 40; Eriksson (2002), 69, 570f.

[265]Aristotle, *Peri ouranou*, 1.4.270b32–271a35; Aristotle, *Mēchanika*, 847b16–848a19.

[266]Heraclitus, 47, fragment 76c; Marcus Aurelius, 4.46.

[267]Frängsmyr (1969), 27, 31.

[268]Agricola, translation, 47, note; Kircher (1665), I, 240.

macrocosmic circulation created existential connections in reality, as expressed in Samuel Columbus's circulations, which may have been inspired by Kircher:

> The water is now still; now can it overflow
>
> Wood, field and mead and land, fine homes and halls also;
>
> Now into earth it sinks, and there its course maintains
>
> As blood in body flows, both to and fro in veins.[269]

Metaphorical thought about the earth as a human body with veins can also be found in Swedenborg's immediate contemporaries. Harald Vallerius wrote a dissertation about the course of water through the globe, how it displays circular movements like the circulation of the blood.[270] The vicar Petrus Gudhemius is thinking along these lines in his work about earthquakes, and Tiselius imagines 'the subterranean systole and diastole' under Lake Vättern.[271] Tiselius also says that water has two movements: an inner motion, with reference to Swedenborg, caused by a subtle fire, and a circulation through the earth.[272] Swedenborg himself drew a parallel in *Camena Borea* between the flow of blood to and from the heart and the flow of water back and forth, and wrote a poem for his father's *Vngdoms regel och ålderdoms spegel* about the ageing of man, in which lukewarm blood flows at a decreasing speed until the source dries up and the wheel stops turning.[273]

The circle metaphor is part of a micro-macrocosmic world-view, a kind of analogical thinking that links the large with the small and the small with the large, which engenders affinity and unity. In a dissertation written under Harald Vallerius about the parallels between microcosm and macrocosm, the author Andreas Pijl defends a heliocentric Cartesian world-view with the archaic idea of the human body as an image of the universe.[274] Just as the heart is the centre of the little world, the sun is the heart of the universe, the centre of the big world. In the same way as the sun leads the heavenly bodies in a constant vortical motion, the heart drives the pulsating blood in a circular movement through the body. Each organ corresponds to a part of the great edifice of the world. Each planet has a different internal organ as its counterpart: the earth is the stomach, Mercury the lungs, Venus the kidneys, and so on. Man is a parallel to the universe, the earth, fire, water, society, geometry, architecture, and music.[275] He arises from a first point, has lines of nerves and filaments, has limbs at different angles, a vertical drainpipe down to the diaphragm, parallel fingers, and a surface covering the entire machine. There are circles, trapezia, a spherical head, a conical nose, cylindrical neck, and a pyramidal heart.

---

[269]Columbus, 17.

[270]H. Vallerius and Ramzelius, 11f.

[271]Gudhemius, 4, § 2; D. Tiselius (1730), 30.

[272]D. Tiselius (1730), 93; *Prodromus principiorum*, 12; translation, 18–20.

[273]*Camena Borea*, edition, 92f; *Ludus Heliconius*, edition, 58f.

[274]H. Vallerius and Pijl, 8; Sandblad, 115.

[275]H. Vallerius and Pijl, 2f, 36–40.

The circulation of the blood fitted well into a geometrical and mechanical world-view. Descartes found that Harvey's idea of the movement of the blood in a closed circle resembled his own idea about the circle of the particles of matter in a plenum, as an example of a purely mechanical process in physiology. For Swedenborg the mechanic there was thus no difficulty in reading the circle metaphor into a strict geometrical iatromechanics. He compares the circulation of the blood in *Miscellanea observata* to the sap in trees and bushes. This analogical connection between blood and sap, he says with a reference to capillary force, provides an explanation why blood circulates more easily in the ramifications of the arteries than in the big arteries.[276] Descartes and many others viewed the circulation of blood as an analogy to the circulation of sap in plants, but caused by the heat of the heart. The circulation system was thus a distinctive feature of living things, as in the circulation of blood, in respiration, the cycle of nutrients in the blood, the circulation of vital spirits by the senses through the blood and nervous fluids.

The idea of linking the structures and processes of the body to geometrical forms is salient in Swedenborg. The view of the human body as a perfect geometry removed the chaotic, incomprehensible organic forms and gave the body order and clarity in its outlines. All living things are tremulations for Swedenborg. Speech, words, and sounds happen though tremulations of the air shaped between the folds of the tongue, sifted below or above the tongue as it twists and turns. In our mouths it is possible to shape 20 or even more tremulations. Throughout the nervous system there can spread a powerful tremulation which may be called '*tremor, shivering, convulsion*'. The tremulation can reveal itself as a pleasant contremiscence in the membrane enclosing the bones (the periosteum) under the membrane enveloping the skull (the pericranium), as a 'pleasantness which plays and titillates in the mind, producing an harmonic motion.'[277] One can also experience a sense of surprise mixed with fear. The tremulation then spreads like cold water over the membranes, and like small waves under the hairs on the head, which then stand on end. The quivering passion spreads everywhere, growing into an even greater trembling. This is palpable evidence that tremulations can be found over the whole body.

There are different degrees of tremulations, just as a local movement can have different degrees of speed. A movement may be so slow that our vision does not notice any change at all, as when one looks at stars and planets for a short time and they seem to stand perfectly still. But motion can also be so fast that it vanishes from our sight, as when a bullet flies through the air. The first degree of tremulation is the *undulation*, which is the most palpable and visible. The larger movements of the body are of this degree. The undulation triggers the smaller tremulations, as a large wheel drives a thousand smaller wheels. If an element starts undulating, for example water, it propels visible waves in circles out to all sides, further and

---

[276]*Miscellanea*, 122–126; translation, 78–80; cf. *Opera* I, 299; *Letters* I, 230; Descartes (1664b), ch. XVIII; *Oeuvres* XI, 124, 132ff.

[277]*Photolith.* I, 137f; *Om darrningar*, 75f; translation, 17.

further out, in increasingly small waves until they level away. The second degree is *tremulation*, which starts where the undulation begins to produce sound and can be heard. Tremulations in the air produce sounds and set the membranes and the eardrum (tympanum) aquiver. A lower speed gives a coarse sound (or darker tones) and a higher one a finer sound until it reaches a speed that escapes hearing. On a clavichord 150 tremulations per second yield high C and 30 or 40 tremulations give low C. The third degree is the finest tremulation, the *contremiscence*, an extremely rapid vibration that cannot be seen or heard. It starts when a tremulation is faster than 200 oscillations per second and can grow to speeds of up to 1,000 or 2,000 oscillations. This motion applies to something other than the five senses, namely, our emotions, 'our motive life-force'.[278] The different degrees are part of a mechanics of motion. Just as a wheel or a pendulum in mechanics can set hundreds of others in motion, so it is with the undulations of the lung, the marrow, and the brain.

In the third chapter Swedenborg describes the nervous system with data taken from Willis, Vieussens, and others.[279] From the brain and the cerebellum comes the medulla oblongata which runs along the spine to the bones and out into the body in increasingly fine ramifications. This entire network of nerves is interconnected in harmony. Every sensory impression leads to a tremulation in the whole nervous system, and they cannot be assigned to a specific place in the cerebellum as Descartes thought, or any particular protuberance or ventricle; no, their home is everywhere at once. So if one trembles, they all tremble. A living sensation is not tied to any place, but free of all coercion, and it passes like lightning through the body. The membranes are a juncture where a great many nerves run together. In a microscope one can see millions of criss-crossing nerve threads. The chief membranes are the meninges, that is to say, dura and pia mater. These are the location of the noblest and most subtle movements in living things. The dura mater is pressed so hard against the skull that it is difficult to separate it without damaging it. The membranes are connected to the bones in the body, so that a tremulation in the membranes and in the nervous system is at the very same instant transmitted to the bones and all the hard parts of the human body.[280] Hardness is necessary for the propagation of the tremulations. Nature has made the skull porous, since porosity facilitates tremulation. The more porous a thing is, the better it plays along with the tremulation of the string. Porous woods such as cedar or spruce amplify the sound of the strings better than others. Nature's most wise God has therefore made a porous skull like the body of an instrument, whereby the membranes acquire their stronger notes from sensory impressions.

---

[278] *Photolith.* I, 141; *Om darrningar*, 79; translation, 22.

[279] *Photolith.* I, 142, 152; *Om darrningar*, 79f, 91; translation, 23, 37.

[280] *Photolith.* I, 148; *Om darrningar*, 87; translation, 32.

Hearing the Music from Within

Thoughts, feelings, and passions tremble. The human body is like a musical instrument with strings and a sound-body, writes the musically minded Swedenborg. The tremulations pass through the membranes and are led out into the skull and the bones of the body. A musical instrument with slack strings cannot produce any sound even if the body is of the best cedar wood. So it is with the body's tremulations too. The strings, like the membranes, must be stretched hard. If they are tensed by blood and lymph they can spread the tremulation to the bones and thus transmit sensory impressions. But if they are slack, nothing happens. The degree of life is determined by the tension of the membranes (meninges). The more they are tensed, the quicker the tremulations are transmitted. The spirit becomes present. But if the membranes collapse, as it were, 'in the same degree do we suffer from absence of mind and of understanding, the body no longer responding to what is quick and prompt.'[281]

Swedenborg describes the circulation of blood ramifying in ever smaller arteries, the vessels enveloping the membranes as 'the finest ramifications and leaves of a tree.'[282] The blood returns in increasingly large branches and thus completes the circle. The lymphatic vessels are like aqueducts between arteries and veins. If we follow our reason we understand that, as the heart drives the blood around, the brain and the medulla also have a power that makes other fluids spread in a circular movement. The heart and the brain are the sources of the circulating in- and outflow of two fluids. The brain and the medulla, like the heart, have a to-and-fro, undulating motion. Moreover, in the medulla oblongata and spinalis there is a nervous fluid called 'humus nervus', which no common sense can deny, since even in the hardest tree there is a sap, a flowing liquid. The nerves ramify out towards the most distant parts of the body, and from there they return to the medulla spinalis. In the same way as the blood has its own 'sources of distillation', lungs, glands, and lymph, which produce the blood before it comes into circulation in arteries and veins, the nervous fluid also has its glands, ventricles, and vessels for its distillation in the brain.[283] He rounds off the discussion by admitting that all this about nervous fluids is just a guess.

The tension of the membranes is crucial. If they become lax for some reason this can immediately be seen on the surface. The senses cannot fulfil their tasks, thought and memory lose their clarity. A person becomes like a fool, 'almost void of life, the vital fire being gradually quenched and approaching a state of quiescence or death.'[284] His father Jesper wrote about this dismal experience at the end of his life in his thousand-page memoirs: 'But my *memory* is vanishing from me. I now

---

[281] *Photolith.* I, 150; *Om darrningar*, 89; translation, 35.

[282] *Photolith.* I, 151; *Om darrningar*, 89f; translation, 35.

[283] *Photolith.* I, 152; *Om darrningar*, 91; translation, 37.

[284] *Photolith.* I, 154; *Om darrningar*, 95; translation, 41.

remember as good as nothing.'[285] Emotions also tremble in relation to the tension of strings and membranes. With *fear* or *terror*, Swedenborg says, the blood is driven to the heart in a jiffy, so that the skin and the muscles are left without blood. A person pales, the membranes become slack and cannot receive the tremulations. Vision loses its acuity, as do hearing and the other senses. Thought and imagination lose their clarity: 'life is in danger'![286] With *amazement* everything comes to a halt, the circulation of blood decreases and the membranes become lax and flaccid. *Fainting* and *stroke* are also caused by laxity or some obstacle in the transmission of the tremulations.

When the blood flows into all the vessels and membranes and makes them tense, it entails other emotions. In *frankness* they are tense and filled with blood, the tremulation can operate freely, the senses and reason are present. Another passion drives the blood out to the outer membranes, 'such as that of *Amor and Venus*.'[287] Then all the membranes in the body become tense and allow the tremulation to put life into everything that ought to feel and live. He had depicted this vibrating, musical, blood-filled mood in a poem intended for declamation at the wedding of a poet and a girl. He urges the poet: 'When you sing your songs, when you strike your strings with the quill, / then see to it that she learns to touch *your string*.'[288] A demure interpretation: musical strings touch the strings of the nerves. *Anger* propels the blood violently and dilates the arteries; the same applies to *fever* and *drunkenness*. Alcohol makes the blood volatile and causes it to expand the blood vessels. If this dilation is too strong, powerful tremulations arise, leading to wild movements instead of smooth, gentle ones. When the right tremulation develops into a different one that does not accord with the usual one, we call it madness. Swedenborg's explanation of diseases and states of mind as being dependent on balanced movements and the correct tension is in line with other iatromechanical explications. Borelli, whom Swedenborg regarded as one of his inspirations, assumes that disturbances in the movement of fluids in the nerves could cause various complications such as fever and cramp. With resilient fibres and movements of ether Hoffmann was also able to explain different diseases. Excessive tension led to cramp and spasms, too weak tension instead meant weakened muscles.

Mechanics cannot be grounded on the unknown, the invisible, or fantasies plucked out of thin air. Geometry, Swedenborg emphasizes, requires a better foundation. If mechanics is to follow its set rules, everything that is unknown and not comparable with geometry must be removed from the theory. One must rely on geometrical demonstrations. Swedenborg seems to be approaching a materialist viewpoint. If it is not tremulations that make life, how could one otherwise explain the complications that arise in those who lose a part of the brain or have been trepanned, or the case of the ossified brain in Bromell's collection? (Fig. 2 of

---

[285] Swedberg (1941), 536.

[286] *Photolith.* I, 155; *Om darrningar*, 95; translation, 42.

[287] *Photolith.* I, 156; *Om darrningar*, 97; translation, 44.

[288] Swedenborg, 'In poetae et puellae nuptias', *Ludus Heliconius*, edition, 128f, commentary 224.

Chap. 1) The brain is necessary to provide the nerves and the membranes with a special fluid. Moreover, it is the membrane or the cortex that has the properties previously ascribed to the inner part of the brain. The cerebral cortex functions in the same way as the bark of a tree. As long as the bark is whole, the leaves and fruits continue to live even if the trunk is broken. If the bark is removed all the way round the trunk, the greenery dies.

The body's tremulations can be compared with strings. If they are slack they undulate without sound; if they are tightened they can immediately produce sound. Or as with war drums: if the skin is loose the sound is not audible very far away, but if it is tightened the speed of the tremulations increases and the sound can be heard at a great distance. Swedenborg also cites an example of a rope with a ball in the middle acting as a pendulum. The longer the rope is, the more slowly it moves. The shorter it is, the faster are the vibrations during the same time. If the rope is tightened the pendulum becomes shorter and the ball vibrates faster. The same thing applies to a membrane that is tensed. Similar musical ideas had been expressed by Stahl, who declared that nerve strings oscillate like the strings in musical instruments. People's characters, Swedenborg claims, depend on the tension of the membranes. He thus modifies the humoral pathology of Hippocrates and Galen, but abandons the idea of fluids and transforms the theory into a geometry of tremulations. In this way Swedenborg translates the more organic metaphor of an earlier doctrine into a mechanical one. Hippocrates' metaphors about the humours proceed from the colour of flowers and from the notion that the soil is for trees what the stomach is for animals: 'As a soil that is manured warms in winter, so the stomach grows warm.'[289] According to humoral pathology, a healthy body has a balance between the four humours, blood, phlegm, yellow bile and black bile. The humours each give rise to one of the four characters or temperaments: sanguine, phlegmatic, choleric, and melancholic. In Swedenborg we find three of the temperaments. A *sanguine* person has thin, volatile blood that flows into all the vessels and dilates them, which makes it easier for tremulations to be transmitted. The senses are then present, mobile, and prompt to act. He finds, sees, and hears faster than others. The life of a *phlegmatic* person is slow, since lymph or serum dominates in the membranes, blood has less room in which to flow and becomes cooler. The *melancholic* has thicker blood which finds it harder to reach into the smaller vessels. Circulation is poorer and decreases in speed. Perhaps this was the answer to the question once posed by Roberg: 'What is the source of the anxiety in melancholy?'[290]

A tremulation needs something hard and tense if it is to be propagated, just as a string needs a sounding box. The hardness quite simply makes sensory impressions more palpable. In a *child* activity is limited, everything is soft and immature. An infant has no stability, its bones are soft and the skull can easily be pressed in by the finger. No tremulation can occur there. The five external senses are scarcely used by them and they cannot perceive anything distinctly. For them sounds, vision,

---

[289] Hippocrates, *Peri chymon*, XI, 82f.

[290] Roberg at the examination by Collegium Medicum in 1697. Dintler, 50.

and everything else is 'like a shadow or a cloud'. This time when all is soft is a
time of oblivion. Not the slightest impression remains for later. Everything that the
senses have perceived has been erased by time and seems as distant as in a dream.
But as soon as the compositions in the body begin to stiffen and are tensed, when the
skull and the sutures join, the body begins to leach its fluids and the nerves become
drier, 'Life then begins to become properly living' and all the senses brighten
up and tremulations can take their course. For *middle-aged* man everything has
reached its proper tension, running promptly and vigorously to full effect, in both
comprehension and expression. The skeleton has acquired its proper dimensions
and hardness, the nerves have stiffened and the membranes have been stretched.
All the senses are in full use, and memory and thought have reached their apex,
their highest level. Tremulations now run best in a man's musical body: 'as soon as
the frame-work is ready, and the whole key-board furnished with taut strings, then
only is it able to convey the sound or the perception, which propels itself by means
of tremulations.' When one approaches *old age* the spinal marrow and the brain
become increasingly sluggish and hard. The humours run more sparsely through
the body and the nerves do not obtain their nourishment. Everything goes slowly,
everything becomes slack and lies in wrinkles and small folds. The membranes can
no longer receive any fine contremiscences, only coarse undulations. Tremulation is
switched off, hearing declines, vision is obscured, and speed disappears, and with
it acumen. Tremulation can also be found in dumb animals. In man, however, the
body attains its ultimate hardness and perfection in a late phase of life. An animal
acquires the nature of its sire and dam after a few weeks.[291]

Tremulations appear most clearly in hearing. There all the organs and membranes
with their cavities, bones, small drums, and spiral coils are shaped to receive and
spread the sound waves from the air through the body. In the ear there is a labyrinth
or a cochlea which, depending on the nature of the sound, amplifies and concentrates
it towards the inner part of the twisted shell. This anatomy shows the mechanism
required by the tremulations. How a tiny membrane in the body can set something
hard, like bone, trembling in sympathy and spread the tremulations through the
whole system can be understood through the art of music. Consider the sound that
can be spread by a war drum or a kettle drum. A violin must have a bridge and
strings bound to something hard. If one strikes a string on a clavichord, it sounds
immediately and plays its tremulations out over everything hard, over bridges and
sounding boards, and lets the tremulation raise its noise and be carried some distance
away. The slightest quiver in a violin can spread over the largest bodies, throughout
the room where the music is played. One can feel with the hand how the walls
vibrate. He takes another visible example of the spread of tremulations from Polhem
and from his own experience of mines. In our mines there are thick cables hanging
in the shafts. If one hangs a weight at one end and hits the cable with a stick it starts
trembling, vibrating from the top down in quick, snaking coils, in waves, just as in
air and water. The rope twitches so violently that a person being lowered into the

---

[291] *Photolith.* I, 162f; *Om darrningar*, 107f; translation, 53–55.

mine on it finds it difficult to hang on. If two men are holding a rope in their hands between them, it is difficult for one to keep his grasp if the other suddenly jerks it. Tugs at the ends spread like tremulations through the body. A long rope that is held horizontally falls in a curve resembling a parabola, and forcing it straight requires a strong and powerful machine. Even then it will not be straight; there will always be an invisible curve.

In the same way as violins and clavichords require a sounding board, so the body's tremulations also need their hard bones and skulls. From dreams it can be understood that sensory impressions consist of tremulations of the skull. Even for the young Swedenborg, dreams and fantasies were living things. In *dreams* one often carries on whole conversations with others, one hears melodies and other sounds which resemble the sounds that enter the normal way. When one tries the next day to remember what one has dreamed, one cannot conceive that it was anything but a real sound. It is also known that, during sleep, the state of the brain is constantly being worked and moved. The sound is not found in the cochlea and the labyrinth; it is internal tremulations in the skull and in other membranes than those in the ear. In *fantasies* we hear complete sounds, coherent speech, 'and often even a musical harmony, so that it sometimes provokes a delusion that a spirit is speaking inside one, ordering and commanding this and that.'[292] A woman had told Swedenborg that she heard hymns being sung inside her, day in and day out, from the first to the last verse, even hymns that she herself had never sung. She visited priests to be cured of this, but the melodies remained in her brain the whole time, as if she were all the time in concert with others. It may have been the starvation artist, the maid Ester Jönsdotter of Norra Åby, 20 miles from Malmö, who had told Swedenborg abut the hymns inside her. Bishop Swedberg, who bought up this divine miracle in his sermons, tells of this remarkable young woman who, while lying in bed for long periods without food or drink, had seen and heard biblical language in a bright white church 'clearer than sun and stars', even though she had never read the Bible.[293] The only explanation, according to the bishop, was that she lived from faith and the word of God which she had in her mouth. Ester was a well-known medical phenomenon. Swedenborg wrote to Benzelius from London complaining that his father had not sent enough money. Swedenborg explained that it is 'hard to live like the wench in Skåne, without food or drink.'[294] There were diverse medical explanations for Ester's ability to lie in bed without nutrition. Block, the provincial physician, thought that she had imbibed nutritious fluid from the air as she lay in this state between 1704 and 1713, while Johan Jacob Döbelius, professor of medicine at Lund University, suggested that she lived off her own soft parts and air. The example of the woman with the hymns being sung inside her, Swedenborg declared, is proof that the sense of hearing consists of internal tremulations in the skull which can spread to the auditory organ. Other such internal

---

[292] *Photolith.* I, 174; *Om darrningar*, 119; cf. translation, 70. Odhner's translation is corrupted here.

[293] Swedberg (1941), 563f.

[294] Swedenborg to Benzelius, London, August 1711. *Opera* I, 217; *Letters* I, 33.

sounds are, for example, buzzing in the ear and rattling teeth. Ester also occurs later in Swedenborg's psychological and neurological studies. He writes about a person who heard tunes being sung in the brain, of hymns that she could not otherwise have known.[295]

Tremulation occurs first in the fluid, then in the membranes, and finally in the bones. The fluid is called 'animale' by others, but in Swedenborg's mechanistic explanation of the vital spirits it is 'animale' because of the trembling movement. When the membranes are tensed against the bones and the blood is at even tension, the fluid flows smoothly and continuously through all vessels and causes the tremulation to spread. This is how we obtain our sensory impressions. Tremulations in the air can cause the auditory organ to quiver. The bang of an exploding bomb in water caused the whole ground and the mountains to quake. It is marvellous the force that exists in a fluid when it can set membranes and bones vibrating, a vibration that gives rise to mobility and vitality in a person.

Vision Extends into the Invisible

Tremulous thoughts continue to vibrate in Swedenborg's later writings. In the manuscript *Generaliter de motu elementorum* (1733) he tries to link the body's tremulations to a theory of particles.[296] There are four movements in the elements: the local movements of particles, as in flowing currents; a movement whereby the whole volume of particles is still while each particle moves in undulations; and yet another movement on the surface of the particle, the tremulation; finally, there is a fourth movement, a pressure or effort in the particle. This doctrine of motions is applied to the movements of the body, the membranes, and the senses. Another manuscript, *De mechanismo animæ et corporis* (1734), describes how analogous movements cause enjoyable, beautiful, and pleasing colours, while divergent tremulations give rise to something unpleasant and sorrowful that disturbs, twists, distorts, and breaks the links of continuity.[297] As an example he mentions how different colours such as green and blue are mixed and result in something unpleasant, that is, brown.

In the second part of *De infinito* from 1734 we find once again his mechanistic ideas about the tremulations of the body and the mind. They show clearly how Swedenborg's theory of tremulations is an attempt to mechanize the idea of vital spirits. He questions the vital spirits that were used as a link between body and mind.[298] Those who used the term denied that there was anything mechanical about the vital spirits. If one thinks about the unknown, but will not accept its similarities

---

[295] *Psychologica*, 108; *Regnum animale* III, n. 509.

[296] *Photolith.* III, 79–83; *Treatises*, 85–93.

[297] *Photolith.* III, 91–102; *Treatises*, 124, 126f.

[298] *De infinito*, 202ff; translation, 167f, 171.

to the known, the result must be that one does not prove anything at all—like those philosophers who use the vital spirits to try to explain the movements of the body, the mind's mastery over it, that the vital spirits run hither and thither in the system to perform their services for their mistress. Others, silently, find the existence of the mind problematic, for they consider it impossible that it could be like a mistress sitting on her throne with the vital spirits standing around her as servants and running off promptly at her command. It is a short and easy distance from doubt and ignorance to denial, Swedenborg observes.

At the same time Swedenborg tried to adapt Wolff's teachings in *Psychologia empirica* to his own tremulation theory. He made excerpts from Wolff's text to see whether the opinions there fitted with his own. It is a strictly mechanistic, virtually materialistic stance that emerges in his critique of the vital spirits. Merely because we cannot see them, we call them spirits. 'If we had the microscopes, we might be able to see the entire structure both of the soul and of the spirit.'[299] But the tremulation actually does not occur in the visible muscle fibres, which consist solely of contraction and remission; it takes place far beyond the bounds of the visible, in the tremulation of the membranes. To obtain knowledge of this invisible phenomenon, metaphorical thought comes to assist: 'As nature operates in the greater so she operates in the lesser. There is no difference. Why take refuge in the unknown just because we do not see? The things which we do not see are infinitely more than those which we see. If we do not see an insect, are we then to say it is [non]existent? that it lacks membrane? that it does not move mechanically, etc.?'[300] Arguments must always proceed from the big to the small, from the coarse to the subtle. What happens in the coarse can also happen in the subtle. From the senses we can deduce our way to the mind, proceed from coarse tremulations to subtle ones, for nature is always like herself. With our senses, then, we can obtain knowledge about the nature of the mind. As regards tremulations, Swedenborg suggests thinking in terms of musical harmony combined with a transverse modality in which different kinds of sensory impressions can be translated into each other. All enjoyable sensory impressions, as of cheerful colours or beautiful sounds, are a harmonious trembling movement in towards the mind, while the disharmony of tremulations causes ugly colours, dreadful sounds, smells, tastes, feels, etc (Fig. 13). The spread of tremulations can explain our ideas, dreams, and fantasies. Associations are tremulations which are caused by similar tremulations. A church can therefore conjure up an image of another church in our mind. If tremulations are given a suitable description and proof, the world will immediately smile in agreement.[301]

Swedenborg's theory of tremulations is permeated by music. The body is a musical instrument that vibrates. Sound has something to do with the very essence of life, it is something palpably physical and has an ability to set a person aquiver.

---

[299] *Psychologica*, 78, cf. 138.

[300] *Psychologica*, 30f, cf. 38–41.

[301] *Psychologica*, 148f, cf. 48f.

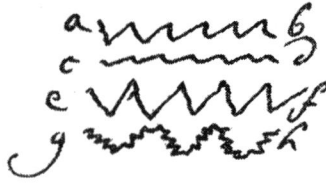

Fig. 13 The speed of tremulations. Long tremors, *ab*. A stronger tremor, *ef*, causes a stronger sensory impression which also absorbs the weaker, shorter tremor *cd*. Faster and slower tremors can also be combined, *gh*, in one and the same membrane, which means that stronger and weaker sensory experiences occur simultaneously. If the tremors are in harmony with each other, a person feels delight and pleasure (Swedenborg, *Psychologica* (1733–1734))

Ultimately, however, sound is not an independent phenomenon. It is possible to see the vibrations and explain them through geometry. Swedenborg's doctrine of the senses is an 'iatrogeometry' in which vision extends towards the other senses. All sensory impressions could be explained by the mechanistic natural philosophers in terms of geometry and mechanics; they could be returned to the fundamental primary properties. Each sense also corresponded to a specific element. In the history of ideas we find a 'hierarchy of the senses' in which they are ranked according to their ability to reveal the world as it is at its very core.[302] The prevailing idea is that vision is the most distinguished, most subtle sense. The domains of vision expand over all the other senses. One can *see* music, *see* sounds, tastes, smells, the sense of touch, and thoughts. To explain sound one could observe water waves and the swinging of ropes in mines, echoes could be described as halls of mirrors and harmonies as mathematical proportions. The body's sensory impressions, thoughts, and feelings could be seen with the assistance of metaphorical thought in waves, in vibrating strings.

Swedenborg's iatrogeometry is a Cartesian reduction of the senses to extent and movement. In his critique of Polhem's comparisons and analogies he seems to want to do without them, but despite this he himself often falls back on analogies. How could one otherwise understand what cannot be perceived with the senses, what is too small to be discovered with the eyes? With the aid of analogy, based on what can be seen in the visible world, such as waves on the surface of water, one could draw conclusions about the invisible world, a world that escapes our sight but nevertheless obeys the same mechanical laws. When the resolution of the microscope and the telescope gives way to obscurity, metaphors and expanded vision take over. Sound became air waves and light became ether waves—and waves became man's sensory movements, nervous impulses, thoughts, emotions, and dreams. Vision expands towards the invisible.

---

[302] Jütte, 61–71.

The Sphere

> But first tell me, [. . .] How soap bubbles can hold together and travel in the air for so long a time.[1]
>
> Swedenborg, *Miscellanea observata* (1722)

Hell Upon Earth

'Behold, a wonderful thing!' With a mixture of terror and rapture, Swedenborg descends into the Falun mine in 1716:

> I glide down from the upper world in a bucket,
> thus hanging, I am brought all the way to the dark shadows of death.
> But, as I moved to and fro hanging in the middle of the air,
> it was pleasant for me to sing holy hymns.
> It was pleasant thus to weigh this poor and fragile life,
> *which all depended on the power of a rope.*
> Behold,—in the recesses of the *mine* the band of Hades hurriedly rushes along,
> with dark faces wondering at me and my followers.[2]

The subterranean beings toiled in the mine, rolling rocks, sweating in the dusty air, throwing wood down the shaft, carrying torches. One went into a recess, another came out of a hole with his face scorched. Some climbed up quivering ladders, others were lowered, huddling in baskets that held only half their body. They followed winding veins through the ground. Some walked around and around in a circle leading stunted horses. The poem is summed up in the concluding words: 'Skill is not inferior to the material, nor is the material inferior to the skill.' This journey down into the interior of the earth describes the fragility of life hanging on a

---

[1] *Miscellanea*, 137; translation, 87.

[2] Swedenborg, 'In Fodinam Fhalunensem', *Ludus Heliconius*, edition, 122–125, commentary 218–221.

D. Dunér, *The Natural Philosophy of Emanuel Swedenborg*, Studies in the History of Philosophy of Mind 11, DOI 10.1007/978-94-007-4560-5_5,
© Springer Science+Business Media Dordrecht 2013

thread, the mine as hell upon earth, but also the union of art and nature. This chapter about the sphere concerns metaphors proceeding from life around the mine, the machines, the processing of the ore, the war industry, but also the round peas and cannonballs that were used in order to understand the structure of subtle matter. The thesis is that abstract ideas rest on experiences of the concrete world. The focus is on Swedenborg's theory of matter, according to which the world is made up of small spherical bubbles.

The experience of the horrors of the mining industry made an indelible impression on him. It became an image, a metaphor for hell, that he carried with him even into the hells of the spirit world. By conjuring up images of the inhumanity of the mines he was able to imagine what hell must be like for a God-fearing person. The mine was conflated with the last judgement and classical mythology in Swedenborg's thought.[3] He saw how the copper mine in Falun extended all the way down to Tartarus and Pluto's realm. In another poem he noticed how the ironsmiths had been deafened by the hammer blows and blackened by the sulphur. 'Just add the wrinkles—and you had the face of Charon', he wrote.[4] The Falun mine as hell upon earth frightened and fascinated many visitors.[5] The miners reminded them of the cyclopes, assistants of Hephaestus in his forge under Etna. Regnard encountered terrible cyclopes of this kind in a grim world. Aubry de La Motraye, who had visited Polhem at Stjärnsund and corresponded with Benzelius, described the great labyrinth where the workers involuntarily dug their own graves. Bishop Swedberg saw the damned being consumed in the blast furnaces. And then there is the best-known account, that of Linnaeus from his tour of Dalarna in the summer of 1734, when the Styx and Pluto's realm made his hairs stand on end. The inferno of the mine recurs time and again in various books on mining, as in what was still the major work on the subject, Agricola's *De re metallica* (1556), which Swedenborg cites so often. This tells of pollution and hazardous gases, of mine slaves and gnomes in the subterranean world.[6]

Stora Kopparberg, the great copper mine in Falun, represented both horror and wealth. It was Sweden's chief source of income, but also a black hole that swallowed huge quantities of wood for the fire-setting method, for the smelting and roasting furnaces. The crater of the great pit, the sterile desert surrounding the mine, the thick, toxic sulphurous smoke from the roasting of the ore led many visitors to think of the volcanoes of Vesuvius and Etna. The destruction of the environment, which also had palpable economic consequences, was discussed in Swedenborg's Board of Mines. In the late summer of 1730 Swedenborg and the assessor Johan Bergenstierna inspected the forests around Falun.[7] At one of the very last meetings

---

[3]*Camena Borea*, edition, 74f; cf. Virgil, *Georgica*, 4.170–175; Virgil, *Aeneid*, 8.443–454.

[4]Swedenborg, 'In Fabros qvi majores ferri massas formant', *Ludus Heliconius*, edition, 122–123, commentary 218.

[5]Regnard, translation, 139ff; La Motraye, translation, xi, 171, 179; Swedberg (1941), 49; Linnaeus (1953), 5, 148–150, 395, 397; Dunér (2008b), 209–233.

[6]Agricola, translation, xxvi, 112, 214, 217f.

[7]RA, Bergenstierna and Swedenborg; presented 16 March 1734, read 16 February 1737.

in which Swedenborg took part, on 30 March 1747, there was a discussion of goats. He agreed with the other assessors that 'Goats are harmful livestock for the forests,' but expressed the opinion that the peasantry should nevertheless have the freedom to let their goats graze on their homesteads until the next session of the diet, to avoid destitution.[8] In the descriptions we see a certain measure of social compassion. Swedenborg and the other visitors could sympathize with the workers' living conditions by evoking biblical and mythological references, how the miners were obliged—like Sisyphus and Ixion—to toil hopelessly and endlessly, tormented by shortage of breath in this earthly vale of tears. But above all, the haunting experience affected the visitors themselves. Swedenborg sang hymns and was reminded of the fragility of life. The opposite of the underground was the garden with its fruit-bearing trees and healing plants. For the mining assessor and gardening enthusiast, earthly life thus became a place between hell and paradise, between the mine and the garden. In metaphorical thought, hell and paradise are here on earth.

Flying in the Air

Swedenborg's theory of matter has its origin in mining and its links to chemistry and engineering. Chemistry was significant for mining because it was used to classify and identify the essential properties of different substances and to improve the processing methods. Mining was high technology for its time in that it mechanized human labour and made the machine power as efficient as possible, as in the devices for hoisting up ore and conveying the energy produced by water power. This introduction to Swedenborg's engineering activity is chiefly intended to show how such knowledge was transferred to natural philosophy. The basic idea is that natural science is applied technology. For Swedenborg and other mechanistic natural philosophers there was no real difference between the mechanism of the microcosm and of the macrocosm, between people's artificial mechanics and the subtle natural mechanics of matter and the celestial bodies. Natural philosophy was modelled on human technology.

During his first voyage abroad, Swedenborg designed amazing inventions. He was no doubt inspired by the fantastic machines of Sturm, Schott, and Wilkins, such as various optical instruments, air pumps, diving bells, and much besides. In London he read Wilkins's *Mathematicall Magick* and perused summaries of *Philosophical Transactions* in which he was able to find descriptions of everything from submarines to Roger Bacon's flying machine. Wilkins's mathematical works are 'very *ingeniousa*', Swedenborg explained.[9] He visited instrument makers and artisans, and as Andreas Hesselius and Göran Vallerius had done before him, he admired the artful timepieces of the London clockmaker Antram, which ran on

---

[8] RA, Bergskollegii arkiv, AI:107:1, 988f.

[9] *Opera* I, 215, cf. 210, 214; *Letters* I, 30, cf. 22, 28; *The Mechanical Inventions*, 9, 20.

light alone.[10] From Rostock, on 8 September 1714, Swedenborg sent Benzelius a list of inventions, one more imaginative than the other. He was busy developing fourteen inventions, including a submarine, various pumps, a crane, sluice gates, a drawbridge, an air pump, an airgun, a 'Sciagraphia', that is, a way of projecting shadows and making sundials, a water clock showing all the mobile bodies in the heavens, and a flying chariot.[11]

'Lo! Daedalus travels through the air, and he laughs from above / at the traps that King Minos has set for him on the earth,' Swedenborg declaims in an introductory poem in *Dædalus Hyperboreus*.[12] It was the science of mechanics and Polhem, 'the Swedish Daedalus', travelling to the heights, but also a flight from the country's enemies. A great deal of space in *Dædalus Hyperboreus* was devoted to technical innovations, with articles about mining machines, speaking tubes, air pumps, and a flying machine. Among other things, Swedenborg describes his improvements to a sophisticated hoisting device that was important for the mining industry. In the second issue there is a description of Polhem's hoist at Blankstöten, the famous *hakspel* with its wooden poles with catches, which was supposed to be more efficient than the old devices, saving leather ropes and thus being cheaper and more durable.[13] A noticeable feature of the mechanical descriptions is the significance he attaches to the geometrical description. Using terms from geometry, such as cylinders, distances, proportions, diameters, radii, centres, and surfaces, he tries to describe what mobile geometry, that is, mechanics, must be like in its perfection.[14] With letters referring to the different parts of the machines on the illustrations, he signals rationality, mathematics, and geometry.

The most fantastic machine is Swedenborg's 'Draft of a machine for flying in the air', which he designed around 1714 and later published in *Dædalus Hyperboreus* in 1716 (Fig. 1). It is original in the sense that it is a kind of glider that is heavier than air, with fixed bearing wings instead of the ornithopters' imitation of flapping bird wings. But the birds can nevertheless show us the way. Swedenborg proceeds from nature's mechanisms, which are then transferred to human mechanics in metaphorical thought, from the gliding flight of birds of prey—the eagles, kites, and falcons above his head—to a human replication. By observing birds, he says, we can invent a machine whose wings can take us up into the air even if we have nothing but the wings of reason.[15] We can learn something from 'eagles and kites, which can lie still on their outspread wings, or can sway in the air.'[16] In the manuscript he

---

[10]*Opera* I, 221; *Letters* I, 43; Hesselius, 106; Rydberg, 287f.

[11]*Opera* I, 225f; *Letters* I, 57f.

[12]*Ludus Heliconius*, edition, 134f; Ovid, *Metamorphoseon*, 8.162–235; cf. Edwards, 45–54.

[13]Polhem, 'Assess. Chr. Polhammars vpfodrings konst wid Blanckstöten', *Dædalus Hyperboreus* II, 25–28; *Polhems brev*, 118; cf. Brückmann, 237, tab. VI.

[14]Swedenborg, 'Then andra opfodrings konsten', *Dædalus Hyperboreus* I, 20–23, cf. 14–20.

[15]*Dædalus Hyperboreus* IV, 80.

[16]*Photolith.* I, 21; *Machine att flyga i wädret*, ed., 39; cf. *Dædalus Hyperboreus* IV, 82; *The Mechanical Inventions*, 23.

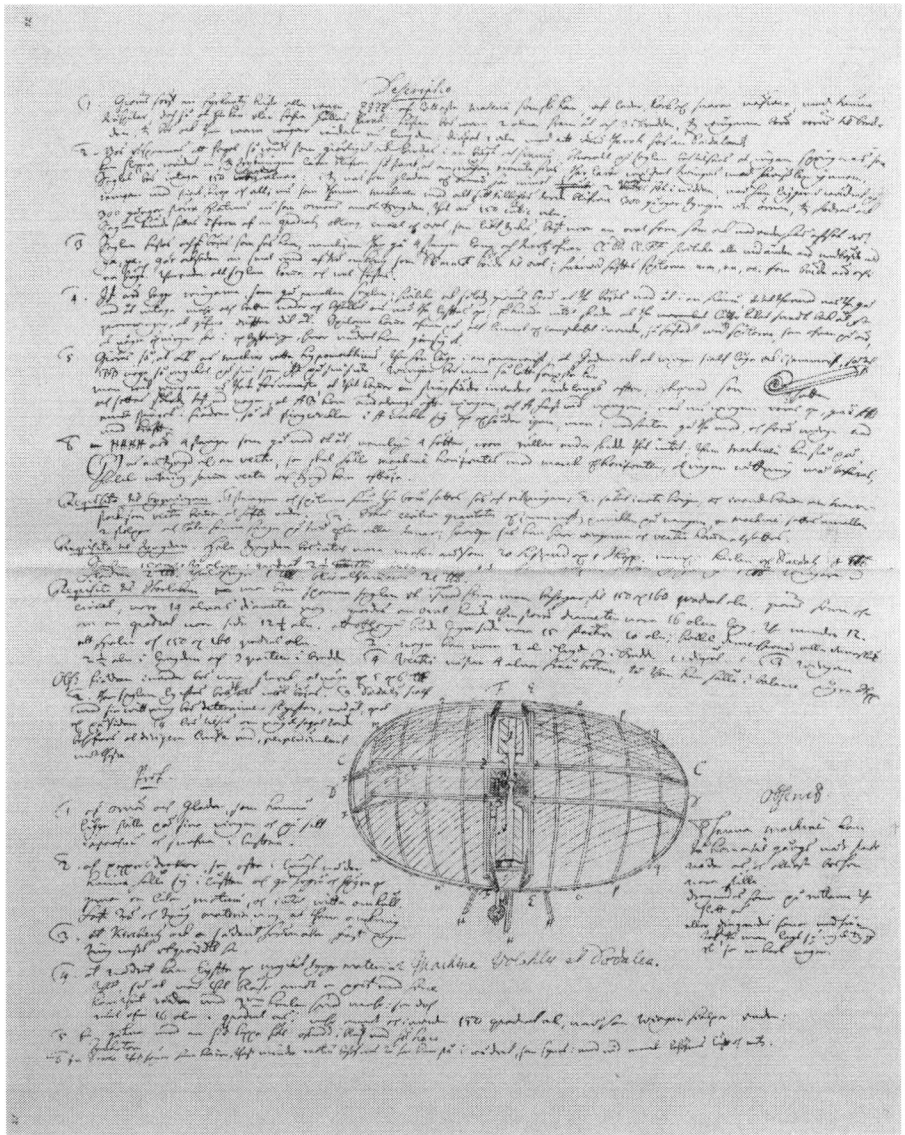

Fig. 1 Machine flying in the air (Swedenborg, *Descriptio machinae Daedaleae sive volatilis* (1716))

goes on to tell of a schoolboy wearing a wide cloak, who was blown by the wind from the tower of Skara Cathedral yet landed unhurt. In *Dædalus Hyperboreus*, however, the schoolboy comes instead from Strängnäs! The sail could be of square, rectangular, circular or, best of all, oval shape, although he does not explain why this should be so.

In an article about air resistance in the same issue of *Dædalus Hyperboreus*, Polhem calculates that if one divides an inch-cube or inch-sphere of copper into 675,269,485,416,752 cubes or spheres, the result would be 'scent', copper fumes with the same lightness as air.[17] Metals can be made to hover and swim in the air. Polhem's notes on air resistance could have given some guidance to Swedenborg, although Polhem explained that much still remains unknown about aerodynamics, not least 'about the strange vortices that exist around all travelling and moving things, such as ships and other things.'[18] In the learned literature, as in his copy of Michael Bernhard Valentini's *Museum museorum* (1714), Swedenborg could read about several different aeroplane projects, such as the sailing aircraft designed in 1709 by the Brazilian monk Lourenço de Gusmão, the 'Passarola', about which there was much talk at the time.[19] Swedenborg himself mentions Kircher's kites, and he refers to Francesco Lana's flying boat from 1670 which Sturm had described. But Swedenborg considered it impossible to try to take off from the ground with the aid of 'large globes evacuated of air'.[20] Swedenborg also cites Fontenelle, whom he may possibly have run into in Paris when he mixed in the circles around the French Academy of Sciences. In *Entretiens sur la pluralité des mondes* (1686) Fontenelle looks at the undreamed-of future prospects of mechanics:

> The Art of Flying is but newly invented, it will improve by degrees, and in time grow prefect; then we may fly as far as the Moon. We do not yet pretend to have discover'd all things, or that what we have discov'rd can receive no addition; and therefore, pray let us agree, there are yet many things to be done in the Ages to come.[21]

Swedenborg was dazzled by the effectiveness of mechanics. Polhem, on the other hand, was not impressed by Swedenborg's strange machine: 'As regards flight or flying artificialities, the difficulty must be the same as making a perpetuum mobile, gold, and other artificialities, although at first sight it seems no less feasible than desirable.'[22] The size is significant. For nature denies one thing: all machines do not retain the same proportion in large as in small. The fact is that the weight increases in proportion to the cube, while the surface increases only in proportion to the square. The dreams of flight in the literature surely inspired Swedenborg, but perhaps they did not directly affect him in his own visions. If anything influenced Swedenborg it was birds.

Swedenborg's flying machine expresses the dream of flying, the freedom and independence of gravity and earth, not needing to be excluded from the upper

---

[17]Polhem, 'Assessor Polheimers wissa anmerckningar om wädrets resistence mot fallande tyngder och areer', *Dædalus Hyperboreus* IV, 69f; *Collegium Curiosorums protokoll* 20 June 1711, 65f; *Opera* I, 254; *Letters* I, 109.

[18]Polhem, *Dædalus Hyperboreus* IV, 77.

[19]Valentini, III, 34–38.

[20]*Dædalus Hyperboreus* IV, 80; *The Mechanical Inventions*, 24; Kircher (1665), II, 479; Sturm (1676), I, 56–66; Söderberg (1988b), 13; Soderberg (1988a), 82f, 92.

[21]Fontenelle, 51f; translation, 63; *Dædalus Hyperboreus* IV, 82.

[22]*Dædalus Hyperboreus* IV, 82; *Polhems brev*, 123.

element. But perhaps it is not least of all the belief that nothing is impossible for mechanics. Yet this need not mean that he believed in the practical feasibility of his own machine. As in many other cases, as with Sturm and Schott, the inventions were not always envisaged as being used; they were instead intended to demonstrate what might be possible, or simply to display the inventor's own mechanical genius. There is nothing to suggest that Swedenborg made any model of his aircraft, much less any practical experiments. There is a risk, he says, that we could be obliged to sacrifice an arm or a leg in an experiment. The machine for flying into the air was signed 'N.N.', perhaps because he wanted to be incognito if it incurred criticism, although a narrow circle of people probably knew who the author was.[23] One could otherwise interpret Swedenborg's fantastic inventions as a kind of advertising, in which he publicly proclaimed his technical genius and offered his mechanical skills to anyone who felt inclined to speculate. Look what I can do! Invest in me!

There is a visionary element in all this inventiveness. At bottom it is a non-verbal mode of thought that is revealed in Swedenborg's technical inventions. Scientific ideas and technical solutions exist first as 'visions', as wordless images in the mind. Then the natural philosopher can attempt to 'translate' the mental image into words or transfer the inner picture to a drawing on paper in order to arouse similar mental images in a constructor who finally turns the idea for the machine into a three-dimensional model.[24] Swedenborg's inventions are based on precisely these mental images. He did not try things out by hand. It was the same with his mechanistic natural philosophy. Just as the engineer thinks in pictures, so too does the natural philosopher. The mechanistic natural philosopher has an inner picture of the geometry of microscopic reality which he tries to convey through text and illustrations. The pictures in Swedenborg's works are attempts to capture his own mental images. They are two-dimensional models which are not primarily intended to show what the world really looks like, but rather seek to communicate an understanding of how it works, how the world machine functions.[25] With mechanical and geometrical visualization he sought to understand nature. It is machines he designs in the mind's creative process. There are undoubtedly personal features in Swedenborg's non-verbal thought. He has his special mental images that colour his scientific theories. The scientific idea is an expression of personality, revealing a personal stance and understanding of the world around him. Swedenborg imagined and visualized nature's hidden structure as bubbles, points, and spirals. Science as visualization is even more clear in Polhem, who pictured the inner structure of matter as cogs, cannonballs, and peas, which is such a palpable visualization that it feels as if one is entering his mechanical laboratory, an ammunition store, or a market with peas and linseed. Traditionally

---

[23]The author is also anonymous in the advertisement in *Stockholmiske kundgiörelser* 2 April 1717, 1f; Swedenborg's anonymous authorship of other works is exposed in Stiernman, I, 45–50, cf. vi–viii.

[24]Ferguson, 828.

[25]Baigrie, 116.

such perceptual processes have not been perceived as true rational thought, and it has been assumed that non-verbal thought is one of the more primitive phases in the cognitive process, inferior to verbal and mathematical thought. As regards Swedenborg's theory of matter, I shall instead try to show that the non-verbal internal mental images are of crucial significance for the creation of new ideas, even that non-verbal communication and perception can serve as a foundation for verbal and mathematical thought.

The Geometry of War

Dædalus Hyperboreus is imbued with gunpowder smoke, cannonballs, and bomb explosions, reminding us that the country, a declining war machine, was engaged in a protracted war. As assessor in the Board of Mines, Swedenborg became part of the military complex. The mining industry was the adjutant of war with the constant demand for iron and copper, and mechanics was dictated by military needs. The martial element in *Dædalus Hyperboreus* is partly a consequence of the restless times, but also a way to demonstrate how natural philosophy could benefit a military power in its practical, concrete strategy. The war also gave rise to interesting mechanical and mathematical problems. The journal proclaims the utility of mathematics and science for artillery and ballistics. Mathematical knowledge was needed for calculating not just bomb parabolae but also piles of cannonballs and the resistance of bullets. Experiments with firearms or ballistics could naturally capture the attention of the warrior king. Swedenborg himself designed a machine gun, an air gun that would fire up to 11,000 shots an hour (Fig. 2).[26] Daedalus placed himself at the service of war, so that he could also gain personal advantage from it. Polhem points out, in a transcript by Swedenborg, that when the kings come to the regiment and immediately become involved in war, it is essential for engineers to watch their interests 'assiduously and milk the cow while two are quarrelling over it.'[27] This passage was later deleted. The rhetoric of war also invaded the language itself, giving it numerous metaphors. Natural philosophers stormed the malicious enemies of the truth, fought on the academic battlefield, met in duels at disputations, triumphed, incurred wounds, or were annihilated. After Charles XII had fallen on the worldly field of battle, the military rhetoric almost entirely vanished from Swedenborg's texts.

In the article 'Ett experiment om watns och snös resistence mot kulor' ('An experiment on the resistance of water and snow against balls', 1716) Swedenborg

---

[26] *Opera* I, 230f; *Letters* I, 65; *Dædalus Hyperboreus* III, 50. Only the figure, without the legend; cf. *Opera* I, 226; III, 278; *Letters* I, 57.

[27] Polhem, 'Instruction för alle våre consuler, comisshafvare och factorer på dhe orther, hvarest dhe vistas' (1716), *Polhems skrifter* II, 66; Swedenborg's transcript, LiSB, N 14a, no. 17, fol. 31–36; *Photolith.* I, 17.

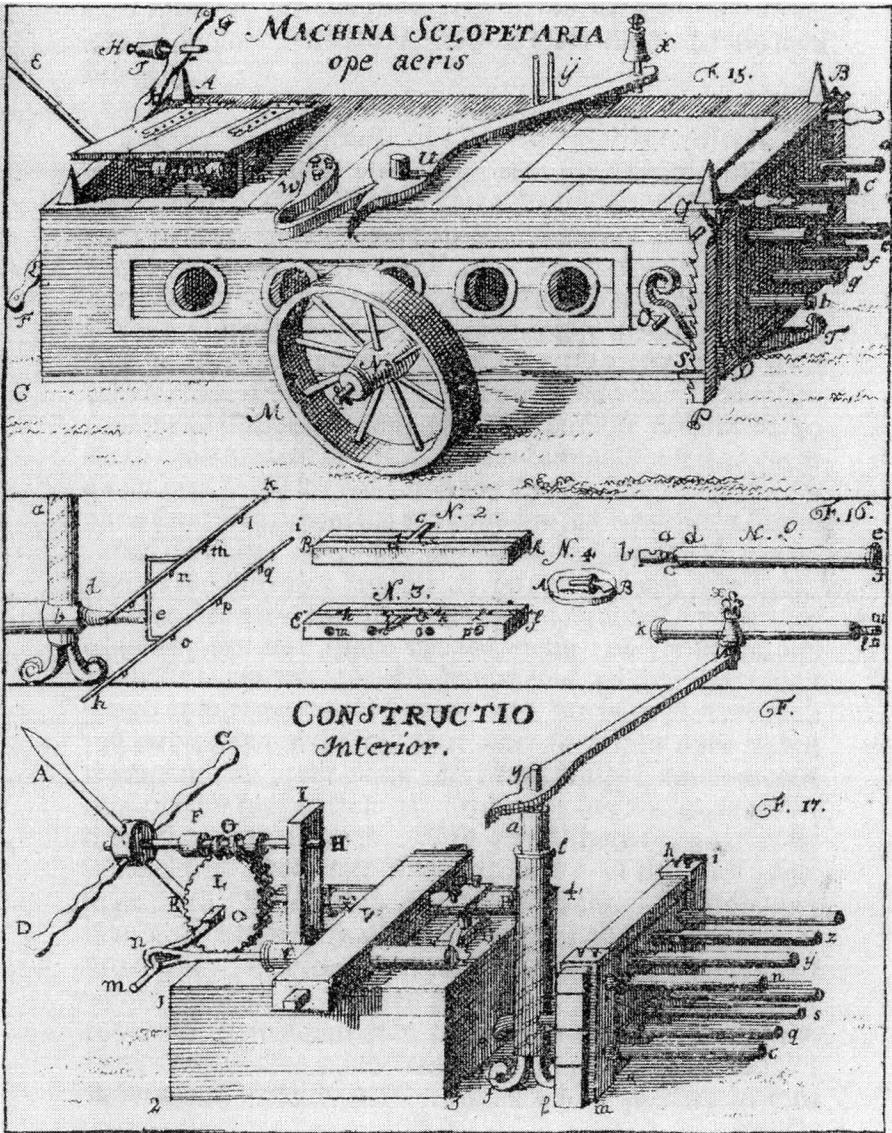

Fig. 2 Machine gun powered by air (Swedenborg, 'Machina sclopetaria ope aeris', *Dædalus Hyperboreus* (1716))

reports on a ballistic experiment he performed together with his cousin: 'Last winter, when little yet level snow remained on top of the ice, to a thickness of 4 or 5 in., I accompanied Joh. Hesselius, Provincial Physician in Västergötland, with a rifled musket.' Swedenborg and Hesselius paced the ice and measured the distances between the bounces of the ball in the snow. From this experiment Swedenborg drew the conclusion that even the lightest substances such as snow and water can exert a noticeable resistance against even the most violent of balls, 'as a thin and trembling leaf can resist a ball, and not permit its progress, unless it makes a hole and breaks its way through.' Moreover, the experiment showed that, the smaller the angle of incidence was, the less power the ball had to break through. The length of the bounces decreased in continuous geometrical proportion, 'as approximately 40, 32, 25, 19 &c.'[28]

In the last issue Polhem, who often cites martial examples and military inventions, poses the question whether balls fall more steeply at the end of their trajectory or follow the parabola the whole time. The latter turned out to be correct, 'all bombs, balls, follow the parabola and nothing else,' as François Blondel and others had also demonstrated.[29] Ballistics is a clear example of the alliance of mathematics with war. The capacity of cannon and firearms had increased, and the officers needed to be able to calculate the angle of the barrel and the distance to the target. Swedenborg's old schoolbook, Bilberg's *Elementa geometriæ*, based on Blondel's art of throwing bombs, no doubt lay close at hand. Also in his library was Daniel Grundell's *Nödig underrättelse om artilleriet till lands och siös* ('Necessary information about artillery on land and at sea', 1705), which deals with topics such as practical ballistics, how a high-elevation shot travels with a violent drive from the cannon and blends with the natural motion.[30] The parabolae of ballistics are naturally included in Swedenborg's *Regel-konsten*. He discusses gun shooting, ballistics, and the power of gunpowder. A ball is projected furthest when the cannon is set at an angle of 45°, whereby the powder provides the force, the speed gives the propulsion, the time determines the altitude, and the ball causes the effect. But had it not been for the air resistance, the ball could have flown 'like a bird through the air', always at the same speed.[31]

The ability of geometry to bring discipline and order also came in handy in the geometry of war. While travelling through Germany in 1733 Swedenborg was fascinated by the marching grenadiers and the mounted gendarmes of Brandenburg. The whole squadron was like a machine put in place, moving instantaneously as the engineer pleased.[32] In battles during this period it was important to arrange the troops in regular formations, with order and discipline, in columns and squares, on easily surveyable battlefields where there was scope for ideal geometry. The infantry was reduced to geometrical units in new formations, in a geometrical ballet moving in time to bassoons, shawms, kettle drums, and trumpets. Another

[28]*Dædalus Hyperboreus* IV, 84–86; cf. Swedenborg, *Pro memoria om några prof och experimenter* (1716). LiSB, N 14a, no. 47, fol. 125; *Photolith.* I, 92.

[29]Polhem, 'Ett prof at wisa bombers och kulors bogskott: giordt och anstelt af herr commercie-rådet Christ. Polhem', *Dædalus Hyperboreus* VI, 2; Polhem (1761), 119, 122.

[30]Grundell, 150–152.

[31]*Regel-konsten*, 116.

[32]*Resebeskrifningar* 3–4 June 1733, 11, cf. 7f; *Documents* II:1, 15, cf. 10.

aspect of the geometry of war is fortification. This art became a kind of applied geometry, a geometrical formalism after Vauban, as earthworks and redoubts were built in regular polygonal forms to withstand the new artillery. With their rules and compasses, mathematically trained fortification officers drew star-shaped, polygonal bastions, well-balanced donjons or fortified towers, casemates where cannon were behind thick walls, safe from bombs. The fortification officer was simultaneously an engineer, an architect, and a mathematician, and besides mastering mechanics and optics he also had to know arithmetic and geometry, especially trigonometry and stereometry.[33] Swedenborg's nephew, Jesper Albrecht Benzelstierna, displayed considerable knowledge of infinitesimal calculus, algebra, and analytical geometry in his examination at the fortification office in 1737. Among other things, he was asked, 'What is the beginning of all things?' He replied:

> The beginning of all things is a thing that does not consist of any parts and is called a point; this, however, does not extend beyond geometry, since in rebus materialibus no such points or monads exist, as all matter can be divided ad infinitum, although not actually but potentially.[34]

This answer suited the fortification office nicely, but did not fit so well with Swedenborg's *De infinito*, a copy of which he had received from the author himself. A few years later Jesper Albrecht drowned during the war against Russia in 1743.

In Berlin in 1733, Swedenborg was struck by the geometry of the buildings. What pleased the eye and exhilarated his senses most of all was the wonderful symmetry and uniformity of the houses.[35] Geometry had an order that contrasted with the uncontrolled, random character of older cities. It was highly visual, which made the discipline aesthetically attractive. Geometrical forms, such as the straight line, the circle, and the polygon, had an aesthetic significance. They were considered more beautiful than the irregular forms. Moreover, as Gabriel Polhem emphasized, the mathematical sciences were useful for the art of building.[36] The carpenter's square was a symbol of symmetry. Baroque urban planning was an extension of fortification geometry and the Renaissance grid plan, which cherished the classical, balanced architecture of Vitruvius. The recently founded town of Karlskrona, where Swedenborg assisted Polhem in his work with the dock, was in baroque style in accordance with Dahlbergh's plan from 1683, but not as explicitly geometrical as his earlier unrealized plan for Landskrona in 1680. Landskrona was envisaged as an ideal, utopian town with two identical churches and a square, proceeding from circular bastions in which the streets were laid out according to a symmetrical grid from a main axis and a canal view out into a distant perspective.[37]

The regular geometry of cities, cultivated landscape, gardens, and fortifications represented order, control, civilization. With the regular forms and symmetry, the

---

[33] Nisser, 149.

[34] KrA, Benzelstierna, fol. 51; Benzelstierna's copy of Swedenborg's *De infinito* is at SLBA.

[35] *Resebeskrifningar* 2 June 1733, 10f; *Documents* II:1, 14.

[36] See G. Polhem.

[37] Snickare, 54f.

formal patterns and the order distinguished mankind and the cultivated landscape from the savage state and drew a boundary between culture and nature. The strict geometrical symmetry and long perspectives of the baroque garden may be seen as a peaceful variant of the art of fortification. They had the same geometrical ideal. Both followed the demands of symmetry and geometrical forms.[38] Swedenborg's natural philosophy is similar in this respect. He tried to describe nature and matter with geometry so as to make them ordered, regular, and easily surveyed.

Nature—A Composite Analogy

With geometry the world could be measured. Geometry was divided into longimetry, planimetry, solidometry, and stereometry, that is to say, the measurement of lengths, plane surfaces, and solid bodies. The latter in particular, stereometry, occupied Swedenborg in several manuscripts and essays. One reason was that trade in commodities such as grain and peas, which were sold in cylindrical or cubic containers of different sizes, gave rise to stereometric problems, and stereometry was of special significance in Swedenborg's geometrical theory of matter. Swedenborg starts the manuscript *Proportiones stereometricæ* (1716) with the weight of different substances in ounces per cubic inch, followed by some stereometric proportions between the volume of a cube and a sphere, and between a rectangle and an oval, inscribed in each other.[39] The inspiration for these stereometric calculations probably came from Stiernhielm and his Linea Carolina or Carl-Staf, the yardstick in which Polhem and the Collegium Curiosorum displayed such great interest.[40] This was because the scales on Stiernhielm's rod could be used to convert a square to a circle of the same area, and showed how cylindrical measures could be converted into cubic ones, and a sphere to a cone of the same volume.[41] It is not impossible that Swedenborg actually used the rod to perform his calculations. Yet in the margin of the tables of the stereometric properties of water, iron, and lead he has noted 'non tamen ille'—but not that. Consequently, these tables are not found in *Dædalus Hyperboreus*. Swedenborg, however, made a fair copy of the stereometric proportions, *Uträkning på watns, medeljärns, blys caliber när tyngden är lika; från unce til marck* ('Calculation of the calibre of water, medium iron, lead when the weight is equal; from ounces to marks', 1716).[42] It contains a table showing the size of a cube, a cylinder and a globe with the same weight of the specific substance, along with a table of weights and stereometric rules. Presumably

---

[38]Dahl, 181; Thomas, 256f.

[39]Swedenborg, *Proportiones stereometricæ* (1716). LiSB, N 14a, no. 52, fol. 131; *Photolith.* I, 100f, cf. II, 44.

[40]*Polhem's Letters*, 46f, 49f; *Polhems skrifter* III, 115, 124, 228.

[41]Ohlon (2000), 194.

[42]LiSB, N 14a, no. 16, fol. 30.

this manuscript was intended for inclusion in *Dædalus Hyperboreus*. It is not there, however, but there are three other articles, about the air pump and stereometry, using these calculations for the numerical proportions of different geometrical bodies.

When Swedenborg describes the conical and cylindrical forms of the air pump, he makes use of these stereometric rules. He compares the volume of a cube and an inscribed cylinder, cone, and sphere. In his characteristic way, however, he skips the mathematical proof: 'Demonstratio. To show all the above would be too circumstantial. I will however give some slight opportunity to those whom it pleases to turn their thoughts to this, to sharpen their minds.'[43] A stereometric essay follows an article about one of the more curious inventions of the Swedish Dædalus, 'Commercie Rådet Polheimers Konstige Tapp' ('The artful tap of Commercial Councillor Polhem'), in other words, a device for saving wine by preventing maids and menservants from deriving 'joy and pleasure at the expense of their lord'.[44] Swedenborg's more sober essay on stereometry in the same issue from 1717 is entitled 'En tafla på Cubers, Cylindrers och Sphærers innehåll, när man tager sidorna i wissa tum' ('A table of the content of cubes, cylinders, and spheres, when the sides are of certain inches'). In this he compares the volume of cubes, cylinders, and spheres of the same height and width in a way similar to that employed in the earlier drafts. For example, a 'cube is related to an enclosed cylinder as 14 to 11, to an enclosed sphere as 21 to 11.'[45] The latter proportion in particular, that between the cube and the sphere as 21 to 11, should be borne in mind. It recurs in many contexts in Polhem's and Swedenborg's stereometric theories of matter. It is precisely this proportion that Polhem cites for the round particles in a cubic structure. Swedenborg goes on to state the proportion of a cylinder to an inscribed cube as 11 to 7, that of a sphere to an inscribed cube as 62 to 41, and finally a proportion known since Archimedes—that between a sphere and a cylinder with the same base and height as the diameter of the sphere—that is, 2 to 3. This was not just a harmless exercise with figures. With these stereometric rules, wine can be saved. The smaller the surface area of a decanter in relation to the volume, the less is the risk of evaporation.

In yet another article, 'Ett lett analytiskt sett at demonstrera så thet föregående som annat dylikt geometrice' ('An easy analytical way to demonstrate both the preceding and other such geometrically'), Swedenborg performs calculations with the aid of analogies or proportions: 'now since equations are merely central points or equal to a concentration of some proportions, their utility does not extend very far in comparison to the benefit if one proceeds from the centre to the periphery, or from equalities to analogies'.[46] Once again he puts forward the same proportions of volume as before between the geometrical bodies. If 'many globes or spheres were to be stacked in hexahedral form, and a square house were made precisely

---

[43] *Dædalus Hyperboreus* III, 63.

[44] Polhem, *Dædalus Hyperboreus* V, 100.

[45] *Dædalus Hyperboreus* V, 116.

[46] *Dædalus Hyperboreus* V, 126.

around them, the volume of the house in proportion to that of the spheres together would be 21/11.'[47] The mental image of the round particles of matter that he later developed proceeds from this concrete image of cannonballs in an ammunition house. This is analogical thinking, metaphorical thought. Swedenborg underlines the importance of demonstrating 'analogies and proportions', since their utility extends further into mechanics as in all kinds of machines, sizes, weights, and movements, and also in the physical proportions to which nature is bound 'and plays throughout her tour, before she comes to equilibrium'.[48] What needs to be clarified is that Swedenborg uses the term *analogia* in a mathematical sense as an expression of numerical proportions, not in the linguistic sense. Swedenborg used the Swedish word *genlikhet* (literally 'counter-likeness'), borrowed from Stiernhielm's *Archimedes reformatus* where it was the vernacular counterpart to the Greek *analogia* and the Latin *proportio*. 'Anima mundi,' Stiernhielm writes, 'the omnific power of the living world, and life-giving spirit' means that the world 'is nothing but a common accord, a convenient proportion'.[49] The world is a harmony of analogies and proportions. Nature is a composite analogy, and everything takes place with proportions in Swedenborg's universe:

> Now as nature plays all her works through proportions before she comes to her equilibrium, one can appropriately say that nature is nothing but a composite analogy; and although one thing may appear to be different from another, it is nothing other than a composition of analogies, which is evident from the harmonic proportion, which is a geometrical and arithmetic counter-likeness together [. . .].[50]

Everything would be easier to understand and would be more natural if analogies were used instead of equations. In nature both geometrical and arithmetic proportions are at work.

One should thus compare things with each other, arrange them in proportions, find analogies between them. 'The art of comparison' is precisely about the rules of proportion, Polhem informs us in *Wishetens andra grundwahl*.[51] Analogies are a matter of considerable interest in Swedenborg's *Geometrica et algebraica* and later in his book about algebra from 1718, where he explains that 'the art of rules'

> deals with everything that has a likeness or counter-likeness of whatsoever nature it may be, that is, everything that concerns an equilibrium or an analogy. There is nothing in the world that does not have a likeness or counter-likeness in itself; all this is comprised in algebra, which seeks out the one and the other in it: her measure is as yet small, but is increasingly beginning to grow, and will, no doubt, through time extend to quite a lot of which we are now ignorant.[52]

---

[47] *Dædalus Hyperboreus* V, 128.

[48] *Dædalus Hyperboreus* V, 130.

[49] Stiernhielm (1644), dedication.

[50] *Dædalus Hyperboreus* V, 132.

[51] Polhem (1716), § 4.

[52] *Regel-konsten*, 2; *Photolith*. II, 30f, 33f.

The world's analogies require mathematical knowledge, and Swedenborg explains: 'Proportio vel regula de Tri' is 'Counter-likeness or the golden numbers she distinguishes through points . . . . E.g. 2.5:4:10. means that 2 is to 5 is as 4 to 10.'[53] 'A euphonious counter-likeness' which is found in the hyperbola is three numbers in geometrical proportion when the difference between the first two is proportional to the difference between the last two, as the first number to the last: 'E.g. 2.3:6. is 3–2.6–3: as 2 to 6.'[54] Naturally, he also considers stereometric proportions in *Regel-konsten*, but as relationships between Swedish terms, such as *omkast* (area), *rull* (a 'roll' or cylinder), *spitsig rull* (a 'pointed roll' or cone) etc.[55] He also treats the weight proportions of water and metals and the proportions between geometrical quantities.

The mutual proportions of things were a constantly recurring theme in Swedenborg's natural philosophy. Everything was part of a universal analogy. Through the concept of analogy, Swedenborg's mechanistic perception of nature takes on an element of Pythagorean philosophy, of a world in numerical relations, a quest for proportions, the conviction that everything is in harmony. In a fragment about infinity from 1733 he describes reality's chain of relationships. Two things form a proportion, proportions between themselves form an analogy, many analogies form a harmony, and a multitude makes up a series.[56] Pythagorean proportions are a prominent feature throughout his works. One of the clearest examples of mathematical analogy can be found in *Clavis hieroglyphica* (1742): 'As the world stands in respect to man, so stand natural effects in respect to rational actions.'[57] As the world relates to man, so man relates to God, and human actions also relate to divine actions. In formal terms this becomes: M.H:E:A, where *M* stands for the world (*mundus*), *H* for man (*homo*), *E* for effect (*effectus*), and *A* for actions (*actiones*). Swedenborg's hieroglyphic key is a kind of universal mathematics with analogies, where scientific statements are transferred to a psychological and then to a theological thesis. Everything in nature, he says, is the 'type, image and likeness of some one among spiritual things, all which are exemplars'.[58] This resulted in the visionary's doctrine of correspondences.

What must be emphasized as regards Swedenborg's theory of matter is his use of analogies and metaphors. Analogy is a part of metaphorical thought. By analogy he really means conclusions founded on similarities between two domains, that is, between two objects, classes of objects, or systems.[59] In traditional logic, analogy is an inductive form of thought which says that if two or more entities are like each

[53] *Regel-konsten*, 9.

[54] *Regel-konsten*, 45, cf. 88.

[55] *Regel-konsten*, 106f.

[56] Swedenborg, *Fragmentum de infinito* (1733). KVA, cod. 65, 669; *Photolith.* III, 168; *Treatises*, 115.

[57] *Clavis hieroglyphica*, 9f; translation, 168; Jonsson and Hjern, 70.

[58] *Clavis hieroglyphica*, 19; translation, 183.

[59] Meheus, 24f, 26; Gentner and Jeziorski, 299–301.

other in one or more ways, then there is a probability that they may also be alike in other ways. If one is faced with a new problem that cannot be solved by deductive means, one can either chance different possible solutions and then test them, or one can utilize similarities to previously solved problems and transfer the information in the source to the target domain. In other words, analogy is not just a pedagogical aid but also a problem solver, a heuristic tool for arriving at suggested solutions based on the structure of the problem rather than its external properties. Based on one's everyday experience, with the aid of analogy one can find explanatory models for the unknown. One compares problems, seeks parallels, favours regularity; one points out similarities and shared abstract properties, and ignores dissimilarities. Structural consistency is preserved in the analogy; the attributes are abandoned while the relations are preserved. In this way analogy can be seen as a system that ties knowledge together. Facts are no longer independent of each other.

A stronger variant of this heuristic analogy is when the analogy also gives reason to accept the solution. Swedenborg often ends up in analogies like this. The fundamental thing for him is that the little is of the same nature as the large, that creation is rational, containing an order that obeys a few simple laws. Analogy is a central intellectual tool that has played a major part in the history of science. We see it in the mechanistic philosophy of the rationalist Descartes, in the empiricist Bacon who saw analogy as something fruitful that could give insight into the harmonious anatomy of the universe, and in Comenius who advocated the method of proceeding from the known and then gradually working one's way to the unknown.[60] Rudbeck the Elder's perception was that man must proceed from his everyday experiences and crafts when describing nature. One can take a watch apart to find out how it is made, but not even with a microscope that discloses what is hidden can one find 'the first point, neither what it looks like nor how it is first tackled and handled in the womb; all this is brought out by the learned through similarities.'[61] In Linnaeus we likewise find analogical thought, with parallels between the mineral, vegetable, and animal kingdoms, and with marriage and the bridal chamber of flowers. Swedenborg was thus certainly not alone in using analogies and metaphors.

In *Nödiga förnufts-öfningar* ('Necessary exercises in reason') Rydelius considers metaphor, or the act of drawing an erroneous conclusion based on shared likenesses between things. He points out something interesting: that metaphor and allegory, unlike the non-metaphorical idea, are 'often much more pleasant to our thoughts, since they touch our imagination, and do so through our hearts', whereas ideas 'touch only the mind, and thus leave the heart cold. As when one calls the world a stormy, roaring sea, it gives a more heart-touching idea of the sea than if it is merely called uneasy, unsteady, and dangerous.'[62] Metaphors, allegories, analogies, similes affect people's fantasy, powers of imagination, and emotions.

---

[60]Nordström, cclxxxviii.

[61]Rudbeck to the Chancellor, Uppsala, October 1688. Rudbeck the Elder (1905), IV, 331.

[62]Rydelius (1737), 275, cf. 273.

The World Machine and the Little Machine

Swedenborg, and above all Polhem, claimed time and again that mathematics and natural science were of benefit to engineering and the economy, that theoretical studies can be of utility for practical life. This is actually a rhetoric to justify scientific activity. There are scarcely any unambiguous examples of science producing new technology from this time, when it was craft skill that led to more advanced engineering.[63] The relation between them was rather the reverse. Engineering inspired science to find different scientific solutions, as is particularly clear in this age of mechanics. Everyday experience of mechanics provided new ideas for an understanding of how nature functioned. Mechanistic natural philosophy was applied mechanics. At many places in the circles around Swedenborg, as with Roberg or in von Hoorn's well-educated midwife, we see how people tried to bridge the discrepancy between theory and practice, attempted to combine occupational skill with academic learning. The marriage of theory and practice was a hobby horse of Polhem's, which also reveals that the relationship between engineering and science concerned not only an epistemological difference but also social differences. Miss Theoria is more noble than Practicus the master builder in Polhem's dialogue, although there would be handsome children if the two were joined, he thought. In another context Polhem writes that the worst thing that can happen in this marriage is that theory will be far too precise while practice will be too dozy, which can lead to strife between them.[64] Swedenborg did not care much about precise calculations and was rather idealistic when it came to practice.

There is a recurrent theme in Swedenborg's theory of matter where technology and theory, artificial and natural technique merge. The laws of mechanics are the same in the little as in the large, in both man's and nature's machines. They ought to share the same regular mechanics. This is where metaphorical thought comes into the light, by transferring the known, everyday knowledge of man-made machines to the unknown mechanics of the world machine. Experiments with pendulums and spheres, or trials with Polhem's model machines can lead us to a knowledge of how the atomic or planetary world must function. Mechanical metaphors could thereby provide arguments for experimenting and justify the use of instruments and tools that otherwise belong to the guild of craftsmen. Since man's machines are completely apprehensible and comprehensible, then so should nature be. This is where mechanistic explanations have their strength. They became intuitively understandable by referring directly to palpable, visible examples from everyday life, where similar machines could be seen. The mechanistic natural philosophy thus encouraged metaphors and analogies, to give a concrete idea of the invisible world. The unknown beyond the senses was conceptualized in mechanical and geometrical terms. Particles became micromechanisms, solar systems became

---

[63]Lindqvist (1989), 176f, 180.
[64]Polhem (1740), 183.

macromechanisms, and the world was a clockwork, a world machine. Theology and mechanics were combined in God as the designer of the world machine, the divine mechanic, the watchmaker, as Polhem said, who can reset his clock any time he pleases.[65]

The metaphor of the baroque is the world machine. The world is a clock constructed by the divine watchmaker. The clock was the universal machine that described movements, measured time, space, and speed, disciplined and synchronized people's work. The world and the clock were synchronous, moving at the same rate. The Creator had previously been described with other metaphors, as a craftsman, a potter, or a builder, as in the Old Testament, in Hesiod's *Theogony*, or in Plato's *Timaeus*. There is constant hammering, drilling, turning, and joining in God's workshop. In *Guds werk och hwila* ('God's work and rest') Spegel compares how God turns the globe of earth on his lathe, round as a ball or a bubble, with the way a peasant builds his hut from turf and beams, moss and birch bark.[66] With the round shape God has denoted his eternity, without beginning or end. For Swedenborg, however, God is no longer a simple artisan. He is a mathematician, mechanic, and scientist.

This agreement between art and nature is an essential foundation for Swedenborg's mechanistic metaphors. He was no doubt strengthened in his conviction of this by Cartesian philosophy, or when reading Kircher drew his attention to the way art and nature coincide like the hours of the sunflower and the day, or the similarity between *arbor philosophica* and real trees.[67] There is an underlying mechanism that we cannot see. In Fontenelle's dialogue about the diversity of worlds the philosopher explains to the questioning marquise that 'all philosophy is founded on these two Propositions. (1) *That we are too short sighted*, or, (2) *We are too curious.*' The true philosopher, he says, 'will not believe what he does see, and is always conjecturing at what he doth not.' As an example he takes an opera, where the wheels and the weights that move the set decoration are concealed behind the scenes. If philosophers at a performance of *Phaëton* were able to 'see the aspiring Youth lifted up by the Winds', what would they say?[68] The one who could reveal the mechanics in Jean-Baptiste Lully's opera, if he had been able to sit in the theatre, is Descartes: pull the curtain aside and show me the world! Swedenborg himself was impressed by the stage machinery at the opera in Verona.[69] The world is a theatrical machine.

For Descartes there is no difference between nature's machines and those made by craftsmen, except that artificial machines must be of sufficient size to be observed with the senses. It is no less possible to acquire knowledge of God's machines than of man's, but it is harder to manipulate them because they are either too big or too small. The large bodies that we can perceive with the senses can serve as models for

---

[65] *Polhems skrifter* III, 4.

[66] Spegel (1685), edition, I:1, 38, 114f.

[67] Kircher (1643); Kircher (1650), II, 414.

[68] Fontenelle, translation, 8f, cf. 11; *Phaëton* (1683), music J.-B. Lully, text P. Quinault.

[69] *Resebeskrifningar* April 1738, 84.

our ideas about the small bodies that evade the senses. We therefore start with our everyday experience of the world, with different kinds of machines, bushes whose branches get entangled, grapes in a dish, eels in buckets, bath sponges or tennis balls, and guess our way to the substructures of things. We surmise that particles of earth are ramified like bushes in order to explain why earth sticks together in solid lumps, or that water consists of eel-like particles which can easily slip past each other in order to explain the flowing property of water. When a body expands, less matter flows into its cavities and vice versa, as in a sponge, which explains how matter can become thinner or denser.[70] If these conclusions coincide with what is really observed, one can assume that the guess is proven or at least probable.

Corpuscular theories are a kind of intellectual model to describe visible phenomena based on invisible, perhaps fictitious reality, grounded in the view that the key to an understanding of nature's processes goes through knowledge of matter and its properties and behaviour. God could have created the world in countless different ways, but only experience can tell us which one he chose. We can thus, in Descartes' opinion, choose any assumption freely, on condition that it agrees with experience. That is also the starting point for Swedenborg. He proceeds from geometrical assumptions about particles, which then have to be tested against real phenomena in nature. If we know the figure of the particles, Swedenborg says, we can calculate all their properties. Locke reasoned in a similar way, that if we knew the mechanics of the rhubarb, the figure, size, texture, and motion of its particles, we could, as with the properties of squares and triangles, work out in advance that rhubarb acts as a purge.[71]

The world as machine, the description of the world as mechanical, means understanding the world as a structure that can be divided into its constituent parts, which together explain the whole. The machine and the world machine are analytical. One can understand them by taking them apart to see how the pieces work and relate to each other. The explanation of the whole of nature with assumptions about the shape, size, positions, arrangement, and movements of the particles has several sources, in the ancient theory of atoms, in the *minima naturalia* of Aristotelian natural philosophy, and in the Cartesian corpuscles.[72] There is a lower boundary for the division of matter, Aristotle says, a least particle of a kind of matter with a specific form. The classical doctrine of atoms, propounded by philosophers such as Leucippus, Democritus and Epicurus, presumed the existence of indivisible atoms in a vacuum. These atoms, Lucretius says, build up the sky, the sea, earth, rivers, the sun, the crops of the fields, trees, living creatures, and so on. The elements, the atoms, the corpuscles function like the letters of the alphabet. They can be combined like modules, and everything in nature can be derived from a number of basic elements or smallest particles of matter. The atoms in the universe are like the letters in language, Lucretius writes, with many letters

---

[70]Descartes (1644), II, § 6–7; *Oeuvres* VIII:1, 43f.

[71]Locke, book IV, ch. III, § 25.

[72]Clericuzio.

common to many of the words 'in my verses' yet in different combinations that differ in meaning and sound.[73] Boyle too compared corpuscles with the letters of the alphabet, letters composing the book of God's work.[74] The theory of atoms was taken up by Gassendi, who tried to wash away the reputation of Epicurean atomism for being materialistic and atheistically tarnished, by emphasizing that the movements of the atoms had their origin in God's activity. One day, Gassendi envisaged, it might be possible to see the atom in the microscope.[75] Corpuscles could thus be conceptualized as letters, as small building blocks, or as an extremely fine dust spreading in space. Atoms, Spole writes, 'that fly around in the air cannot be seen at all in an open place, in sunshine, or by the light of a fire; but if we admit a beam of light to a dark room, we see numerous such atoms in it.'[76] Was it dust he saw in a camera obscura? The same metaphor is found in Lucretius' *De rerum natura*, where grains of dust whirl in the light of the sunbeam, as in an everlasting battle. The atoms of Democritus and Epicurus are called 'sun-motes' (*solegrand*) by Forsius, as when the sun shines through a window or a hole and one can see countless sun-motes hovering back and forth, up and down in the air.[77]

Descartes' theory of particles and vortices in *Principia philosophiæ* was the natural starting point for Swedenborg and the circle of the Collegium Curiosorum.[78] No indivisible atoms can exist, Descartes says; all matter, irrespective of how small it is, can be divided into smaller and smaller parts. The indefinitely extended space is set in motion by God, and in this whirling mill matter is crushed into small parts. The differently shaped particles rub against each other. Edges and angles are worn down and rounded to become like sand and stones in a river. The first element is fire, which lacks a definite size or figure and is the most subtle and penetrating substance in the world. Fire builds up the sun and the fixed stars. The second element, the ether of space, consists of small, round spheres, and can explain the movement of the planets, gravity, and light. Earthy matter, the third element, is of a coarser nature, building up our earth, the planets, and the comets.

Peas and Cannonballs

It was with metaphorical thinking that Swedenborg formed a perception of matter. He also found inspiration and ideas in other natural scientists, reshaping what he used, not infrequently beyond recognition, and he reinterpreted them through his

---

[73]Lucretius, 1.823–824; translation, 51.

[74]Brooke, 132.

[75]McMullin, 38.

[76]Spole and Odhelius, 4; translation, 480; Lucretius, 2.114–118.

[77]Forsius, 52f, cf. 73.

[78]Descartes (1644), III, § 46–52; *Oeuvres* VIII:1, 100–105; Descartes (1664a), ch. VIII; *Oeuvres* XI, 51; Hoffwenius (1698), VII, 30, § 15.

own understanding and for his own purposes. In many ways his theory of matter is closest to that of Polhem. For Polhem everything is mechanics, everything can be explained with matter and motion. Without these concepts 'all our speculations are vain and to no avail.' Since the figures of the particles are of such 'disproportionate fineness towards our eyes', one must use reason.[79] Because our weak senses do not allow us to venture far out from land, across the infinite main and deep into eternity, we must confine our senses to what corresponds to their proportions. The rest we leave to God's own wisdom. Subtle nature's mechanical structure is, for Polhem, not unlike an inventor's or craftsman's mechanical devices. The theoretical and practical mechanics are the same. They differ solely in that the former consists of infinitely fine matter, while the latter is of coarse stuff in the service of man. It is thus essential to find out the geometrical forms of matter and thereby calculate all its mechanical effects. Knowledge of nature is derived from the forms of geometry. Small particles could thus be handled as if one could really see them. The same kind of analogical thought was embraced by Swedenborg, although he arrived at a partly different micromechanics. But both Polhem's and Swedenborg's theories of matter proceed from the round figure, the sphere.

Sometimes their thinking is so similar that several manuscripts have been confused.[80] As Polhem's secretary, Swedenborg made transcripts of his manuscripts. One example of a work written in Swedenborg's hand, but composed by Polhem, is the dialogue between the kinswomen Chymia and Mechanica on the essence of nature, *Discours emellan Mechaniquen och Chymien om Naturens wäsende* (1718).[81] These personifications of chemistry and mechanics discuss, among other things, the roundness and smoothness of particles, and the angular figures of salt. Chymia asks Mechanica how stones have become round. It is an effect of the water rubbing the stones against each other as it flows and washes with a strong and steady movement, Mechanica answers. In a notebook Swedenborg had also written 20 cryptic lines of excerpts from Polhem, probably from the years 1716–1718.[82] These unnoticed lines in Latin contain physical experiments and notes on topics such as the way spheres move in water and in pendulums. Towards the spring of 1720, Swedenborg also had a collection of Polhem's books in his home: '*Councillor of Commerce Polhem's* books must be somewhere among my papers; as soon as they are found, they should be sent back', he wrote to Benzelius.[83] He may be referring to bound collections of Polhem's manuscripts.

---

[79]Polhem to Benzelius, 19 November 1710. *Polhem's Letters*, 41, cf. 7; *Polhems skrifter* III, 186, 316, cf. 3, 65, 327.

[80]Polhem, *Copia af en instruction och fullmacht som igenom sådant tilfelle är inhemptat som förtalet vtwijsar* (1716). LiSB, N 14a, no. 17, fol. 31–36; *Photolith.* I, 7–18; *Polhems skrifter* II, 61–67; Polhem, *Om sättet för handelns och manufacturernas uphjelpande* (c. 1716). LiSB N 14a, no. 37, fol. 102–104; *Photolith.* I, 68–73; *Polhems skrifter* II, 72–74; *De causis rerum*; *Discours emellan Mechaniquen och Chymien*; Dunér (2002), 3–27.

[81]Polhem, 'Discours emellan Mechaniquen och Chymien om Naturens wäsende' (1718), *Opera* III, 248f; *Polhems skrifter* III, 161f, 165f.

[82]Swedenborg, *Ex Polhemio*. KVA, cod. 86, 277.

[83]Swedenborg to Benzelius, 29 February 1720. *Opera* I, 300; *Letters* I, 231.

A manuscript that was long attributed to Swedenborg, but which has been shown to have been composed by Polhem, is *De causis rerum*.[84] The manuscript also exists in a further two versions, one complete in a transcript by Jacob Troilius from 1711, and a fragmentary copy in Polhem's hand. There is a great deal to suggest that it was part of plans by the Collegium Curiosorum for the publication of Polhem's essays. In letters to Benzelius in the autumn of 1710 Polhem tells of his ideas about the compression of air, the equilibrium of the planets, the rise and fall of the barometer, the difference between liquid and solid matter, the floating and sinking of fish bladders, all topics that recur in *De causis rerum*.[85] Polhem has also made additions to Troilius's transcript, which can be interpreted as a sign that it was authorized and approved by him. *De causis rerum* in Swedenborg's hand is in part a summary and in part a pure transcript, probably of Troilius's version, but also of another transcript by Troilius, Polhem's *Tanckar om Barometrens stigande och fallande* ('Thoughts about the rise and fall of the barometer').[86] Swedenborg's version of *De causis rerum* was presumably written in 1716 as a digest of Polhem's thoughts about the causes of things, probably intended for the second issue of *Dædalus Hyperboreus*. There is no obvious reason to assume that it was written in 1717, as was previously believed. Several letters from the spring of 1716 discuss the experiment in *De causis rerum* about the oscillation of a sphere under water.[87] At the last moment, however, Polhem wanted it to be dropped. His reason is not stated; perhaps he no longer agreed with himself.

De causis rerum gives good insight into the foundations for Polhem's natural philosophy of the causes of things. Its contents were thus undoubtedly known to Swedenborg, besides all the other manuscripts he must have read and all that was said during their conversations. Polhem begins with 'On the first beginnings and creation of the world', where he attempts to understand and explain what Moses really meant by his account of the creation. He cannot get it to make complete sense and tries to find explanations in natural philosophy. What kind of water was above and under the firmament? Moses must have meant that both the heaven and the sea consisted of water,

> chiefly because he says of the flood that *the windows of heaven were opened* and poured out water: but thus and much besides seems to be a metaphorical manner of speaking, as is otherwise stated, e.g. that it rained as if the skies were open. Or it did not rain but gushed down, it poured down, etc.[88]

The earth and the sea were held together like a round ball or sphere and hovered freely in the air, 'like the egg yoke within the white'. It is not until the second chapter about the equilibrium of the planets that Swedenborg begins his excerpts.

[84]Polhem, *De causis rerum* (before 1711). KB, X 517:1, fol. 176–197; Troilius's transcript (1711), KB, X 521, 1–29; Swedenborg's transcript (1716), LiSB, N 14a, no. 30, fol. 54–55; *Photolith.* I, 24–27; *Opera* III, 231–233.

[85]Polhem to Benzelius, 10 September 1710. *Polhem's Letters*, 7f, cf. 11f; *Opera* III, 231f.

[86]Polhem, *Tanckar om barometrens stigande och fallande*. KB, X 521, 34–37; cf. LUB, Polhem.

[87]*Opera* I, 244, 247, 250, 259; *Letters* I, 84, 89.

[88]KB, X 521, 6.

The core of Polhem's theory of matter can be found in *De causis rerum*. His theory combined a corpuscular natural philosophy with Cartesian mechanics and a micro-macrocosmic world-view, where Plato's regular polygons and the classical idea of the perfect sphere are the models for the subtle geometry of microparticles. Here he is still on Descartes' side regarding the non-existence of the vacuum. There is no space or extension that does not contain matter. Universal matter therefore cannot be of a form that prevents movement, such as cubic and tetrahedral forms, but must be round 'as all heavenly bodies, stars, sun, moon, earth, and planets are round'.[89] Particles, in other words, are imagined by analogy with the heavenly bodies. And perhaps the spaces between these are also filled with even smaller spheres. These spheres fill infinity, and each sphere is filled with other particles of ether, air, water, and these in turn of even smaller, and so on to infinity, to both minimum and maximum. Universal matter becomes infinite, or rather indefinite in smallest and the largest, 'and thus no particle of matter can be envisaged as being so small that it cannot be infinitely smaller'.[90] It is like an infinite regression, where one can never come to the end, any more than to the diagonal in a square or quadrature of the circle. In a fragment of *De causis rerum* Polhem postulates three things: (1) That we humans cannot imagine a thing being so small or so large that nature cannot make it even smaller or larger to infinity. (2) Nothing in nature can be grasped that does not have a certain figure, round or square, however small or large it may be. (3) Particles which are further apart occupy more space than those of equal number which are more closely packed.[91]

This central idea of Polhem's about the particles of matter being round was taken over and elaborated by Swedenborg. Why the sphere attained this significance in Swedenborg's theory of matter can be understood against the background of general metaphorical thinking. We can see spheres around us, as Swedenborg explains the 'globus' or 'sphæra' in *Regel-konsten*, 'A globe; when smaller a ball and marble; when even smaller, hail or shot.'[92] The entire mass of the sphere could be imagined as assembled in a single point, at the centre of gravity, and knowledge of the planets and the star sphere could be transferred to the microcosm. Another important reason for the supremacy of the spheres as adduced by Polhem is of a classical kind. The sphere is the body that describes the largest possible volume with the smallest possible surface area, in the same way as 'the glass blown at glassworks seeks a round and globose form, because it has the greatest internal volume and the smallest enclosure'.[93] The sphere had a special significance as being without a beginning or an end, with all the radii converging at a single point. Since the sphere is the most perfect, beautiful, and harmonious of forms, the earth must be round and

---

[89] KB, X 517:1, fol. 196v; cf. *Polhems skrifter* III, 74, 313; Forsius, 85.

[90] KB, X 517:1, fol. 197v.

[91] KB, X 517:1, fol. 178v.

[92] *Regel-konsten*, 6.

[93] Polhem, 'Mechanica naturalis eller naturens konstiga sammanhang framstelt under små frågor och svar', *Polhems skrifter* III, 251, cf. 68.

the universe spherical, according to the Pythagoreans. The idea that all bodies in the universe seek this form, from the celestial sphere to a drop of water, found no obstacle in the Copernican system, and did not lose its regular spherical perfection until Galileo's chiaroscuro passed over the mountains and valleys of the moon and the sun was disfigured with dark spots. During his student days Swedenborg had borrowed Milliet Dechales' guide to world mathematics, which explained how one notices that the earth must be round when a ship vanishes beneath the horizon or when a tower sinks in the sea.[94] In Swedenborg's time, however, the question was no longer whether the earth was round or not, but whether it was pointed or flattened at the poles. As a Cartesian he preferred the former.

The classical idea of the perfect circle and the sphere, which was embraced not least by Plato and Aristotle, recurs in Harald Vallerius, Polhem, and Swedenborg.[95] That the sphere is the most uniform, simple, convenient, and appropriate form for motion, whereas angular particles lose their angular shapes and become spherical through the constant milling movement of the universe, could be read in Hoffwenius's *Synopsis physica*.[96] Hoffwenius goes on to write that the smooth, polished bodies, for example the particles of water, move more easily and reflect light better, in contrast to rough surfaces which consume light. This identification of the capacity of particles for movement with their round form can also be found in a work to which Swedenborg often refers, Hiärne's *Actorum chymicorum Holmensium parasceue* (1712). Hiärne considers the differing opinions about the figure of water particles— as cylindrical, oblong, and other shapes—but concludes that it is absurd to question the roundness of water particles.[97] Wilhelm Homberg, a Dutch chemist, born in Java, whom Hiärne might have met in Stockholm, reasoned along similar lines when he assumed that the liquid property of mercury was due to its corpuscles being made up of very smooth spheres which rolled easily against each other.[98]

Roundness became a universal principle for Swedenborg and Polhem, a micro-macrocosmic line of thought according to which the round shape ranges between the two infinites, from the infinitely small to the infinitely large, from the first matter to the celestial bodies. The particle world and the planet world are bound together. The figures of the small parts are as unfathomable and enigmatic as the figure of the whole universe, but since the round figure is the most appropriate for motion, one can conclude that they must be round.[99] Celestial bodies or planets can serve as a model for the particle world. Like them, the particles describe circular, polar movements, and rotate around their axis in eternal motion. This micro-macrocosmic idea is noticeable in Forsius, for whom the perfect round figure in the element of water means that water envelops the earth, like the parts of a thing, and this

---

[94] Milliet Dechales, III, 195.

[95] Aristotle, *Peri ouranou*, 2.4.286b10–287b22; H. Vallerius and Melander, 10.

[96] Hoffwenius (1698), V:8, 17.

[97] Hiärne (1712), 18.

[98] Principe (2001), 549; Lindroth (1946–47), 93.

[99] *Polhems skrifter* III, 4, 44ff, 66, 187, 227.

is how it is in everything.[100] Since drops of water are round, water as a whole must be round too. Since the parts of earth, such as mountains, sand heaps, and grains of sand, are round, the whole element of earth must also be round. Newton likewise transferred knowledge of the planetary world to the invisible particle world, concluding by analogy with his law of gravitation that the small particles of bodies also exert an attraction, a force that can work at a distance; as with the earth, so also with drops of water.[101] The roundness of particles thus comes from the metaphor of the celestial bodies, and Polhem also cites cannonballs, peas, bladders, soap bubbles, fish roe, and the blood corpuscles that Leeuwenhoek found in his microscope: 'The inquisitive Dutchman Leeuwenhoek has observed through his subtle microscopes the figure of blood down to its particles and found it to consist entirely of round globules floating in a clear liquor just like fish roe in water.'[102] The round form, according to Polhem, is most appropriate for motion and the liquid state, for transparency, refraction, compression and expansion, gravity, life, tremulation and sound for hearing, movement towards storm and tempest, hot and cold, clean and wholesome, rotten and deadly.

Particles as cannonballs serve as a heuristic model of reality, a general idea that provides a foundation for the continued investigation of matter. The models proceeding from cannonballs, peas, and bubbles steer the continued thinking about matter in a specific direction, but also function as a kind of self-confirmation of the model which presupposes what is to be found. The models also have a certain pedagogical role in drawing the general outline, while the theory provides the details. First one constructs a mental model of a problem. One then examines the model in search of the conclusions that follow from the premises. Finally, one checks the conclusions by trying to find counter-examples. If no counter-examples are found, the conclusion is correct. Apart from performing the final check, Swedenborg follows this way of thinking in models.

Polhem took his concrete examples from the nearby reality around him. Sweden in the 1710s was an agrarian country harried by plague, crop failure, and war. The amount of peas in a barrel was thus not a negligible problem. The heaped measure to which Polhem refers had been formally abolished long before, but its great economic significance meant that it was replaced by a system called *kappmål*, with a special vessel used to measure the added volume to compensate for the difference between heaped and level measure. The fact that the size of the heap above the brim could vary considerably, depending on the shape of the measure and also on the properties of the dry goods, caught the interest of Polhem the theorist of matter when he set out to explain fluid properties in *De causis rerum*.[103] How universal matter can be fluid is demonstrated by coarse and palpable things like peas and

---

[100]Forsius, 75, 142, 171.

[101]Newton (1706), 337f.

[102]Polhem, 'Pulspendel bequemlig för dem som eij hafva tijd vara siuka' (*c.* 1720), *Polhems skrifter* IV, 13; Leeuwenhoek, 1–3, 496.

[103]KB, X 517:1, fol. 178r; KB, X 521, 16; cf. *Polhem's Letters*, 15; *Polhems skrifter* III, 178, 197.

linseeds. Why the heaped portion on a barrel of peas is not as large as that of grain was explained by the round figure rolling more easily and being more fluid than elongated figures. Linseeds give a smaller heap than peas, even though the seeds are not round, but this is due to their smoother surface, so that if peas were as smooth as linseeds they would make an even smaller heap, and the smallest heap or none at all if they were entirely without friction. Concrete reality provided the models; from what was concrete, visible, and known one could draw conclusions about what was abstract, invisible, and unknown. This is not just a pedagogical device. He does not stop at a didactic analogy. Mechanical laws are the same in both the smallest and the largest. The mechanics of peas differs from that of particles only as regards size. The roundness and the smooth surface of water and air particles explains their fluid property, in the same mechanical way as with peas and linseeds in a barrel.

Polhem's observations of the association between peas and the fluid state were brought up by the experimental physicist Mårten Triewald in his lectures on the new natural science, delivered at the House of the Nobility in 1728–1729. Triewald had the same opinion: the particles in liquid matter are round, slippery, and smooth. Speculating about peas and linseed does not immediately and necessarily mean thinking in a popular or non-academic way. It is first and foremost a metaphorical mode of thought. Lucretius explains fluid substances as consisting of smooth, round atoms, unlike the atoms of solid substances which are held together with hooks. A handful of poppy seeds, he writes in *De rerum natura*, moves just as easily as water flowing down a slope.[104] Locke too speculated that the particles of matter are 100,000,000 times smaller than a mustard seed.[105]

So what is it that prevents air, water, mercury, and other liquid substances from also consisting of such exactly round and smooth particles, Polhem asks himself. The only thing that could argue against it is air pressure, with its capacity for compression and rarefaction; air could be compressed as much as 40 times, as Boyle had shown. Through experiments with the air pump, Boyle found that air corpuscles must be elastic and springy, perhaps like Cartesian whirls with an inner rotation, like spiral watch springs of metal, like wool or sponges, which could explain the ability of air to be compressed and expanded.[106] Therefore, Polhem goes on, Descartes and others have assumed that air particles are 'ramosæ and plumosæ', that is to say, branching, spring-like, or—as Polhem writes elsewhere—'like creaking springs in watches', soft and flexible.[107] The disadvantage of the theory is that it cannot explain transparency and refraction. There is, however, a way to get round this without abandoning the idea of round and smooth particles, yet simultaneously explaining the capacity for expansion and compression.

---

[104]Lucretius, 2.453–454.

[105]Locke, book I, ch. XIII, § 22.

[106]Boyle (1669), 154f, 187–189, 193.

[107]KB, X 521, 17; KB, X 517:1, fol. 178v; *Polhem's Letters*, 14f; cf. *Polhems skrifter* III, 196, 226, 250, 301; Descartes (1644), IV, § 33ff, 46, 57; *Oeuvres* VIII:1, 238f; Hoffwenius (1698), XII:8, 56.

This is how Polhem arrives at his spherical theory, that the structure of constellations of round particles can explain phenomena. That round balls can be put together in forms of varying sparsity and density is something that 'anyone can easily investigate, especially those who stack cannonballs in ammunition houses'. If they are placed so that eight balls form a room, with four under and four over, the form is at its sparsest, a structure that Polhem calls 'hexahedral' or 'cubical'. If one instead lets four balls form a room, with three below and one above, the form is denser, in other words a tetrahedral structure (Fig. 3). If a vessel is filled with balls in these different formations, then there are 10 balls in the loosely packed form as against 28 in the tightly packed form. The particles of particles and so on to infinity can be pressed together in the same way, from cubic to tetrahedral structure, 'so that the whole universe can in this way be squeezed together into a mathematical point'.[108]

Polhem returns time and again to the cannonball metaphor, making a distinction between three stacks of cannonballs: tetrahedral, octahedral, and hexahedral structure, representing three of the five regular Platonic solids. A form between solid and liquid matter is the octahedron consisting of four balls in a square with one ball on top and one below. It resembles Plato's combination in *Timaeus* of Pythagoras' mathematics, Democritus' atomism, and Empedocles' four elements of a geometrical particle world consisting of the stable, immobile cube (earth), the icosahedron (water), the octahedron (air), and the tetrahedron (fire) with the smallest number of parts, most mobile, lightest, and with the sharpest corners.[109] The fifth regular, equilateral polygon, the dodecahedron, according to Plato, corresponded to the whole universe, but later this regular solid came to stand for the fifth element, ether. The relation between the intervening space and the balls touches on Kepler's problem, that of finding the way to pack spheres most densely.[110] Placing cannonballs in regular stacks, as in tetrahedra or octahedra, and calculating how many balls there were was one of the mathematical problems of an artillerist (Fig. 4).[111] Swedenborg too was interested in pyramids of cannonballs. How do you work out the number of balls in a stack? The answer to the question was Swedenborg's essay 'En lett vträkning På kulors samman leggningar i Triangel-Stapel' ('An easy calculation of the packing of balls in triangular stacks') in the last issue of *Dædalus Hyperboreus* from 1718.[112] He solves the problem with the aid of the rule of proportionality. It is sufficient to know the base to work out the number in the stack.

An image that Polhem often uses to describe the relative movement of round particles invokes cogs. When matter is loosely packed, no solid form can arise,

---

[108]KB, X 521, 18f; cf. *Polhem's Letters*, 15f, 62f; *Polhems skrifter* III, 6, 68, 74, 79, 186, 188, 227, 321.

[109]Plato, *Timaeus*, 55D–56B; Cromwell, 51–57.

[110]Aste and Weaire, 29ff.

[111]KrA, Rappe, 123; see also KrA, *Artilleribok*, XVI:47.

[112]*Dædalus Hyperboreus* VI, 4–6.

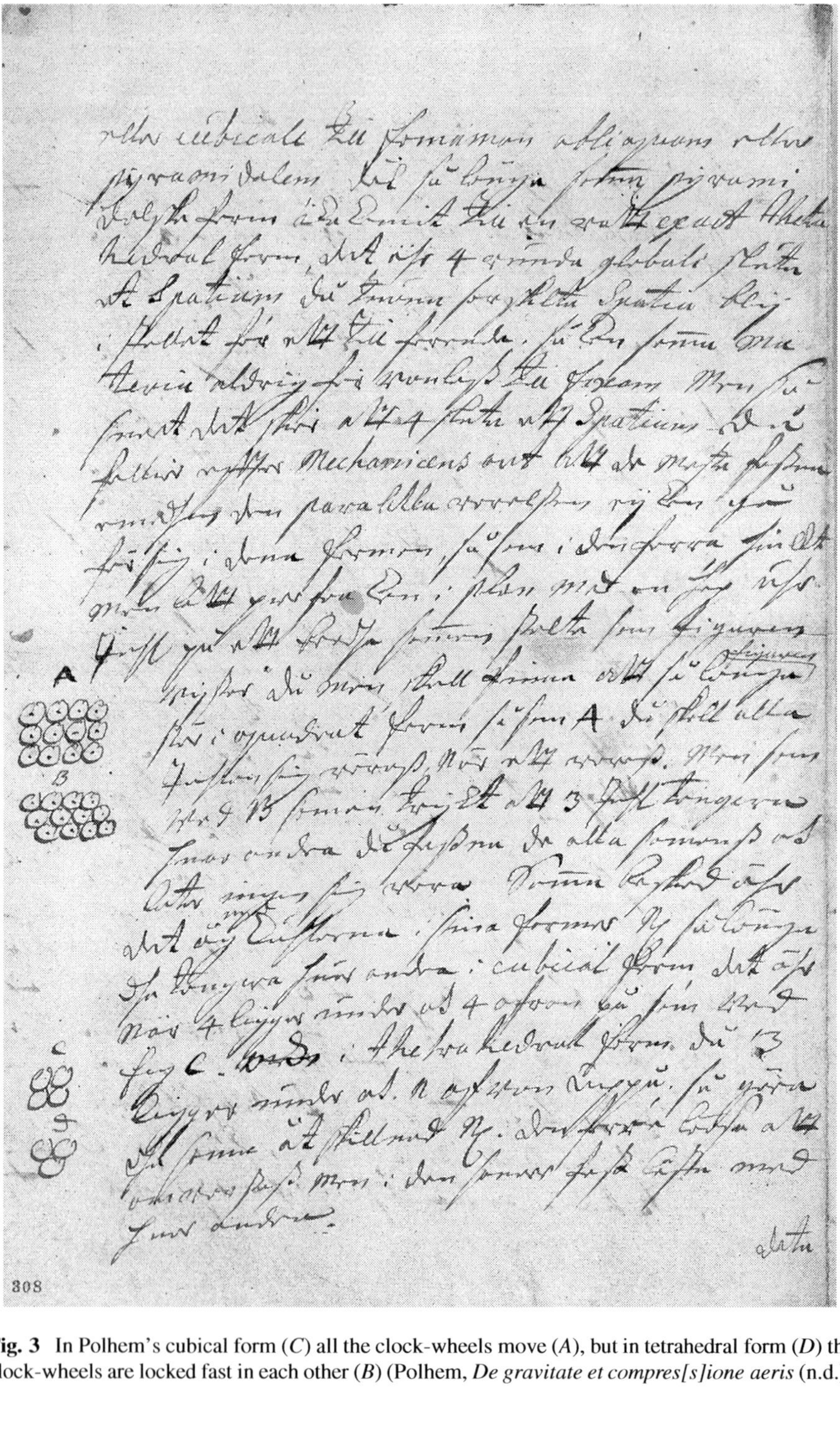

Fig. 3 In Polhem's cubical form (*C*) all the clock-wheels move (*A*), but in tetrahedral form (*D*) the clock-wheels are locked fast in each other (*B*) (Polhem, *De gravitate et compres[s]ione aeris* (n.d.))

Fig. 4 Cannonballs stacked in pyramids, the first of them in a tetrahedric structure and the second octahedral. Major General Niklas Rappe passed the time during his captivity in Moscow by writing *Åtta böcker om artilleriet* ('Eight books about artillery' (1714))

since the motion is uniform, without deadlock or friction. This can be tested with watch wheels or cogs, simple machine elements reminiscent of the models in his mechanical alphabet. If one places four wheels together, Polhem explains in *De causis rerum*, and then rotates one of them, the others will follow without difficulty, even if there are several thousand of them. But if one places three wheels in a triangle, they deadlock and form solid matter. Rest is like glue or assembly, while movement is like oil or separation in the particle machinery, which causes matter to become liquid and fluid. Polhem finds the numerical proportion 10 to 28 for the weight of water as against different solids such as glass, crystal, diamond, grapes, marble, alabaster, quartz, flint, coral, slate, granite, and so on. But, he says, 'This proportion is not so mathematically exact, but a little more or less, just as materials can differ. But if one goes to the purest materials such as diamond and gold, the proportion is tolerably correct.'[113]

The substances salt, sulphur, and mercury arise in veritable chemical battles in Polhem's *De causis rerum*. Fire, glass, and salt on one side against water, oil, and sulphur on the other:

> In sum, these conflicting parties have been thronging so long and tussled until they || have mortified and killed each other, their bodies having become earth and pumice, almost identical, although it could serve somewhat better as a wall against both, for pumice withstands both fire and water.[114]

Polhem, in other words, combines the Paracelsian *tria prima*—mercury, sulphur, and salt—with a mechanistic corpuscular philosophy. Fire, being lighter, seeks to move upwards from its centre, while water, which is very heavy, wants to move down towards the centre of the earth. Water, 'or large oceans, which are so terribly deep', seek their way down in the earth which is packed with fire, inevitably resulting in 'such a dreadful tumult and boiling, or more precisely battle, against one another'.

---

[113]KB, X 521, 21; *Polhems skrifter* III, 5, 46, 189, 201, 253, 267, 301, 322f; *Polhem's Letters*, 15, 57.

[114]KB, X 521, 13f; KB, X 517:1, fol. 194r; up to '||' is Troilius's transcript, after which the text is in Polhem's own hand.

What was once water is pressed together by 'this hard squeeze' into solid matter, first salt, then sulphur, and finally mercury. He arrives at a highly unorthodox conclusion which it is difficult to see as anything but a materialistic outlook. Matter is alive as long as it has salt and sulphur, 'but after death the body becomes pumice, and the soul quicksilver and air'.[115]

Polhem explains the rising and falling of water through an experiment with fish bladders which he lets balance in water so that they neither sink nor float to the surface. If the water becomes slightly warmer, the bladders rise; if the water becomes colder they sink. Water particles function in the same way. If the ether, when heated by the rays of the sun, expands them, the water particles rise in the air. As long as the lower altitude of the atmosphere permits this upward movement, the weather will be rainy, but if the altitude of the atmosphere is greater there will be sunshine. In his version of *De causis rerum* Swedenborg passes on to another manuscript of Polhem in a transcript by Troilius, *Tanckar om Barometrens stigande och fallande*. Among nature's hidden secrets, the rise and fall of the barometer is not the least. There is a link between the changing air and the changing barometer, 'now clear, now rain, now wind', depending on the varying gravity of the air.[116] Both have their origin in the changing height of the air sphere. To investigate pressure, Polhem performed experiments with fish bladders in water and with small glass bubbles the size of hazelnuts or peas, to see how they balance in water. This could also be done with urinary bladders or with a glass bottle submerged in water, 'if it is sunk too deep it will break so that all that is left is the stopper and the cord'.[117] With these experiments he sought to discover how water can balance in the air and rise, how the air becomes viscous from salt, earth, sulphur, and so on, 'like soapy water which can be blown with straw pipes and released in the air, to hover and quickly perish, a depiction of vanitas'.[118] When the air pressure is lower, that is to say, when the atmosphere is lower, the water particles can become slightly larger than usual. They then rise to a certain height, where they assemble and form bodies that become new clouds that fall again in drops, 'as when one empties a whole bucket of water from the top of a tower or down a mine, no drop reaches the ground, but all are transformed together into a thin mist which one cannot see, much less feel'.[119] The different height of the air spheres, he says with reference to Kircher, is a result of the battle between fire and water in the earth's crust. 'And it seems as if nature herself works like a chemist, using the earth as a furnace, caverns and openings as retorts and flasks, and the very air above as the recipient.'[120] Just after this, Polhem compares the earth to a living body, where the central fire is the heart, the air is the blood, cavities are the liver, dry land is the arteries, wet land and the waters are the veins.

---

[115] KB, X 521, 24f.

[116] KB, X 521, 30.

[117] KB, X 267:1, fol. 90r; cf. *Opera* III, 232f.

[118] KB, X 517:1, fol. 182r; cf. *Polhems skrifter* III, 198.

[119] KB, X 521, 39.

[120] KB, X 521, 41.

A Sea of Bubbles

In January 1718 Swedenborg wrote to Benzelius to tell him that he was working on an essay in which he sought to prove, with the aid of geometry, that air and water particles must be round. It would be good, he thought, if he could have some opinions about it from Roberg, 'who is himself subtle in all that is minute and subtle', but the judgement of Johan Vallerius would also be valuable, if he 'would lay aside a little his own and his father's *Cartesium*'. In fact, a portrait of Harald Vallerius shows him resting his arm on a thick book by Descartes, holding a pair of compasses in his hand, prepared to measure the world on a firm foundation. One can thus discern a certain revolt against dogmatic Cartesianism, although Swedenborg's future theory of matter on the whole does not depart from basic Cartesian principles. His intention was to proceed with this topic in a 'large book' as scholars in foreign lands do with their speculations, 'but since here one has not the facilities for so large a publication, the mouth must adapt itself to the foodsack'. The great utility of the geometrical approach would be that it would be easier to ascertain the nature of air and water. If one knows the geometrical properties of the particles one can also deduce their other properties: 'for if one finds the correct figure of the particles, one will in turn get to know all the properties which belong to that figure.'[121] If one can identify the geometrical forms of particles, then, one can deduce and draw conclusions about their other properties. He hopes that there is a solid foundation for this, since he does not want to publish anything that is not better grounded than what he wrote in the previous issue of *Dædalus Hyperboreus* about stereometric rules. 'If the post is to be so atrociously raised, as is talked of, one will likely have to take leave of his friends and relatives', he writes anxiously and sends the manuscript about small round particles with Rudbeck as courier. I believe it 'will be well-received abroad', he says self-consciously, and it is to be dedicated to Abbé Bignon.[122]

Swedenborg says that he wants to penetrate everything to do with fire and metals. The method is to use chemical experiment and experience, derived from learned men such as Boyle, Becher, Hiärne, and Lémery, in order to search for the smallest constituents of nature which can be compared to geometry and mechanics. Yet it was not necessary to perform experiments of his own, as he wrote to Benzelius in May 1720:

> It seems to me, that an endless number of experiments is a good foundation to build upon, in order to make use of the labour and expenditure of other men, that is, to work with the head, over that on which others have worked with their hands. From this could come a multitude of deductions in *chymicis, metallicis*, fire and all their phenomena.[123]

---

[121] *Opera* I, 281, cf. 280; *Letters* I, 178f; cf. UUB, Roberg; J. Klopper's portrait of Vallerius in Dunér (1910).

[122] *Opera* I, 283f 178; *Letters* I, 181ff.

[123] *Opera* I, 304; *Letters* I, 237; *Bokwetts Gillets protokoll*, 24; cf. *Letters* I, 245.

He sought the inner geometry of matter. He would have this printed abroad, in two works on the natural principles of round particles, with geometrical and experimental demonstrations of the internal mechanics of water particles, salt, acids, saltpetre, oils, and lead, while also presenting his ideas about colours and fire, along with various technical inventions and his solution to the longitude problem.[124]

He presented his theory of round particles, 'the bullular hypothesis', in 1721 in an introduction to the principles of natural things, *Prodromus principiorum rerum naturalium*, which in many ways can be regarded as an elaboration of the same theme that Polhem had followed on the spherical nature of matter. Swedenborg proceeds from the sphere as the figure of motion, assuming that the world is built up of small round 'bullæ', that is, bubbles, and he applies this hypothesis to the elements and chemical substances, such as salts, acids, oils, saltpetre, lead, and other metals. Until now, all knowledge of the invisible had been concealed, but, Swedenborg declaims, perfect geometry has now ascended the holy mountain. For what are physics and chemistry, if not a special mechanism, a geometry with variations in positions, figures, weights, and movements in the particles?[125] Geometry and simplicity was what he sought.

Swedenborg applies geometry to matter in a way that is very similar to Polhem's stacks of cannonballs. He places bubbles in different formations with specific numerical proportions between matter and intervening space. The idea of combining balls in different formations is in line with mechanistic natural philosophy. Boyle, for example, whom Polhem often mentions on the subject of air pressure, believed that material things could be explained in terms of their geometrical form, motion, and how the particles are arranged and bonded in clusters of differing structure.[126] In *Micrographia* Hooke proposes different positions for globular particles, with solid substances having particles in a regular figure and liquids having spherical particles.[127] The mechanistic corpuscular theories seem to proceed from a metaphor that the world is made up of small building blocks which can be combined to give different structures with different functions. Swedenborg piles small bubbles on each other, like Polhem with his peas and cannonballs. But unlike Polhem's basic system of two different ball formations, Swedenborg identifies eight different formations in which round particles can be arranged (Fig. 5).[128] First comes the vertical position, corresponding to Polhem's cube, followed by three different triangular positions. In the fifth position, fixed triangular pyramidal, one globe rests on three others, which is the same as Polhem's tetrahedral structure and is similarly characterized by the maximum degree of rest and cold. The sixth, the fluid triangular pyramidal position, has the same structure as the preceding one, but the balls are at

---

[124]*Acta literaria Sveciæ* (1721), 209–211; *Neue Zeitungen* 8 May 1722, 418–420; *Letters* I, 253f; *Bokwetts Gillets protokoll*, 50.

[125]*Prodromus principiorum*, edition 3; translation, 1.

[126]Boyle (1965), 192f, 202.

[127]Hooke, 85f.

[128]*Prodromus principiorum*, edition, 8–13; translation, 8–15.

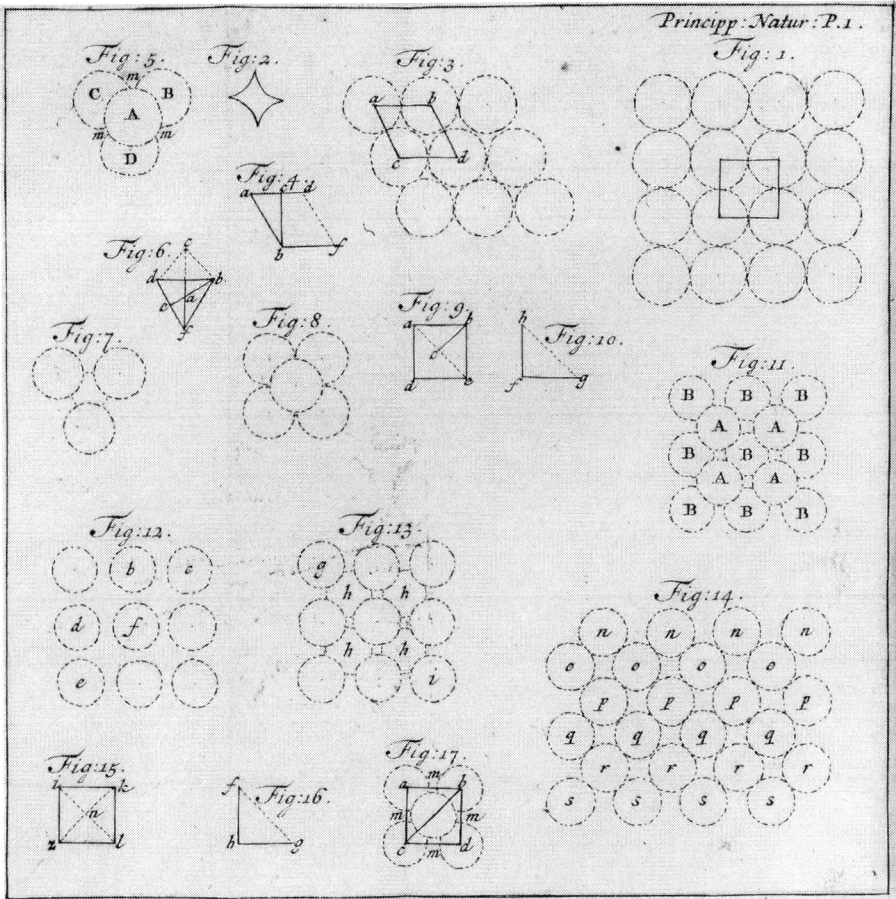

Fig. 5 The formations of round particles. The vertical position (*1*), with cubic particles in the intervening spaces (*2*). The triangular position (*3*). The fixed triangular pyramidal position (*5*). The fluid triangular pyramidal position (*7*). The fixed quadrilateral pyramidal position (*8*). The fluid quadrilateral pyramidal or the natural position (*11–13*) (Swedenborg, *Prodromus principiorum rerum naturalium* (1721))

a certain distance from each other. Polhem's octahedral formation corresponds to Swedenborg's seventh position, fixed quadrilateral pyramidal, which is formed at very high pressure. Finally, we have the fluid quadrilateral pyramidal or natural position, that is, a position occupied by water particles, where one globe rests on four others but separated by a distance.

Swedenborg's particle structures can be described as a kind of crystallography. Substances are divided and acquire their properties from geometrical forms, with the spherical parts making up tetrahedra, hexahedra, and other crystalline figures. With a geometrical world-view, of course, the geometrical forms of crystals were

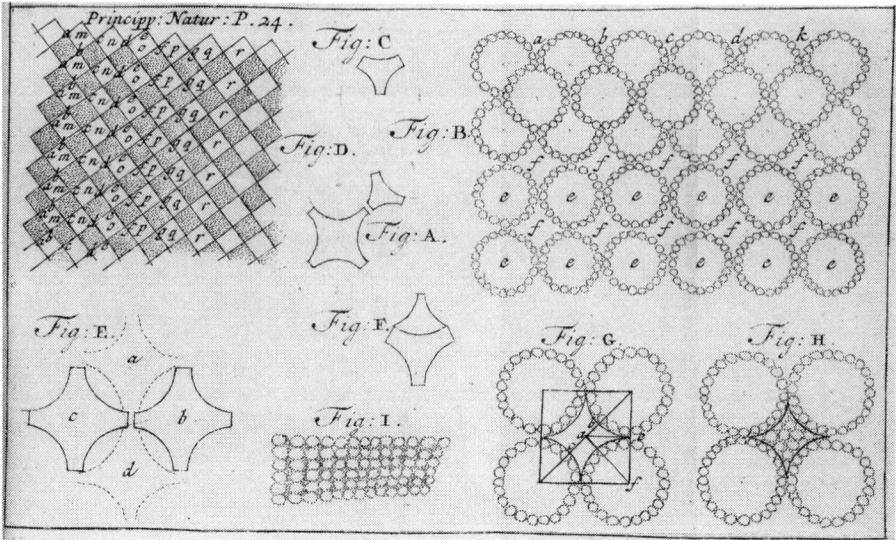

Fig. 6 Angular salt particles. At the bottom of the sea, the particles are in the fixed quadrilateral pyramidal position, (*B*). The interstices of the first kind are cubes with hollowed sides, (*A*). Another kind of intervening particle is the cavo-triangular type with four hollowed sides and four points, (*C*). The number of triangles is twice as large as the number of water particles, (*D*) (Swedenborg, *Prodromus principiorum rerum naturalium* (1721))

of interest. For example, Boyle found that saltpetre forms long, hexahedral crystals, and Huygens was fascinated by the remarkable figures of crystals, salts, minerals, and plants, with their regular angles, ordered polygons, hexagons, and pentagons. This regularity, according to Huygens, comes from the small, invisible, equal-sized particles of which they are composed. He was particularly amazed by Iceland spar, with the remarkable double refraction which had been noticed by the Dane Rasmus Bartholin. Huygens sought to explain the optical properties of Iceland spar as consisting of spheroids, that is, rotated ellipses arranged in regular pyramidal structures.[129] In his classification of the kingdoms of nature, Linnaeus, who had read Swedenborg's books about the mineral kingdom, likewise distinguished minerals through their basically geometrical figures, through truncations and angles.[130] The physical and chemical properties were thus subordinate in this theory to the geometrical forms of the crystals, wholly in keeping with a mechanistic view of the world as consisting solely of primary properties, form and motion.

In the space between the round particles, where Polhem sometimes says he finds God's dwelling or the vacuum, Swedenborg discovers other kinds of particles: angular and pointed, hollowed-out cubic and triangular particles (Fig. 6). As regards

---

[129] Huygens (1690), 92–94.

[130] Linnaeus (1748), 220f, tab. VIII; Linnaeus (1907), I, 4f & II, 37–44; cf. Lesch, 89.

the existence of the vacuums, Swedenborg is not entirely clear. He seems to avoid a vacuum in the sense of space without matter, the vacuum that Polhem would later propose. When he was assistant to Polhem, the latter was likewise hesitant about the vacuum and was still close to the Cartesian idea of plenum, whereby elements enclose each other. He gradually progressed to conceiving the real existence of the vacuum and its significance in providing scope for motion.[131] Although Swedenborg distinguishes between *spatium vacuum* and *spatium plenum*, the meaning of the former, 'empty space' comes closest to 'free space' as a designation of the space between the round particles, while 'full space' is the space occupied by the globules.[132]

Salt is an example of a substance with these intervening angular particles which are formed, in Swedenborg's view, through high pressure on the bottom of the sea, when the water changes from fluid quadrilateral pyramidal to fixed quadrilateral pyramidal position.[133] The spaces between these water particles at the bottom of the sea are where salt and other substances arise, of two different triangular kinds, cubes or triangles with concave sides. A complete salt particle consists of a cube and eight triangles. Polhem also believed that the pressure on the sea bed caused angular salt particles; quite simply, if one pressed fresh water hard it would give salt water. Triewald, in his lectures on natural science, refers to Polhem's idea that salt particles occupy the 'secret holes' between the water particles.[134] For Swedenborg, the angular forms of salt particles fit exactly in between the round water particles. The sharp angles of salt particles are what gives salt its biting taste. Round shape, in contrast, gives rise to a sweet taste since these particles roll freely on the papillae of the tongue—another idea he shared with Polhem and many of his contemporaries. Furthermore, according to Swedenborg, the cube is alkaline and the triangle an acid. The pointed tips represent the acid in salt, which causes the sourish taste perceived by the papillae. When water evaporates in rising bubbles, the salt cannot accompany it and thus remains where it is. Sea-fire, phosphorescence at sea, can be explained by the movement of salt water in the sea, which spreads a weak fire or luminescence, a sign that the salt particles are being broken down by the motion. With reference to Swedenborg's theory of water and salt, Tiselius says that fresh water dries more easily in clothes than salt water, precisely because salt has rough, sharp, and pointed particles which stick longer to clothes.[135]

For Swedenborg, then, geometrical forms and motion could explain sensory experiences such as colour, smell, taste, and touch. The corpuscular theories stress this persistent difference, that substances which are agreeable to the senses, things

---

[131]Polhem, 'En kort manuduction och inledning till philosophiam naturalem' (1716), *Polhems skrifter* III, 177; Frängsmyr (1969), 92; Hård, 57.

[132]*Prodromus principiorum*, edition, 8f; translation, 9.

[133]*Prodromus principiorum*, edition, 24, 28–30, 44, 47–49; translation, 28, 34f, 54, 58–60.

[134]Triewald (1736), II, 6; *Polhem's Letters*, 42; *Polhems skrifter* III, 153, 166, 179f, 200f; IV, 26, 47.

[135]D. Tiselius (1730), 93.

that caress the senses, are made of smooth, round particles, while bitter, acrid, and unpleasant tastes are caused by substances with sharp and rough particles. Honey and milk, Lucretius wrote, roll delightfully on the tongue, but wormwood and centaury 'screw the mouth awry with their nauseating savour'.[136] Similarly, the horrific rasping of a screeching saw cannot be caused by the same smooth atoms as the tones that a musician conjures from a lyre by touching the strings with his light fingers. In mechanistic philosophy this is a recurrent theme in the distinction between primary and secondary properties or qualities. Primary qualities are the properties possessed by things in themselves, such as extension, form, solidity, motion, number, and size. They are objective properties, independent of our senses. Secondary qualities are instead subjective, dependent on sensory perception. Properties such as taste, smell, heat, hardness, sound, and colours are thus not inherent in things but arise in our perception of the primary qualities. This idea was embraced by all the leading philosophers, from Galileo to Descartes and Locke.[137] The same was true on the Swedish mechanistic stage, as with Polhem and Roberg. Harald Vallerius discussed the atoms of Democritus and Epicurus and the particles of Descartes in a dissertation where he also considered how the surfaces of bodies affect taste, smell, and hearing.[138] A dyer or fuller can get used to the stench of his work, Spole claimed, because the pores of his sense of smell have become adapted to the form of the 'dreadful particles'.[139] The sour taste of unripe fruit was due to the form of the salt particles, with their sharp points that cut into the nerves of the tongue, said Rudbeck the Younger.[140]

This division had ontological consequences. It tells us what mechanics, the mechanistic natural philosophy, believed to be the real, that is to say, the fundamental, primary properties such as geometrical form and motion. Everything else could be reduced to these two basic properties. At the same time, the subject, the feeling person, became active, not a passive recipient of impressions. The geometrical forms of reality must be interpreted, reshaped as sensory impressions. Secondary properties became dependent on the observer and thus no longer a part of objective reality, or they could lead to sure knowledge of what the world is really like. Some people prefer the squealing of pigs to the harmonies of the organ, as Mersenne exemplified the subjectivity of the senses.[141] Swedenborg's mechanistic, Cartesian programme was in many respects an endeavour to explain how nature's innermost, fundamental constituents could give rise to the secondary qualities. The mechanistic challenge was thus: identify the form, size, number, motion, and structure of material things. Qualities were geometrized. The division between primary and

[136]Lucretius, 2.399–401; translation, 72.

[137]Also Mersenne, Hobbes, Boyle, Newton; Descartes (1641), III; Locke, book II, ch. VIII, § 18; cf. Berkeley's critique of primary qualities. Berkeley, § 9f.

[138]H. Vallerius and Melander, 60–70.

[139]Spole and Odhelius, 8f; translation, 480.

[140]Lindroth (1989), II, 433.

[141]Mersenne (1625), book III.

secondary qualities, in my opinion, is grounded in metaphorical thought. It involves a cross-modality, a human ability to 'translate' one sensory experience into another. One can talk of 'a round taste', 'a sharp sound', or 'a cold gaze'. From this one can understand the thinking behind Swedenborg's idea about the sharp taste of salt and other reductions of qualities to geometrical, visible, or tactile properties.

Swedenborg follows the geometrical principle according to which, if one can determine the different positions and forms of bubbles, one can draw conclusions about their properties, as for example concerning weight, colour, and taste. Geometrical nature is built up of particles of matter in the form of small bubbles, *bullæ*, round particles consisting of surface and nothing else, and capable of expansion and contraction. Water, air, ether, fire, and light consist of these bubble-like particles, which differ only in size. On the surface of every bubble, moreover, there are smaller particles, whose surfaces in turn have even smaller particles, and so on down to mathematical points (Fig. 7:18–19 in Chap. 4). Water particles, which thus consist of a series of these bubbles, are relatively large and have a surface covered with smaller bubbles, whereas fire particles are very small, moving bubbles with a surface consisting of mathematical points.[142]

It follows from geometry, in the opinion of both Swedenborg and Polhem, that the fluid property is due to the roundness of the particles, and the rounder they are, the more suited they are to the fluid property. Round particles touch each other only at points and thus encounter less resistance in their movement. Solid states and connections are due to points of contact obstructing the movement. In round bodies the radii from centre to surface are equal and there are no protruding parts to impede the forward motion. The round form of the globe is the form of movement; quite simply, roundness is equal to motion, 'rotunditas = motus'.[143] Swedenborg quotes much experimental evidence of water globules that can be proved geometrically, such as the undulations of water, how water forms slow waves in exact circles, how drops of water are pressed equally by the air on all sides to form a sphere, how water rises from the earth and lakes as vapour, which is visible to the naked eye as round bubbles. And if one empties a barrel of water down the mine in Falun, the water is dispersed into vapour, so that a person standing on the bottom 200 foot down does not feel a single drop. One can also experience the force of water in a steam engine or when water is transformed into ice and can then burst a cannon.[144]

By combining all these particles—round, triangular, and square—Swedenborg seeks to explain the true structure of the entire particle world. Acids penetrate into pores and have mechanical properties like wedges which can split other particles into bits.[145] Saltpetre likewise has an angular figure. The idea of corrosion by sharp particles is a typical geometrical metaphor. Acids which corrode and attack metals are pointed and can drill their way through the figures of other substances, as

---

[142] *Prodromus principiorum*, edition, 105; translation, 134.

[143] *Prodromus principiorum*, edition, 17; translation, 19.

[144] *Prodromus principiorum*, edition, 20–23; translation, 23–27.

[145] *Prodromus principiorum*, edition, 63, 68f; translation, 80, 86f.

Hoffwenius explained, for example. Chemical properties can thus be understood through metaphors from mechanics and geometry. For Block the sour particles of salt were hard and long with sharp edges, while alkaline water particles were flat and round.[146] The theory of acid and alkali, the opposites involved in chemical processes, for example, in the creation of metals, was a popular thesis developed by Sylvius (de le Boë). A work that was particularly influential, and that Swedenborg often cited for the geometry of chemical substances, is Nicolas Lémery's *Cours de chymie* (1675). Lémery claims that natural things cannot be better explained than by assigning to their parts the figures that correspond to the effect they cause. Sulphur, or the oil that Lémery alludes to, consists of supple, ramified particles that get caught up in other particles, which results in the special viscosity of oils. He also assumes the existence of small, round particles of fire and provides mechanistic explanations of the neutralization of acids and alkalis. In his geometrical chemistry, acids consist of sharp-pointed particles, whereas alkaline particles are porous like pincushions. These could vary, with larger or smaller pores, sharper or blunter needles, and so on. When the needles are stuck into the pincushion they are neutralized.[147]

It is with this kind of metaphorical thought, based on the mechanics of geometrical forms, that Swedenborg builds up his advanced particle mechanics. All properties could in principle be explained through geometry and motion. Weight, for example, depends on the position and relative number of different particles. Oil particles are round and consist of a subtle matter within, and salt flakes of different kinds on the surface.[148] Since oil particles are round and of the same size as water particles, they display similar properties. They are fluid and mobile, they are in the same position, have the same undulations, and so forth. As for metals, he rejects the Paracelsian idea that they are compounds of salt, sulphur, and mercury. No sure—that is to say, geometrical—knowledge of metals had yet been attained, according to Swedenborg. His path is that of geometry. By ascribing forms, positions, movements, and other geometrical attributes to metals, one can arrive at knowledge of the mechanics of the invisible. A solution of lead is sweet, which indicates that lead particles are round. Lead ore is grey in colour, which shows that its particles are in vertical position with large holes between them absorbing light. Practical geometry and mechanics are the very origin of variations in the particle world. But there is a risk of illusions here. In this little world, objects are mere mental phenomena. Reason and the inner eye are the only organs of vision at our disposal. But since everything has laws, which means that the experiments fall under the command of geometry, the mechanics of the invisible can rest on geometrical calculations and can share in the infallibility of mathematics. Everything that I state follows from the power of geometry, says Swedenborg.[149]

[146]Lindroth (1973), 172f.

[147]Lémery (1675), 10; Lémery (1713), part I, ch. I, § 43–52; Powers, 167.

[148]*Prodromus principiorum*, edition 90f, 93; translation, 115, 117.

[149]*Prodromus principiorum*, edition 104f, 114, 121, 132; translation, 133f, 146, 154, 168.

The foundation in natural philosophy of the bullular hypothesis is presented by Swedenborg in *Miscellanea observata*. The method proceeds from geometry. With the aid of geometrical skills, it is possible to detect the forms of particles that cannot be perceived by the senses. He postulates a number of axioms that are necessary for the revelation of nature's secrets of particle mechanics. First, nature is assumed to prefer the simplest solution, which means that the particles must have the simplest and least artificial form. Moreover, the origin of nature is said to be identical to that of geometry. Nature's particles have their origin in mathematical points, in the same way as lines and bodies in geometry. There is nothing in nature that is not geometrical, and vice versa. Thus there is nothing geometrical that is not natural. The third axiom is motion and the fourth is experiment through which nature should be investigated. The principle of simplicity means that there can be no simpler particles than round bullæ. This inescapable fact is obvious from geometry. For the sphere has no angle, but a single endless angle that sums up all angles. In this geometrical body all radii are the same length. There is just one surface and it is perfectly smooth. It is simply the most perfect figure without discrepancies, and since it is the most uniform figure it also has the most uniform movement. This form is the simplest possible. At the start of time there were only mathematical points without forms, dimensions, movements, or geometrical properties, and these were set in motion by the prime mover, God.[150]

But the bullular theory is only a hypothesis, since all assumptions about invisible things raise many doubts. 'I cannot assert, I shall only imagine or suppose,' Swedenborg writes. Yet in the mean time experiments can show whether my assumptions are in harmony with the truth or not. Since fires burn worse on high mountains where the air is thin, he assumes that there is more fire at the bottom of the atmosphere. In our sublunar world there cannot be fire without air. This can be experienced when fire burns in mines. A powerful wind arises and is sucked in towards the fire and the people standing by. It can also be tested by the air pump. Moreover, sulphur, oils, and other substances contain fire particles. The properties of the elements are in geometrical and mechanical harmony with the hypothesis of round particles. The diameters of the particles are in a geometrical ratio of 30 to 1. If air is 1, then the diameter of ether particles is 1/30, and that of light particles is 1/900 that of air. Let us picture, he says, a sea filled with such round particles and imagine that they are sufficiently large to be visible. Let us then apply geometry if it proves that its mechanics is the same as in the mechanics of the elements.[151]

If there is more fire inside than outside a particular substance, the fire forms a bubble, which has been proved by thousands of experiments, as in boiling water where fire transforms water into myriads of bubbles, or in steam, slag from furnaces, bubbles in glass, and other chemical experiments.[152] If there is more fire outside than inside a substance, it is shaped into a globe by the fire, as can be seen from

---

[150] *Miscellanea*, 133–136; translation, 84–86.

[151] *Miscellanea*, 91, 146, 153; translation, 58, 92, 96.

[152] *Miscellanea*, 64–67, 77; translation, 41f, 49.

the metal that drips down into the fire, drops of glass, water, and mercury which form exactly spherical globes in both air and vacuum. In a similar way as liquid in square stones forms circles, fire penetrates hard bodies. Swedenborg's idea rests on the belief that the roundness arises through an evenly distributed pressure, like a balance between inner and outer forces. This was how Descartes explained the roundness of water drops as the evenly distributed pressure from without of the ether spheres.[153] It is for a similar reason that the earth has acquired its round shape. The idea that steam consists of small, hollow bubbles, filled with a lighter substance that lifts the bubbles up into the sky, had been put forward by both Halley and Leibniz. Nils Wallerius wrote about it in the *Acta* of the Scientific Society, and in the Swedish translation of William Derham's popular *Physico-Theologie* (1736) we read that the water particles driven aloft by heat 'are water bubbles, which can be seen with the eyes through a microscope, when they fly around in a sunbeam' in a dark room.[154] Swedenborg's bubbles are based on a metaphor whereby knowledge of vapour and water drops is transferred to the invisible world. In *Miscellanea observata* Swedenborg writes in connection with icicle-like stalactites in caves, that when we treat the invisible, we can only arrive at a conclusion by reasoning with the aid of analogies. They can assist in revealing the mysteries of nature. The conclusion can be adopted until experience teaches us the opposite. When it comes to the invisible, we have only cogitation and geometry available to us in our investigations.[155]

The Power of the Water Bubble

The bullular hypothesis links Swedenborg in an interesting way with what was a completely new technology for Sweden, the steam engine. In a letter from 1714 he mentions a machine that raises water with fire and that can be used where there is no running water, in other words a steam engine that wins mechanical energy from thermal energy.[156] This is the first time a steam engine is recorded in Swedish sources. It may have been a steam engine of the Savery type, which Swedenborg could have read about or perhaps even seen in England.[157] A number of years later, in 1720, Swedenborg wrote once again to Benzelius, enclosing an excerpt from a letter from an auscultator in the Board of Mines, Henrik Kalmeter, who was then in Newcastle. Kalmeter describes a remarkable pump machine, an invention 'powered by fire and water'. To describe something that no one in Sweden had previously seen or heard about, Kalmeter was forced to use metaphors in order to understand

[153]Descartes (1644), IV, § 19; *Oeuvres* VIII:1, 211f; Hoffwenius (1698), XI:5, 47f.
[154]Derham (1760), 73; Halley (1694), 468–473; N. Wallerius, 339–346; cf. Rüdiger, 121f, 332; Beckman, 206.
[155]*Miscellanea*, 169; *Opera* I, 176, 179; translation, 106, 128, 132.
[156]*Opera* I, 226, cf. 231; *Letters* I, 57; Lindqvist (1984), 119.
[157]Savery, 228.

and convey what he had seen. The pump goes like 'a kind of churn or drum like that by which one makes butter, made of metal, which is so tight that no air can press in at the sides of the piston which goes inside the drum.'[158] The butter churn works in principle in the same way as the steam engine's almost 2-ton cylinder, with a piston moving up and down.

In *Prodromus principiorum* Swedenborg gives his explanation of the phenomenon of steam power in terms of his theory of particle physics. He describes a number of experiments with water globules, empirical experiences that can all be proved and explained geometrically. The 26th experiment concerns the steam engine. Water converted into steam has so much power that it can set a whole machine like a wheel in motion.[159] Steam power is one of the empirical phenomena that the bullular hypothesis must take into account. The round particles should therefore be able to provide a physical explanation as to why the steam engine functions. A passage in *Miscellanea observata*, about the strong force and vigorous motion of the small bullular particles, starts with four empirical facts. The third observation draws attention to the fact that steam can set a whole machine wheel in movement, as has been tested in both large and small machines.[160] From these factual circumstances one can draw the conclusion that if air consists of bullae then it possesses a very great resistance. Ether has greater resistance, and particles of light even greater, because the strength of the bullular particles increases in proportion to the reduced diameter, since the surface area increases in proportion to the diameter. It is thus the very geometrical form of the bubbles which contains the force that sets the wheel of the steam engine in motion.

As assessor in the Board of Mines, Swedenborg appeared in the capacity of an expert in the evaluation of the advantages and disadvantages of the new technology in April–May 1725. One of his verdicts on the Newcomen machine was that it was designed with such proportions that was 'based not only on the laws of mechanics, but also on those of physics'.[161] The steam engine was thus also based on physical, scientific laws. In addition, Swedenborg's memorandum includes calculations of the diameter of the cylinder (Fig. 7). His geometrical method sought to find the diameter of a circle with an area twice as large as that of a circle of known diameter. But Swedenborg's method gives only a determination of the theoretical value. It was in fact a geometrical method for finding the root of the number 2. When Swedenborg then writes that 'the diameter is always made slightly larger' because of the friction, this shows a characteristic side of Swedenborg: his fascination with theory, with drawing geometrical circles with a compass, rather than the practical execution. Swedenborg is a theorist, not so much a practically minded mining

---

[158]Kalmeter, in Swedenborg to Benzelius, Stockholm, 29 February 1720. *Opera* I, 300; *Letters* I, 231f; cf. *Bokwetts Gillets protokoll*, 17; Swedenborg also obtained information about blast furnaces in England from Kalmeter in *De ferro*, 155–158.

[159]*Prodromus principiorum*, edition, 21; translation, 25.

[160]*Miscellanea*, 139; translation, 88.

[161]RA, Bergskollegii arkiv, EIV:169, I, 760, 765; Lindqvist (1984), 166–169, 177f, 344.

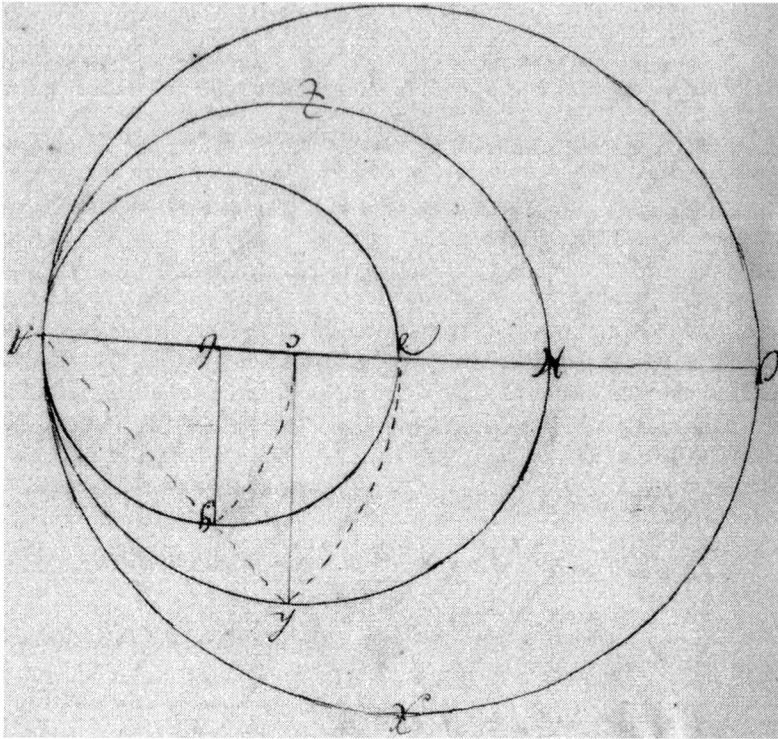

Fig. 7 Swedenborg's geometrical method for calculating the radius of a circle, the area of which is to be twice as large as the area of a known circle. The circle with radius *ch* is twice the area of the circle with radius *bg* (Swedenborg's statement about the Newcomen machine in the Board of Mines (1725))

engineer as a scientifically and mathematically schooled official. One can also see it as an assertion of theoretical knowledge and physics, at the expense of practical mechanics, as way to take command over the new technology. The steam engine requires a man of science; an artisan is not enough.

Three years later, in 1728, Triewald set up the first Swedish steam engine at the Dannemora mines. According to his own calculations, the machine, which was used for pumping out water, could raise as much as 15,120 barrels of water daily, and could do more work in a day than 66 horses.[162] In his description of the fire and air machine at Dannemora, he compares the air pressure to how three strong men on either side of a door without lock or catch try either to open it or to prevent the others from pushing it in. He also describes the same experiment that Polhem often cited, showing that an ox bladder dilates as a result of heat. The heat gives the

---

[162]Triewald (1734), 26, 33.

air a great expansive force. Triewald comes close to Swedenborg's explanation of steam power in terms of bubbles. The steam released into the cylinder is a damp, heated air, 'because each and every air particle is surrounded by an incomparably thin water membrane or bubble.'[163]

Triewald had definitely been impressed by Polhem's mechanical genius, but it is less certain how much he had been guided by Swedenborg's bullular hypothesis. The review of *Prodromus principiorum* in *Acta eruditorum* was not entirely flattering: Swedenborg's description of the geometry of water particles is unclear because he uses terms that he does not explain properly.[164] Two writers who really do cite Swedenborg as an authority in the physics of water particles are Pehr Martin and Daniel Tiselius. In his work about Vättern from 1723, Tiselius regards water particles as small, long, smooth motes (atoms) which 'wriggle and twist about and penetrate things'.[165] This is in agreement with Descartes' assumption. Water consists of two kinds of particles according to Descartes; both of them are long and slender, but whereas fresh water particles are soft and pliable, those of salt water are hard and inflexible. But Tiselius also says that he eagerly awaits the publication of Swedenborg's treatise about the innermost nature and properties of water. Pehr Martin reminds Tiselius, in a letter published among the previous comments on Vättern, that there are many theories of the structure of water particles: that they are rectangular, slippery, cubic matter ordered like intestines or concave parabolae. But the prevailing opinion at the moment is that they are round, as assumed by the best mathematicians, and in Sweden by Polhem and Swedenborg.[166] Fresh water has a viscosity that allows it to illuminate and retain the air in its pores, writes Tiselius with reference to Swedenborg's essay on Vänern, since it contains a fattiness that can be distended, 'as can be seen from soapy water, in which one can blow large bubbles.'[167] Tiselius says that he follows Hiärne, Bromell, and Martin in the opinion that water consists of particles which are flexible, rectangular, eel-like, elusive, and easily penetrate the pores of different bodies.[168] Plato, Aristotle, and all the Peripatetics, when they saw drops of water becoming shaped in a spherical figure, said that everything in nature is formed in the same way. 'Water particles are round with a hard crust and in the middle of the particle a hole or cavity,' says Tiselius with reference to Swedenborg, 'so nothing else can follow but that there must be an elasticity (as one can see from the blows of the round marbles with which children play, that one blow propels the other by knocking it on) and a tension in the water.'[169]

---

[163]Triewald (1734), 23, cf. 17f, 22.

[164]*Acta eruditorum*, February 1722, 84.

[165]D. Tiselius (1723), 4, cf. 6.

[166]Martin (1730), 11f, cf. 90f.

[167]D. Tiselius (1730), 55; *Om Wennerns fallande och stigande*, § 9; *Opera* I, 40.

[168]D. Tiselius (1730), 91f; H. Vallerius and Martin, 17; Descartes (1644), IV, § 19.

[169]D. Tiselius (1730), 97, cf. 110; *Prodromus principiorum*, edition, 14f; translation, 16.

The Vapours Rising Over the Mountain

As extraordinary assessor in the Board of Mines, Swedenborg plunged into the science of mining. There was a considerable amount of knowledge for him to acquire, such as the properties of metal, the processing of ore, and the location of mines. His two first essays on mining science dealt with blast furnaces and how to find new mines. He submitted these two works to the Board of Mines in the autumn of 1719, and as he says to Benzelius, they won the approval of his colleagues. One of the essays, *Nya anledningar till grufwors igenfinnande, eller några än oopfundna grep till at opleta grufwor och skatter, som i jorden diupt äro gömda* ('New guidelines for finding mines, or some as yet undiscovered methods for searching out mines and treasures concealed deep in the earth', 1719), is a rather strange piece, one might think, to come from a mechanistic philosopher.[170] In it he considers a number of ways to find veins of ore under ground by means of visible signs in the landscape, the bedrock, and the vegetation. He particularly emphasizes one phenomenon as a sure sign of underlying ore deposits. In Sweden this is called *brånad* or *bergvittring*, often described as *effluvia* or *exhalations*, that is to say, a kind of strong sheen emitted by rock rich in ore, as a fire-like vapour or steam. Swedenborg is an interpreter of signs, a baroque semiotician who reads and interprets signs in nature, draws conclusions from nature's symptoms in order to reveal the inner causes to which the signs refer. The world consists of signs which we must learn to decipher. Everything has a significance, referring to a more profound internal meaning. Observations of reality take on an uncommonly great significance. One can notice three things in this text, as in virtually all his early scientific works: the semiotic interpretation of reality, the metaphorical thought, and the mechanistic explanations. Before this Swedenborg's brother-in-law, Lars Benzelstierna, had defended a dissertation under Elvius which considered the question of how to find new veins of ore, and like Kircher he mainly used three different ways: reading signs, using tools, or pure chance.[171] One book that was fruitful for Swedenborg's *Nya anledningar till grufwors igenfinnande* was Vallemont's *La physique occulte*, which describes ten different methods for discovering mines, taken from Agricola, Cardano, Glauber, and Kircher.

There are great gains to be made, says Swedenborg, from a more reliable way of finding the treasures that lie hidden in the ground. Often these finds of ore have been discovered by chance, as when someone happens to kick away the topsoil or when a bull digs into the earth with its horns. Such events were the source of Sweden's prosperity. But under the wild forest, under sand, clay, and soil, huge unknown treasures may still be concealed. The divining rod, a cleft hazel twig that points downwards when held over a vein of ore, is an invention of the curious world,

---

[170]LiSB, N 14a, no. 73, fol. 171–177; *Photolith.* I, 106–119; *Opera* I, 294; ; *Letters* I, 221; Hj. Sjögren, 436–443.

[171]Elvius and Benzelius.

he says, which can be banished from the learned world as superstition, as there is no proof of its reliability. But there is another phenomenon that is perfectly real: 'Otherwise it is plainly true that all the spaces and caches where metal rock, streaks of ore, and other treasures are to be found have above them a stream of vapour, so that a powerful light shines thence at night and is spread a long distance.' In the darkness one can see a light as of fires, but it vanishes as one comes closer. This fire is a clear sign of the fumes that arise from veins of silver, copper, and perhaps iron. Sulphur and volatile salt fly up like a current 'and make a semblance of a flying dragon, from which one can also judge that there are chambers where treasures of silver and gold are hidden.'[172] The divining rod could perhaps be explained by this vapour, which causes the joints in the fingers to become paralysed and slack, and the spirits in the blood sluggish, so that the divining rod must fall and point towards the streak of ore.

There are other signs to look out for, for example, whether there are unusual plants, trees, and mosses at deserted mines. The reason is that the effluvia affect and change the properties of grass and herbs, as Agricola and Vallemont had also pointed out.[173] The metals can form new plants and give everything that a herb needs. This can be seen from the way silver and gold are dissolved, and when something is blended with it 'a plant is immediately formed into a bush with small twigs and branches, which is called arbor philosophica.'[174] Swedenborg's reference is to the alchemical experiment with the tree of philosophy, in which mercury is added to silver nitrate, which then takes on a bush-like shape, also known as 'Arbor Dianae' or 'Arbor Lunae', the moon tree or silver tree. In a lecture on the growth of metals Magnus von Bromell explains how to create '*arbor Lunae*. If aqua fortis [nitric acid] and Luna [silver] are mixed with mercurio vivo [mercury], up shoots a tree, or at least a moss.'[175] In *Miscellanea observata* Swedenborg talks of artificial trees from metals, like underground trees, leaves, flowers, and fruits which have their origin in the forms and positions of the mechanism of their particles.[176] This is a mechanistic variant and reinterpretation of Paracelsus's idea of the water tree with its roots growing down towards the centre of the earth and with branches bearing fruit of the earth's substances: salt, stones, and minerals.[177] Swedenborg goes on to tell of the Arbor Lunae and how one winter when he was on his way to Brunsbo he saw how the ice crystals on the windows made beautiful 'ice roses'. Artful nature and the mechanics of particles create the hexagonal and stellate forms of snow. Although water particles are round, in mechanical combinations they can result in hexagons, spirals, and so on.

---

[172]LiSB, N 14a, no. 73, fol. 173r.

[173]Agricola, translation, 38; Vallemont (1696), 329f.

[174]LiSB, N 14a, no. 73, fol. 175r.

[175]Bromell in ch. 'De generatione metallorum'. KB, X 601.

[176]*Miscellanea*, 127–131; translation, 81–83; *Prodromus principiorum*, edition, 88, 136f; translation, 112, 174f.

[177]Frängsmyr (1969), 39.

If one looks around one can also see how every stone has its distinctive mosses, green, white, red, and brown, and a moss that 'trails in serpentines and whirls and erodes all around it like ringworms.'[178] Moreover, the vapour can turn trees and branches entirely into stone. The sulphur and salt of the effluvia, which have flowed for a thousand years through the soil, through humus and clay, must also have left visible traces. One can also see how frost, ice, and snow lie more or less thick over mines than at other places. This is a sign that Agricola mentions, and he explains that the ground over veins of ore is not white from frost because the veins give off a warm, dry exhalation that prevents the damp from freezing.[179] If you lift stones, Swedenborg says, 'you notice how many millions of insects creep and crawl under them, where they have their nests and their republic'.[180] If our senses had just been sufficiently sharp they would have been able to find what is in the depths. Swedenborg sums up: if God had given us 100,000 times finer senses we could have found, merely by the smell and the faint glimmer, how the effluvia poured out of the rich veins of ore. But since we have not been granted this, we must find other ways using our minds. It takes accuracy and 'subtle observation.[181] We must be good chemists, with a complete knowledge of hidden nature.

Effluvia and exhalation above ore-rich mountains were an accepted phenomenon among miners and were considered a sure sign of ore below ground. At the great copper mine in Falun, the weathering of rock had long been observed as a way to find new deposits of ore. Aristotle explained the origin of metals in terms of two exhalations, one of them moist, the other dry and smoky, rising from the centre of the earth as a result of the heat of the sun. In the underground these exhalations are brought together to produce stones, minerals, and metals, and when they rise into the air they cause thunder and lightning, hailstorms, and other meteorological phenomena.[182] In mining and alchemy it was taken for granted that there is a link between the metals and the planets. The steam or effluvia was drawn up from the depths, rising through the veins and combined to make ore through the influence of the planets. Several of Swedenborg's sources on the science of mining mention this phenomenon that miners can see at night: Basilius Valentinus, Kirchmaier, Kellner, and Hellwig.[183] In *Physica subterranea* Becher talks about the existence of a metallic smoke over mines, about subtle metallic corpuscles rising from them.[184] Becher also compares the circular movement of this metallic smoke with the steam caused by heat, how the exhalations rise from the depths of mines and return again in a constant circulation, as in a chemical distillation. Balthasar Rößler,

---

[178]LiSB, N 14a, no. 73, fol. 175r.

[179]Agricola, translation, 37.

[180]LiSB, N 14a, no. 73, fol. 176v.

[181]LiSB, N 14a, no. 73, fol. 177r.

[182]Aristotle, *Meteōrologikōn*, 3.6.378a13–378b6; cf. Principe (2000), 37.

[183]Kirchmaier (1687), 42; Kirchmaier (1698), 13; Kellner, 44; Basilius Valentinus, II, 67–70; Hellwig, 9.

[184]Becher (1703), I, 70f, 92–98.

in a work frequently cited by Swedenborg, *Speculum metallurgiæ politissimum* (1700), considers both weathering and the divining rod as ways to find streaks of ore.[185] *Witterung*, Rößler explains, is what miners call both a natural heat that ore gives off when it has become perfect, and a steam that is sometimes emitted by the rich mine tunnels, and in the daytime can burst out like fire.

In Sweden Hiärne tried to collect more data on the phenomenon when he asked his informants the question: 'Is there rock from which there has at times been some remarkable warmth, violent heat, or weathering and fumes?'[186] Swedenborg often referred to Hiärne, although he says in a letter to Benzelius that he finds him 'to be very little grounded in the things on which *chymie* should be built.'[187] Hiärne's Paracelsism was probably alien to Swedenborg. For him the proper foundation of chemistry was mechanics and geometry, not the kind of '*puerilia* and crudities in *Mathesi*' that Hiärne busied himself with.[188] However, his statement here also includes a certain indignation over Hiärne's defamation of his father the bishop on the language issue. Polhem was curious about physical knowledge of 'the brightly shining rock-fires'; the same curiosity was displayed by Bromell and by Erik Odhelius, who defended a dissertation in Brussels about effluvia from metals.[189] What was new in a way about Swedenborg was his attempt to incorporate effluvia in a mechanistic explanation. They cannot be of a mystical, immaterial nature, but must be understood as a type of matter. A predecessor here was Boyle, who likewise assumed the existence of exhalations above mines and tried to show that their nature was not incompatible with contemporary scientific theories.[190] The effluvia could in fact explain many natural phenomena, such as electricity and magnetism, which ought not to be impossible to discern in a microscope.

The divining rod was also used to search for ore in the mining districts of Sweden. The divining rod, which was supposed to go back to Moses at Mount Horeb, is mentioned in most books on mining science as an implement for discovering mines, as in the works of Agricola and Basilius Valentinus.[191] Paracelsus, on the other hand, did not believe in it, while Boyle was convinced that it worked. The ironworks owner Otto Dress regarded it as pure superstition in the description of how to find ore that he submitted to the Board of Mines, a text that Swedenborg would also use in his *De ferro*.[192] The year before Swedenborg finished his essay about new ways to discover mines, a dissertation was published in Uppsala about the divining

---

[185]Rößler, ch. XXVII, § 5; § 12, & ch. XXXI, § 8.

[186]Hiärne (1706), 228.

[187]Swedenborg to Benzelius, Brunsbo, January 1718. *Opera* I, 278; *Letters* I, 172.

[188]Swedenborg to Benzelius, Brunsbo, 5 October 1718. *Opera* I, 287; *Letters* I, 198.

[189]Polhem to Societas Literaria et Scientiarum, Stjärnsund, 17 April 1728. *Polhems brev*, 174; Bromell (1739), 72; Odhelius; J. G. Wallerius and Roman, 4f; J. G. Wallerius and Scheffel, 15–19.

[190]Boyle, 'De temperie svbterranearvm regionvm ratione caloris et frigoris', in Boyle (1680), 16; Vallemont (1696), 130; Kargon, 101.

[191]Agricola, translation, 38, note; Basilius Valentinus, II, 87–111; Exodus 17:6.

[192]Dress, 48f; Carlborg, 33; C. Sahlin 1931, 11.

rod, examining its use for metal detecting.[193] Swedenborg was more doubtful about the divining rod, but he still tried to adduce a mechanistic explanation for it. In his library one can find a rather strange French work that may very well have inspired him, namely Vallemont's *La physique occulte*.[194] Vallemont attempts a mechanistic explanation of the motion of the divining rod by looking at the corpuscles of matter. The corpuscles flow through the hands, penetrate the divining rod, and make it bend when one walks over ground where water vapour rises from springs or exhalations from mines (Fig. 8). It can also be used in criminal investigation, for the divining rod bends above the bodies of murder victims and criminals in hiding. It is, quite simply, a valuable invention, Vallemont observes, because it is the only pointer we have to lead us to underground treasures or the places where nature produces metals.

Swedenborg's *Nya anledningar till grufwors igenfinnande* was read aloud by Benzelius at a meeting of Bokwettsgillet in 1720.[195] In the ensuing discussion Roberg mentioned the red moss found on the stones around Salberget. He also expressed a wish that Swedenborg would say whether Swedish mines also display strong effluvia or exhalations like those seen from the looser rock in Germany. Was it the case that metals give off a palpable exhalation where they are produced in the depths of the earth, and is it not more uncertain whether there is such a steam where the metals are already complete? Roberg recalled having seen at Dannemora mine a kind of immature vitriol bursting forth like rime frost.

Swedenborg also discusses the effluvia of metals in *Miscellanea observata*.[196] There is a special fire from the sun or the centre of the earth that enriches stones with metals. But he would not contradict those who assume that there is an influence of particles from the rays of the planets. There may thus be an actual connection between metals and the planets. Swedenborg, however, suggests a different origin for metals: water. Mines are always full of water, gold is often found in the sand of rivers and iron ore in lakes. There may thus be a special liquid that arranges particles in specific forms and positions. Water functions in this way as a vessel transporting metallic effluvia and particles to their womb. This idea of metallic effluvia and exhalations that Swedenborg embraces is a striking example of metaphorical thinking. The exhalations or *brånad* is a metaphor proceeding from the mist and steam rising from the earth and from water when they are heated by the sun.[197] It can also be seen as a product of confusion with the sulphurous fumes and toxic gases at mines and smelting houses. The poisonous environment and the lethal fumes of the mines are also generally discussed in the literature on mining, as in Vallemont or Agricola, who mention a dangerous pestilent air in mines that causes lungs to rot.[198]

---

[193]Hermansson and Maehlin, 10–24.

[194]Vallemont (1696), 60, 93, 127, 333; cf. Valentini, III, 68–71, 113–228; cf. *Principia*, 235; translation II, 27.

[195]*Bokwetts Gillets protokoll* 5 February 1720, 13.

[196]*Miscellanea* IV; *Opera* I, 166–168, 172f; translation, 117–119, 124f.

[197]Cf. Forsius, 113, 119.

[198]Vallemont (1696), 124; Agricola, translation, 6.

Fig. 8 A French gentleman walking with a divining rod, 'la baguette divinatoire'. Thick exhalations rise from the mineral-rich rock, causing the divining rod to turn downwards (Vallemont, *La physique occulte* (1693))

Pag. 106.

The Geometry of Heat

Swedenborg's suggestions as to how mining could be improved are based on a belief that theoretical science could provide the answers after the empirical evidence was assembled. In the summer of 1719 Swedenborg put into print some thoughts about fire and hearths, as he explained to Benzelius, by collecting knowledge from 'smiths, charcoal burners, roasters [of ore], smelting masters, etc.'[199] Swedenborg's *Beskrifning öfver swenska masvgnar och theras blåsningar* ('Description of Swedish blast furnaces and their blasts), which was submitted to the Board of Mines in autumn 1719, was one of the first in a series of works on mining science. The aim was to ascertain the nature of fire and its effect in all kinds of furnaces, in blasting, smelting, metal manufacture and oar roasting. At this time there was a great interest in the nature of combustion and various ways to achieve greater efficiency, such as the tiled stove, which gave more economical heating. Mining consumed enormous quantities of firewood, which thinned out the forests. If one could save some wood, a great deal could be gained.

Swedenborg follows his characteristic metaphorical thinking, progressing from the large and perceptible to the small things beyond the reach of the senses. The idea was to scrutinize the nature of fire on a grand scale 'and anatomize it in a large subject', and then draw conclusions about its effects and properties on a small scale.[200] The manuscript is permeated by an eye for geometrical structures and forms. He gives accounts of the structure of the blast furnace, the slope, width, and height of the form, but he also provides sensory descriptions and investigations of the colour of the slag, the properties of fire and smoke, demonstrating the need for a furnace master's professional skill and experience. Swedenborg the geometrician explains that, the narrower the figure of the blast furnace is, the greater is the area in relation to the volume, which thus means more cold, but the broader the figure is, the more round or square in form, the less is the area in relation to the volume, and consequently less cold and more heat, 'which also has a secure foundation in geometry.'[201] The best effect as regards fire and heat is when a blast furnace has the form of a hyperbolic flame over a wax candle.

In 1721 Swedenborg published an essay about iron and fire, *Nova observata et inventa circa ferrum et ignem*, in which he puts forward the theory that fire consists of bubbles with mathematical points on their surface.[202] Fire particles are small and can penetrate all hard substances. They resemble the particles of water and air, which differ only in weight and volatility, so that air and water tend to pull downwards, while fire rises. Swedenborg sent a copy of *Prodromus principiorum* to Boerhaave in Leiden.[203] Boerhaave, in his *Elementa chemiae* (1732), had expressed a view that agreed with Swedenborg's. Fire, as he saw it, consisted of small, hard, smooth, and highly mobile spherical particles, which penetrate all bodies.[204] Fire, Swedenborg went on, moves in spirals like shells when it rises through a hole, and when it ascends it makes a sound, in the same way as when water sinks.[205]

---

[199]*Opera* I, 292; *Letters* I, 215.

[200]RA, Swedenborg, [3f]; *Beskrifning öfver swenska masvgnar*, edition, 201; it was to be included in *De ferro*, the first 70 pages. *Letters* I, 215.

[201]*Beskrifning öfver swenska masvgnar*, edition, 229, cf. 231.

[202]*Nova observata*, edition, 185; translation, 199f.

[203]British Museum, dated 21 October 1721. *Letters* I, 256.

[204]Boerhaave (1732), I, 165f, 327f; Frängsmyr (1969), 237.

[205]*Nova observata*, edition, 181, 189f; translation, 194, 204f; *Miscellanea*, 97; translation, 61.

The undulations of water and fire and many other properties are similar, which demonstrates that water and fire are concordant in their elementary nature, differing only in weight and lightness. He also shows how to design a furnace which in eight to ten days consumes only the amount that ordinary furnaces need in one or two days. Swedenborg believes that the hearth can be of a lesser depth, but the reviewer in *Acta eruditorum* says that the rule can perhaps be applied in Sweden, where the wood is placed upright, but not here in Germany and other countries where the wood is laid parallel to the hearth.[206]

Cold and chill, as Swedenborg used the words, became a positive entity, a true property and not a lack of heat; this idea was also expressed by Polhem, Block, and other contemporaries. In *Acta literaria Sveciæ* for 1722 Swedenborg writes about the heating of rooms, describing cold as a separate identity, how stones spread cold, how chill radiates, is exhaled and evaporated from the walls of stone houses.[207] 'Sit upon a stone, and in less than a quarter of an hour you will feel great cold.' For timber houses the opposite applies, with the example concerning blast furnaces. The bigger the timber walls are in relation to the volume of the room, the warmer the room is. Wood, unlike stone, conserves heat. A square room is thus warmer than a round one, a rectangular room is warmer than a square one, and the more oblong it is, the warmer a room becomes. Now there are no descriptions or explanations of any independent entity called cold in Swedenborg's writings. In his theory of matter he is, if anything, in agreement with the Aristotelian idea that cold was a non-being, an absence of heat, or the Cartesian explanation of cold as an absence of movement in the particles of matter. Swedenborg speaks of fire as particles shaped like bubbles and heat as the motion of the particles. The example of the heating of rooms shows how language and abstract natural philosophy have drifted apart. Swedenborg's abstract scientific understanding of heat has distanced itself from the intuitive, preconceived meaning that is built into our language. In language cold and heat are categories of existing things or objects.

A Mineral Cabinet Without Stones

Prodromus, the precursor of the principles of natural things, was the introduction to a large-scale project that Swedenborg intended as a 19-part work on metallography and mining science. In 1722 he announced a prospectus for subscriptions to this plan, promising that publication would begin towards the end of 1723.[208] This never

---

[206]*Acta eruditorum*, May 1722, 264; *Miscellanea*, 85–101; translation, 54–64; Swedenborg could read about the use of fire in stoves, smelting furnaces, tiled stoves, etc. in a book he owned, Leutmann (1723).

[207]*Novæ regulæ*, 282–285; LiSB, N 14a, no. 148:8; *Photolith.* I, 188f; translation, 154f.

[208]Swedenborg, *De genuina metallorum tractatione* (1722). KB, Okat., subskr., Pist., Fal.; published in *Nova litteraria*, 1720–1722 in Supplementum actorum eruditorum, Leipzig, August 1722, 126–128; *Neue Zeitungen* 21 December 1722, 1008.

happened. The extant manuscripts of this gigantic work deal with salt, silver, vitriol, sulphur, and pyrites, along with magnets. The papers consist almost exclusively of collected material and excerpts from other authors, and were no doubt used as his own reference work and notebooks, to which he made continuous additions. But the ambitious work was also to contain investigations of the nature of fire and blasting, experiments about metal processes, flux, the growth of minerals, and much besides. At the start of 1724 Swedenborg reported to Benzelius and Bokwettsgillet on the progress of his major work on minerals, especially the part about copper: 'I hope in time to increase this my mineral cabinet or collection, just as others add to their stones.'[209] When Benzelius asked him later in the spring if he was not going to apply for the chair vacated by the late professor of astronomy, Nils Celsius, Swedenborg declines:

> My *affaire* [occupation] has now been *Geometrica*, *Metallica* and *Chymica*, and it is a far cry between them and *astronomica*. To abandon that with which I think to perform a good use, would be indefensible. Besides that, I have not the *donum docendi* [the gift of teaching], as my Brother knows, by reason of the *naturella difficultate* of speech. I hope, therefore, that the Academy does not put me in nomination.[210]

Geometry, metals, and chemistry were now the task of the stammering assessor. Not long afterwards, on 15 July, he finally became a full assessor in the Board of Mines.

In that year, 1724, he was working on the part about the separation of silver from copper.[211] He also wrote about vitriol, different kinds of sulphates, above all oxidized sulphurous ores of copper and iron which formed beautiful dark blue and green crystals (Fig. 9). In *De victriolo*, Swedenborg tells about Leeuwenhoek, who was able to observe in his microscope how sea-blue copper vitriol is dissolved in rainwater.[212] Copper vitriol commonly occurred in mine water and had a number of uses. On the practical side, Swedenborg agitated to have a vitriol works established.[213] But the manuscript about vitriol mostly consists of notes and excerpts that he had collected over a very long time from other mineralogists and chemists.[214] Among other things, he cut out a picture from Agricola's *De re metallica* describing how vitriol water is collected in basins (Fig. 10). One man is carrying a pail of vitriol water from a mine, walking out into the strong sunshine towards the square bath. In hot regions or in the summer it was poured into outdoor basins, where it

---

[209] Swedenborg to Benzelius, Prästhyttan, 14 February 1724. *Opera* I, 312; *Letters* I, 328; *Bokwetts Gillets protokoll*, 98, 100–103, cf. 107.

[210] Swedenborg to Benzelius, Stockholm, 26 May 1724. *Opera* I, 313; *Letters* I, 333f.

[211] *De secretione argenti a cupro*; Swedenborg cites authors such as Agricola, Barchusen, Becher, Boyle, Ercker, Glauber, Kellner, Hellwig, etc.; he also used Löhneyß (1690).

[212] *De victriolo*, 405.

[213] In 1719 Swedenborg submitted a letter to Queen Ulrika Eleonora about the establishment of a vitriol works at Stora Kopparberg. In her reply she ordered the Board of Mines to consider the matter. Ulrika Eleonora to the Board of Mines, Stockholm, 14 November 1719. RA, Bergskollegii arkiv, kungl. brev 1717–1721, fol. 150 1/2; Hammarström, 63.

[214] References to Agricola, Barchusen, Ercker, König, Lémery, Willis, Wolff, and to Brandt's distillation of copper vitriol in 1727. Brandt (1741), 49–63; *De victriolo*, 433–446.

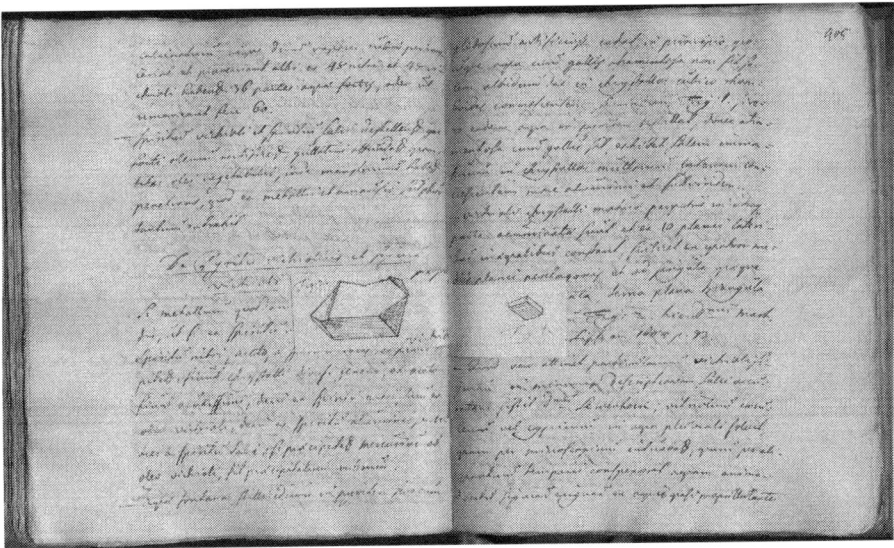

Fig. 9 White salt in cubic-rhomboid crystals (*1*), and a pentagonal vitriol crystal (*2*) (Swedenborg, *De victriolo*, drawing from Lister's essay in *Acta eruditorum* (1684))

was condensed by the heat of the sun. In cold regions and in the winter the vitriol water was boiled in rectangular lead cauldrons. Agricola writes that 'the thickened solutions congeal and adhere in transparent cubes or seeds of vitriol, like bunches of grapes.'[215] In Agricola's illustrations there is a special vision that involves peeling off the soil and seeing the mine galleries in cross-section, in order to display what one wants to show, not what it looks like in reality.

Around 1727 he wrote about sulphur and pyrite. In the manuscript *De sulphure et pyrite* he mentions an unusual event that had happened two years previously in Stockholm.[216] A sulphurous rain fell on the town in 1725. The weather as a whole had been odd that year, with constant rain and cold. Bishop Swedberg saw divine signs and omens in the unusual weather and the endless rain. Yet the dangers dispersed and Swedberg was able to breathe out: 'And God in his grace has heard our prayer: and the fine summer weather is now starting again.'[217] In *De sulphure et pyrite* Swedenborg has collected a great deal of information about the properties of sulphur.[218] Like the other books, however, it contains notes written on different occasions. Among other things he added a drawing copied from a report from 1694

---

[215] Agricola, translation, 574; *De victriolo*, 3f.

[216] *De sulphure et pyrite*, 86f; *De sulphure* refers to *De cupro* on p. 23, to *De victriolo* on p. 111, to *De ferro* on pp. 248, 251, 261; references to Agricola, Boerhaave, Henckel, Hoffmann, König, Lémery, Hellwig, Valentini.

[217] Swedberg (1941), 446; Swedberg (1725).

[218] *De sulphure*, 22, 40f, 105, 232f, 260.

Fig. 10   Vitriol boiling (Agricola, *De re metallica* (1556))

by Samuel Buschenfelt (Fig. 11).[219] Swedenborg himself visited the mines in Harz in 1722. In Grotta del Cane in Pozzuoli near Naples, he goes on to say, there are sulphurous fumes so toxic that they can suffocate animals and humans alike, a vapour that extinguishes fire and light, and prevents muskets from being fired. He also touches on subjects such as volcanoes and pyrotechnics, and talks of the terrible fire of Etna, which casts out stones like a barrage of bombs that devastate the treeless surroundings.

The salt of life is indispensable. Salt was something Sweden had to import. It was a challenge for Swedenborg, Polhem, and Roberg to find a way to extract salt in the cold North and thus avoid having to pay high prices for Portuguese salt. Under the hot southern sun salt could be obtained from evaporated water, as Agricola describes. Swedenborg cut out a picture from *De re metallica* showing the

---

[219]Buschenfelt, *Den äldre fadren Buschenfelts marchscheider Relation 1694. tilhörige Ritningar.* KB, L 70:54:2, p. 366.

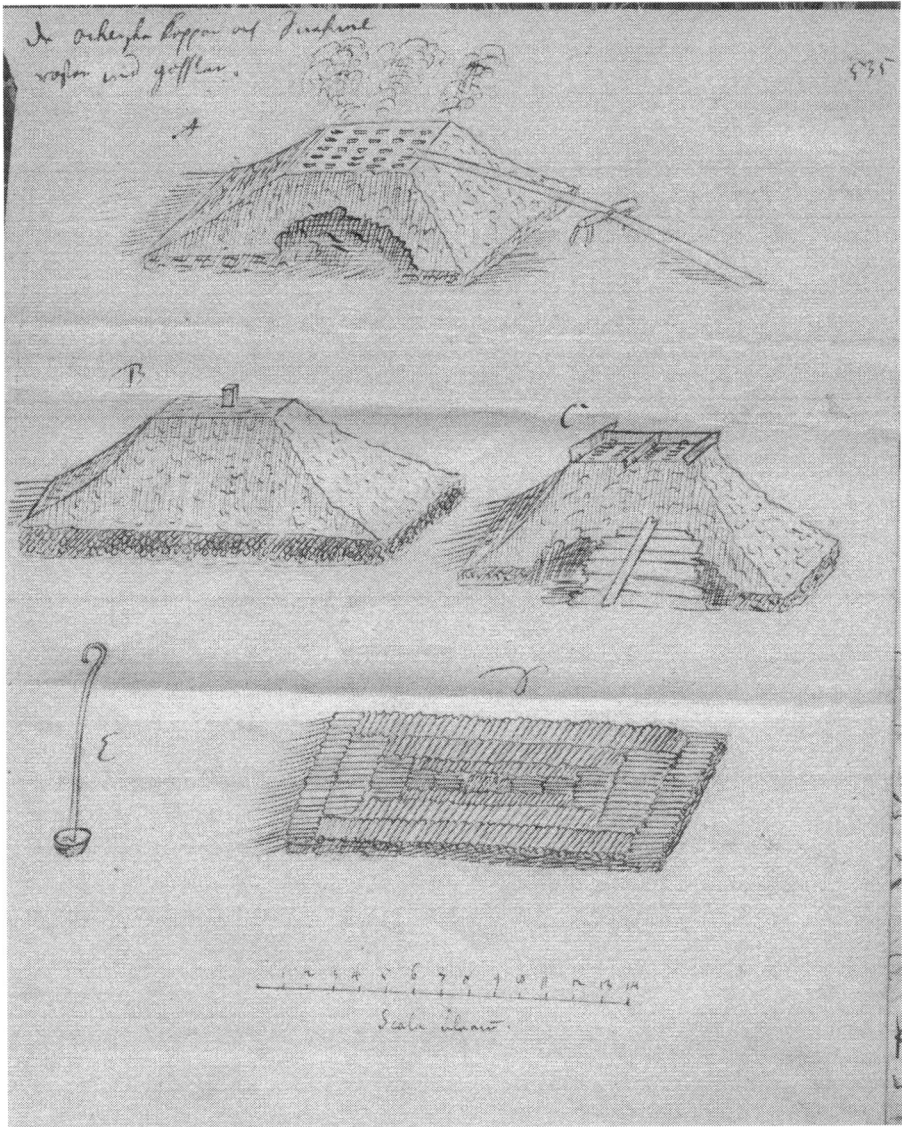

Fig. 11 Drawing in Swedenborg's manuscript on sulphur and pyrite of 'the Oker copper and sulphur roasters at Goslar', copied from a report from 1694 by Samuel Buschenfelt (Swedenborg, *De sulphure et pyrite* (*c*. 1727))

manufacture of salt from sea water.[220] Merchants stand on the beach in the cooling shade of palms. Ships lie at anchor, others are on their way out over the sea, laden with the precious salt. But what could be done in our sun-deprived country? One

---

[220]*De sale communi*, ms, 83; Agricola, translation, 547.

method that Agricola suggests is to boil seawater in large vessels. But this would be a further drain on the forests so badly needed by the mining industry. Roberg has a different idea. In *Dædalus Hyperboreus* for 1716 Roberg put forward his 'thoughts about the manufacture of salt in the Northern countries'.[221] Since the Nordic sun is far too weak and boiling far too expensive and require too much firewood, he found another way by exploiting Sweden's plentiful supply of cold. His plan was to extract salt with the aid of the low temperatures in Sweden, by freezing, letting the ice and frost separate the salt. Swedenborg suggests in particular Gullmarsberg on the shores of Gullmarsfjorden in Bohuslän, as a suitable place for a salt works.[222] But this project was never successful.

For several years, at least up until 1728, Swedenborg collected a multitude of data from a number of different authorities, about the nature of salt and its production. He wrote all this down in the manuscript *De sale communi*. One passage deals with the geometrical forms of common salt. When water evaporates one can obtain elegant salt crystals in quadrilateral pyramidal forms, Swedenborg writes. Other salts too can display geometrical forms, such as cubic parallelepipeds or various mineral salts whose cubes can be seen in a microscope (Fig. 12). Swedenborg refers not only to the crystallography of Martin Lister and Basilius Valentinus but also to Leeuwenhoek's study of the sweat that drips from a human face, both after a person has drunk wine and when he has not drunk any. In the face of the latter Leeuwenhoek could see cubic and pyramidal salt-like forms in his microscope.[223]

The manuscript *De magnete* (*c.* 1722–1729), according to the title page, may have been intended for publication in London in 1722. The majority of the notes may also be from that year, but he was to make additions for several years to come, including entries from a work by Pieter van Musschenbroek from 1729.[224] It was not until 1734, with *Principia rerum naturalium*, that he used these notes about magnets for a printed work. In *De magnete* he has numerous excerpts from a great many authors about different experiments with magnets and static electricity. He makes notes about a range of fantastic phenomena that can be achieved with a magnet, and a number of experiments with the aid of Hauksbee's electricity machine which generated static electricity by rubbing against rotating glass spheres or discs (Fig. 13).[225] Swedenborg viewed magnetism through a mechanistic filter. These phenomena are explained as a kind of mechanical effluvia or flows of particles in a set of metaphors analogous to his thoughts about exhalations and steam.

---

[221]Roberg, *Dædalus Hyperboreus* II, 28–30; Roberg, *Doct. Robergs upgift om saltets tillvärkande genom köld och blåst. 1706*. LiSB, N 14a, no. 58, fol. 140; *Opera* I, 266f, 268; *Letters* I, 145f, 149.

[222]*Miscellanea*, 104; translation, 66; Swedenborg, *Memorial om salt siuderiers inrättning i Swerje* (1717). LiSB, N 14a, no. 41; *Photolith.* I, 74–78; *Letters* I, 140–143; cf. *Kongl. Maj:ts nådige öpne bref, och privilegium*, 1717; *Underrettelse om docken*, [5–8].

[223]*De sale communi*, ms, 289–294; edition, 130–135; translation 1992:2, 131–135; Leeuwenhoek, 425–429.

[224]*De magnete*; Musschenbroek (1729); cf. Musschenbroek (1731); refers to Gilbert, Henckel, Hoffmann, Löhneyß, Rößler, Valentini, Wolff, and others, and on p. 224 to Ask's dissertation about prospecting for iron mines using magnets. Burman and Ask.

[225]*De magnete*, 237–240, 251–264.

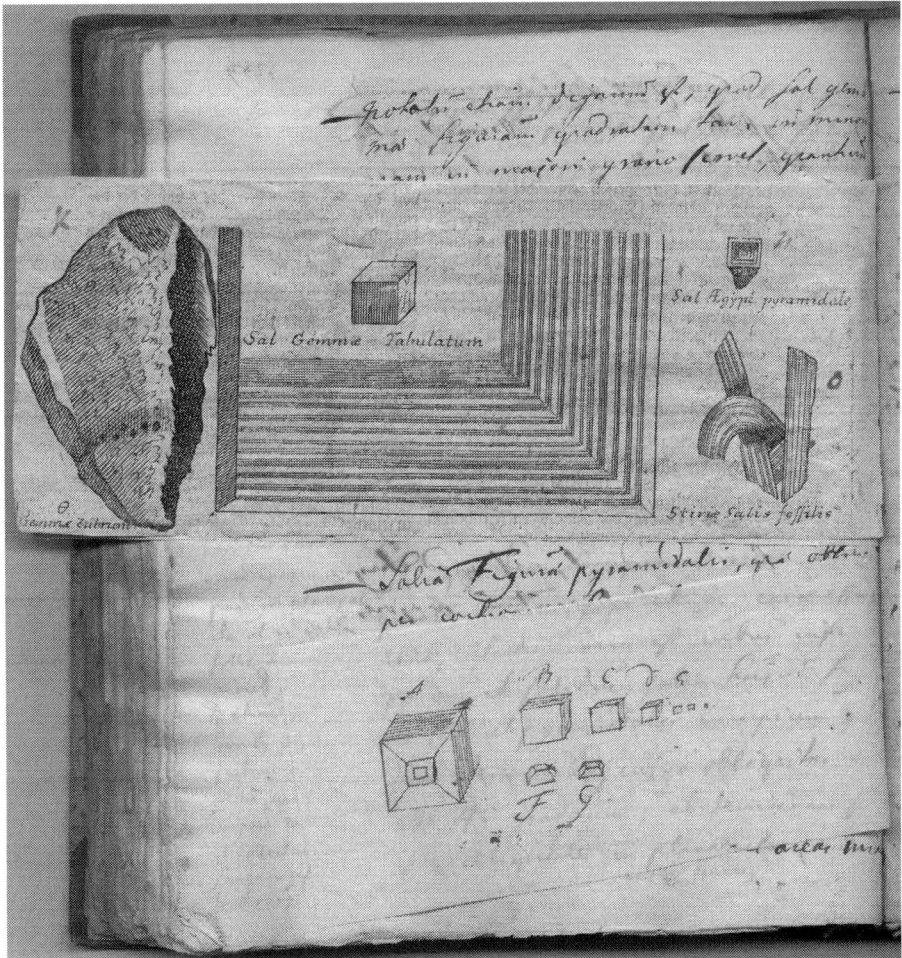

Fig. 12 The geometry of salt crystals. *K*, red rock salt. *L*, cubical salt. *M*, layered rock salt. *O*, icicles of mined salt, and *N*, a salt that is always shaped like small pyramids and is therefore called Egyptian Pyramidal salt. *A* to *G* are crystals obtained from sea water, rock salts, or salt water from springs (Swedenborg, *De sale commune*. Cutting from Basilius Valentinus, *Chymische Schriften* (1694), and Swedenborg's own drawing based on Lister's essay in *Acta eruditorum* (1684))

For his chapter on the attraction of the magnet and the transfer of forces, Swedenborg cut out a remarkable picture from Valentini's *Museum museorum* (Fig. 14). It shows a terrible basilisk with a poisonous gaze. Under the bar that it holds in its mouth hangs a magnet, and above it three keys which are held together by the magnetic force. To the left are three iron spindles and to the right a band of rings, likewise held together by magnetism. In the background is a window through which the viewer's gaze is drawn out into a mountain landscape. According to Valentini, the picture illustrates how the strength or force of the magnet can

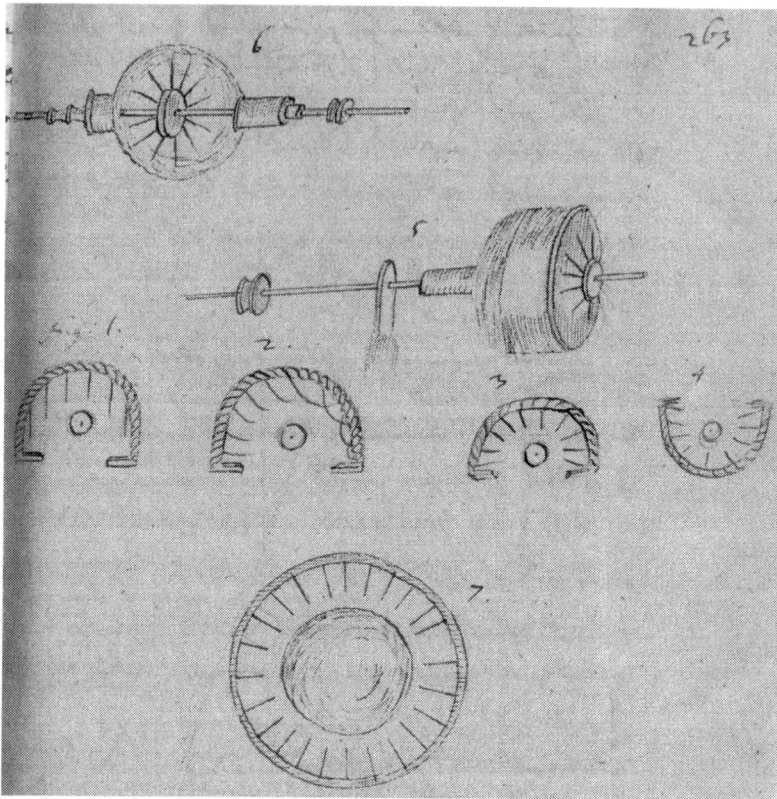

Fig. 13 Electricity machines (Swedenborg's drawing in *De magnete*, based on Hauksbee's *Physico-Mechanical Experiments* (1709))

be increased or diminished, and can be renewed when it is lost.[226] Swedenborg explains in *De magnete*: 'From the figure alongside one can also see how the magnet shares its power of attraction with different iron objects, such as many keys, or rings or pins, since one can hang in the other. This is best illustrated by the adjoining figure.'[227] The picture actually occurs later in Swedenborg's *Principia*, although in a completely new guise (Fig. 15). The occult alchemical atmosphere in Valentini's picture has been translated into a stripped geometrical-mechanistic drawing. The keys in *Principia* describe how the attraction of the magnet passes through iron, according to a surprising experiment by Derham in which the key *BK* is attracted more strongly to *C* than to the magnet *A*.[228]

---

[226]Valentini, III, ch. XVI, 85f, & ch. XXX, 94–96.

[227]*De magnete*, 79.

[228]*Principia*, 222f; translation II, 2f; Derham (1705), 2139.

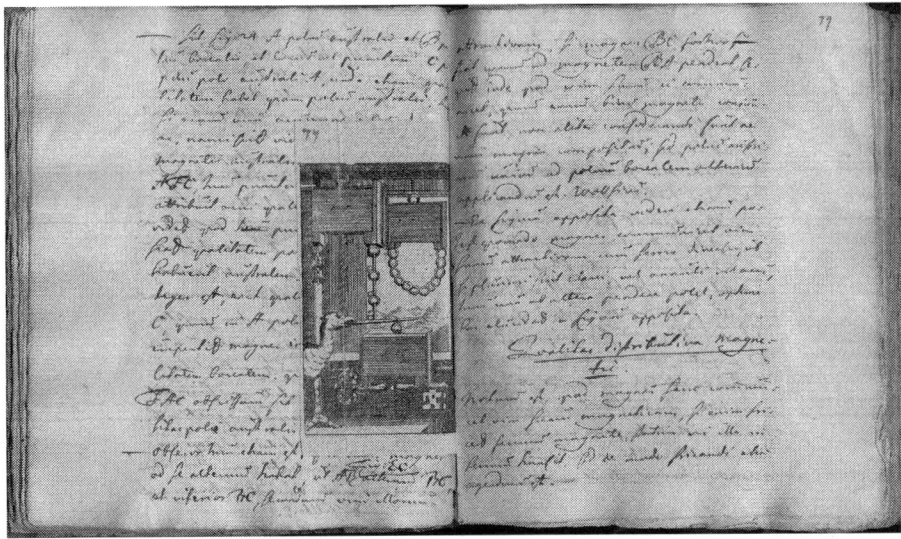

Fig. 14 The magnetic basilisk (Swedenborg, *De magnete*, cutting from Valentini's *Museum museorum* (1714))

Fig. 15 Magnetic experiment with keys (Swedenborg, *Principia rerum naturalium* (1734))

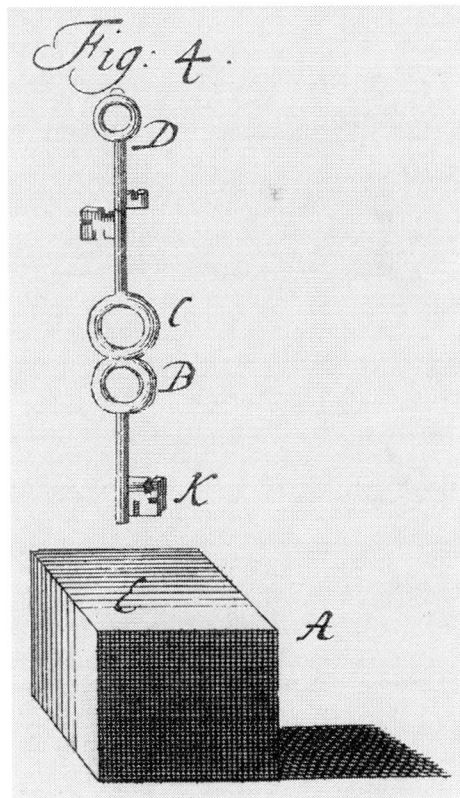

The Fruits of the Volcano

Swedenborg's studies in the mineral kingdom ended up being largely incorporated in *Opera philosophica et mineralia* (1734), the three volumes of which deal with the principles of natural things and with the two most important commodities of Swedish mining: iron and copper. This work, which sought to go through the entire mineral kingdom, leaves an undigested, unrevised impression, as if it were still in the hearth of the furnace or left behind, glowing red on the anvil. Yet he nevertheless casts loose from the safe haven and sets sail on the hazardous ocean of science: 'it seems to me as if I were on a little vessel with small, ragged sails, trying to travel over the whole great sea of the world, so that I have every reason to be afraid,' in the words of Virgil, that '*The seamy craft groaned under the weight, and through its chinks took in a marshy flood.*'—'In any case I nevertheless prefer to run the risk and sail out of the harbour rather than, seized by despondency and discouragement, abandon the entire enterprise.'[229]

Swedenborg dedicated the second volume, *De ferro*, to the Swedish king's brother Wilhelm, landgrave of Hessen-Kassel: if his book 'may seek protection under the shield of such a mighty prince, its dark colour through the gleam of the shield's rays is transformed into the brighter shimmer of silver.'[230] *De ferro* is about smelting methods, assays, and chemical methods concerning iron, and it relies in large measure on his excerpts from Swedish and foreign sources.[231] Among other things, he gives a description of a new kind of rolling and cutting mill which attracted some attention and was translated into French.[232] But he was not interested in doing his own laboratory experiments: I have 'not yet had the time or opportunity to let everything go through new fire with new and toilsome work.'[233] He wants to reveal the secrets of the professionals, their hidden knowledge, to bring it into the light, make it public and accessible to Latin-reading scholars. In this way it is an antonym of 'arcana', the secrets concealed in alchemical texts and carefully guarded by guild ordinances. It is a social struggle between the man of science and the craftsman, a contest between theory and practice, between natural philosophy and experience. He wants to get at the countless secrets belonging exclusively to 'smiths, smelters, and similar occupations, a crowd of simple people with sooty faces like the cyclops, of whom one would not have reason to expect anything

---

[229]*De ferro*, preface; Virgil, *Aeneid*, 6.413f.

[230]*De ferro*, dedication.

[231]E.g. from L. G. Schultze's report to the Board of Mines, where Swedenborg is also mentioned. Schultze, 7; Grönwall and Saxholm; passages from Swedenborg's *De ferro* have been translated into French: Bazin, 17, 24, 27–31, 36; translation.

[232]*De ferro*, 252f; 'Traité du fer'; Rinman, 133f; before this, on 11 April 1723 he had submitted to the Diet a humble memorandum about rolling mills of a similar kind to what he had seen in Liège in the winter of 1721. A rolling mill according to Swedenborg's proposal was installed at Vedevåg in Västmanland in 1726–1727. C. Sahlin (1934), s. 149–155; *Letters* I, 306–310.

[233]*De ferro*, 239.

brilliant or ingenious.' But their knowledge lies entirely within the practical sphere, relying on 'experience and lessons derived from real life'. Their knowledge should therefore serve 'to be preferred to or at least equated with much scientific insight'.[234]

With this aspiration to expose secrets, he cannot be content with the goals and methods of alchemy. He describes the activity of alchemists in metaphors of disgust, darkness, and the theatre. He is revolted by the unmentionable alchemy: 'I have touched lightly on the whims of the alchemists, scarcely wishing to touch them except with the outermost part of my lips—*that weeds may not choke the gladsome corn.*' Instead of coming in clear light the alchemists come 'with shadows, instead of with real meat and blood with masks, instead of reality with stage scenery.'[235] Earlier in *Miscellanea observata* Swedenborg had explained alchemy and transmutation as being impossible, precisely because they conflict with the assumption of particle theory that each metal has its distinctive particles.[236] Metals taste different. One has a plain taste, another a sweet taste, others are repulsive, such as mercury, or bitter as silver. This variation in taste must be due to the form of the particles, whose tips cause different taste sensations on the papillae of the tongue. The same conclusion can be drawn on the basis of other properties such as colour, compactness, fusions, which all depend on the primary geometrical properties. It is thus beyond human ability to destroy the forms of these metal particles and transmute them into new forms, for example, to gold. This is because gold consists of heavier and more homogeneous and stable particles than other substances. They are also bigger than other particles, with a diameter up to ten times greater than the particles of water. This rejection of alchemy was brave in view of the fact that *Miscellanea observata* was dedicated to his superior, the alchemist and president of the Board of Mines, Gustaf Bonde.[237] Swedenborg, one might think, ought to have known of Bonde's alchemical interest even though he could not know exactly what Bonde was up to when he tried to make gold from his own excrement.

Alchemy most probably lacked the rationality and clarity that Swedenborg expected in studies of nature. One can view his reaction against alchemy as part of the exclusion of that pursuit from science. Mechanistic natural philosophers, as rational scientists, wished to be above the manual craft of laboratory workers and to combine chemistry with the exalted geometry that they adored—the standard bearer of rationality and objectivity. Critical opinions about alchemy had also been expressed by Polhem, who observed that the only divine gold-making was economic activity.[238] But Swedenborg's attack on alchemy is in large measure pure rhetoric. His library contained a not insignificant number of alchemical works such as those by Pseudo-Geber, Schweitzer (Helvetius), Barchusen, and Kertzenmacher.[239]

---

[234]*De ferro*, preface.

[235]*De ferro*, preface; Virgil, *Georgica*, 1.69; cf. Edenborg (2002), 76.

[236]*Miscellanea*, 117–122; translation, 75–78.

[237]Fors, 239–252.

[238]*Polhems skrifter* III, 291–295.

[239]Geber; Schweitzwer; Barchusen, 481ff; Kertzenmacher.

The laboratory skills of the alchemists taught him a great deal, yet he wanted to belong to a more modern, demystified, rational science. Contemporaries noticed that Swedenborg's *Opera philosophica et mineralia* had points in contact with alchemy. Mr. Swedenborg excludes the alchemists and says that he will not even speak their name, says a Russian academician, yet we still find Becher, Stahl, Kunckel, Boerhaave, and others in these works, and one can see from the palpable traces of their writings that he does not regard alchemy as heresy.[240] At the archives of the Russian Academy of Sciences in Saint Petersburg there are some documents that describe Swedenborg's contacts with Russia.[241] In 1734 the Russian Academy of Sciences in Saint Petersburg had received a parcel of books sent by Swedenborg, including *Opera philosophica et mineralia*.[242] The Academy then appointed a group to see if anything in it might be of utility for Russian mining. In the assessment of Swedenborg's *De ferro* the reviewer observed, apart from the alchemical element, that this enterprise was far from being parochial, and if the author feared that his boat might be too fragile for such a huge ocean, he had at least been bold enough to attempt the sailing.[243] The secretary of the Russian Academy of Sciences wrote back to Swedenborg, thanking him for his valuable studies of nature's hidden secrets, and invited him to correspond with the Academy.[244]

Swedenborg did not find any direct theoretical guidance that was to his mechanistic taste in the alchemical books, apart from a fair amount of detailed descriptions of various chemical experiments. Yet one can see that the terminology and patterns of thought in alchemy entered his mind, as it were, through the back door. A recurrent concept in the history of chemistry and alchemy is the perfection of substances, that is, a wish to transform substances into their most pure and perfect state. This was a desirable goal for a mining assessor who wanted to extract iron and copper of high quality. In addition, we find the attempt to ascertain the correct proportions between the constituent elements or principles. For Swedenborg it was easy to read both the perfection and the proportions into his mechanistic philosophy. In addition he could possibly find in the alchemists the ancient wisdom and the analogical thinking that he would be searching for, especially in his later works. Alchemical images and symbols occur in his writings, as in *The Dream Book*, where he sees his own spiritual purification as an alchemical process to achieve gold—goodness.[245]

---

[240]SLBA, S2 Ac12, 5; German original in Helsinki University Library, Petersburger academiens handlingar, no. 63, vol. 13; *Green Books* V, no. 583; in Swedenborg's library: Kunckel von Löwenstern (1716), III, ch. XLI, 563–625 (Swedenborg owned the edition from 1722); Becher (1726); Stahl (1720).

[241]*Materialy dlya istory Imperatorskoy akademy nauk*, tom 2 (1731–1735), 507–511.

[242]*Green Books* V, no. 569.14.

[243]SLBA, S2 Ac12, 2; RAN, razryad V, opis 1–C, no. 6, concerns Swedenborg's *Principia*, 'Praesent. in acad. scient. d. 5. Novbr 1736'.

[244]Russian Academy of Sciences to Swedenborg, St Petersburg, 28 December 1734. RAN, Ausgehende Briefe 1734–1735, fond 1, opis 3, no. 19, letter no. 12; *Letters* I, 465.

[245]*Swedenborgs drömmar*, 12–13 April 1744, 26; Bergquist (1988), 156f.

De ferro places great emphasis on the professional skill and experience of miners. Observations on the spot take on particular significance, as during blasts in a furnace when the sparks spread as pulverized charcoal is added, and 'it looks like when a furious gust drives a cloud of yellow sand over a sandy plain.' The fire from a blast furnace is like the sand whizzing over a field. In his descriptions he uses metaphorical thought; everything reminds him of something else, everything is understood and imagined in terms of other phenomena or has parallels in classical literature. When the mass of charcoal has settled, 'one can hear a dull noise and din, just as when a violent south-westerly storm rushes through the forest and the wind sounds its coarse bass. This is because the fire rolls through the powerfully heated furnace and strives upwards as in the volcanic eruptions of Etna.' It is as in Virgil's *Aeneid*, where Aeolus in his huge cave rules the tumultuous winds and noisy storms. The blast furnace is a volcano, from which the flames rise like a billowing fire from the innards of Vesuvius's crater. But on clear days with sunshine these fiery masses are less distinct, resembling 'evaporations from the earth, as one can see rising on a hot summer day from fields and meadows, perceptible through a gaze from the side.'[246] Fire as vapour.

The great pit in Falun is not infrequently described as a volcano in travel accounts from the time, as a Hekla, Vesuvius, or Etna. The terrified travellers found several similarities between volcanoes and mines, not just that they resemble each other in appearance and both lead down into the underworld. They also reek of sulphur. Swedenborg also mentions these three famous volcanoes in a discussion at Bokwettsgillet in 1724 about the northern lights and meteors. Swedenborg claims that neither the sulphur of the volcanoes nor that of the Falun mine can cause these phenomena.[247] Kircher makes an analogy in a work that was important for Swedenborg, *Mundus subterraneus*, between the alchemist's laboratory and the fire-filled subterranean chambers of the geocosmos. The alchemist's furnaces, stills, retorts, flasks, and circulation vessels, according to Kircher, resemble the underground fire-chambers, caves, metal veins, and rivers. Kircher, who had himself lowered on a rope into the crater of Etna in 1637, saw the heat of the volcano as a counterpart to the alchemist's furnace, its smoke as alchemical decoctions, and the stench reminded him of the sulphurous fumes in his own laboratory.[248]

The account of iron also includes a description of bog ore and the ore found in lakes. In Småland Mars loves the bottoms of lakes, which bulge with their wealth of iron, like a permanent treasure chamber, a profuse field that constantly provides new offspring and new harvests of iron, 'just like a field that is verdant every other year with a new crop', and the bog 'is the mother spring from which the juice secreted in the mire flows incessantly into the lake.' Iron is growing grain. Ore is like round balls, uneven-grained sand, in which each particle is usually close to a round shape

---

[246]*De ferro*, 59; Virgil, *Aeneid*, 1.52f.

[247]Swedenborg to Benzelius, *c.* 20 March 1724. *Opera* I, 305f; *Letters* I, 329; *Bokwetts Gillets protokoll* 27 March 1724, 104; cf. Burman, 567–570; *De sulphure*, 105.

[248]Kircher (1665), preface; Baldwin, 47.

and resembles the grains of 'barley and wheat or even beans.'[249] Iron blossoms and forms crystal vegetation with juices, shoots, branches, and foliage in subterranean cavities and grottoes. In the microscopic world he finds, like René Antoine Ferchault de Réaumur, geometrical forms in the fracture surface of pig iron. He sees rays, stars, round balls, leaves, and twigs (Fig. 16).[250] In a letter from 1724, Swedenborg writes about Réaumur's work on steel, that 'so far as I have glanced through it, it is very fine and *curieus*.'[251] Swedenborg's descriptions of the mineral kingdom reveal how he proceeds from what he has already learned. There is a preconceptual, geometrical vision based on his mechanistic and geometrical world-view. If the world is geometry, then it must also be possible to find geometrical forms when one observes the subtle world in a microscope. He searches for geometry when he admires a very beautiful and rare ammonite with six spiral coils. But iron ore can occur in nature not just as geometrical forms such as cubes, he reads in Emanuel König's *Regnum minerale* (1686). They can also be like siliquae, beans, lupins, peas, shells, exotic fruit, peppercorns, like elegant iron balls of wrinkly coriander and cinnamon sticks of iron.[252] The unknown is described in terms of the known. Everyday experience of the rich pleasure that a fine dinner gives to the eyes is transferred to the rougher and more alien world of iron ore.

In the Bride-Chamber of the Mineral Kingdom

In the introduction to the third volume about copper, *De cupro*, Swedenborg is amazed at all the variety of substances to be found under ground.[253] The subterranean realm is like a mother who produces rich treasures and gives nutrition to an infinite variation. The shimmering caves, walls, halls, columns, and portals of the Falun mine must make a traveller think that he has come to Venus in her bride-chamber. Metals are living entities that are born, germinate, and grow in the fertilized womb of the underworld. This metaphor is not uncommon in the alchemical literature, as in Basilius Valentinus whom Swedenborg cites not infrequently. This meant that metals grew in Mother Earth, fertilized by the heavens, like a human foetus.[254] As living creatures, metals inhale life-giving spirit from the sun's rays, which are invisible but can be discovered by the divining rod. Metals are engendered in the earth's veins, born of saltpetre, sulphur, and mercury, through the heat of the earth and the influence of the sky, Forsius wrote.[255] Venus, copper, and

---

[249] *De ferro*, 115–117, 287, 294.

[250] *De ferro*, 218; Réaumur; cf. Granström, 367.

[251] Swedenborg to Benzelius, Stockholm, 20 August 1724. *Opera* I, 315; *Letters* I, 343.

[252] *De ferro*, 290f; cf. König, 16–22.

[253] *De cupro*, preface; translation I, ix.

[254] Basilius Valentinus, I, 264f; cf. Kunckel von Löwenstern (1716), I, ch. VI, 70–80; Debus, I, 95f.

[255] Forsius, 175.

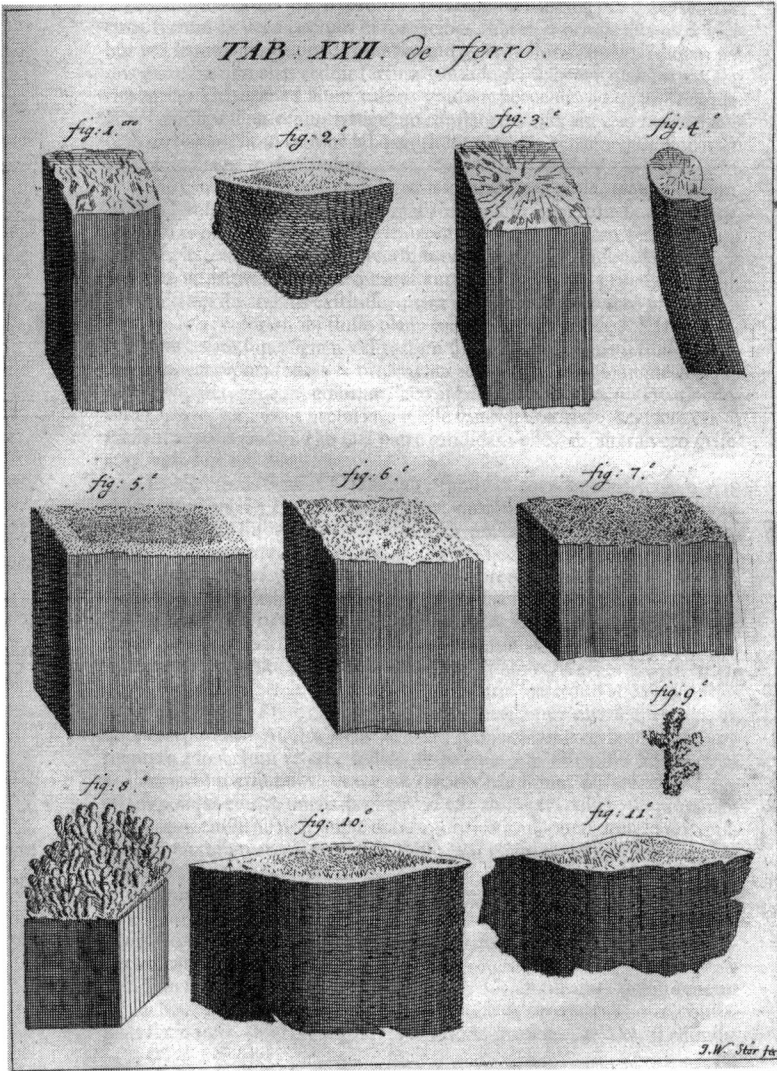

Fig. 16 Irregularities in the fracture surface of pig iron. Fracture surface with rays and 'playing figures' (*3*), with spoke-like radiation (*4*), with 'bright images of stars scattered here and there' (*6*). Figure (*8*) is a detail of figure (*5*) viewed through a magnifying glass, showing that it is composed of infinitely many leaves. Figure (*9*) shows a leaf detached from figure (*8*) to illustrate how each leaf consists of very small grains over and under each other (Swedenborg, *De ferro* (1734), after a picture in Réaumur's *L'art de convertir le fer forgé en acier* (1722))

the blue planet have connections in chemistry. There are similar correspondences as regards iron. The iron bath that removes slag is described by Swedenborg in *De ferro* as when 'the divinity of iron (Mars) shows his body naked and his limbs exposed to the goddess of the hearth (Vesta).'[256]

Swedenborg transfers the metaphor of the womb to a mechanistic mode of thought, similar to the way the mechanist Boyle assumed that metals grew under ground. What Swedenborg tries to do is to explain geological processes in mechanical terms and make them accord with biblical history. The great flood was not just a punishment, he explains in *De cupro*; it also brought something good. Through God's providence man was granted ores, which were discovered when the flood receded. Before the deluge flooded the entire world, the surface of the earth was perfectly smooth and unbroken, without pointed mountains or deep valleys, without lakes or oceans. Through the violent force of the deluge, however, cracks arose in the crust. The earth became deformed, uneven, and ragged. The ore that had previously been hidden in the lowest recesses of the globe, which otherwise would have lain forever hidden deep in the subterranean shadows of Tartarus, was now brought up into the daylight.[257]

For the descriptive sections he has, as usual, read many different works. The account of the smelting process, for example, is a translation of a transcript which he owned of Natanael Ekman's report on the smelting works at Stora Kopparberg.[258] Swedenborg the geometrician writes that the smelting process for copper in Falun did not satisfy what was required, 'and was not, so to speak, in accordance with geometrical rules.'[259] When it comes to refining he has a word-for-word translation of accounts by his step-brother Anton Swab and Lars Harmens.[260] Swedenborg gives a fairly detailed description of mining in places where he himself had never set foot, such as the Russian copper mines in Siberia.[261] In *De sulphure* Swedenborg had previously referred to the Russian mining pioneer Tatishchev when citing data about a kind of pure sulphur in the mountains at Kazan.[262] This was no doubt one of the conversation topics, alongside the decimal system of measure, that Swedenborg discussed with Tatishchev when he visited Sweden in 1724–1726. Tatishchev was thoroughly acquainted with Siberian mining. He had been dispatched by the tsar in 1720 to discover new finds of ore in Siberia. Among other things, Swedenborg recounts in *De cupro*, he found copper ore by the rivers Utka and Iset, at the town of Kungur, in Perm by the river Kama, at the monastery of Piskoy, and he founded the metal city of Yekaterinburg. In the same year Tatishchev met Swedenborg's cousin, Peter Schönström the Younger, a cavalry captain with an interest in history, who

---

[256] *De ferro*, 50.

[257] *De cupro*, preface; translation I, xi.

[258] RA, Ekman; Lindroth (1955), II, 16f.

[259] *De cupro*, 42; translation, chapt. II.

[260] KB, Swab, fol. 143; edition, 19; RA, Harmens; Lindroth (1955), II, 348f;

[261] *De cupro*, 133f; translation I, 134f.

[262] *De sulphure*, 107.

was a prisoner of war in the salt-producing town of Solikamsk.[263] They had met at least once before, albeit fleetingly, on either side of the front line at the Battle of Poltava in 1709. Tatishchev was wounded, Schönström was captured, and Charles XII fled. Schönström too had an interest in mining. On his arrival back in Sweden after peace came, he wrote a report on the Siberian mines, of which Swedenborg also obtained a copy.[264] The Board of Mines was concerned about competitors popping up on the market for raw materials, and discussions were held about sending some people incognito to pursue industrial espionage in Russia. The Russian mines were increasingly viewed as serious competitors. Russia took a comparable interest in Swedish mining. That was one reason for Tatishchev's sojourn in Sweden. On the tsar's orders he was to study the reputable Swedish mining industry, and also to learn about the organization and state of manufacture and coinage. In addition, he was supposed to recruit Swedish professionals for the mines in the Urals, and to find apprenticeships for young Russians at Swedish mines. At a meeting of the Board of Mines in February 1725, when Swedenborg was present, Tatishchev handed over a map of Russian mines in the Urals, the same map that Swedenborg would include in *De ferro*.[265] Tatishchev can almost be called a political spy, for his mission concerned not only mining; he was also, according to a report from 1726, given the task of ascertaining 'the Swedish government's overt actions and concealed intentions.'[266]

In his studies of manufactures, Tatishchev also visited Stjärnsund in 1725, where he was fascinated by a pond made of wooden planks, labour-saving and easy to repair; it had been made by a famous 'mechanicus', in other words Polhem.[267] Tatishchev also made the acquaintance of Benzelius, with whom he discussed matters of editing, but also a question that tickled the imagination more: the Siberian mammoth. The subject had arisen when prisoners of war returned home to Sweden with tales of a remarkable animal found in the soil of Siberia.[268] In the spring of 1725 Benzelius asked Tatishchev, as an expert on Siberia, to write an essay about the mammoth for publication in *Acta literaria Sveciæ*.[269] Tatishchev asked, is the mammoth an underground animal, Behemoth in the book of Job, an elephant washed away with the deluge, or what? This is in fact the only scholarly work by Tatishchev

---

[263] Küttner, 112.

[264] RA, Bergskollegii arkiv, AI:77:1, 1f.

[265] RA, Bergskollegii arkiv, AI:71, 1223; *De ferro*, 164–167; The original map is preserved in RA, Kommerskollegium, Gruvkartor, markområden, utländska.

[266] Tatishchev to Empress Catherine I, 17 October 1726; cited in Küttner, 119.

[267] Küttner, 124–126.

[268] Kagg, *Baron Kaggs beskrifning på ett vattendjur i Obijströmen* (Dec. 1722). LiSB, N 14a, no. 54; *Bokwetts Gillets protokoll* 14 December 1722, 79.

[269] Tatishchev, *Beschreibung von Mamontowa Kost oder Mamontz Knoche* (12 May 1725). LiSB, N 14a, no. 93, fol. 237–240; Tatishchev, 36–43; Tatishchev to Benzelius, Stockholm, 20 January 1726 and 20 February 1726. Benzelius (1979), II, 294f; *Bokwetts Gillets protokoll*, 122; Hag (1979–1981), 66–68; Küttner, 149f.

that was published during his lifetime, which means that Swedenborg's data on Siberia in *De cupro* probably came both from Schönström and as oral information or documents personally communicated by Tatishchev. This excursus on Siberian metallurgy and palaeontology reveals an important point: oral communications, conversations that have died away for good, and have easily been ignored in research, were of greater significance than we might first suspect, concealed behind the extant written documents.

In *De cupro* Swedenborg goes on to discuss how balls were cast in France and Leeuwenhoek's description of the particles left in the ashes in a furnace where balls have been cast. Thousands of small pipes could be assembled in a single drop of water. To find the right mixture of metal for the manufacture of mirrors, he cites data from Johann Rudolf Glauber. He writes about how the sun's rays are collected through a magnifying glass and can display wonderful forms: cylindrical, pyramidal, parabolic, etc.[270] From copper one can also obtain pigments for paints. Swedenborg examines different processes for producing ultramarine and azure paint, and continues to the topic of how artists prepare paints.[271] Sweets can be dyed with copper paint and then eaten without risk. Painters can safely draw their brushes between their lips with this paint, which they cannot do without risking their lives when using other paints.

How geometrical vision structures reality and orders it in geometrical forms is evident from a section about different kinds of pyrites. On his visits to the Falun mine or other copper mines, Swedenborg must have noticed that the pyrites which frequently occur in copper ore are often shaped like cubes, or else take the forms shown by ordinary quartz crystals, appearing as hexagonal pyramids and prisms. Nature orders herself in a remarkable way according to the rules of geometry. Here Swedenborg proceeds from the German mining councillor Johann Friedrich Henckel and his division of sulphates by their geometrical form. Henckel's *Pyritologia* (1725) tells the whole story of pyrite, of how it is mined, and its subterranean causes. In Henckel Swedenborg found both a friend and a metallurgist with the same basic geometrical outlook. They met in Dresden on 30 August 1733, and together they went to the mining councillor Johann Wolfgang Trier to see his collection of minerals and ore samples. A large fossil of a 'marine cat' found in a copper mine particularly fascinated Swedenborg. He asked to borrow a copperplate which he then included in *De cupro*.[272] Henckel in turn was interested in obtaining information from Swedenborg about Swedish mineralogy for his planned dictionary of practical mineralogy.[273] In *Pyritologia* he also refers to Swedenborg and writes that he has made a laudable attempt in *Prodromus principiorum* to proceed from geometry and to discover the nature of natural bodies in agreement with hydrostatics. But, he says, it still seems to me to be

---

[270]*De cupro*, 373f, 376; translation III, 361, 364; Glauber, IV, 64–70.

[271]*De cupro*, 464f; translation III, 449.

[272]*De cupro*, tab. 39; *Opera* I, 324; *Letters* I, 453; *Resebeskrifningar*, 57; *Documents* II:1, 72f.

[273]Henckel to Swedenborg, Freiberg, 21 November 1732. *Opera* I, 322; *Letters* I, 450.

too early to draw such conclusions.[274] What had not previously been known is that Henckel further praised Swedenborg and his books about iron and copper in a book printed at the same time and by the same printer in Dresden and Leipzig as Swedenborg's *Opera philosophica*. Moreover, in *Idea generalis de lapidum origine per observationes experimenta & consectaria succincte adumbrata* (1734) Henckel refers to several Swedish works which may well have been recommended to him by Swedenborg.[275]

Colour cannot have been a secure point of departure for Swedenborg in the investigation of metals. In a letter from this time Swedenborg replies to questions from an unknown addressee about the reason why metals change colour.[276] One and the same phenomenon, such as a particular colour, can be created in many different ways. One must therefore, he says, observe the same rules as in algebraic analysis, that is, proceed from propositions and facts from which one can draw conclusions. One such secure point of departure for Swedenborg is evidently the geometrical form. When it comes to pyrites, Agricola had previously classified them primarily by colour.[277] Some were silvery, others yellow, gold, lead-grey, ashen, iron-coloured, and so on. This division was rejected by Henckel, who believed that all pyrites contain iron. The classification advocated instead by Henckel, and followed in turn by Swedenborg, consists of a number of classes based on the mechanist's eye for the geometrical forms of the pyrites (Fig. 17).[278] Henckel identifies pyrites that are round, spherical and hemispherical, radial, discoid, oval, sausage-shaped, comb-shaped, angular, tetrahedral, hexahedral, cubic like small dice, oblong, rhomboid, cellular, or shaped like honeycombs. Moreover, some are octahedral, decahedral, dodecahedral, decatesserahedral like dice with 14 sides, prismatic, trapezoid, or irregular, coin-shaped, tubular, symmetrical, symmorphous or regular, lithoxyloid like fibres, helical, cylindrical, lenticular, vortical, pyramidal, conical, stellate, and so on. It seems to be only words that are lacking to describe all the geometrical forms of pyrites. With pictures, in Henckel's opinion, one can arrive at a better understanding than with letters.[279] The knowledge then reaches the eyes as well as the mind. Henckel's classification of pyrites was evidently of great value for Swedenborg, for he cites it no less than three times, in *De victriolo*, and presumably after this in *De sulphure*, and finally he included it in the published *De cupro*.[280]

---

[274] Henckel (1725), 1006f.

[275] Henckel (1734), 13, 55, 58, 73; Henckel refers to *De cupro*, § 17, fol. 168ff; Roberg and Sunborg; Hiärne (1712); Bromell, in *Acta literaria Sveciæ* (1727).

[276] Swedenborg to N.N. (*c.* 1734/1735). *Opera* I, 335f; *Letters* I, 457.

[277] Agricola, translation, note, 112.

[278] *De cupro*, 411f; translation III, 395f; Henckel (1725), 119–214; cf. *Acta eruditorum*, June 1726, 271.

[279] Henckel (1725), 55f.

[280] *De victriolo*, 404–409; *De sulphure*, 107–109.

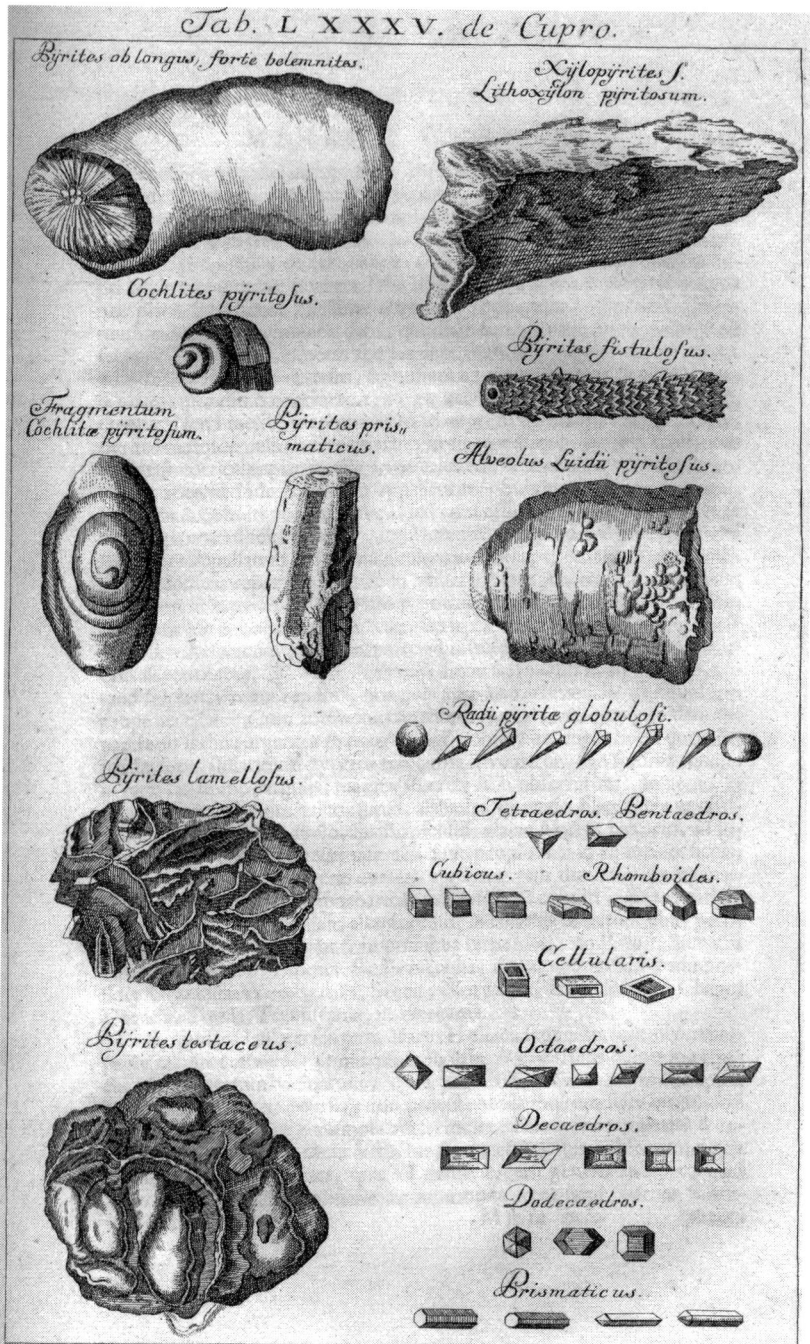

Fig. 17 The geometry of pyritology (Swedenborg, *De cupro* (1734), after a picture in Henckel's *Pyritologia* (1725))

Vanitas Bubbles of Soap and Water

Swedenborg's theory of matter is characterized by concrete imagery and the typical analogical thinking of his time, when technology and the everyday world's concrete objects and shapes had their counterparts in the particle world. The pure forms of geometry—bubbles, drops of water, cannonballs, and peas—could show the way to an understanding of subtle matter. With the aid of known reality he could draw conclusions about the unknown. But it was something more than just a didactic device. Ultimately it was based on a conviction that the laws of mechanics applied to the little and the large world alike. The difference between cannonballs and round particles was merely a difference in size. The deductive method frequently used by Polhem, and not least by Swedenborg, for exploring the unknown is by analogy. If peas, for example, display a number of properties such as roundness, smoothness, and mobility, one may conclude that water, which shares one of these properties, mobility, also has the other properties, roundness and smoothness. Polhem's ideas about round particles and their formations were probably an important source of inspiration for Swedenborg. But the two men differed slightly in their ways of approaching the problem: one of them went about it like a practical mechanic, the other like a theoretical mathematician. With bubbles, cannonballs, and a barrel of peas, they tried to capture the mechanics of the microcosmic particle world. These were spherical bodies which simultaneously reflected a time, an agricultural country at war, where the mining industry financed arms manufacture and scientists acted in the service of war. Peas in heaped measure gave clues about the reasons for the properties of liquid matter. Cannonballs in ammunition houses suited as an image of the structure of subtle matter.

The sphere is something that is volatile, unstable, easily set in motion, while the right angle is honest, clear, and stable. Knowledge rests unsteadily on a sphere, but stands firm on the imperturbable cube. A sphere rolls away, but a pyramid stands its ground: 'I myself shall be thy pyramid,' Swedenborg wrote in a poem in 1709 while mourning at the grave of Eric Benzelius the Elder.[281] He would be a stately gravestone like the pyramid of Cheops at Memphis, as Peringskiöld puts it, which with 'its size and artful working makes one quite perplexed with wonder.'[282] Victoria, the personification of victory, stands in Swedenborg's *Camena Borea* with one foot on a globe representing the world and destiny, while the other foot is on the move in the air, ready to fly away to place a wreath of victory on the approaching person.[283] The sphere may be the earth or the celestial sphere, but also the irrational caprices and swift changes of fate. In emblematics the inconstant Madame Fortuna stands on the easily rolling sphere of unstable, unpredictable life.

---

[281] Swedenborg, 'Patriæ planctus et lacrimæ, …' (1709), *Ludus Heliconius*, edition, 48f; Helander (2004), 457.

[282] Peringskiöld, 177.

[283] *Camena Borea*, edition, 44f.

Soap bubbles denote the fragile life that can break at any moment. 'Life is a beautiful bubble,' Linnaeus writes in *Nemesis divina*.[284] Swedenborg's world is a world of soap bubbles, small bubbles with thin film of even more bubbles, smaller ones, all the way down to mathematical points. Hovering, perishable soap bubbles, symbols of the vanity of vanities, are in harmony with the geometry of the elements—Vanitas bubbles of soap and water.

---

[284]Linnaeus (1968), 101f; translation, 90.

The Point

We see the spider construct her webs in a geometrical manner, drawing radii from
the centre, and connecting them together with polygons and circles; and when it is
finished, she places herself in the middle, and lies in ambush for her prey.[1]

Swedenborg, *Principia rerum naturalium* (1734)

The Spider in the Polygonal Web

The world is geometry, a finite universe that obeys geometrical and mechanical
laws. In the world geometry, the spider follows nature's laws of mechanics and
spins its web in polygonal and circular forms, the beaver builds dams according to
geometrical principles, the bird makes its nest based on the figure of the circle,
and bees shape their honeycombs in hexagons. There is a natural mechanics,
Swedenborg writes in *Principia rerum naturalium* (1734), in which the senses are
shaped in harmony with the elementary mechanics of the world. Geometry applies
to everything that is finite and bounded, but there are things that are not geometrical
or mechanical: the infinite, providence, love. This chapter focuses on the point,
the mathematical and natural point. Swedenborg conceives the world as one large
geometry in motion, where mathematical points outline the parts of the machine, the
perfect geometrical figures. The world geometry rests on a number of metaphors,
that *matter is geometry*, that *mechanics is geometry in motion*, that *a point is a spot*.
It also concerns the hypostatization of geometry, making the abstract real. It was
believed that geometry is found in the world, not only in our imagination of it. It is
the world of the geometrical spider.

---

[1] *Principia*, 14; translation I, 24; cf. Swedenborg, *Argumenta quaedam in Principia rerum
naturalium* (1733); edition *Opera* II, 205f; *Treatises*, 110–112; a summary of the italicized
passages in the printed *Principia* can be found in *Ex Principiis rerum naturalium meis* (1734),
edition *Opera* II, 207–262; translation.

D. Dunér, *The Natural Philosophy of Emanuel Swedenborg*, Studies in the History
of Philosophy of Mind 11, DOI 10.1007/978-94-007-4560-5_6,
© Springer Science+Business Media Dordrecht 2013

A true philosopher, Swedenborg says, is one who can arrive at the real causes and the true knowledge of the invisible mechanical world beyond the senses.[2] He should then be able to think a priori from the first principles about the world and its phenomena, in physics, chemistry, metallurgy, and all other subjects that follow mechanical principles. As if from a central vantage point, he can look out over the whole world system and its mechanical and philosophical laws. Nature's mechanical world is like the spider's web. At the centre of the web, the natural philosopher waits like the spider, drawing radii that meet at the centre and linking them through circles and polygons. When the spider sits in the middle and places its feet on the radii, it feels the movement of every thread, every particle that falls into the web. It is ever-prepared to rush out and snare its prey, the sought-after knowledge. The spider can immediately feel and know everything that happens on the periphery, what it is and when it comes. From its centre it can measure and survey the whole infinite periphery, the whole world, with a single glance, while the fly, with great toil and effort, increasingly nestles itself in the web and becomes a victim of its own wisdom and philosophy. Nature is a spider's web with infinite radii from a centre, and it is bound together with infinite circles and polygons, so that nothing can happen in one of them without it being instantly conveyed towards the centre and then thrown back over the whole web.

While spending a few days at Karlsbad in Bohemia in the middle of August 1733, Swedenborg wrote about the geometrical spider.[3] In May he had gone down to Dresden to have his *Opera philosophica et mineralia* printed. The spider's web captivated him. Swedenborg's spider metaphor expresses philosophy's dream of order in chaos, the dream of a world in which everything follows set laws, everything is palpable, and deductive reality can be surveyed. Nature's most diverse phenomena are bound together through the threads. The parts of the world machine grasp each other. Swedenborg is searching for a synthesis, a Wolffian idea of the nexus, cohesion, and perfection of all things, in which the smallest units in the particle world are combined with the largest structures of the universe.[4] Nature becomes a baroque garden with mathematical pleasures, where the spider combines abstract science with practical engineering. The spider is a geometrician, an architect, an intelligent little being with skills in engineering and art, equipped with finer senses than man. It perceives the slightest vibration in nature's global web. As in Kircher, the spider is also musical (Fig. 1).[5] Her regular strings are tuned in harmony.

The spider represents a geometrical ideal of science. The whole of reality is regarded as a uniform deductive system. Based on a few axioms, the natural philosopher could derive all theorems and thereby reveal the cosmic plan. If he is at the centre of the web he can survey all of reality, even the most distant phenomena. With the aid of a limited number of initial premises, such as God's providence,

---

[2] *Principia*, 19; translation I, 32.

[3] *Psychologica*, xi; *Treatises*, 113.

[4] Wolff (1731, 1737) edition, § 10.

[5] Kircher (1650), book VI, 441f; cf. Kircher (1665), II, 369.

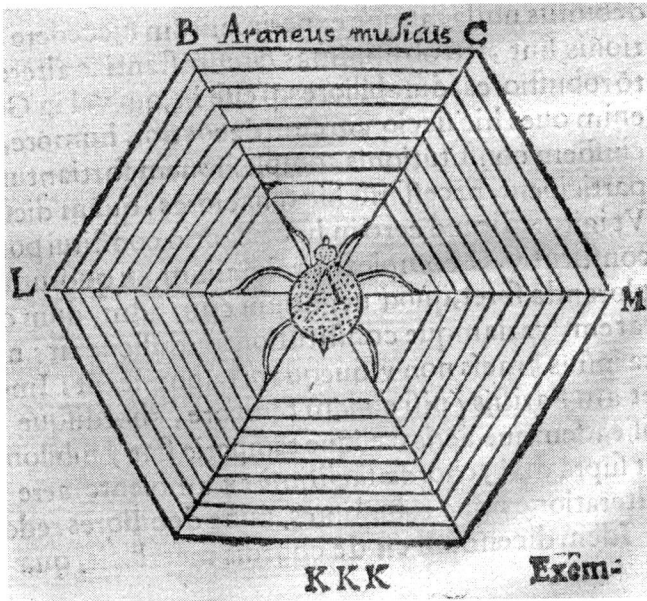

Fig. 1 The musical spider pursues geometry (Kircher, *Mvsvrgia vniversalis* (1650))

the mathematical point, the spiral movement, and a few others, Swedenborg tries to accomplish a universal synthesis, in which the whole of reality is summed up in one theory, the particle world with the solar vortex, the microcosmic man with the macrocosmic universe, the creation with the last judgement. The result is a deterministic system, in which the geometrical-mechanical laws by logical necessity give rise to the geometrical particle world. In *Principia* he puts forward his explanation of the parts of the world machine, the creation of our geometrical world from mathematical points, the spiral motion of particles and the vortical movements of the planets: everything from the smallest invisible to the first visible.

The spider metaphor concerns the relationship between art and nature, between artificial and natural mechanics. The human weaver imitates the spider. As in so many other cases in Swedenborg's works, one can find threads going back to Ovid. In *The Metamorphoses* Minerva is challenged by Arachne to show which of them is more skilled at weaving.[6] When Minerva sees that she is surpassed in skill she resentfully rips up Arachne's artful weave and turns her into a spider. This was a popular motif that could express how art imitates nature, a Swedish example being Stiernhielm's *Peplum Mineruæ* (1652). The way the spider joins lines became an expression of the manner in which the baroque sought links between things, 'friendships, harmony, and kinship'. If 'you stop for a moment and consider the

---

[6]Ovid, *Metamorphoseon*, 6.5–145.

wisdom with which its sets up its loom, how it draws out its finest threads while calculating distance, angles, and centre', how it stretches the threads, strengthens the web with new lines, and weaves it all into a geometrically perfect polygon inscribed in a circle, 'will you not find, I wonder, that Arachne, this geometrician and architect, seeks to emulate Minerva, that is to say, Reason or God, creator of the world? O goodness! O wisdom! In the meanest little animal I see God!'[7]

Many works from the time are crawling and swarming with insects: buzzing bees, busy, social ants, and cunning spiders lying in ambush in dusty corners. With insects one could picture the proper society and the proper way to attain knowledge. For Francis Bacon the empiricists are like ants, merely collecting and consuming, while the rationalists are like spiders, merely creating threads out of their own bodies.[8] The true philosopher should be neither ant nor spider, neither storing material unchanged in the chamber of his memory nor solely relying on his own spiritual powers. No, the true philosopher should be a bee who sucks the juice from the flowers of garden and meadow, processes the nectar and transforms it by his own ability into sweet honey. In the workshop of the true philosopher, the work is done in the manner of the bees, who rationally digest and order what they have collected. Jonathan Swift varies the theme of insects in *The Battle of the Books* (1710), where a conversation between the spider and the bee represents the struggle between the Moderns and the Ancients.[9] The spider of the Moderns is swollen with an infinite amount of flies, spins entirely from the innards of its modern brain, and constructs a web that risks falling into oblivion, neglected and concealed in a dusty corner. But we, the Ancients, says Swift, promise nothing beyond our own wings and our voice. What we have created comes from endless toil and searching out every corner of nature. Instead of dirt and poison we fill our beehives with honey and wax, and bestow on mankind the noblest of all things, sweetness and light. For Polhem, bees are not only good mechanics and chemists. They are also good mathematicians who use the most perfect figure when they make their 'six-sided chests which, with the least building material, give the most room, and the greatest rooms or chests can be contained in smallest place.'[10]

The bee occurs in other contexts in Swedenborg. In a poem he is disturbed by a bee falling on the paper where he had written the name of a nymph. It then flew off with yellow legs to its honeycomb.[11] Among his notes taken from Jan Swammerdam's *Biblia naturae* (1737–1738) in 1743, he digresses in a contemplation of the form of government among bees, which serves as an analogy for human kingship.[12] A famous example in the genre is Bernard de Mandeville's

---

[7]Stiernhielm, *Samlade skrifter* II:2:1, 7, cf. 6, commentary 222f; cf. I:1, 27.

[8]Bacon (1620), I, § 95.

[9]Swift, 11.

[10]Polhem, 'Dispu[ta]tion om olijka regementzformer' (after 1718), *Polhems skrifter* II, 177.

[11]Swedenborg, 'Lusus extemporalis ad amicum qvendam Oxoniae 1712', *Ludus Heliconius*, edition, 98f.

[12]Swedenborg, *Johannis Swammerdamii Biblia naturæ* (1743). KVA, cod. 53, 167–175; *Photolith.* VI, 231–237; Swammerdam, II, 498.

The Fable of the Bees (1714), whose political philosophy was adopted by Anders Nordencrantz, with whom Swedenborg incidentally had political disputes in the diet in 1761, and by Anders Johan von Höpken.[13] Swedenborg's spider metaphor makes an interesting departure from the admiration of the honey-sweet little bee that was so commonly expressed by Enlightenment philosophers, not least by Swedish lovers of science. Swedenborg's spider is 'baroque' in the totality of its control and its creation or order. It is 'modern' in that it makes its own secretion from its abdomen.

Swedenborg pays tribute to the wise spider, as Aristotle, Seneca, and Pliny had done.[14] The life of the spider reminds Spegel of the marital household.[15] Bishop Swedberg had instead chosen the ant as a moral example. Young people should not lie too long sleeping away the best time of the day, but learn from the busy ants to gather during the summer of youth what they need for the winter of old age.[16] Usually, however, the spider is not just a useless rationalist but also frightening, dangerous, and treacherous, as in Carl Johan Lohman, who had preached to Charles XII on the day he died at Fredrikshald. For the king's funeral on 26 February 1719 he wrote in a poem:

> *Our most chaste King* is dead. Thou sloth and lust and laxity,
> In these three spiders who do weave the web of death,
> Could one of you have poison placed in *Carol's* roses?[17]

Swedenborg concludes a manuscript about the anatomy of the brain with notes about the tarantula.[18] These contain largely the same information as found in a book in his library which he used in his studies of magnetism, Kircher's *Magnes siue de arte magnetica* (1643).[19] There we can read about tarantism, a life-threatening disease caused by the tarantula, a wolf spider from Taranto in southern Italy. Those who are bitten by the tarantula go mad, starting to smile and weep, sing and dance. Kircher depicts this dangerous spider and prints the music of the tarantella dance, which counteracts the effects of the poison and cures the madness. Scholars were fascinated by the tarantula. In Sweden Göran Vallerius defended a dissertation under his father, *De tarantula* (1702), which is based on Kircher. The soul could also be like the spider in the web in attempts to describe the nervous system, as in Linnaeus for whom the soul is between the medulla oblongata and the cerebellum, as it were between stalk and root, 'and sits like the spider in the web, with its hands in the cerebellum and its feet in the medulla.'[20] But the spider's web also expressed

---

[13]Swedenborg, *Ödmjukt memorial* (1761). KVA, cod. 56; *Documents* I, 511–515; Bergquist (2005), 345–351; Högnäs, 93, 98.

[14]Aristotle, *Tōn peri ta zōia istoriōn*, 8.38.622b23–24, 8.39.622b27–623b2; Seneca, *Ad Lucilium epistulae morales*, 121.22f; Pliny, 11.28.79–84.

[15]Spegel (1685), edition, I:1, 365.

[16]Swedberg (1709b), 158–160; Prov. 6:6.

[17]C. J. Lohman, 'Vyrdsam åminnelse', Cited in Nyström, 228.

[18]Swedenborg, *De tarantula*. KVA, cod. 55, 878; *Photolith.* V, 627.

[19]Kircher (1643), 755–765; cf. Schott (1657), II, 236–250; Kircher (1667).

[20]Cited in Broberg (1975), 132; cf. Linnaeus (1739), 14.

vulnerability. In Augustine the soul is ripped apart like the spider's easily destroyed web.[21] The soul and the body as spider and web can be found in Heraclitus, Chrysippus, the Stoics, and the Sceptics. Fate woven in a net by Lachesis occurs in Swedenborg's *Camena Borea*.[22] The three Parcae or Fates are Clotho who spins the thread of life, Lachesis who twines it, and Atropos who finally cuts it off. In *Principia* the spider's web is an image not just of life but also of the connections of the body down to the smallest parts, the cohesion and continuity of all things in the world.[23] 'Life hangs by a thread, cobweb fine, nothing more fragile,' Linnaeus noted among his divine retributions.[24]

World anxiety, the dread of the inexplicability, boundlessness, and chaos of the world, is a characteristic of man's quest for knowledge, unease in the face of the unexplained, the alien, of creation, the last judgement, death. How are we to orientate ourselves in this world which gives us such a sense of irresolute anxiety, desolation, and vague uncertainty? This disorientation engenders a desire to bind words and the world in firm rules, in a mathematical language. Elusive reality is too painful in its anarchy. It must be subjected to order, captured in a language. Man wants to make sense of life, to turn a chaotic reality into something logically coherent and meaningful, to interpret reality in the form of fiction in order to make it manageable and interpretable. It is a quest for discipline and order, a resistance to chaos. The pattern we find in the world is coloured by the cultural context, and its cultures may be seen as habits that decrease the unpredictability of the world, of human thoughts and actions.

In the creation of a world-view, Swedenborg was seeking simplicity, order, regularity, harmony, unity, and totality. Searching for knowledge means assuming that the world is ordered, not a random collection of separate phenomena, and that it is possible to acquire knowledge of the world. The world is intelligible in principle, at least for God himself. Natural philosophy and mathematics are a part of man's endeavour to bring order into the chaos of reality, to organize meanings in the world, to make it meaningful and to find his own place in nature, or perhaps more correctly to try to distinguish himself from it. Important distinctions in science are those between true and false, present and absent, real and unreal. The 'real' is what we need to grasp in order to be realistic, that is, to be able to function successfully, to achieve goals, to arrive at a working understanding of the situation in which we find ourselves.[25] Man has a tendency to look for causes in the world. To understand how the world functions and to predict the future, we assume that there are purposes. Observing and describing the world as geometry was a way to oppose chaos, the irregular, the indescribable.

---

[21] Augustine, 7.16.

[22] *Camena Borea*, edition, 44f, 48f, 62f.

[23] *Principia*, 13; translation I, 22.

[24] Linnaeus (1968), 101f; translation, 90

[25] Lakoff and Johnson (1999), 109.

The Point That Delineates the World

Geometry proved to work well for solving practical problems, and it became a successful way to describe the world around us. 'If we cast our eyes about us, far and wide, high and low, the all-measuring *Geometria* is at hand, the mother of a thousand arts; it swings over everything that is high, bounds the widest, reaches the longest, and scoops out the abyss,' writes Stiernhielm.[26] Geometrical observation of reality had ontological and metaphysical consequences. It gave rise to the idea that the world was in itself geometrical, a kind of materialized geometry, not least in the mechanistic world-view. The world was a geometry, created according to geometrical principles, built up of purely geometrical forms. Not only could nature be described in geometrical terms, it was geometrical in its very essence. That God had created the world according to mathematical proportions in dimension, number, and weight could be demonstrated, for instance, through some lines in The Wisdom of Solomon in the Old Testament Apocrypha: 'For the whole world before thee is as a little grain of the balance, yea, as a drop of the morning dew that falleth down upon the earth.'[27] Spole wanted the sermon at his funeral to be based on that passage.[28] In Ecclesiasticus (The Wisdom of Jesus Son of Sirach) we read that

> All wisdom cometh from the Lord, and is with him for ever. Who can number the sand of the sea, and the drops of rain, and the days of eternity? Who can find out the height of heaven, and the breadth of the earth, and the deep, and wisdom? Wisdom hath been created before all things.

'It is useful to read Sirach frequently,' Swedberg told the young students of the Västmanlands-Dala nation in Uppsala.[29] Wisdom and the mathematical structure of the world have their guarantor in the Omniscient Mathematician. In mathematical textbooks and treatises, God was described as a geometrician, as on the title page of a book that Swedenborg bought in London in 1710, Girolamo Vitali's *Lexicon mathematicvm astronomicvm geometricvm* (1668). God measures with his compasses against the earth, with the triangle of the trinity on the head and a cupid in the sky bearing the motto *In pondere, et mensvra*, 'In weight, and measure'.[30] In the same way, Roberg underlines in his article about salt production that things consist of mixtures of the elements in specific measures, spaces, and weights, 'according to which Nature arranges everything.'[31]

Swedenborg and the mechanistic natural philosophers lived in a world geometry, where man interpreted the world using his geometrical inner eyes, his geometrical

---

[26] Stiernhielm (1644), dedication.

[27] Wisd. of Sol. 11:22f; cf. Job 28:25, Isa. 40:12, Ecclus. 1:2f; cf. Triewald (1734).

[28] Spole (1946), 31f.

[29] Swedberg to the nation, 21 November 1703. *Constitutiones nationis Dalekarlo-Vestmannice*, 14; Ecclus. 1:1–4.

[30] Vitalis. UUB.

[31] Roberg, *Dædalus Hyperboreus* II, 30.

concepts, and thereby constantly saw geometrical forms in nature. Wherever Swedenborg turned his gaze, towards layers of stone, fossils, metals, lakes, or heavenly bodies, he saw geometrical figures. Geometry gave him the ability to describe the reality around him, just as the other concepts of mathematics, such as quantities and proportions, were one of the few intellectual devices available for mastering the uncertain, imprecise sensory experience. Geometry, which in his natural philosophy stands for order, clarity, reason, and perfection, has the function of structuring the chaotic world of the senses. The fact that the world is mathematical and geometrical is simultaneously an expression of man's desire that the world should resemble his own rationality.[32] By understanding reality in terms of geometry we can think about the world. To see the pure, simple geometrical forms instead of the complexity is a way to make the world palpable and comprehensible.

The Swedish mathematician Gestrinius asks in a dissertation about the point from 1627 whether there is actually any point in the world, what it is, what use it has, and what are the most important points. There are, he replies, starting points, finishing points, tangential points, intersection points, mid points, central points, points of balance....[33] The point had a particular significance in mathematics. But what is a point? The answer was to be found in Euclid. The point in geometry can be described as the undefined entity that represents an object with a location but no extension. Euclid starts his *Elementa* with the definition: 'A point is that which has no part.'[34] From the point one can then form the line, the circle, the triangle, the sphere, and other geometrical figures. This geometrical creation was central to Pythagorean and Neoplatonic philosophy.[35] Things are numbers, 1 is a point, 2 is two points forming a straight line, 3 is three points making up a triangle or surface, 4 is four points forming a tetrahedron or body. Together this makes 10, the perfect number. One can also let a point move back and forth from one end to another and thereby form a line. A line in lateral motion forms a surface and a surface in motion back and forth forms a body. If the same point moves at high speed in a circular orbit around a centre, a circle arises. If a semicircle moves around its axis it creates a spherical surface. Precisely this geometrical movement that Aristotle and Proclus had described is outlined by Swedenborg in a draft of his *Principia*.[36] Yet the Kabbalah also has the creative point. The first character in the Hebrew alphabet, 'aleph', is a point in space that encloses all points. The Italian Renaissance humanist Pico della Mirandola designated the divine as the indivisible point where the circular movement has its beginning and its end.[37]

---

[32] Hoffmeyer, 38, cf. 36.

[33] Gestrinius and Schomerus; Dahlin, 83f.

[34] Euclid, 1.

[35] Aristotle, *Peri ouranou*, 1.1.268a7–28; Philo, 16.49–52; Diogenes Laertius, 8.24f.

[36] *Opera* II, 200; *Treatises*, 102; *Principia*, 28; translation I, 48f; Aristotle, *Peri psychēs*, 1.4.409a4f; Proclus, 97.6 ff.

[37] Pochat, 95.

The abstract point is conceptualized, understood, through metaphorical thought. The point is a spot as when one can see a person, a bird, or a house far, far away on the horizon. One perceives them as small dots in the field of vision, without extension, with only a location, but in fact they have substantiality. The point is also the centre of gravity, where all mass is envisaged as being assembled in a geometrical point. The spot, as something ostensibly non-substantial, can contain a whole world that the microscope and the telescope have allowed us gaze into. The smallest dot grows to an enormous mass in a microscope, Pascal writes, as infinitely divisible as the firmament.[38] Hooke regarded it as natural to begin, as in geometry, with the point, and in his *Micrographia* he first directed his microscope at a point on a printed page, a point that was grey like a splatch of London dirt but contained valleys and hills like the earth.[39] With the minerals, nature began to practise geometry in regular figures, triangles, squares, tetrahedra, cubes and so on, Hooke wrote, and in plants we find even more complicated structures and mechanisms. Swedenborg's ideas about the mathematical point that traces out the world proceed from this kind of metaphorical thinking, in which the everyday experience of spots in the field of vision and points on a paper is transferred to the tiny particles that cannot be seen by the eye and to the enormous cosmic movements.

It is from the mathematical point that the whole of Swedenborg's particle world is conjured up; it encapsulates all the subsequent stages of creation. Swedenborg's mathematical point delineates the finite world; as with a pen, points are drawn to geometrical figures. The creator is an artist, a great geometrician, Kepler wrote.[40] Swedenborg's God is an infinite geometrician who creates a world according to geometrical laws, and who through his boundless wisdom can design nothing but the most perfect geometrical world, with the most perfect, beautiful, and unlimited geometrical figures. This is in line with Leibniz's 'lex melioris': our world which God has created is the best possible. The all-powerful, all-good God has infinite freedom to create the most perfect world, and he does so too. In Plato's *Timaeus* we likewise find the idea that the origin of the world from God assures its beauty and perfection in circles and spheres.[41] When God began to engage in spatial geometry, He quite simply created matter. In Swedenborg's natural philosophy the act of creation is a kind of extended Euclidean geometry. At the same moment as the world is created, geometry is also born. The genesis of nature and geometry became one and the same, or if you like, the substantial relationships of physics coincided with the spatial relationships of geometry.

This equation of natural points with mathematical points is a metaphorical way of thinking, whereby knowledge of concrete points is transferred to abstract ones. The world and geometry arising from one point is a metaphorical idea that occurs in Proclus's commentary on Euclid and in Giordano Bruno's atomic theory, which

---

[38] Pascal (1567/58), translation, 440.

[39] Hooke, 1–3, cf. 154.

[40] Kepler, ch. II, 49.

[41] Plato, *Timaeus*, 32C–34B, 53B–C.

proceeds from a point that is impossible to observe and can only be grasped by thought. Stiernhielm's identification of the mathematical with the physical point in his speculations about the vacuum is an example of metaphorical thinking that need not necessarily have been borrowed from Bruno or Kepler, as has been assumed.[42] Apart from the fact that this identity is a general metaphor, it is more likely that the inspiration comes from the frequently cited Proclus, especially when Stiernhielm talks about how three points form a triangle, that is, a surface, and if one adds a fourth the result is a regular body, the tetrahedron.

Swedenborg's point follows from the basic Cartesian metaphors, that *matter is extension*, that *mechanics is geometry in movement*, or as Newton put it, that geometry is perfect mechanics.[43] The circles and straight lines of geometry must be drawn with the aid of motion and are therefore mechanical. Geometry is the attribute, figure, and space of every finite thing, Swedenborg explains in *Principia*, in other words, the property that something finite must have in order to continue being itself, while mechanics is *modus*, the way or state in which the finite exists, acts, or is acted upon. Mechanics is *acting geometry*.[44] The Cartesians, as Hoffwenius says following Clauberg, differ from the Aristotelians in that they proceed from the mathematical point and not from being.[45] Clauberg declares that God first created a point, and when this point was set in motion, it formed line, surface, and body. Swedenborg indirectly mentions Clauberg's ideas about extension in some ontological notes.[46] Otherwise the interest Swedenborg showed in the point as the beginning stemmed from his own circle of acquaintances. Harald Vallerius wrote a dissertation about it, and in his manuscripts Polhem often returns to the idea that the world is immanent in a point: 'the whole universe can be packed into a mathematical point at least rationally.'[47]

The idea of the point as the starting point of geometry and nature is put forward by Swedenborg in 1722 in *Miscellanea observata*. Nature's principles should be the same as those of geometry, and the natural particles can ultimately be derived from mathematical points.[48] This idea that particles have their origin in mathematical points was criticized in a review in *Historie der Gelehrsamkeit unserer Zeiten* (1722).[49] The reviewer of Swedenborg's *Miscellanea observata* remarks on the many typographical errors and the lack of a pure, elegant, and

---

[42]Stiernhielm, 'Om tomrummet', *Samlade skrifter* II:2:1, 21–31; cf. Nordström, cccxxxf; Lindborg (2000), 166.

[43]Cf. Dear (1995), 211, 213; Dear (2006), 15–38.

[44]*Principia*, 121; translation I, 207.

[45]Hoffwenius (1698), VII:5, 26; Descartes (1664a), ch. VII; *Oeuvres* XI, 39f; Lindborg (1965), xiii, 63, 160f, 340, 347.

[46]Swedenborg, *Ontologia* (1742). KVA, cod. 54; *Photolith.* VI, 333; *Ontology*, n. 50; Wolff (1730); 2nd ed. (1736), II, part I, ch. II.

[47]H. Vallerius and Mozelius; *Polhems skrifter* III, 227, cf. 62f, 68.

[48]*Miscellanea*, 132f; translation, 84.

[49]*Historie der Gelehrsamkeit* (1722), 315–327.

clear style. The more one tries to delve into his sentences and the connections between his thoughts, says the reviewer, the more impossible it is to agree with him. Either it is incomprehensible or it should be understood in a completely different way. The mathematical principles scattered here and there indicate that he had studied mathematics, but it seems as if he assumes that it is unnecessary to explore in any depth the true nature and limits of these theorems. So it seems that he is not aspiring for the reputation of being a master of higher geometry. Swedenborg's theory of the elements and his physical principles are incomprehensible. The reviewer criticizes Swedenborg's concept of the mathematical point, the way he confuses the points of geometry and physics, that the geometrical line consists of an infinite number of points, that an infinite number of lines form a plane, that an infinite number of planes placed on each other form a body, and that he presupposes motion in the points. These ideas are beyond comprehension. There is no line consisting of points, no surface made up of lines, no body consisting of surfaces, as Bonaventura Cavalieri has demonstrated, says the reviewer. Mathematical points are indivisible signs that exist only in the imagination. When one says that the motion of a point forms a line, this line exists only in the mind. Noticing mathematical figures is just a way to remove the imperfection in the bodies that exist in nature. We wonder how this author has arrived at this extraordinary and incomprehensible opinion, declares the reviewer.

In *Neue Zeitungen von gelehrten Sachen* (1722) this critique is cited, with the addition that Swedenborg's thoughts are so incoherent that one cannot possibly understand him or agree with him.[50] The mathematical principles are not applied with care, which shows that he has not had much practice in higher geometry. Moreover, Swedenborg has a completely false conception of the mathematical point, which he equates with the first element. His other statements, furthermore, do not agree with each other. Swedenborg responded to this harsh condemnation in *Acta literaria Sveciæ* (1722) by simply choosing not to refute the criticism against his mathematical points. He says he will do it some other time, God willing. It is not worth challenging an anonymous reviewer to combat: 'But who they are and what they are like is of no matter, but since they are anonymous and without leader or law, in order that they may safely ambush wayfarers, they will have to forgive us for considering that it is incompatible with our honour to challenge them to fight.'[51]

A Grain of Dust at the Equator

Swedenborg continued to ponder on the point. In a manuscript about the principles of natural things entitled *The Minor Principia* he has distanced himself some way from his original particle physics in *Prodromus principiorum*. Instead of a rather

---

[50]*Neue Zeitungen* 1 August 1722, 616.
[51]*Acta literaria Sveciæ* (1722), 356; cf. Nordenmark (1933), 54.

stiff, Polhem-inspired theory of matter as stacks of balls, bubbles, and angular intervening particles, he has developed a more dynamic theory in which the motion of the point has a central role. It is probably this smaller *Principia* that occupied him in November 1729 when he wrote to the secretary of the Royal Scientific Society in Uppsala, Anders Celsius.[52] Swedenborg says that he has spent ten years collecting everything about metals and the mineral kingdom, but first has to put his excerpts in order and then print the work abroad. He wants to demonstrate nature's principles on the basis of reason and experience.

In *The Minor Principia* Swedenborg sides with those who wish to show that nature is completely mechanical, that everything in the invisible particles takes place geometrically. The only difference between our mechanics and particle mechanics is the size, since nature acts in the same way in the smallest things as she does in the largest. This way of thinking rests on an analogy from the visible world. Because large and visible things follow the rules of geometry, small and invisible things should do so too. There is no need to take refuge in any occult properties, for natural philosophy has the same origin as geometry. In geometry the mathematical point is the first unit from which line, surface, and body are formed. Geometry arises from a formless, weightless point, as from something unknown which can only be described ambiguously in words—it cannot be imagined.[53] God is the prime mover, infinity without geometry, who through His own endless motion in the infinitely small space gives rise to the point in which geometry has its origin. That something finite can come out of something infinite is demonstrated by infinitesimal calculus, the endless movement of a point in geometry that forms a line, or in physics where matter can be divided to infinity.

It was motion that created the point, and the figure that is most appropriate for motion is the sphere. Through infinite motion back and forth a line is formed, in the same way as an infinite series of points forms a line. It is as when a lead ball rotates at high speed and then seems to form a continuous surface. It would be even clearer if the circle were formed through an infinite or near-infinite speed (Fig. 2). If you imagine a grain of dust at the Equator, it follows the earth's motion at a speed of 64,800,000 cubits during a day. If you further imagine that the point is in such motion in a circle, but in an orbit that is just a thousandth of a cubit, then with the same speed as the grain of dust it would move around its circumference 1,800,000,000 times. In that case one could say that the point is everywhere in the circle. Swedenborg uses metaphors in his thinking. He envisages points as small spheres, grains of dust, and oscillating lead balls.[54]

Swedenborg arrives at the assumption of the existence of the vacuum. Neither previously nor subsequently has he been so explicit about his belief in the reality of the vacuum. We must assume that movement occurs in a space where there is no

---

[52]Swedenborg to A. Celsius, Stockholm, 27 November 1729. *Opera* I, 321; *Bokwetts Gillets protokoll* 12 December 1729, 174.

[53]*Opera* II, 4; *The Minor Principia*, 2.

[54]*Opera* II, 9, 33; *The Minor Principia*, 9, 36f.

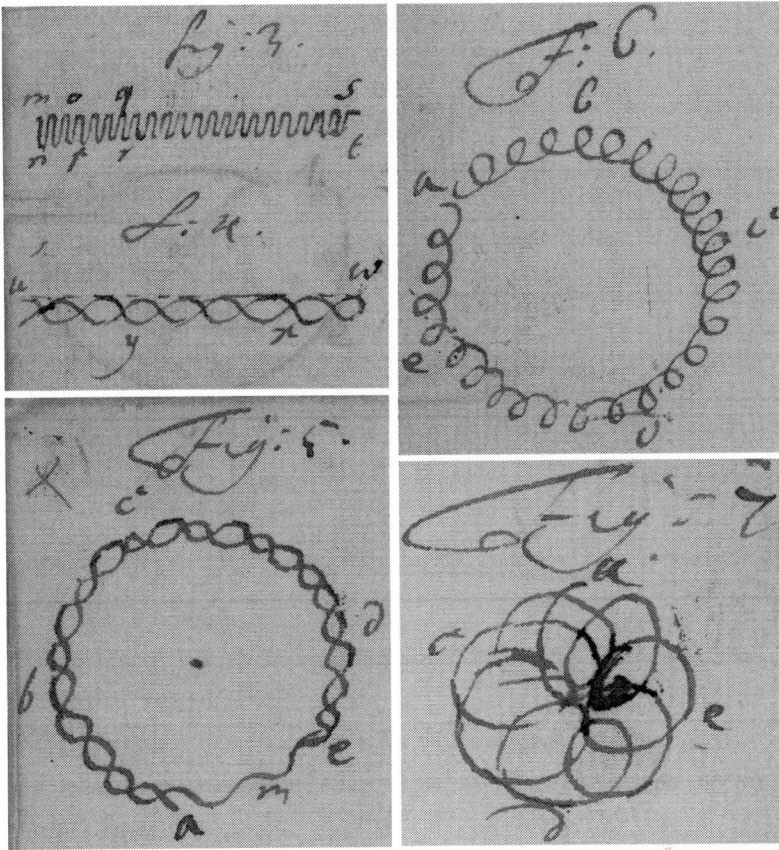

Fig. 2 The point draws. A point moves from *m* to *s* in an up-and-down movement, forming a plane surface (*3*). An extremely fast undulatory movement from *u* to *w* by a single point forming a plane surface (*4*). The movement of the point from *a* to *b* and *e* can also form a circular plane or ring (*5–6*), or an entire circular area where the point moves from *a* to the centre at *b*, and out to the circumference at *c* (*7*) (Swedenborg's manuscript *Principia rerum naturalium* (*c.* 1729–1731))

resistance, where it can approach infinite speed. In other words, we must imagine that this motion happens in an empty space or vacuum.[55] Elsewhere he writes that 'The space in which the point moves must be regarded as a perfect vacuum, free from any matter whatsoever.'[56] In this space there is nothing but the first movement of the point, the mechanism which we are incapable of discussing because it existed before the origin of any mechanism.

Swedenborg also admits the unintelligible motion of the point: 'I confess that these things are somewhat obscure, and that it is very difficult to form a conception

---

[55] *Opera* II, 10f; *The Minor Principia*, 10f.

[56] *Opera* II, 20; *The Minor Principia*, 21.

of motion in a circle and of motion from place to place along a line in a vacuum, where place cannot be conceived.' No explanation can be given with the aid of geometry, since there is no matter to which geometry can be applied. We must have recourse to a motion that has no similarity to the movements in the elementary state, but includes all kinds of movements. In other words, we must arrive at the assumption of circular motion. The movement of the natural point is eternal and can only be stopped by the prime mover, God. This motion occurs in a vacuum where nothing but the first points are driven by a continuous and endless movement, which means that one cannot speak of up or down; there is thus no location or space anywhere. The constant circling of the points will finally lead to their striking each other. And he says: 'This cannot at all be reduced to calculation; for if I wished to deal with every particular, and submit everything to calculation, time would hardly permit of this. Before I submit the theory to the test of experiment, I desire to deal merely with the general argument.' Swedenborg develops a kind of transcendental geometry beyond the powers of human conception. Geometry cannot be proved, nor is that necessary. Geometrical proof, Swedenborg remarks, is performed by those who wish to make easy things difficult, to turn light to darkness, to tie everything up in riddles and knots, in the belief that the learned world will esteem their teachings.[57]

He continued to work on the manuscript about the principles of natural things at least until 1731, and it then served as a basis for the version that he had printed in Dresden and Leipzig in 1734. *The Minor Principia*, however, displays significant differences from the printed version. We do not find central concepts such as finite, active-passive, conatus, or nexus. Something led him to put his original thoughts into a completely new conceptual apparatus. What intervened was his reading of Wolff's work, which gave him cause and inspiration to update and edit his manuscript. He saw that Wolff's ideas agreed with his own, saw his own ideas confirmed in the German scholar's work. The new concepts that Swedenborg would use were not unique to Wolff; they are found in a number of other thinkers, and they are primarily based on metaphorical thought. One cannot immediately assume that Swedenborg is directly dependent on Wolff. It is above all a matter of his having adopted terms and concepts.

On a visit to the secretary Rüger in Dresden in July 1733, Swedenborg was able to see Wolff's *Cosmologia generalis* (1731). This is a work that 'endeavours to build up elementary nature on purely metaphysical principles', he notes, adding that it 'rests on an extremely wise foundation.'[58] Proceeding from metaphysical principles in natural philosophy was the very desire that Swedenborg himself nourished. In his *Cosmologia generalis* Wolff builds up a world system, not based on observations, but—as in the particle theory in *Principia*—on ontological principles, with order as a fundamental principle. Wolff, of course, was not unknown to him before this summer day in Dresden. There is no evidence of any personal contacts between them, but we know that Swedenborg intended to send a letter to Wolff about

---

[57]*Opera* II, 11, 15f, 28; *The Minor Principia*, 11, 16, 31.
[58]*Resebeskrifningar* 10 July 1733, 22; cf. Jonsson (1969), 47.

his hydrostatic observations.[59] Swedenborg had read Wolff's mathematical works during his first study trip, and he describes them as very useful and clearly written.[60] In the 1720s Wolff's philosophy made its entrance in Uppsala, where it was advocated by scholars like the astronomer Celsius, but also by the mathematician Klingenstierna. The lecturer Carl Runcrantz tells of Wolff's significance in Uppsala in the 1730s, how Wolffian philosophy opened 'a gladder, wider view. As in the most beautiful garden, there was a contest here between nature and art. Everything of one kind in its place: everything was measured, clipped, and tidy; all in order, all in system.'[61] Why it was Wolffianism that had such an impact and not the philosophy of Rydelius was due to the actual method, according to the historian Lagerbring. Wolffian philosophy employed the mathematical method 'which undoubtedly is highly convincing, when the description, Definitio, which is laid as the foundation, is true and indubitable.'[62]

What captured Swedenborg's interest in Wolff was the geometrical method, the clear concepts, and the systematic thinking. Swedenborg's own natural philosophy and metaphysics are a kind of systematic philosophy in which things are placed in a coherent, ordered system. Systematic thinking in philosophy has its model in Euclid's axiomatic systematization of geometry. Philosophy is envisaged as an architectural structure, a metaphorical, cognitive system in which everything is connected, nothing is unbound or isolated. It is an organized structure with interrelating principles which display links between the parts of an ordered whole.[63] The real system, as the system-builder Leibniz imagined it, is built into nature and guaranteed by God and the perfection of creation. The system gives unity in the chaotic variation of reality. It is this that lies behind universal mathematics with its systematization of concepts and propositions, and Swedenborg's quest for fundamental, universal principles. Wolff's geometrical-scientistic aptitude and his emphasis on logical-deductive methods in the investigation of the world machine was a style of thought in Swedenborg's taste. Their similar modes of thought agreed in the cultivation of reason, the search for syntheses, system-building, geometrism, and rational theology. He found something of himself in Wolff's work. When Swedenborg read *Theologia naturalis* (1736–1737) he recognized himself. Reading it constantly turned him in towards himself, a confirmatory, self-reflecting rediscovery of his own thoughts. He saw Wolff's book 'in which he seemed to touch me, but without name.'[64]

---

[59] Swedenborg to Benzelius, Stockholm, 26 May 1724. *Opera* I, 314; *Bokwetts Gillets protokoll*, 108; cf. Jonsson (1983b), 164; Sandels, 15; Swedenborg is mentioned in a letter to J. C. Wolf, 13 July 1736. Benzelius (1983), 126.

[60] Swedenborg to Benzelius, Greifswald, 4 April 1715. *Opera* I, 229; Wolff (1715).

[61] Runcrantz, 21; cf. Frängsmyr (1972), 104.

[62] Lagerbring, IV:3, 49.

[63] Rescher, 29f, 39.

[64] *Resebeskrifningar* 20 July 1736, 64; *Photolith.* III, 53; Wolff (1736–1737), II, 610ff; cf. Lamm (1915), 48; translation, 48; Jonsson (1967–1968), 35–37.

Immediately after the discovery of Wolff's cosmology, Swedenborg began to outline a preface to *Principia* in which he compares his principles with Wolff's *Ontologia* and *Cosmologia*.[65] In a notebook from this time Swedenborg has entered a long series of definitions of philosophical terms, above all from Wolff's *Ontologia* but also from *Cosmologia* and *Psychologia rationalis* as well as Scipion Dupleix's *Corps de philosophie* (1636).[66] He strews this philosophical terminology in *Principia* and *De infinito*, as he had never done before. He acquired a new language through Wolff. In an appendix to *Principia* Swedenborg praises Wolff's *Ontologia* and *Cosmologia*.[67] Through them Wolff has shown the way to the first principles through rules and axioms. Swedenborg wrote that a reading of Wolff's works had confirmed his own opinions, although he had written down his principles two years before first consulting them, that is, in 1731. Looking at the history of his own ideas, he acknowledges the great debt of gratitude that he feels to Wolff's work, so much so that if someone were to compare them, he would find that his thoughts agree almost entirely with the metaphysical and general axioms in Wolff.

Nature's Labyrinth

The aspiration for wisdom is a characteristic feature of mankind; this is how Swedenborg opens his *Principia*. We strive towards a knowledge that is beyond or above our senses, a knowledge deeper than what is offered through the senses. There are three ways to true philosophy and knowledge of the mechanics of the universe: experience, geometry, and faculty of reasoning. With these we can arrive at the outermost limits of human wisdom. Philosophy should be understood here as knowledge of the mechanics of our world, everything that is subject to the laws of geometry, everything that can be revealed by experience with the assistance of geometry and reason. Three realms are under the mastery of geometry and the mechanical laws of motion: the mineral kingdom, the vegetable kingdom, and the animal kingdom; but there is also a fourth, the kingdom of the elements. On the immense ocean of natural phenomena

> I should not venture to spread my sail, without having experience and geometry continually present to guide my hand and watch the helm. With these to assist and direct me, I may hope for a prosperous voyage over the trackless deep. Let these be therefore my two stars to guide and enlighten me on my way; for of these it is that we stand most in need amid the thick darkness which involves both elemental nature and the human mind.[68]

---

[65] *Opera* II, 197; *Treatises*, 97.
[66] From Wolff (1731), § 293–301 & Wolff (1730), § 866–873. KVA, cod. 88, 276–278; *Photolith*. III.
[67] *Principia*, 452; translation II, 366.
[68] *Principia*, 2; translation I, 4.

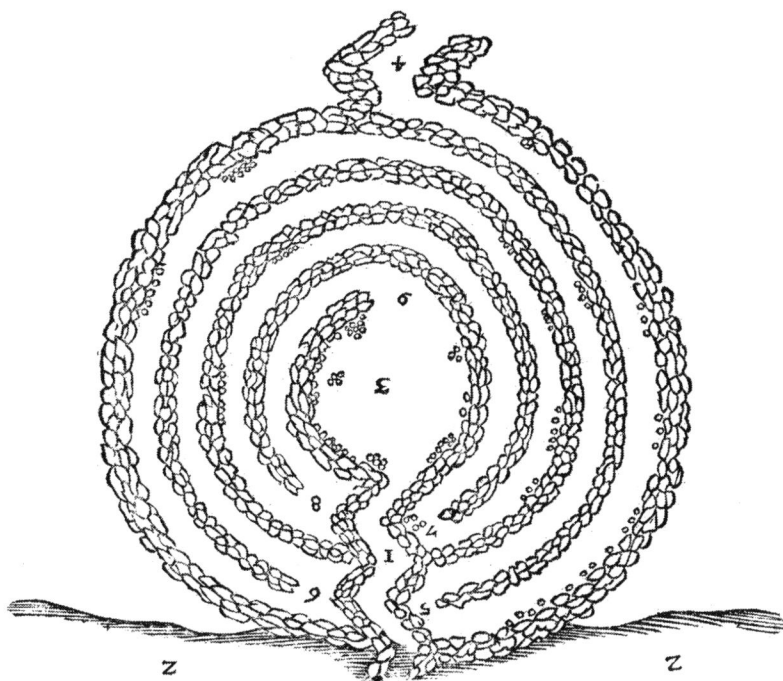

Fig. 3 A 'Troy Town' labyrinth on the island of Öland, with just one way out and no side tracks (Rudbeck the Elder, *Atland eller Manheim* (1679))

Experience is all the knowledge of the world that we acquire through the senses. The assembled knowledge of the millennia has now reached so far that we no longer need to postpone our investigation of nature's secret and invisible activity.

In the study of elusive nature, the investigator is cast into a dark labyrinth, where he easily gets lost in dark passages and dead ends. Nature is a maze with intricate paths which he wants to explore with impatient apprehension.[69] It would be fruitless to try to walk around its convolutions and note down all the directions. You would end up following your own footsteps in a circle. If you wish to find the shortest route through the labyrinth towards the exit, you must refrain from meaningless exploration of all its meanderings. Instead you should stand at one of the crossroads of the labyrinth, try to familiarize yourself with its contours based on the snaking paths you have already trodden, and turn back occasionally to retrace your steps. When you have finally got to know the concealed plan, after patient, methodical work, you can leave the solution behind you and walk fearlessly into the labyrinth. Swedenborg's labyrinth is a maze, not the linear Cretan labyrinth without sidetracks, with just one way to the centre and one way out. It is not like the Troy Town depicted by Rudbeck in his *Atlantica*, where one is forced to follow one path and cannot get lost (Fig. 3). Nature is instead a multicursal maze with countless routes that

---

[69]*Principia*, 4; translation I, 6.

constantly require you to choose paths, which sometimes turn out to be dead ends. A net, like the spider's web, can likewise be seen as a labyrinth, where each point is tied to all the other points.

Among his classical excerpts, presumably intended for his dissertation, about orbiting, travelling in circles, Swedenborg noted some lines from the myth of Daedalus and Icarus: 'As soft *Maeander's* wanton current plays, / When thro' the *Phrygian* fields it loosely strays; / Backward and forward rouls the dimpl'd tide,/Seeming, at once, two different ways to glide.'[70] Ovid compares King Minos's labyrinth, which Daedalus built, to the glittering waves of the Maeander, making their uncertain way up towards the source and back down to the embrace of the sea. Nature as a labyrinth is a metaphor that is varied in many scientific accounts in an effort to capture the drudgery and disorientation of scholarly research. Thinking is like walking in the dark, in a maze, in a wild forest, a roadless land, making a voyage over a sea with no landmarks. Falsity and deceit are a labyrinth with wrong trails, as in *Camena Borea*, with thousands of crossroads and thousands of paths.[71] Everything that is difficult, incomprehensible, impenetrable puts man in a maze. Thinking, life, nature, even the Bible is portrayed as a labyrinth, a forest, an ocean.[72] The world of things is a labyrinth, writes Bacon, where one must, like Theseus, follow Ariadne's thread to find the way out. With the aid of the senses and the light of experience we can open and secure a paved road for reason without abandoning ourselves to 'the darkness of traditions, or to the eddy and whirl of arguments, or to the waves and windings of chance and casual unregulated experience.' The method used by Swedenborg—reading through other people's statements and adding one's own reflections—is considered by Bacon to have 'no foundation and simply spins around on opinions.' In that sense Swedenborg is far from being an empiricist in the spirit of Bacon. Science which has lost sight of experience, according to Bacon, has got lost as in a labyrinth, 'since a properly organised order takes one through the woods of experience by a steady path to the open country of axioms.'[73]

Man is in darkness, having lost his place. 'He looks for it everywhere restlessly and unsuccessfully in impenetrable darkness,' wrote Pascal.[74] Descartes wandered alone in the dark, slowly and cautiously, afraid of falling. He wanted to be staunch in his opinions and imitate 'those travelers who, finding themselves lost in a forest, should not wander off by turning first this way and then that, and still less stop in one place but rather always walk as straight as they can in one direction, and not change it at all for slight reasons'.[75] The chiaroscuro of reality is dispersed by that natural light, reason. Without knowledge, without experience, a method, wrote Hooke, we are left to wander in nature's complicated meanders, in labyrinths of ungrounded

[70]KVA, cod. 37, 85; Ovid, *Metamorphoseon*, 8.162–164.

[71]*Camena Borea*, edition, 142f.

[72]Eco (1986), 149.

[73]Bacon (1620), I, § 82, cf. § 61; translation, 67f.

[74]Pascal (1999), 8.

[75]Descartes (1637), III; *Oeuvres* VI, 24; translation, 43.

opinions.[76] Nature is for Linnaeus like a palace with countless inaccessible rooms, concealed and containing precious treasures in the inner rooms. He had not yet tried to penetrate it, 'but when I peeped in through an open window, looking into a hidden chamber, I observed something . . . .'[77]

One of the solutions to nature's labyrinth, leading out to wisdom, is experience, in which all sciences have their origin. Experience is the means to wisdom. But nowadays, Swedenborg remarks, those who are called the wisest are those with the most experience, the greatest skill in experiments, those who perform demonstrations with eloquence and the ability to captivate listeners with a pleasant and melodious voice. Perhaps he had a popular demonstrator like Triewald in mind? But experience itself is not wisdom. Or take, for example, a historian or scholar who has ploughed through piles of books, learned about the vicissitudes of the ages and the lives of heroes—this does not make him wise. He must be able to see analogies and similes in history, and must go on to penetrate, with rational thought, all the way down to the cause of things.[78]

As a natural scientist Swedenborg himself was scarcely an indefatigable empiricist who amassed experience. He wanted immediate results, to take a short cut through reason to the truth. Yet in an epistemological sense there are empiricist tendencies in *Principia*, in that he emphasizes that all knowledge builds on experience.[79] Man must rely on experience, the senses, and assembled human knowledge, since he lost his intuitive knowledge of the world during the deluge and can no longer attain immediate knowledge from the mind. There is a risk, however, in the pure collection of facts, he thinks, since this can obscure what is essential for the researcher. At the same time, he could not accept radical rationalism with its sceptical attitude to the experience of the senses.

The other means leading to wisdom, with the ability to reveal and dissolve invisible nature, is geometry and rational philosophy, with which one can digest experiences through analysis, reducing them to laws, rules, and analogies. This is done by gathering a multitude of observations in one's memory and then drawing conclusions about the unknown things on the basis of their similarities and analogies to what is known. Geometry accompanies the world from its very beginning, from its first boundary to the last. So it is with mechanical principles, although they can be different in different worlds, with elements that are formed and arranged differently. He who claims otherwise resorts to occult properties and conceals his ignorance. Both the largest and the smallest have the same mechanical laws. Even if the particles of the elements are invisible, they are still geometrical and mechanical. The same applies equally to what is within sight and what is above the sphere of vision. We see mechanics and geometry in the solar vortex, the circling of the satellites, in the circulation of the blood or the alternating movements of the lungs,

---

[76] Hooke, 1, 87.

[77] Linnaeus and Ramström, § 1; Broberg (1975), 105.

[78] *Principia*, 8; translation I, 13f.

[79] *Principia*, 4f; translation I, 7–9.

like a pair of bellows. We can investigate these mechanical and hydraulic machines. The same mechanics, with lungs, hearts, and senses, can be seen in the smallest animalcula under a magnifying glass to the largest whales. They have wishes, pleasures, satisfactions, loves, reproduction, and emotions through their vital spirits. Nature resembles herself, with the same mechanical principles, in the smallest finite existence and in the largest. Geometry is always the same. The same proportions exist between large numbers as between small ones. A 100 billion is to 500 billion as 1 billionth is to 5 billionths. But the smaller mechanics is more pure, more regular than the larger and more composite. Everything hangs together. Means and ends are linked to each other and constitute life itself and the essence of nature. Man's senses are also constructed in accordance with the elementary world, a world in geometrical harmony.[80] His eyes determine whether something in the finite world or in nature has harmonious proportions or not, whether a garden's ornamentation follows the rules of art. It is the mutual analogy and harmony of these finite parts that constitutes the beauty of the world.

Reason consists of knowing various facts about the world and being able to put them in order and in context, to see their analogies and to bring out by analogy a third or fourth previously unknown truth. The third road that leads to a true philosophy is the faculty of reasoning, the ability that distinguishes mankind. The path from the senses to the mind can be compared to the path taken by light through a dense cloud towards the eyes.[81] He who lacks experience and a capacity for thought fumbles in the dark. One such metaphor of light and dark is illustrated in the frontispiece of Johann Kunckel von Löwenstern's book about the art of glass, *Ars vitraria experimentalis* (1679), a book that Swedenborg owned (Fig. 4). The experimental fire takes two different paths. The light of truth radiates from the clear, powerful sun through the burning glass of reason which is being held by Mens, the mind. The focused ray ignites the natural light that is held up by Experience in her triangular-patterned skirt. The mind also holds the jug of wisdom, from which she pours being or the thing down into the cup of science. The little schoolboy, intellectual labour, stands with his funnel hoping to catch a few falling drops. Waiting in the background is the trial by fire, examination. The other side is ruled by the darkness of uncertainty, with gloomy moonlight and impenetrable cloud. There reality is obscured by unreason, which can see hardly anything through its blindfold. Unreason is holding the extinguished lantern of folly and has the inflated udder of vacuity in her lap. The mistakes of fantasy are tugging at the skirts of vanity as she fills the udder with air. On her lap lies an open book with a list of mistaken predecessors. Fruitless labour, standing on the uncultivable soil, tries to seize the mistakes of fantasy with a pair of tongs. On the dark sky in the background, the winged ass of planlessness flies among the non-existence of the dark mountains. At the bottom is the cornucopia of truth, and to the right truth rides past Christ towards the horn of poverty.

---

[80] *Principia*, 14; translation I, 23f.

[81] *Principia*, 16f, 24f; translation I, 28–30, 42; cf. *Psychologica*, 10f.

Fig. 4 The light and dark of thought (Kunckel von Löwenstern, *Ars vitraria experimentalis* (1679))

Thinking as a walk in an unknown landscape shrouded in darkness is a metaphor intended to conceptualize abstract thought. If you exert yourself a little, Lucretius writes, 'one thing will become clear by another, and blind night will not steal your path and prevent you from seeing all the uttermost recesses of nature: so clearly will truths kindle light for truths.'[82] *To know is to see* is a central metaphor for

---

[82]Lucretius, 1.1114–1117.

Descartes. One of his main questions was to find out how we can obtain *clear* and sure knowledge, how we can see and grasp ideas plainly and clearly. Knowledge is guaranteed by the natural light of the mind, which illuminates mental objects or ideas, in the same way as an eye in full daylight can see every detail of physical objects. The difficulty, Descartes says, lies 'in correctly recognizing which are the things that we conceive distinctly.'[83] Swedenborg follows the metaphor of light representing human thought and notes in *Psychologica*, following Wolff, the proposition that the light of the mind is the clarity of perception, and the darkness of the mind is obscurity. A clear perception is called distinct, while the opposite is confusion.[84] Nature as light and dark is a metaphor that constantly accompanies Swedenborg's thought. The researcher cannot go beyond reason, beyond his finiteness, he writes in a manuscript about the infinite, the indefinite, and the finite.[85] If he tries to go further, his thought will be shrouded in darkness, even with a careful examination of natural things. He may become blind and dream of murky nights. So even in finite things there is a thick darkness that enshrouds, proliferates, and brings reality into disorder.

The Janus Face of the Mathematical Point

Since the world consists of finite things and obeys the laws of mechanics and ultimately of geometry, like all finite geometrical things it has its origin in a point. Through a divine act of will a movement in 'the natural point' is created in the infinite. Mechanistic philosophy corroborated divine activity. It is a voluntaristic conception of God that permeates Swedenborg's creation, according to which God creates the world through an act of will like a person's soul which by free will causes the body to move. There is an organic analogy between human and divine will. If one wants to ascertain the divine will, there is good reason to perform experiments. God may have made the world in many ways. But which is the right way? Swedenborg's chosen path is not to find out by experiment which solution God chose; instead he tries to think like God, which solution God as the chief mathematician ought to have chosen.

The point is the first thing to enter in the creation, just as the point is the first thing in geometry. Since all things in the finite world are necessarily geometrical, this means that, at the very moment when the world was created, geometry was born as well. The world and geometry come from the same seed. Swedenborg's point is analogous to the point in geometry. Like the latter, it is indivisible and lacks extension, but it occupies a specific location in space. The only difference, according to Swedenborg, is that the point in the creation of the world is called 'the

---

[83]Descartes (1637), IV; *Oeuvres* VI, 33; translation, 53.

[84]*Psychologica*, 12f; Wolff (1732b), § 35f, 38f.

[85]Swedenborg, *Fragmentum de infinito* (c. 1733–1734); edition *Photolith.* III, 173; *Treatises*, 121f.

natural point', while the point in the creation of geometry is called 'the mathematical point'.[86] The natural point is the same as the mathematical point or 'Zeno's point' after the Eleatic philosopher. Swedenborg's Zeno points show how independently he reads Wolff, from whom he could very well have borrowed the concept. He changes the content of Wolff's *Cosmologia* to suit his own purpose, for Wolff emphasizes that nothing material can arise from Zeno's points.[87]

The finite world is created through the mathematical point, that is to say, it arises immediately, via the point, from the infinite. The natural point thus stands between the infinite and the finite. Swedenborg compares this intermediate position to the two-headed Roman god Janus about whom Ovid wrote in his *Fasti*, the doorkeeper and the symbol of entrances and exits, whose opposing faces allow him to look out on two worlds at the same time.[88] The Janus face looks simultaneously at the unknown infinite and at the finite geometrical-mechanical world of which man can be part through the point. The mathematical point is a doorway into nature. We are admitted into a geometrical field, we come into the finite universe with its specific laws and geometry. And in this geometrical world, human reason can extend itself: 'man instantly begins to have a knowledge of himself, to perceive that he is something, that he is finited, mechanical, nay, even a machine.'[89] The point, in other words, stands between the intelligible finite world, and the unintelligible infinite world. It was not until ten years later that Swedenborg took the step over the limits of the unknowable and entered the infinite geometry of the spiritual world.

In the creation of the world God thus implants motion in the point, or more exactly an effort, *conatus*, at motion.[90] Hence there are two basic principles in Swedenborg's particle system: the point and its inherent effort to move, which proceeds from Descartes' dichotomy of extension and motion. Motion is what gives rise to both figure and space. The term conatus appears in the printed version, presumably following studies of Wolff's Leibniz-inspired philosophy, but it is also found in Hobbes and others.[91] Conatus can be regarded as a metaphorical concept in which human drives and endeavours are transferred to matter. Man strives towards certain goals, and in the same way there is also a kind of will in nature towards specific goals. Leibniz explained the relationship between conatus and motion as corresponding to the relationship of the point to space, as the beginning and end of motion. He described static force as a dead force, the start or the finish of a tendency towards motion, that is, a conatus. Living force or energy relate to dead force as impetus to conatus, as a line to a point, or as a plane to a line.[92] The conservation of the living force, *vis viva*, was for Leibniz the same as the eternity of

---

[86] *Principia*, 29f; translation I, 51; cf. *Opera* II, 201f; *Treatises*, 105.

[87] Wolff (1731), § 215–218; *Oeconomia* I, n. 592; Jonsson (1969), 48.

[88] *Principia*, 31; translation I, 53; cf. *Oeconomia* II, n. 116; Ovid, *Fasti*, 1.89–144; Pochat, 89–103.

[89] *Principia*, 31; translation I, 54.

[90] *Principia*, 33; translation I, 57.

[91] Wolff (1731), § 166; Jonsson (1999), 30.

[92] Westfall, 136; Blay, 104f.

God's creation. In Leibniz and Wolff we also find the idea that material substances have their origin in metaphysical points to which a force is allocated. The substantial atoms, Leibniz writes, might be called '*metaphysical points.* They have something vital, and a kind of *perception*, and *mathematical points* are the *points of view* from which they express the universe.'[93]

The natural point is pure motion in the universal Infinite, a movement that cannot be imagined through any geometrical laws.[94] But it is still nothing, even if it does not have any analogies with finite movements. Since there are composite, extended, and mixed things, there must also be simple, non-extended, pure things. The point has no extension and no parts, it is indivisible and without figure. It can only be imagined through analogy with finite things with geometrical lines; it is so close to infinite and so distant from geometrical and finite comprehension. In a manuscript of *Principia* Swedenborg writes that analogies end in the point.[95] When God placed the point, it was thus not a full stop; it marked a start.

---

[93]Leibniz (1695); translation, 745.

[94]*Principia*, 32; translation I, 55.

[95]*Opera* II, 198; *Treatises*, 99.

The Spiral

Its curvature twists more and more in a way like a snail, although always further
from its central point. Is drawn thus.[1]

Swedenborg, 'Om en boglinia eller curvâ, hwars skärlinier thet är secantes giöra altid
med boglinien lika wincklar' ('On a bent line or curve, whose cutting lines or secants
always make equal angles with the bent line', 1718)

Helical Lines

The spiral curves, twists in expanding circles from a centre, as in the coiling of
the snail, in the whirling waters of the maelstrom, in the vortices of the universe.
For Swedenborg the spiral is the most beautiful and complete of all geometrical
figures. It is return, it is perfect motion in the particle world and in the firmament.
Its movement leads one's thoughts to disorder, and can be perceived as chaotic,
suggestive, alive, and expressive. In a broader sense the spiral is a curve that winds
around a centre, a pole in the plane, while simultaneously distancing itself from
this point of departure. The present chapter is about this class of gentle lines in the
particle world, in the cosmos and the soul, about whorls, whirls, vortices, helices,
scrolls, serpentines, and meanders, but also the related circle.

In Swedenborg's thought the spiral takes on a number of properties which cannot
be explained by the geometry of the form; these properties include perfection,
dignity, beauty, reason, and goodness. Throughout the history of ideas the forms
and figures of geometry have often been connected with something outside themselves, with something over and above their strict geometrical properties, obtaining
associative properties and being assigned symbolic meanings. Certain forms have

---

[1] *Dædalus Hyperboreus* VI, 14.

D. Dunér, *The Natural Philosophy of Emanuel Swedenborg*, Studies in the History
of Philosophy of Mind 11, DOI 10.1007/978-94-007-4560-5_7,
© Springer Science+Business Media Dordrecht 2013

been regarded as more beautiful and noble than others. The spiral is one such simple symbol, like the cross, the circle and the pentagram, with a greater semantic potential than others, with greater variation and indeterminacy. The forms of geometry are of ideal objectivity, they are the same always and everywhere for all observers. The geometrical forms in nature and on paper point towards a transcendental reality of ideal, eternal, perfect forms. Three forms in particular— the point, the circle, and the sphere—came to represent perfection, infinity, eternity, harmony, wholeness, and divinity. In Swedenborg it was instead the spiral that underwent this apotheosis. The aim here is therefore to find our way to the meaning of the spiral, the cognitive incentive for its supremacy, how Swedenborg's spiral may have had its foundation in metaphorical thought, in the mathematics and physics of the time, the aesthetic and perfection of forms in science, metaphysics, and the visual arts.[2]

In Swedenborg's early essays from the 1710s the spiral is a geometrical figure consisting of concrete curves on a sheet of paper, curves which can be exactly described with the aid of a mathematical formula. The spirals he drew seem iconic in that they show graphic similarities to the ideal forms of mathematics. What he was interested in was the denotation of the spiral, or its core meaning, its objective significance, in other words, nothing but the purely geometrical and demonstrable properties of the spiral. In *Geometrica et algebraica*, among mathematical descriptions of different curves and conic sections, such as the cycloid, the parabola, and the hyperbola, which he studied in Reyneau's *Usage de l'analyse*, he also has a section about spirals. He tries to grasp the mathematical properties of the spiral, which almost whirl out over the paper through the text. The spiral is mechanical and infinite, he explains (Fig. 1).[3]

The spiral has, above all, two geometrical properties that Swedenborg noticed. It is mechanical, and it is infinite or unchanging. The Archimedean spiral's lack of fixed points, its property of being mechanical, mobile, and dynamic recurs in Swedenborg's spiral. Archimedes' spiral combines two constant movements, that is to say, if a point travels with uniform motion along a straight line, while this line is simultaneously rotating at a constant speed around one of its ends, which is held fixed, the point will describe a spiral in plane.[4] The logarithmic or equiangular spiral, or as Jakob Bernoulli called it, 'spira mirabilis', the marvellous spiral, was first described by Descartes in 1638. Swedenborg was probably the first to describe the logarithmic spiral in a printed Swedish text. He does so in an article 'Om en boglinia eller curvâ, hwars skärlinier thet är secantes giöra altid med boglinien lika wincklar' ('On a bent line or curve, whose cutting lines or secants always make equal angles with the bent line') in the last issue of *Dædalus Hyperboreus* from 1718. He also explains how to draw one of these lines in a geometrical and mechanical way, as Polhem had shown him. This equiangular spiral can be of use

---

[2]See also Dunér (1999), 51–91; 2nd ed., 7–77; translation, 1–58.
[3]*Photolith.* II, 76–78, cf. 39; Reyneau, II, passim, 545–879, see especially 593.
[4]Archimedes, *Peri elikōn*, 44.17–23.

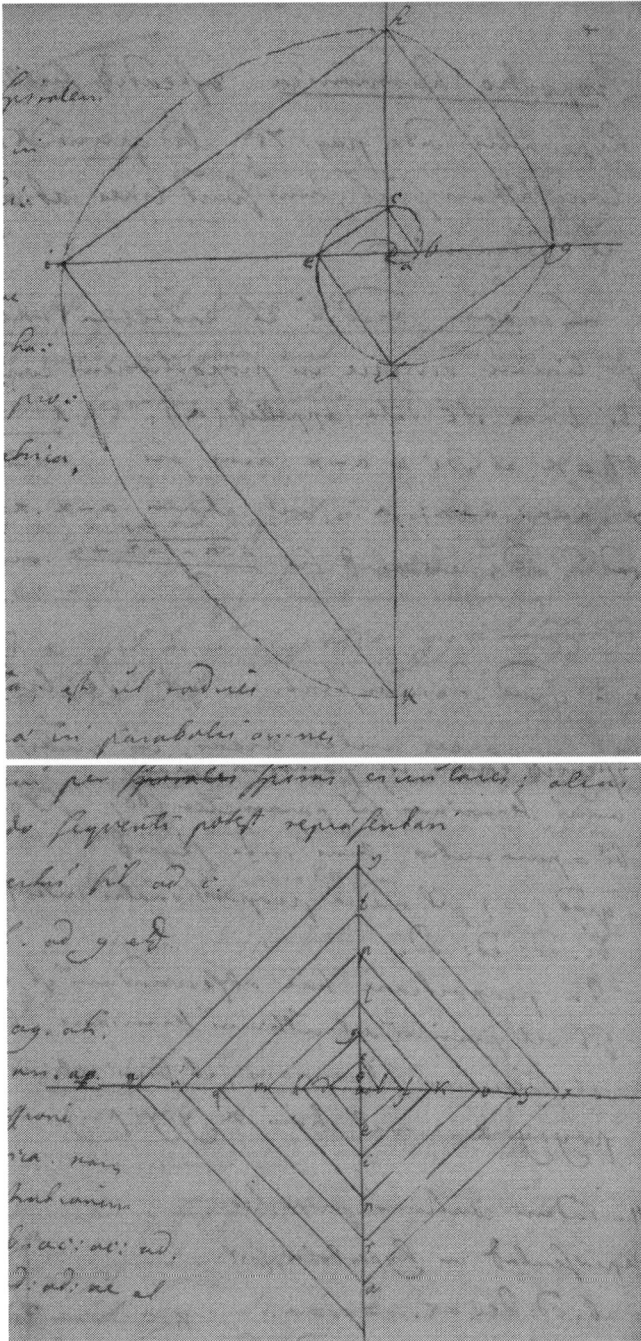

Fig. 1 Logarithmic spiral, the equiangular spiral in geometrical growth (Swedenborg, *Geometrica et algebraica* (1713–1714))

in the manufacture of a special kind of shears for cutting sheet metal and for paper mills which always have to cut evenly. Yet despite the references in the text, the article lacks illustrations. In the autumn of 1718, however, he had sent a figure of a spiral to the king, but that one seems to have disappeared into eternity.[5]

Swedenborg does not go so deeply into the geometrical properties of the logarithmic spiral in *Dædalus Hyperboreus* as he had done in *Geometrica et algebraica*. The logarithmic spiral has precisely the properties that Swedenborg constantly invokes in his natural philosophy in *Principia*. It is infinite, the angles are always the same, it is unchanging, its curve grows and shrinks infinitely without changing form. The logarithmic spiral therefore looks the same all the time, regardless of which part of the spiral one considers. In each little part of the logarithmic spiral one can thus see the whole. This also coincides with the golden section and the Fibonacci sequence, 1, 1, 2, 3, 5, 8, 13, 21, 34, 55, 89, 144... The mathematician Jakob Bernoulli had a spiral inscribed on his gravestone in Basel with the words *Eadem mutata resurgo*, 'Although changed, I arise the same.'[6] For Leibniz too, the spiral is a symbol of resurrection. There is a metal spiral on his coffin under which are the words *Inclinata resurget*, 'The bowed will rise.'[7] In Swedenborg's spiral we repeatedly find the logarithmic spiral's property of infinity, the peculiar quality of being self-reflecting, with the part and the whole merging. No matter which part you observe, you see the whole.

Swedenborg's interest in mechanical curves was typical of the age. After Descartes presented analytical geometry in 1637 there was flourishing mathematical activity with curves until the middle of the eighteenth century. With Descartes' method, which established a link between geometry and algebra, the concept of motion was introduced to geometry, creating a new way of representing and analysing curves. Mechanical curves such as Galileo's cycloid, Pascal's limaçon, and Roberval's sinusoidal could only be described approximately and incompletely. Fermat, Torricelli, Bernoulli, and many others worked with spirals turning in space, such as the helix and the loxodrome. This led into a geometry that lay beyond finite algebra and required a new mathematics, an infinitesimal analysis, a 'hidden geometry' as Leibniz put it.[8]

The interest in curves spread to natural philosophy and art. The spiral fascinated Swedenborg and many others, not just because of a mathematical fascination, but particularly because of the beauty of the spiral, its wealth of variation, mutability, infinity, and movement. The dynamic, mobile, and infinite curves of the baroque contrast with the classical timeless and finite Euclidean geometry, with statics and Apollonius's conic section, and with the immobile and angular Platonic bodies of the Renaissance, with their symmetry, straight lines, and closure.[9] In Greek

---

[5] *Opera* I, 287.

[6] *Acta eruditorum* (1706), 44.

[7] Vogler, 251.

[8] Gandt, 204; Mahoney, 704.

[9] Bense, 26; Spengler, 19; translation, 15; Hallyn, 224; Rosengren, 180f, 184, 186; Netz (2005), 251–293.

Fig. 2 The pillar of the
Temple of Solomon. Sveden,
Dalarna (Photo by the author)

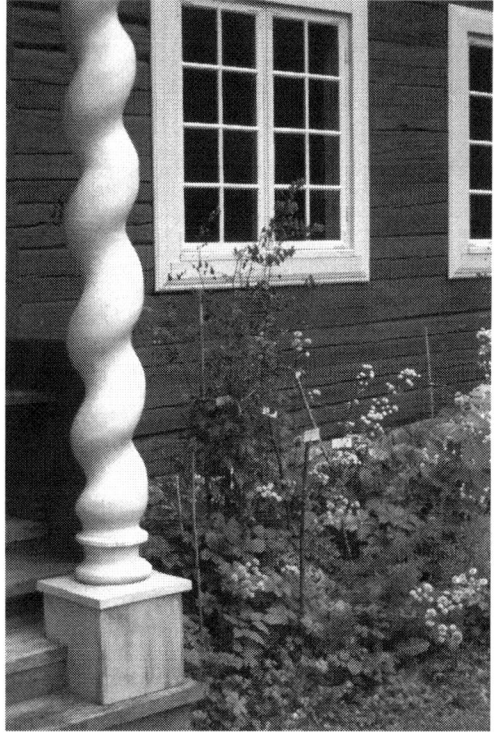

Fig. 2 The pillar of the Temple of Solomon. Sveden, Dalarna (Photo by the author)

geometry, studies were confined to the circle and the straight line, with the aid of their physical counterparts, the compasses and the rule, by which the harmony, simplicity, and aesthetic of geometry were to be preserved.

The Archimedean spiral was of interest not just to mathematics, kinematics, and physics, but also to the visual arts and architecture. In mathematical textbooks such as Wolff's *Der Anfangs-Gründe aller mathematischen Wissenschaften* (1710), which Swedenborg read in a Latin translation in Greifswald in 1715, there are detailed explanations of how to construct the volutes of an Ionian capital, based on Andrea Palladio.[10] Ionian capitals are like the waves and the force of the wind on classical buildings, or like the serpent in Gothic architecture. Swedenborg probably walked through the beautiful entrance of his father's childhood home, Sveden outside Falun, where Linnaeus married his Sara Lisa. The doorway is flanked by two helical Solomonic columns which, like those he had seen in Saint Mark's Cathedral in Venice in 1738, were supposed to have stood at the entrance to the Temple of Solomon in Jerusalem (Fig. 2).[11]

With the baroque, things were no longer timeless and in the present; they were constantly changing and becoming, continuous metamorphoses, a potentiality in the

---

[10]Wolff (1710); 4th ed. (1732), 387ff; Wolff (1715); *Opera* I, 229; Cook, 17f.

[11]*Resebeskrifningar* 19 April 1738, 85.

aspiration to perfection. It is not an immutable Platonic static-geometrical symmetry that characterizes the world but a dynamic harmony with power and movement, change, openness and infinity, writhing ornamentation and crooked lines. But the idea of the perfection and simplicity of the cosmos was not abandoned. Simplicity instead came to mean general validity rather than absolute Platonic symmetry. God has ordered the universe in such a way, Leibniz thought, that it always chooses the simplest and most perfect solutions. The curving, swelling works of baroque architecture were decorated with meandering volutes and cartouches. The ellipse summed up both the church interior and the planetary orbits. Both the small rooms of man and the huge universal space of God arched over people. Longitudes and latitudes, on the surface of the earth and on the celestial vault, were calculated by means of spherical geometry. Swedenborg's natural philosophy was perfectly in line with the swinging volutes of the baroque, an infinite dynamic in movement. No resting circles, but instead oscillating spiral movements. It is a geometrical particle world of 'infinitesimal' boundlessness, where motion draws the figures, without straight lines and angles.

With such concepts of the spiral, people began to see it in living nature. Swedenborg saw spirals everywhere around him, in the shell of the snail, the tissue of the brain, in the blood, the heart, the muscle fibres, the cochlea of the ear. The logarithmic spiral is not uncommon in nature. On the shell banks at Uddevalla where he had strolled, or the petrified shells in Aachen which he had drawn in *Miscellanea observata*, he could observe the logarithmic spirals of molluscs (Fig. 6 in Chap. 4). In palaeontological and conchological works he could study shells arranged in symmetry, variation, and contrast, or in his friend Bromell's beautiful collection of stones and *Lithographiæ svecanæ*. The chambered nautilus, *Nautilus pompilius*, which was depicted in books on his shelf by Gottlieb Friedrich Mylius and Bromell, was a desirable item in collections of natural history and was used during the baroque for magnificent cups.[12] Swedenborg himself owned an antique 'Credence of a shell', a goblet of silver studded with jewels.[13] Roberg noticed how hair formed whirls or 'vortices' on the head, 'lying as if mown, as the life-spirits employ their speed as the foetus grows, and the nerves run away, the outward ends of which, it seems to me, are the hairs.'[14] The coiling of hair, these offshoots of the nerves, was caused by the vital spirits. A recurrent theme in this chapter is how Swedenborg imagines invisible reality based on the visible, how concrete forms, shells, eddies, spirals on a sheet of paper are transferred to abstract reality. Spirals are eddies, like Scylla and Charybdis in the Strait of Messina, the maelstrom in the Lofoten Islands, the enormous whirlpool at the North Pole that Kircher describes in *Mundus subterraneus*, or like Kircher's birds circling as they rise in helices from the centre of the earth.[15]

---

[12] Mylius, I, 53; Bromell (1726–1727), II, 65–75.

[13] July 1770. KB, Swedenborg's manuscripts, biography of Swedenborg, in 'Swedenborg', no. 12.

[14] Roberg (1718), 32.

[15] Kircher (1665), III, 148, 160, & I, 19.

The Circle of Time

The circle is eternal return. Swedenborg comes back time and again to the return of movement to its starting point, the cyclic recurrence of these, the way history repeats itself.[16] It is in accordance with the thoughts of the Pythagoreans about the sphere and the circle as perfect figures, ideas echoed through history by Plato, Aristotle, Proclus, and others. The circle is perfect in that it has neither a beginning nor an end, each point on the circumference is equidistant from the centre, and the circumference is the line that encloses the largest possible area in relation to the length of the line. The circle remains the same and is best suited to the movement of the sphere. There is a longing for immobility in the classical world-view with its spherical celestial bodies in circular movements. Swedenborg develops it into a world of spiral bubbles in spiral motion and whirling spheres, constantly in movement.

Matter, motion, and time go in circles. Time always returns to the same point:

> After clouds and fog are ended,
> Shines the sun in all its splendour,
> After chill of winter cometh
> Springtime and the balmy summer.[17]

The 12-year-old Swedenborg, in opus number 1 in his long list of works, writes about the return of the day and the seasons for the wedding of his cousin, Beata Hesselia, to Johan Kolmodin in the vicarage of Folkärna in May 1700. Time is a cosmic cycle, through the orbit of the earth around the sun, with the seasons repeated in a circle, or the peregrinations of the heavenly bodies, the annual tasks in the fields, the ebb and flow of the sea, the course of rain and rivers, the stages of life and regeneration. The preacher in Ecclesiastes starts with the vanity of the eternal, meaningless cycle, as generations come and go, the sun rises and sets, the wind goes towards the south and turns to the north. All rivers run into the sea, yet the sea never becomes full. There is nothing new under the sun.

The understanding of time is related to other concepts in metaphorical thought, to movement, space, and events. Time flows like a river, the wheel of fortune rotates with nature's cycle, the movements of the planets, rise and fall, life and death. In metaphorical thought, abstract time is translated into something spatial, a straight line or a returning circle. Augustine, who would later be significant for Swedenborg, has a presentiment of something similar in his psychological interpretation of time as something inherent in the mind, in thought. Past, present, and future which never are, but exist as memory, attention, and expectation.[18] In Christian eschatology time ran like a straight line between creation and doomsday. Through the linearity of time, history had a direction, something ahead and something behind, an incipient

---

[16]*Ludus Heliconius*, edition, commentary 14f, 229; cf. *Camena Borea*, edition, 142–145.

[17]Swedenborg, 'Post nubila Phœbus', *Opera poetica*, 1.

[18]Augustine, 11.18, 11.28; translation, 267, 277.

notion of progress. But circular time was advocated later on by the historian and philosopher Joest Lips (Lipsius), and it recurs in Vico's cyclic perception of history.[19] Only the circle can recall the past.

Cyclic time had its classical advocate in Pythagoras, known for the theorem that Swedenborg explains in *Regel-konsten*: 'All right-angled triangles have the property that the squares of two sides is the sum of the square of the third.'[20] The Pythagorean theorem became the very symbol of geometry in heraldry. When Polhem was raised to the peerage in 1716 he placed it on his coat of arms, at the suggestion of the numismatist Elias Brenner.[21] People also took an interest in Pythagorean symbols, which it was assumed that Pythagoras had borrowed from Egyptian wisdom. Schefferus, for example, wrote about Pythagorean philosophy and the relation of Egyptian hieroglyphs to allegories, and Jesper Swedberg commented on some Pythagorean metaphors and analogies by Demophilus in his doctoral dissertation in 1682.[22] Swedenborg also mentions Pythagoras as a philosopher at two places, albeit not especially crucial, in his dissertation.[23]

Swedenborg's *Festivus applausus* begins with a meditation on cyclic time. He paraphrases Ovid's *Metamorphoses* and talks about Pythagoras's quick-witted disciplines who saw that everything under the sky moved in circles that wound their way like the river Maeander.[24] The sun, the sky, and time return to their beginning and their end. The earth is carried around in the same orbit, the seasons return to their harvest, the ages of man go from shade and winter to new youth. From this, Swedenborg continues, they drew the conclusion that everything under the path of the moon moves in circles, as if to depict the pattern of heaven as the origin of all things. They believed in the return of the ages, that scenes are constantly repeated on the theatre of the world. In his natural philosophy Swedenborg developed these parallels between microcosm and macrocosm, the small circles on earth and the big ones in the universe, the analogies between the ages of man and the seasons, the movements of the universe as the pattern for everything. It is the perspective of the eternal cycle, with future and past, rain and drought, summer and winter, sun and shade, life and death. In the very last work he wrote before he moved to the subjective, relative time of the spiritual world, he begins the creation story *De cultu et amore Dei*:

> Walking once alone in a pleasant grove to dispel my disturbing thoughts, and seeing that the trees were shedding their foliage, and that the falling leaves were flying about—for autumn was then taking its turn in the revolution of the year, and dispersing the decorations

---

[19]Lindberg (2001), 18; Nordin, 39–41.

[20]*Regel-konsten*, 66f.

[21]*Polhem's Letters*, 122.

[22]Schefferus (1664); Micrander and Swedberg; Swedberg (1941), 57; Forelius and Gardmann; Ellenius, 130f, 275.

[23]*Selectæ sententiæ*, 7, 15; translation, 6, 12.

[24]*Festivus applausus*, 4. KVA; edition, 54f, commentary 92f; *Camena Borea*, edition, 144f; cf. Ovid, *Metamorphoseon*, 15.199–213.

of summer—from being sad I became serious, while I recollected the delights which that grove, from spring even to this season, had communicated, and so often diffused through my whole mind. But on seeing this change of scene I began to meditate on the vicissitudes of times; and it occurred to me whether all things relating to time do not also pass through similar vicissitudes, namely, whether this is not the case, not only with forests but also with our lives and ages; for it is evident that they, in like manner, commencing from a kind of spring and blossom, and passing through their summer, sink rapidly into their old age, an image of autumn. Nor is this the case only with the periods of men's individual lives, but also with the ages or eons of the world's existence, that is, with the general lives of societies, which from their infancy, integrity, and innocence, were formerly called the golden and silver ages, and which, it is now believed, are about to be succeeded by the last or iron ages, which will shortly moulder away into rust or the dust of clay.[25]

Swedenborg's thought is influenced by the park through which he is walking. The autumn and the changing seasons make him think of the cycle of time, and through metaphor he transfers his experience of the shifting world around him to the cosmic course and abstract thought. The circle metaphors are a central part of the cosmic understanding. There are circles hidden in chaos, which bind and link. Circles are a perpetual motion machine which go on unchanged. The vortical theory and Swedenborg's spirals are a continuation of the basic idea of the circle metaphor. For Linnaeus, everything is linked in a chain. 'All things revolve in a circle. It is as at a market: at first sight one sees only a huge mass of people wandering hither and thither, yet each one of them must have a home of his own, from which he is on the way and to which he shall return.'[26] Coming back to the same place means moving in a circle. Eternal return is the circle.

The Force of the Moon

'Why, BENGT, do you turn your attention and your thoughts to the constellations above, leaving the mortals? Why do you survey the stars? Are you perhaps led by love for the celestial fields?' With a poem full of baroque variations, Swedenborg pays tribute to his friend Bengt Bredberg when he is to defend his dissertation about the stars in November 1707.[27] 'Do you wish', Swedenborg asks insistently, 'to investigate the gleaming stars with your swift intellect and to measure sagaciously with mental power the whole of Heaven?' Many questions, but he himself provides one answer which may explain why people look at the stars: 'But I want you here to rejoice at the beautiful appearance of the heavens.'

The immense spaces above us, the beauty of the billions of sparkling stars over our heads, filled him with lifelong amazement. The sky was the backdrop for a

---

[25]*De cultu*, n. 1.

[26]Linnaeus and Wilcke.

[27]Swedenborg, 'Doctissimo et ornatissimo juveni D. Benedicto Bredberg, eruditum De asterismis laborem', *Ludus Heliconius*, edition, 46f, commentary, 37, 142f; Elvius and Bredberg.

remarkable spectacle enacted by comets and meteors, lunar and solar eclipses. Poets, theologians, and astronomers all stood gaping. Celestial movements and phenomena were read into life, transferred to other thoughts by metaphor. Think of your creator in youth, wrote Swedberg, 'before the sun and the moon cease to shine'.[28] Judgement awaits us after death, Swedenborg responded in his poem in *Vngdoms regel och ålderdoms spegel*, so you should not let your joys extend too far; 'but they should remain confined within a circle and a border.'[29]

When Swedenborg tells us in *Festivus applausus* of how the moon showed itself again after an eclipse, he expresses the matter in no less than nine different variations on the same thought. In his poetry Swedenborg sought to achieve the *variatio* of the baroque through parataxis (that is, coordination of several clauses), shedding light on the same phenomenon from different angles and creating a steady flow of thoughts, ideas, and images on the same theme.[30] He builds up a poem with historical parallels and analogies. The lunar eclipse is a metaphor for Charles XII's captivity in Bender under the crescent moon of the Ottoman Empire. With astrology as a poetic tool Swedenborg says: 'will you not then refuse to believe that Heaven does not, by a secret flowing, influence our actions, our lives and the vicissitudes of fortune? I have myself no doubts about this.'[31] Astrology in the strict sense cannot be found in Swedenborg's astronomical works, but there is a mode of thought based on the same metaphorical grounds, the micro-macrocosmic idea, but also the obvious fact that the heavenly bodies affect earthly life, like the rays of the sun and the ebb and flow that follows the phases of the moon. The whole of Swedenborg's mechanistic natural philosophy actually seeks to provide purely mechanical explanations such as pressure and motion in a continuum. Occult forces working at a distance had no explanatory value. What heaven has to say is merely that God is an omnipotent and omniscient Creator. Swedenborg had distanced himself from the kind of interpretations embraced by his father, who read more concrete messages into the sudden celestial phenomena. Swedberg wrote to the priest he had sent to the Christina congregation in Pennsylvania, Andreas Hesselius, to tell of 'the dreadful and terrible *celestial sign*' that had been seen on the night of 6 March 1716. Hesselius replied that this 'would be like *the gleaming of swords with their points facing down*, and many quickly moving and unusual flashes. May God let it not mean something evil!'[32] Swedenborg wrote a letter, no longer extant, about the northern lights in 1716, which was discussed in Bokwettsgillet in 1721.[33] The aurora borealis was often explained at this time, for example by Triewald, as being due to the reflection of moonlight on the ice crystals in the atmosphere.[34]

---

[28]Swedberg (1709b), v. 5, introduction and p. 144; Eccles. 12:2.

[29]*Ludus Heliconius*, edition, 52f.

[30]Helander, 30.

[31]*Festivus applausus*, edition, 78f, commentary 37, 131, 139.

[32]Swedberg (1941), 345; Swedberg (1732), edition, 85f; cf. *Bokwetts Gillets protokoll* 25 September 1724, 111.

[33]*Bokwetts Gillets protokoll* 1 December 1721, 61f; cf. *Opera* I, 210.

[34]Lindqvist (1984), 269; Lindroth (1967), I:1, 496.

There was occasion to study a total eclipse of the sun on 22 April 1715. Swedenborg wrote to Benzelius wondering whether Elvius was going to study the great eclipse.[35] Elvius had been waiting for the chance. He had observed a lunar eclipse in 1706, and in his almanac for the same year there is an article about how to observe solar eclipses without damaging the eyes. Both Elvius and Johan Vallerius went out that day and saw the disc of the sun darken as daylight was screened off. In *Dædalus Hyperboreus* the following year Swedenborg inserted a simple table with Johan Vallerius's observations from this unique occasion.[36]

Whirls and Voids

'My head must spin,' she said laughing, 'it is good to know what vortices are.'[37] Fontenelle's thoughts about the plurality of worlds during his evening walks in the garden, like those of Swedenborg himself, followed Descartes in the tracks of his vortical theory. The Cartesian conflict had died down. Aristotelian natural philosophy had been forced to concede, with its circles and its crystalline, concentric spheres representing an immutable, eternal celestial order, in contrast to the notion of rectilinear movement upwards or downwards that dominated under the moon, on the imperfect, mutable earth.[38] Swedenborg adopted Cartesian vortices, a whirling movement that applies both above and below the moon, to the biggest and the smallest alike. But the idea that astronomy was a kind of mathematics, that planetary astronomy was an extension of spatial geometry, an astronomy based on circles, homocentric spheres, and epicycles, as with Eudoxus, the Aristotelians, or in Claudius Ptolemy's *Almagest* (*c.* 140 AD), or later in Kepler's ellipses, was adopted as a basis for attempts to capture the chaotic movements of the heavens. In the book by Zahn that Swedenborg owned, he could read on a plate illustrating a Tychonian geocentric solar system of concentric circles: *Memento rerum omnium circulum esse*, 'Remember that all things are a circle.'[39] *Circulus Æterni Motus*, 'The circle is eternal motion', can be read on the title page of Becher's *Physica subterranea*.[40] The idea of the perfect circle also made an impression in heraldry and allegory. In a key work about allegory, Giovanni Campani's (pseudonym Cesare Ripa) *Iconologia* (1593), the circle and the zodiac are the very sign of the Perfect, signifying reason

---

[35] *Opera* I, 229.

[36] J. Vallerius, 'Solens stora förmörkelse som åhr 1715 den 22 april sig tildrog, blef på följande sätt observerad i Upsala af math. profess. Johan Vallerius', *Dædalus Hyperboreus* II, 40; *Opera* I, 213; Elvius (1722), 319f; LiSB, N 14a, no. 69–70, fol. 162–166; J. Vallerius, UUB, fol. 4r; Nordenmark (1959), 146f.

[37] Fontenelle, translation, 112.

[38] Aristotle, *Physikēs*, 8.8.261b27–262a12; Aristotle, *Peri ouranou*, 1.2.268b15–26.

[39] Zahn, introduction.

[40] Becher (1703).

and regularity.[41] With a pair of compasses, an allegorical female figure inscribes an eternal, perfect circle. The movements of the firmament are a divine geometry. Swedenborg does not let go of the idea of a perfect order established by an infinite geometrician. The world is a reflection of perfect figures. The doctrine of the perfection of the circle lived on as an important basis of his thought, immortalized as a spiral.

Geometry established for science which figures one should look for in nature. It had to be the most perfect geometrical figures, the circle and the sphere, as also for Copernicus and Galileo.[42] It was essential to eliminate the irregularities and reduce phenomena to regular circular movements, to aspire to symmetry, simplicity, and harmony in planetary movements. There is an underlying structure in reality, unique, true, and necessary.[43] But during the first half of the seventeenth century the supremacy of the circular movement declined, since it could no longer be taken for granted; it had to be explained. If the circle cannot describe the paths of the planets, then the less perfect ellipse can do so for Kepler. Natural motion no longer followed the form of the circle, as according to Descartes' second law of nature became rectilinear. A body travelling in a circle tends to move out from the centre in the tangent of the circle, as when a stone is slung away by its rotation.[44] The idea of natural motion has to do with everyday human experience. Teleological explanations were read into people's actions and movements, and transferred with the aid of metaphor to hylozoistic explanations, that the objects and processes of matter have human, spiritual properties.

For Descartes all movements in the world were circular in a universal perspective, but not necessarily in an exact circle. When a body leaves a place, another comes in its place, a third in the place of the second, and so on until the last has left its place for the first.[45] No vacuum is needed for movement, nor can it exist. Through the implanted movement in the ether-filled universe, vortices of matter arise causing matter to be crushed to bits. In the centre of the vortices, stars are formed from the first element, fire, and around them the planets are transported in the whirlpool of the ether, like boats on a river, which could explain why they all moved in the same direction and in the same plane. The success of Descartes' vortical theory lay not in its mathematical exactitude but rather in its instant visual strength. The whirls of the universe could be clearly visualized and imagined in mental images, as macrocosmic cyclones and eddies.

The planets need not follow exactly circular paths in Swedenborg's vortical theory, being more or less elliptical, but it is striking that he builds up his theory heedless of the fact that he does not relate it to or harmonize it with Kepler's laws

---

[41]Ripa, 391f, cf. 41, 109, 500.

[42]Galilei (1632), translation, 36f; Hallyn, 105, 121; Blay, 38, 40; McAllister, 163–181.

[43]Galilei (1613), first letter.

[44]Descartes (1664a), ch. IV; *Oeuvres* XI, 16ff; H. Vallerius and Moell, 27–30; Hoffwenius (1698), III, § 10.

[45]Descartes (1644), II, § 33; *Oeuvres* VIII:1, 58f.

and Newton's law of gravitation. Swedenborg owned and had at least read parts of Newton's *Philosophiæ naturalis principia mathematica* (1687). In October 1710 he wrote from London to Benzelius telling him that he read Newton every day, and that he hoped to meet him and hear him.[46] His reading also included Newton's variant of calculus and in all probability his optics too. Yet he was never convinced by the theory of gravitation. If Swedenborg was a 'good' mathematician one would think that he ought to have been 'forced' with mathematical necessity to accept the theory of gravitation, especially since he highly admired the objectivity and certainty of mathematics. A good mathematician, one might think, ought to capitalize on his knowledge and use it in his astrodynamics. In Swedenborg's cosmology, however, mathematical calculations were only of marginal, illustrative significance. They were irrelevant to him, just tiresome groundwork in the counting house which did not need to encumber the text. The aim was rather to create a uniform synthesis, a coherent explanation of the whole universe and creation.

It need not be the case that it was mathematics itself that was an obstacle to him. Swedenborg could, with a little will power, make his way through advanced books in mathematics. The problem with Newton was instead on the cognitive and metaphysical plane. It was perfectly clear to Polhem that Newton's theory tended towards occult explanations and distanced itself from the continual clashes of pure mechanics.[47] Action at a distance was absurd. Many prominent mathematicians—such as Leibniz, Johann Bernoulli, and Fontenelle—preferred Cartesian vortices as late as 1752.[48] Swedenborg thus cannot be chided for being totally behind the times. Newton's natural philosophy was not popularized on the continent until the publication of Voltaire's *Eléments de la philosophie de Newton* (1738), incidentally a book that was on Swedenborg's shelf. And indeed there were unschooled laymen who knew no more than the four rules of arithmetic but wholeheartedly adopted Newton's teachings. The fault did not lie in the mathematics.

One reason why the theory of gravitation was so absurd for the Cartesians can be found in Swedenborg's correspondence with the members of the Collegium Curiosorum. Elvius, who owned a copy of Newton's *Principia* as early as the 1690s and mentioned the theory of gravitation in a dissertation from 1703, was sceptical.[49] He asked Swedenborg what learned people thought about Newton's *Principia* in London 'inasmuch as they seem to be pure *abstraction* and not *physicae* [physics], namely as to how one *corpus planet.* [body of the planets] shall *gravitate* to another, etc. which seems to be unreasonable.'[50] Swedenborg gave a straight answer to the question: no Englishman doubts it on patriotic grounds, since they are blind when it comes to their own countrymen. Harald Vallerius mentioned Newton and his *Principia* for the first time in Sweden in a dissertation from 1693, but he

---

[46] *Opera* I, 207, 210; *Letters* I, 21.

[47] *Polhem's Letters*, 101; *Polhems skrifter* III, 343, 424ff.

[48] Hildebrandsson, 14; Gaukroger, 159f; cf. Kragh, 67–77; Shank.

[49] Elvius and Frondin, 19; Nordenmark (1959), 138.

[50] Elvius to Swedenborg, Uppsala, 28 July 1711. *Opera* I, 212, cf. 216; *Letters* I, 26.

did not care to abandon his basic Cartesian outlook, and his son Johan lectured on Newton's physics.[51] Yet it is Polhem who expresses himself most drastically. Newton may have been a great mathematician but he was somewhat childish in the way he complicated everything for the sake of his honour. Wallis has a touch of the same disease: 'But for those who diligently wish to sharpen their brains they are good whetstones both of them, that is, sharpening first with Wallis and then with Newton.'[52] Newton's theories 'should be combed on the fine hackle'![53] Polhem said that he could explain everything in a simpler way. The simple explanation is the pressure of the ether. There is no need for occult antipathies and sympathies, just pure mechanics.

On the Eternal Spring in the Age of Winter Cold

Everything is perishable, nothing lasts forever. 'Since this is the case with earthly things, one can well conclude as to things superterrestrial, and that even there everything progresses toward a goal and a [state of] rest,' wrote Swedenborg in 1717 in *En ny theorie om jordens afstannande* ('A new theory about the retardation of the earth').[54] In this manuscript he puts forward a number of reasons for his opinion that the earth has been moving increasingly slowly since the beginning. Ether effects a resistance in the earth's orbit, and air in the motion around the axis, which makes the years and the days longer and longer. He bases this on metaphorical thinking: there are similarities between the mechanics of the universe and the tools of man. The earth rolls and spins on her axis or her central pole, as when one takes a ball and throws it into the water, when a point on a wheel rotates around its axis or the spindle on a spinning wheel. Another reason why the movement around the axis must have been faster in the past can be detected in the roundness of the earth. The irregularities have been rubbed off and subsided in spherical form. 'If one takes a lump of clay and rolls it around its axis in water, it is formed into an oval.'[55] For a Cartesian, then, in contrast to a Newtonian, the earth is oval owing to the resistance of the medium. For Polhem the earth ought to become cylindrical since it rotates around its axis as in a lathe. The motion around the poles cannot be the reason for its round figure. The oval figure is due instead to the action of the sun and the central fire, he explains.[56]

There is also written evidence for the theory that the world is slowing down, as when the Bible and classical poets talk of an eternal spring in primeval times. When

---

[51]H. Vallerius and Moell, 7; Spole and Alinus, 43, 46; Sandblad (1944–1945), 114.

[52]Polhem to Elvius, Stjärnsund, 31 May 1712. *Polhem's Letters*, 88.

[53]Polhem to Benzelius, 19 April 1712. *Polhem's Letters*, 84.

[54]*Opera* III, 271; LiSB, N 14a, no. 34; *Photolith.* I, 28–65; translation, 43.

[55]*Opera* III, 272f; translation, 45.

[56]*Polhem's Letters*, 26f, cf. 11.

the year and the days were shorter, the antediluvians, those who lived before the Deluge, could reach an age of 700–800 years. So Methusalah's 969 years actually correspond to only $121^{1/8}$ of our years. Perhaps with the support of Peringskiöld's genealogical tables in Charles XII's Bible, Swedenborg calculated that in the time from Adam to the Deluge, the age of the antediluvians had decreased by 100 years in 1,656 years. 'Let, therefore, a triangle be constructed, the area of which [describes] the course which the earth would cover until its end'.[57] If the earth had not been flooded, it would now have had an age of 18,212 years. In this middle of this proportional calculation, time seems to have run out for Swedenborg. He set aside the manuscript that he had commenced, either out of fatigue or to turn to some other task that was calling for his attention. Time is short in this age of iron and winter cold, when the sea is shrinking, temperatures are falling, and the human lifespan is decreasing.

Yet he never abandoned the idea. He resumed his attempt to prove the conclusion that the motion of the earth has slowed down. Around 1718–1719 he wrote two versions about the motion and position of the earth. In the first version, in manuscript, *En ny mening om jordens och planeternas gång och stånd eller några bewis at jorden löper alt sachtare och sachtare: at winter och sommar, dagar och dygn til tiden blifwa lengre och lengre in til werldsens sista tid* (1718), he puts forward 'A new opinion about the motion and position of the earth and the planets or some proof that the earth is moving more and more slowly: that winter and summer, days and nights are becoming longer and longer until the end of the world'. The increasing distance of the earth from the sun has made it poorer and less fertile. But this has not been noticed, and no wonder, for 'we have not been aware that our earth rotates like a ball each day: indeed our vision often deceives us in small matters; one does not notice that the ship is moving forward or backward.' If one sails around the earth one does not know that one has travelled in a circle, if one does not return to the same place again or finds that one has lost a day. 'Behold, thus are we deceived by the light of our own senses, and are blind with them, and believe an untruth to be true.'[58] We therefore cannot trust our senses. It requires reason and geometry for Swedenborg to make us truly able to see.

Swedenborg published his theory in 1719 in *Om jordenes och planeternas gång och stånd* ('On motion and position of the earth and the planets'). A theory about the origin of the earth from chaos ought to follow geometry and be compared with the opinions of Descartes and Newton. But that would be beyond the scope of the present work, he says and hastens on to other things. Geometry is wholly fundamental for him, but he does not have the energy to deal with it in this connection. Instead he puts forward a number of reasons for the decreasing speed of the earth based on metaphorical thought. He is part of a Rudbeckian tradition in

---

[57] *Opera* III, 282; translation, 56.

[58] 'En ny mening om jordens och planeternas gång och stånd eller några bewis at jorden löper alt sachtare och sachtare: at winter och summer, dagar och dygn til tiden blifwa lengre och lengre in til werldsens sista tid', *Opera* III, 285, cf. 303.

which ancient myths have something to say. Rudbeck can explain, he writes, why Canaan and the golden land can be compared 'with our mountains and iron-cliffs, its bright and delightful summer with our chilly and cloudy winter'.[59] God's word, Ovid, and other pagan poets such as Homer, 'True Grandfather of Poets', and Plato's Atlantis tell us that the earth once had 'a glorious air, a constant spring, a golden age, an Eden and Paradise, an earth inhabited by *Atlas*'.[60] There was a heaven here on earth. Life flourished in the eternal spring.

Methusalah's abnormal age is due to the fact that the earth used to move faster. He lived as long we do now, but he experienced more summers and winters. With the decreasing speed of the earth's rotation, the generations after us will regard 20 or 30 years as a high age. Swedenborg seems to assume the existence of an absolute time, that the biological clock ticks independently of the movement of the heavenly bodies and the length of days, solar time and the solar year, as if for God there was an absolute universal time. He is thinking in metaphors as a variant of the old idea of *mundus senescens*, the ageing earth. Our globe would age and die like a human being. In the past there was a golden age, according to the myth related by Ovid and Hesiod, but now we have descended into a poor age of copper, iron, and clay.[61] Invoking the support of Rudbeck's *Atlantica*, as well as Pufendorf, he seeks to show that Sweden was once a marvellously abundant land, like a Canaan, Savoy, or Italy, like a Florence or Mantua.[62] Swedenborg's theory is also able to explain why the great island of America is inhabited. During the eternal spring, when the air was still and mild, as in the month of April in today's Sweden, the first inhabitants walked across Lapland and Greenland to America. Then winter spread itself over the world. Physical experiments could demonstrate that the thesis was correct, as when one rotates a thermometer slowly over a fire, the alcohol rises, but with greater speed to a smaller peak.[63] Another experiment can be performed with the vibrations of the pendulum. Just as an ordinary cart wheel makes as many revolutions during a mile regardless of whether one is travelling fast or slow, there are always 365 days in a terrestrial year irrespective of the speed of the earth's rotation. In the printed version there were no illustrations to explain the experiments, but in a copy once owned by Moræus and then by his son-in-law Linnaeus, Swedenborg's own drawings are appended. In his childhood Linnaeus had been amazed that the early ancestors lived to such an age, but Swedenborg's theory explained everything to him.[64]

What the Bible says about the signs that will precede the last days is compatible with the retardation of the earth, with reference to the description of the destruction of Jerusalem in Luke 21:25: 'And there shall be signs in the sun, and in the moon,

[59]*Opera* III, 301.

[60]*Opera* III, 306, cf. 287; Homer, 7.112–132; Virgil, *Eclogues*, 4.

[61]Ovid, *Metamorphoseon*, 1.89–150; Hesiod, translation, 117–120.

[62]*Opera* III, 312, cf. 293.

[63]Cf. *De cultu*, n. 17, note i; *Principia*, 447; translation II, 359f.

[64]Linnaeus's copy of Swedenborg's *Om jordenes och planeternas gång och stånd*, 24/25; *Green Books* 171.11; Photostat copy UUB, D 27; Linnaeus (1958), 17.

and in the stars; and upon the earth distress of nations, with perplexity; the sea and the waves roaring.' The earth will be destroyed, Swedenborg warns, the human race will dry up, barrenness and famine will spread. There will be signs visible in sun and moon, in air and sea, '*dreadful eclipses shall come*, causing the peoples anxiety and fright', stars shall fall down, 'The sea and the wave shall roar' with storms and rain.[65] This could be interpreted as showing that Swedenborg at this time read the Bible literally, not embracing a doctrine of correspondences with levels of meaning, and that he believed the creation, the age of the antediluvians, Noah's flood, and doomsday to be real events. But he is thinking metaphorically all the time. He transfers the particular history in the bible to a universal history with the aid of metaphor. To understand the hidden mechanism of the solar system he uses physical and biblical examples.

Of special significance for Swedenborg's cosmogony and geogony was Thomas Burnet's theory of the mundane egg in *Telluris theoria sacra* (1681), which was in his library.[66] In a notebook started in the 1710s he made annotations about Burnet's cosmogonic world egg alongside data from David Gregory's astronomy about the distance of the planets from the sun, and something about Kepler's theory and the hypotheses of Epicurus, Descartes, and Leibniz.[67] The surface of the earth, according to Burnet, was perfectly smooth, regular, and uniform before the deluge, without mountains or oceans (Fig. 3). The earth arose from chaos, a fluid mass in which everything was mingled in confusion. Heavier matter sank towards the middle, while lighter matter swam up. The oilier and lighter substances were separated from the heavier ones, floating on the surface, like cream and milk, oil and water.[68] Stiernhielm envisaged the beginning as from a black sphere, a dark chaos. 'What is meant here by the dark sphere?' asked the innocent Simplicius in one of Stiernhielm's dialogues from the 1660s. Astraeus replied: 'It represents the darkness in the beginning', the darkness of which Moses spoke in Genesis.[69] Chaos and the idea of the cosmos are based on the conceptual pair of cognitive dialectics, that cosmos presupposes chaos, that order presupposes disorder. The order that we have must come from disorder.

Swedenborg's metaphorical thinking on the theory of the slowing earth is linked not only to Burnet and Polhem but also to Newton's *Principia*. What is most obvious, however, is that he proceeds from the Cartesian vortices. Since all particles in Descartes' vortex theory seek to move in straight lines, this means that the stronger or heavier particles will describe the largest circles. There is thus a centrifugal force, as mathematically studied by Huygens. According to Descartes, an orbit arises through a balance between the planet's centrifugal tendency and an

---

[65]*Opera* III, 319, cf. 296.

[66]*Opera* III, 297, 319f; Jonsson (1999), 27.

[67]KVA, cod. 53, 165–186; Burnet (1694), II, ch. X, 137–140; Burnet (1697), II, ch. VIII, 184f; Swedenborg owned Gregory, 2nd ed., 1726.

[68]Burnet (1694), I, ch. V, 21; Burnet (1697), I, ch. V, 38.

[69]Stiernhielm, 'Dialogus, interlocutoribus Simplicio, and Astræo', *Samlade skrifter* II:2:1, 51.

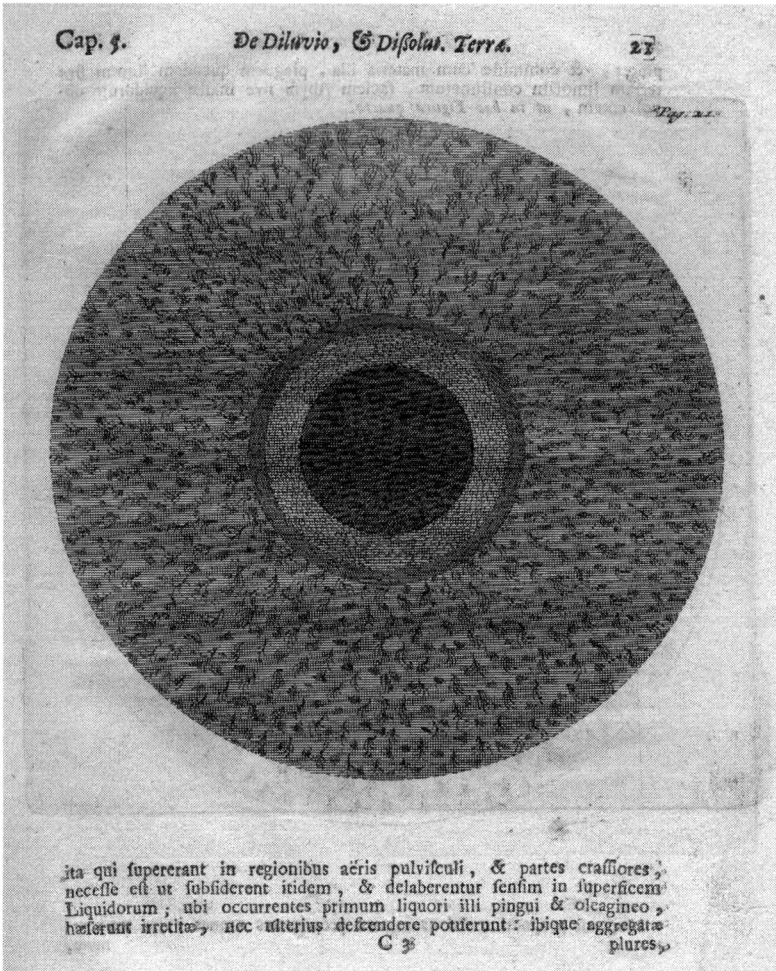

ita qui fupererant in regionibus aëris pulvifculi , & partes craſſiores ,
neceſſe eſt ut fubfiderent itidem , & delaberentur fenfim in fuperficem
Liquidorum ; ubi occurrentes primum liquori illi pingui & olcagineo ,
hæferunt irretitæ, nec ulterius defcendere potuerunt : ibique aggregatæ
C 3   plures,

Fig. 3 Things are separated in the beginning as cream is separated from milk, or oil from water (Burnet, *Telluris theoria sacra* (1694))

opposite pressure from matter in the vortex.[70] Centrifugal and centripetal forces can thus explain planetary orbits. In other words, the spiral path of the earth is due to its centrifugal force being greater than the opposite force of the ether.

The idea of the retarding motion of the earth, with a spiral path proceeding from the sun, was something that Swedenborg had read and transcribed several times from Polhem. One of Polhem's favourite examples, which he varied countless times in order to illustrate the centripetal and centrifugal forces of the planets

[70]Cf. Schuster.

in the solar vortex, is an experiment with a sphere that is rotated under water.[71] Polhem may have derived the inspiration for this experiment from his teacher Rudbeck the Elder, who had just devised a model to describe the movement of the heavenly bodies by revolving a globe in water.[72] The movement of the planets is understood in a hydrological metaphor; knowledge of the movements of water is transferred to the matter-filled vortices of the universe. In Swedenborg's transcript of Polhem's *De causis rerum*, this particular example of a sphere rotating under water is mentioned.[73] The fact that the earth and the planets hover freely in the air and move in circles is not the least of God's great and wonderful works. It can be concluded from the experiment that the motion of the earth and the planets around their axes is caused by the resistance of the ether, which balances them in their movement around the sun, in a way similar to the effect of the water on the submerged sphere. Gravity can thus be explained by the pressure of the ether, 'in the same way as the woodwork in the mill pond.'[74] If the lead ball is heavier than water, it tries to escape from the centre, whereas a wooden ball which is lighter than water pulls in towards the centre. If the sphere were as light as the water it would stay in its circular orbit. Just as one can make a hollow lead ball filled with air float on water, it may be the same with the subterranean fire which is lighter than the ether and brings the earth into equilibrium with its surroundings. Polhem envisages the whole space of the universe as being filled with an infinitely fine matter in constant whirls in both the smallest and the largest.[75] In Polhem's dialogue between chemistry and mechanics, of which there is also a transcript by Swedenborg, we again find this analogy between the sphere in the vortex of the water and the earth in the vortex of the ether. Our globe is held in equilibrium in the ether by the lighter fire in its interior. Miss Mechanica compares this with the way a hollow lead ball under water can be kept floating. If the earth were heavier than this equilibrium with the ether, it would travel out in a spiral, further and further from the sun, and if it were lighter it would be drawn in towards the sun and destroyed.[76] Yet Swedenborg later questioned the idea of the central fire in *Miscellanea observata*.[77]

In November 1719 Swedenborg heard rumours in the capital that the earth had come 25,000 Swedish miles (about 166,000 English miles) closer to the sun.[78] He was amazed that such a jump could take place in just a year or two, and he seriously doubted the statement. He told Benzelius he thought it was unreasonable. Moreover,

---

[71] *Dædalus Hyperboreus* I, 7; *Polhem's Letters*, 8f, 10f, 13, 24, 90, 118f.

[72] Spole and Humerus; Eriksson (2002), 238.

[73] *Opera* III, 231; KB, X 517:1, fol. 180–181; also in *Ex Polhemio*. KVA, cod. 86, 277.

[74] Polhem to Benzelius, 30 September 1710. *Polhems brev*, 10f.

[75] Polhem, 'Tankar om elementernass uphof och warelse efter mechaniska principier 1 Om wädret'. KB, X 517:1, fol. 227v; Polhem, 'Om wäderparticlarnas mechaniske structur naturaliter'. KB, X 517:1, fol. 240.

[76] *Opera* III, 250; *Polhems skrifter* III, 164; cf. *Polhem's Letters*, 23.

[77] *Miscellanea*, 161f; translation, 101.

[78] Swedenborg to Benzelius, Stockholm, 3 November 1719. *Opera* I, 290f; *Letters* I, 214f.

the claim was in opposition to his own theories. This gave him occasion to comment on the motion and position of the earth. The stronger the movement and rotation in the solar vortex, the further the planets are thrown out from the centre, but a weaker movement causes them to be drawn inwards. 'And it is well known in what proportion the vis centrifuga, in travelling outward and inward, increases according to the speed. Of this, Isaac Newton treats in his *Principia*.' Swedenborg read Newton with Cartesian eyes, reading vortices into the theory of gravitation. He had probably read a section in Newton's *Principia* about the circular motion of bodies in resistant media, where Newton demonstrated that if the force of gravitation varies in inverse proportion to the cube of the distance, instead of to the square, then the planets would not follow an elliptical orbit, but would be hurled away out from the sun in a logarithmic spiral course.[79] There is no risk yet that the earth will be swallowed up by the sun, Swedenborg assures us: 'If the sun grows larger and larger before our eyes, then first would be the time to entertain fear because of it, and to commend oneself to God's hands.'[80] Benzelius also wondered about another thing he had read, namely, a statement that the abode of the damned is in the sun.[81] 'I think just the opposite,' Swedenborg replied. The sun is, if anything, the abode of the blessed. For it is at the centre of our planetary system, in which all vortical movements have their origin. It has the most splendid light and magnificence, whereas darkness and terrors are furthest away from it. The sun consists of the most subtle particles,

> almost devoid of composition, and so put off the denomination of matter, and also of form, weight, and many other properties possessed by compound particles. And it would also seem likely that in this finest, must be the finest essences. A God, an angel, a thing which, moreover, has nothing materiale in its being.

Swedenborg is led towards the idea of God and the sun being brought together in a metaphor in his assurance: 'God has his seat in the sun.'[82]

Swedenborg could thus explain gravitation in mechanical terms, as did Polhem and other Cartesians, as being dependent on the pressure of the atmosphere or the 'ether'. In *Miscellanea observata* he expresses this equilibrium in terms of heavier bodies falling downwards and lighter ones rising. This Cartesian understanding of gravity becomes odd when he concludes his reasoning by referring to an authority—Newton, the 'star of the learned world'.[83] Swedenborg did not read Newton's Newton, but rather Newton with a Cartesian pre-understanding. One reviewer points out that Swedenborg must be a very special admirer of Newton, since it cannot be unknown to him how little value Newton attached to concocted physical theories based solely on the discoverer's own fantasies.[84] It is not permissible for a natural philosopher merely to create things from his own brain.

---

[79] Newton (1687), II, sect. IV, prop. XV, theorem XII.

[80] Swedenborg to Benzelius, Stockholm, 26 November 1719. *Opera* I, 293; *Letters* I, 220.

[81] Benzelius had been reading *Neue Zeitungen* 2 August 1719; *Letters* I, 218f.

[82] *Opera* I, 294; *Letters* I, 221f.

[83] *Miscellanea*, 160; translation, 100.

[84] *Historie der Gelehrsamkeit* (1722), 315–327; *The New Philosophy* (2003), 565.

From Centre to Circumference and Back

The vortex of the solar system becomes a source of metaphors in itself. *The particle world is a planetary system.* Swedenborg transfers the structure of the macrocosmic vortical theory to the corpuscles of the microcosm. And in the middle is man in the mesocosm. Knowledge of the sun, the stars, and the planets that we see can be transferred to an understanding of the subtle matter beyond the bounds of the visible. Closely akin to Polhem's theory of spheres, he tried in *Miscellanea observata* to construct a particle system based on the simplest figure, the sphere, and the angular particles lying between spheres. But then he suggested another possibility. 'If then we choose to devise hypothetically particles of other forms, sinuous, serpentine, spiral, or fibrous, I confess that the notion will be deservedly ingenious, the inventors worthy of great praise, and their inventions cordially to be welcomed, if they can be referred to mechanical principles.'[85] He himself would succeed in this brilliant achievement. A few years later, in the second draft of the principles of natural things, known as *The Minor Principia*, the spiral makes itself felt as the figure of matter, the universal and perfect form of motion.

The basic idea is that the natural point moves in a circular spiral that creates a spherical surface. Polhem had previously speculated about this in a manuscript that Swedenborg contemplated publishing in *Dædalus Hyperboreus*. In his *Mechanica naturalis* (after 1716) Polhem mentions a spiral movement which first revolves around a cylinder, but which the ambient pressure finally turns into a spherical form, 'round bubbles like the so-called bullæ vanitatis that boys like to play with, or else bubbles on rough water.'[86] Malebranche had entertained similar thoughts when he replaced Descartes' globules with small elastic vortices which would explain heat and light. The elasticity of the small vortices was produced by the centrifugal force through their rapid internal circular motion.[87] One of the reasons for Swedenborg's choice of the spiral as the figure of matter may have to do with the fact that this form facilitated the explanation of the compressibility of substances. Air can be compressed, and perhaps its particles were like spiral springs that can be squeezed and expanded. The spiral shape possesses a mechanical force. Polhem the mechanic wrote about springs, about the choice and treatment of spring steel for locks and clockwork.[88] Swedenborg utilized the mechanical force of the spiral spring in his manuscript about the flying machine. The machine flying in the air would have a spiral spring, attached to a pair of 'oars' or wings, the force of which would make the aeroplane rise and propel it when there is too little muscle power.[89] (See the picture on page 211)

---

[85] *Miscellanea*, 156f; translation, 98.

[86] *Polhems skrifter* III, 254, cf. 241; cf. *Opera* I, 281; *Letters* I, 152; *Polhem's Letters*, 128, 276f.

[87] Taton and Wilson, 218.

[88] Polhem, 'Observ: om fjädrar', *Polhems skrifter* I, 183–185.

[89] Holmer, 96–98; Löfkvist, 108.

The spiral motion that forms a spherical surface, as Swedenborg writes in *The Minor Principia*, can be more easily understood through the imagination than with a figure. Geometry recognizes only what is material and is restricted to the finite. Since geometry does not treat such infinite circular movements, it is difficult to embrace it in words. Nothing other than an infinite motion or infinite qualities could have given rise to the natural point, an infinite motion in an infinitely small space. An infinite first movement from the centre to the circumference and back to the centre in a perpetual spiral means that the first natural point is spherical. This spiral sphere has two poles, an equator and an ecliptic, like a small planet. The first movement may be called the most perfect circular movement, the circle of circles, for in a point an infinite number of circles are driven in all directions. It is the motion of motions. Swedenborg assumes the existence of two different kinds of points (Fig. 4). The first point has a spiral motion from centre to circumference and back to the same centre. This is the first particle. The other kind is a spiral motion that does not return to the same centre but to a different centre nearby, and so on. This is the second particle. It is futile to calculate this spiral. And he says: 'I have considered it sufficient to deal with a particular aspect of the subject—of primary origin—rather than to enter that field in which every detail and every law would have to be explained. Time would not suffice for the purpose, nor is the investigation necessary.'[90]

The third particle consists of points or particles of the first kind on the surface and variable points or particles of the second kind inside (Fig. 5). Using Wolffian terms, in the printed *Principia* he would later call these passive and active, but those are concepts that he does not yet use. Here in *The Minor Principia*, he just says: 'I might more fully prove this from geometry, but since it is quite obvious, there is no need to waste time in useless demonstrations.'[91] Particles of the third kind have the same movement as the variable points, which expand and contract like the heart and lungs, enclose themselves and undergo constant changes. If a small body moves in a circular flow at a long distance from the centre, it can be perceived as something at rest in its own movement, like a sailor on a ship, a child in the womb, inhabitants of a globe, even if the body really moves in a circle. The spiral motion on the surface leaves the area at the poles free, where no points move. The polar cones are fountains and sources from which matter flows, like the heart and the ventricles through which blood passes in and out, and the lungs where the blood can pass in and out through its own small vessels.[92] Heavier matter makes its way towards the centre in a globe, while light matter can go out and in through the cones. This idea of the polar cones of particles is transferred by Swedenborg from Descartes' vortical theory, where subtle matter pours into the vortex through the poles, moves towards the centre and then out along the equator, subsequently to acquire new vortical poles. Swedenborg's theory is thus a Cartesian vortex theory applied to the little world.

---

[90] *Opera* II, 13, 36–40, 163; *The Minor Principia*, 13, 40–42, 45, 198.

[91] *Opera* II, 43f; *The Minor Principia*, 49f.

[92] *Opera* II, 46, 55, 64, cf. 123, 161; *The Minor Principia*, 53, 63, 75, cf. 147f, 196.

Fig. 4 The two points. In the
first point the spiral moves
from the centre, *a*, and on to
c, d, e, f, g, h, i, k, and back to
a. In the other point, however,
the spiral moves from *a* to *b*,
c, d, e, f, g, h, i, k, l, and on to
the centre, *b*, and not *a*
(Swedenborg's manuscript
Principia rerum naturalium
(*c.* 1729–1731))

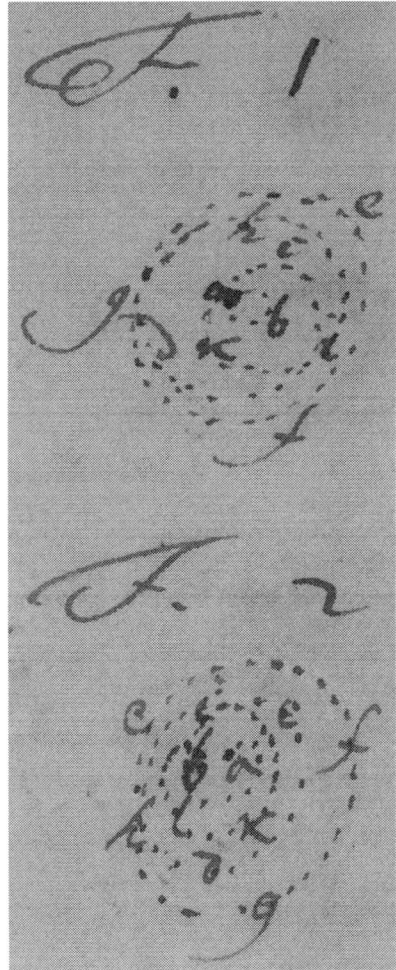

The universe is finite. It must have boundaries, as it cannot extend without limits.[93] All movements proceed from the sun, which consists of extremely thin matter. The solar ocean is like a wide sea, where solar matter flows through the circumference of the sun out into the vortex, but flows back in through the poles, as when air is drawn into a pair of lungs. In a similar way the third particle has a vortex resembling the sun's vortex. Swedenborg thinks more like an astronomer than a mining assessor when he constructs his theory of matter. The particles cannot be combined at the poles, since the vortical matter would then be able to flow in from one pole to the other. This is also what Descartes explains in *Principia philosophiæ*, that the poles of the vortices never touch each other. If the vortices moved in the

---

[93]*Opera* II, 70; *The Minor Principia*, 82.

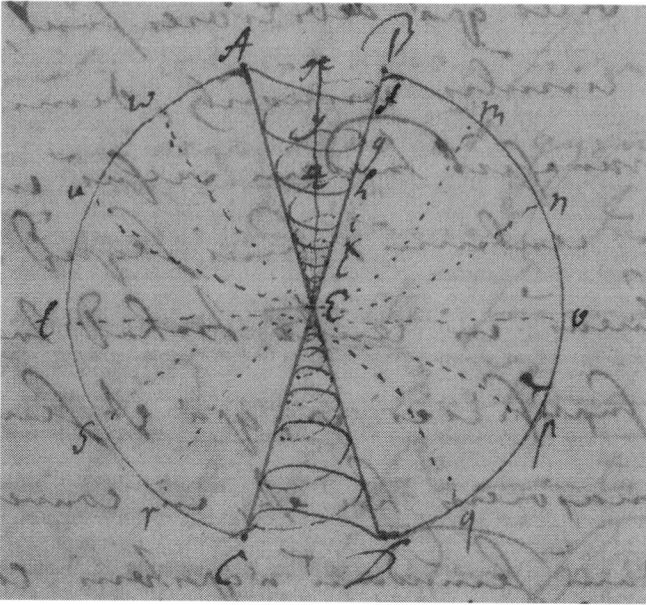

Fig. 5 The third particle. Points move in spirals on the surface *B, m, n, o, p, q, D,* and *C, r, s, t, u, w, A.* In the polar cones *AB* and *CD* the points move in increasingly narrow spirals towards the centre to an exact circle (Swedenborg's manuscript *Principia rerum naturalium* (*c.* 1729–1731))

same direction they could be combined, and if the movements are in the opposite direction they would counter each other. Instead, Descartes thinks, the pole of one vortex touches the equator of another vortex, which allows a constant exchange of fire between the vortices. The first matter is first forced to the centre of the vortex, the star, and then out through the poles of the vortex and afterwards sucked in at the equator by another vortex.[94] Swedenborg therefore assumes that the particles of the third kind can only be combined at the ecliptics. All the poles have the same direction towards such a distant point that the polar axes can be considered to be parallel. Implicit in this is that there is an end point, but Swedenborg tends towards a boundless universe, where it is not possible to determine any fixed point that can be regarded as the end point; instead the extension continues seemingly to infinity.

Through pressure the particle is reduced, the surface shrinks to a central globe and forms a particle of the fourth kind. At even greater pressure close to the sun, the fifth particle arises. As an example he takes a paddle spinning around in water. At first it moves sluggishly, but once some speed has been built up it goes easily. So it was at the start of time, before matter began to move in a vortex. The sun was then surrounded by a crust which darkened it, which caused the true chaos that was exploded, thus creating things in the solar vortex. The sixth particle consists of the

---

[94]Descartes (1644), III, § 65–71; *Oeuvres* VIII:1, 116–125; Hoffwenius (1698), VIII:9–12, 33–35; *Opera* II, 79, 83; *The Minor Principia*, 93, 98.

Fig. 6 The sixth particle
(Swedenborg's manuscript
Principia rerum naturalium
(*c.* 1729–1731))

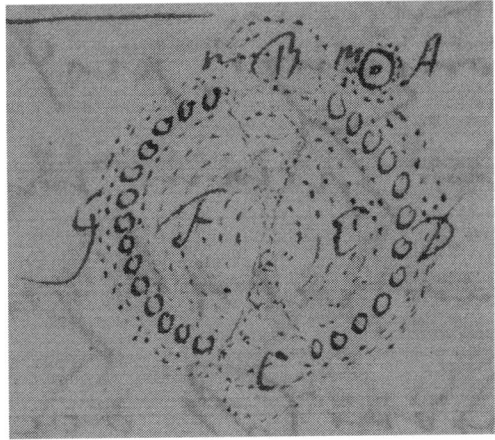

fifth particle on the surface and the second one on the inside, and it resembles the third and has the same movements, but is larger. 'These things can be geometrically proved, as all the rest can,' says Swedenborg, 'but if we were to give ourselves up to proving each detail, this work would be increased enormously, and perhaps matters otherwise clear would become obscure.' The movement of the sixth particle also takes place in spirals, with two poles through which the particles of the second kind flow in and out (Fig. 6). The solar vortex consists of particles of the third and fourth kind, while the earth's vortex consists especially of particles of the sixth kind with particles of the fourth kind between them.[95]

There is a rapid vibration movement on the surface of the third, fourth, and sixth particles. The vibration theory is of great significance, for it explains unknown properties in light and other phenomena. By vibration he means a movement on the surface of the particle, without the particle as a whole moving. The centre of the particle is at rest as in a blown-up bubble (Fig. 7). An undulation is a greater vibration where the centre too is in movement. A special undulation takes place in the solar vortex and in the planetary vortices because of the sun's motion. The sun at the centre of the vortex functions like the heart or the lungs in a person. Undulations stand for phenomena such as vision, light, colours, and heat, and have the same properties as a rectilinear pressure. This pressure or undulation can be transported over long distances but is nevertheless forced to decline gradually. The particles lie like an unbroken chain in the whole universe. Light arises in the sun and extends as an undulatory pressure in straight lines to our vortex with the aid of particles of the third and fourth kind, which press on particles of the sixth kind, that is, the ether particles.[96]

Compression of the sixth particle produces the seventh particle, and further compression forms the eighth particle. A combination of the fourth particles on

[95] *Opera* II, 94–97, 100f; *The Minor Principia*, 112–116, 120f.
[96] *Opera* II, 126, 1230, 136, 145; *The Minor Principia*, 151, 156, 163, 174f.

Fig. 7 Vibrations on the
surface. *BCA* is the surface of
a particle of the third, fourth,
or sixth kind. If a small point
is struck so that it starts to
vibrate, the vibration will
immediately spread like a
wave around the particle, over
the whole surface, while the
centre is at rest
(Swedenborg's manuscript
Principia rerum naturalium
(*c.* 1729–1731))

the inside and the eighth on top forms a bubble that constitutes a new particle, the ninth, which is the same as the air particle. Through compression of the ninth particle a tenth arises, the water particle. At the start of the formation of the earth it consisted entirely of water, but the water in the ocean then started to form crusts. The earth was an ocean in primeval times, a deep sea that extended all the way down towards the centre of the earth, 500 or 600 miles down. All hard materials have their origin in fluid particles. All elementary particles are spherical, extremely mobile, and sensitive to movement. They are arranged in a structure where one rests on four others, whereby one particle receives pressure from four others. Since calculations of the relation between the particles and the intervening space are tedious, Swedenborg says, 'it is sufficient here to adduce the results of a calculation without the details.' In the natural position the proportion of the space occupied by the particle to the intervening space is *almost* 5 to 1. The particles must be spherical and be part of an exact structure. If they had another form instead, such as cylindrical, sinuous, or eccentric, the pressure would be divided unevenly.[97]

Finally we come to a particle that we can see with the naked eye; all the previous particles are beyond the scope of our vision. This takes place when the 10th particle expands to the 11th particle, or the steam particle, with a surface of water and with the fourth and seventh particles inside it. We see the truth in our theory from the things that we can discern with the eyes. These steam particles can be seen when we look at an evaporating surface from the side. When this 11th particle rises to higher regions, where the atmospheric pressure is less, it expands. Finally the bubble bursts to become rain, which falls as drops of water. The raindrops are particles of the 12th kind. The elements aspire to evaporate and become spherical, as when lead is exposed to fire and becomes lighter than air and is propelled in the air as if it had been given wings. Water mixed with soap becomes a multitude of bubbles.[98]

---

[97] *Opera* II, 181f, 184, 186; *The Minor Principia*, 221, 224, 226.

[98] *Opera* II, 187, 190; *The Minor Principia*, 227f, 231f.

Impossible Figures

There is no other work by Swedenborg in which the spiral plays such a prominent part as in *Principia* from 1734. The spiral is the fundamental figure of motion in nature, the recurring sign that links all the mechanical parts of the universe, combines the smallest units in the particle world with the largest structures of the universe. Since nothing imperfect can be created by an infinitely perfect God, the inherent motion of the point must necessarily be the most perfect of all movements. This conatus or effort at movement must also describe the most perfect figure, even though it actually does not describe anything spatial. If this form of motion is to be perfect and uniform, it must resemble the circular figure, without end or angle. The most perfect figure of motion in the point must be the perpetually circular motion, that is to say, the eternal, unbroken motion from centre to periphery and back from periphery to centre.[99] The perpetually circular motion is a spiral, the most perfect of all figures, an infinite circular motion in which every revolution around the centre forms a circle in the progression from centre to periphery. It is circular in all its dimensions, has no limits, ends, or angles. Like the Archimedean spiral, Swedenborg's is a combination of a circular and a rectilinear movement. Both describe an infinite number of circles in the motion from the centre. Moreover, the circle shares a property with the logarithmic spiral. It is equiangular, although with an angle of 90° where the degree of growth is zero. Circles, spirals, and vortices revolve around a centre. Swedenborg's spiral is a continuation of the doctrine of the perfection of the circle. It is a perpetuated circle, an eternalized circular movement that describes all possible circles, in all dimensions.

This infinite internal spiral motion in the point can only be understood by analogy with the geometry of finite things, Swedenborg emphasizes.[100] It is a supersensual geometry. Neither the centre nor the periphery in the spiral figure of the point can be imagined geometrically, nor can any extension be assigned to it. One is forced to use paradoxical expressions. This spiral actually does not proceed from *one* centre to *one* periphery, but from *several* centres to *several* peripheries and from the peripheries back to the centres. There is thus an infinite number of centres and peripheries, where centre and periphery become one and the same. Geometry applied only to finite things, but since the point is neither finite nor infinite, it cannot be understood geometrically. The geometry of the point is impossible to demonstrate, because the point has not yet been made finite. It does not yet have an end, it has not become 'something', but is only a 'becoming'. It cannot yet be analysed with finite terms, merely imagined with the intellect. Despite that, Swedenborg declares, geometry has its place in this reasoning. This is where geometry has its origin—it is the genesis of geometry. Geometry can neither express this conatus to motion nor describe its figure, but only as an analogy. Since it cannot

---

[99] *Principia*, 36; translation I, 63, cf. 82; cf. *Treatises*, 103f.

[100] *Principia*, 37; translation I, 64f.

be demonstrated with the aid of geometry, we take refuge in the principles and axioms of rational philosophy. We are at the limit of knowledge, and science can never reach beyond that limit. This both non-finite and non-infinite point contains everything in the finite world.

The inner spiral motion of the point produces motion around the axis of the point. There also arises a progressive, inner movement, that is to say, the motion of the individual spirals around their poles in the point. If there is no external resistance, a further movement arises from the combination of these. It is the local movement that describes the figures of surfaces and particles. The three movements of the point are in analogy with the macrocosmic movements. The motion around the axis is the earth's motion around its axis, the progressive motion is the movement in the interior of the earth, and the local motion is the earth's movement around the sun. These ideas cannot yet be geometrically demonstrated, investigated, explained, or dissected. In these phenomena, so extremely small, experience cannot come to our assistance, no phenomenon or movement can be understood with our senses unless we first turn to finite and more complex things. We have no experiment which can confirm or prove the theory; we can only observe that in nature there is nothing that can move with greater freedom and ease than spiral motion.[101] It is the figure of figures, an eternal and spontaneous movement that includes and overcomes all forces in nature. It produces all the geometry and mechanics in the world.

Motion is in itself nothing substantial, but can resemble a substance if something substantial is set in motion. If a body moves in a line or a circle, the motion creates something resembling a line or a circle, although nothing other than the moving body itself is substantial. If the body describes a spiral movement from centre to periphery in all directions, something that resembles a sphere will arise. But nothing other than the actual point is substantial, and this substance, if it moves at high speed, causes all the space it describes to be perceived as a substance. In a similar way, the point is also the origin of geometry, through which the geometrician forms lines, surfaces, and bodies. Take a ball, D, which is tied to a string and swing it around a centre, C (Fig. 8).[102] If the ball moves at very high speed, this revolving motion resembles a circle. Although there is only a substantial body (the ball), its circular movement creates the illusion of a substantial circle. In the surface of particles there is nothing substantial apart from the fluid substance itself. It is a physical truth, in Swedenborg's opinion, that nothing finite can exist without motion. The movement of Swedenborg's point thus forms the extension of space. Unlike what he had said in *The Minor Principia*, he assumed in the printed version of his *Principia* that space is filled with ubiquitous active points.[103] There is no vacuum.

---

[101] *Principia*, 39; translation I, 68.

[102] *Principia*, 71; translation I, 124.

[103] *Principia* 79; translation I, 139.

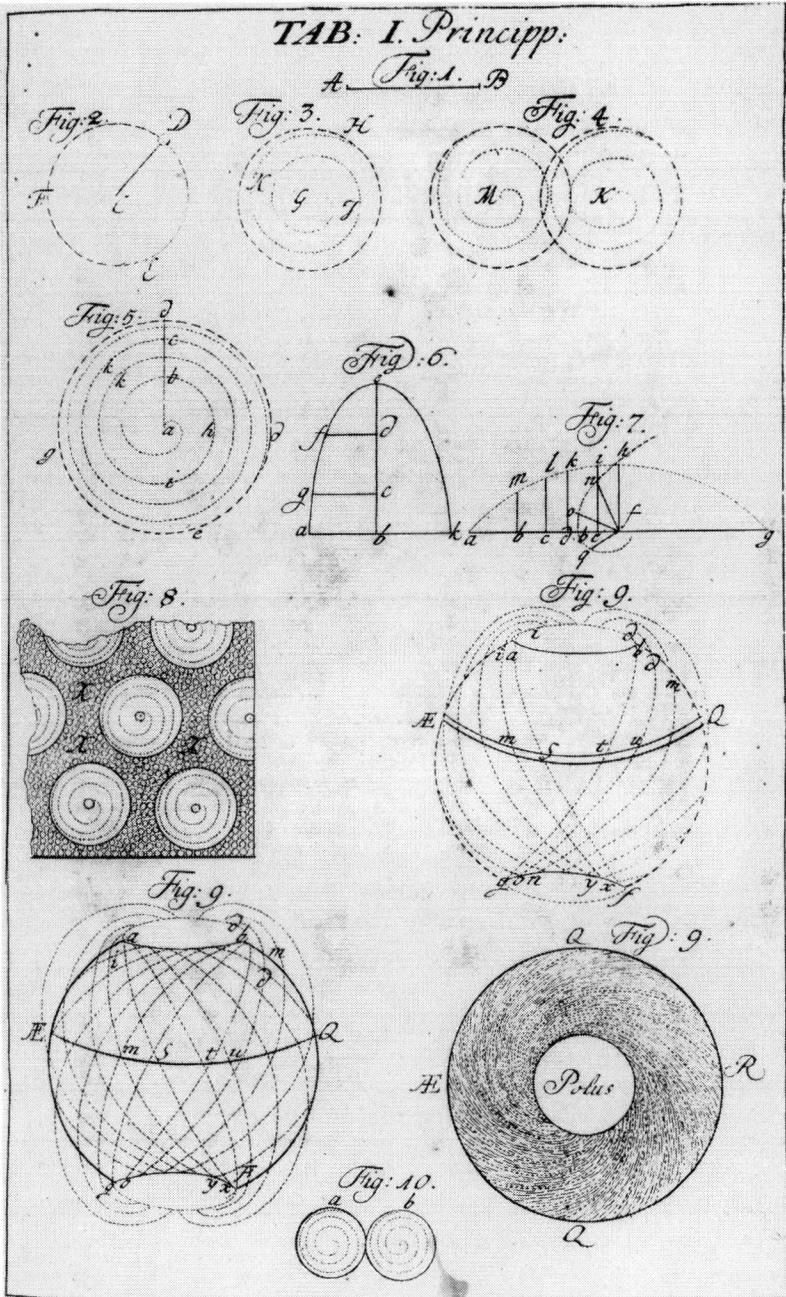

Fig. 8 The geometry of the spiral. The circular path of a ball (2). A spiral movement from the centre *a* to the circular circumference *deg* (5). Comparison between the spiral and the parabola (7). A section from the microcosm (8). The apparent plane formed by a substance in spiral movement, where *ÆQR* is the equator (9) (Swedenborg, *Principia rerum naturalium* (1734))

The Microcosmic Spiral Motion

The points in turn describe a spiral motion that resembles the inner conatus to motion of the points. This is the first 'finitum' where geometry begins and receives its limits, a term for the first substance that now appears in *Principia* after Swedenborg's reading of Wolff.[104] This first finite has two opposing poles shaped like cones in towards the centre, and like the macrocosmic world it has an equator, ecliptic, and meridians. The microcosm, in other words, has the same order as the macrocosm. The solar vortex moves in spirals like the first finites, with equator, ecliptic, and meridians and other circles belonging to heaven and earth, like imitations on a large scale of what happens in the small and simple, or the simple and small imitates the movements of the composite and large.[105] The whole visible world is full of proof that the parts are in agreement with the circles of the universe. The spiral and the motion around the axis, the progressive and the local movement are principles that recur throughout the series of units, from the first finites to the largest constellations of galaxies. Without mechanics, nothing can be established with certainty. One can see the force and strength of mechanics in a lever, its movement in the sloping plane, the eternal force in the eternal lever, eternal movement in the eternally sloping plane.[106] All the possibilities of mechanics can be found in the spiral.

The first finites are 'passive', they are immobile and resistant, while finites in local motion in perfect circles forming a surface are 'active' and thus mobile. There are three degrees: finite (passive), active, and the combination of the two into an element. Without these passive and active principles, no elementary particles could arise. There is thus no distinction between active and passive units at all in *The Minor Principia*; this is not introduced until the printed version. Presumably he had seen in Wolff how these concepts could be applied to the particle world. But he is not wholly dependent on Wolff in this idea. The conceptual pair of active-passive is a typical example of metaphorical thought. It is a metaphor that proceeds from human actions in activity and passivity, acting and resisting. Such metaphors appear more or less independently of each other in many contexts and in several of Swedenborg's sources. Lémery divides finites into active and passive, and Christoph von Hellwig, in a book that Swedenborg used, *Der accurate Scheider und künstliche Probierer* (1717), asserts that all things in the whole world consist of two principles, active and passive.[107] In human thought there is a constant division of the world into opposites: light and dark, man and woman, right and left, up and down, active and passive, and so on. The opposites become reference points on the basis of which other observations can be added. The active ones pass above our most subtle senses, Swedenborg says, 'for in things so minute the senses can have no experience.'[108]

---

[104] *Principia*, 43; translation I, 75; cf. *Treatises*, 105; Wolff (1730), § 796ff; Wolff (1731), § 44f.

[105] *Principia*, 49, 53f; translation I, 86, 94.

[106] *Principia*, 60; translation I, 105.

[107] Hellwig, 2.

[108] *Principia*, 77; translation I, 136.

From the movements of the first finites there arises the second finite, whose figure is larger, but not as perfect. The actives of the first and second finites, which make up the solar ocean, have the same quality, but the latter describes larger circles and thus moves more slowly. From the combination of passives and actives, or more exactly the second finites and the actives of the first finites, is formed the first elementary particle, the universal element that extends between the eye and the sun and the stars, through the heavens. It has a surface that is held in equilibrium through the motion of the actives and the resistance of the passives, between external and internal forces (Fig. 9). Under pressure, the elementary particle forms cones from the poles to centre, and through compression become a new finite. The first elementary particle, which builds up the solar vortex, contains everything preceding it, such as the point, the first and second finites, and the actives of the first finites. It becomes a microcosm resembling the macrocosm of the solar system. The elementary particle forms a vortex, not unlike the macrocosmic solar vortex, in which the polar axis of the elementary particle corresponds to the polar axis of the zodiac and its equator corresponds the zodiac of the solar vortex. The universe is filled with this first element, as we can conclude from the fact that we can see the stars in the heavens with our eyes, and this presence cannot be effected without contiguity. Through this element we can contemplate the most distant stars and the reflected light of the planets.[109]

The force of the point creates similarities to itself, multiplies itself in itself. To take an example, Swedenborg says, if the first point is 1, then 100 points make up the first finite, 10,000 points the second finite, and the fifth finite would consist of 10,000,000,000 points, and so on through multiplication of itself. There is a series of increasingly complex classes of finites, actives, and elementary particles. The more classes of particles arise, the more enriched, beautiful, and perfect the world becomes. The same geometrical laws apply to them all. The difference is merely a question of the size and speed of the circles. The second elementary particle is the magnetic element, consisting of third finites on the outside, and the actives of the first and second finites on the inside. The first and second elements constitute the solar vortex. They are bubbles like water rising as steam over a fire. On the earth's journey from the sun to its set orbit, new degrees of particle compositions are formed, making air, fire, water, steam, and the surface of the earth. Elementary particles of the third order correspond to ether, the fourth air, and the actives of the fourth and fifth finites are fire.[110]

The elementary world is like itself in both the least and the greatest, in the microcosm and the macrocosm, in the least particle as in the universe. In every part of the world there is a world concealed. In the microcosm is the macrocosm. The least is the model for the greatest, the greatest is an image of the least, like Leibniz's monads, each of which reflects the whole universe.[111] Swedenborg's world is part

---

[109] *Principia*, 93, 97; translation I, 163, 168.

[110] *Principia*, 108, 111, 114, 421, 425f; translation I, 185, 191, 196, & II, 317, 323, 325.

[111] Leibniz (1714), § 56, 62f, 65.

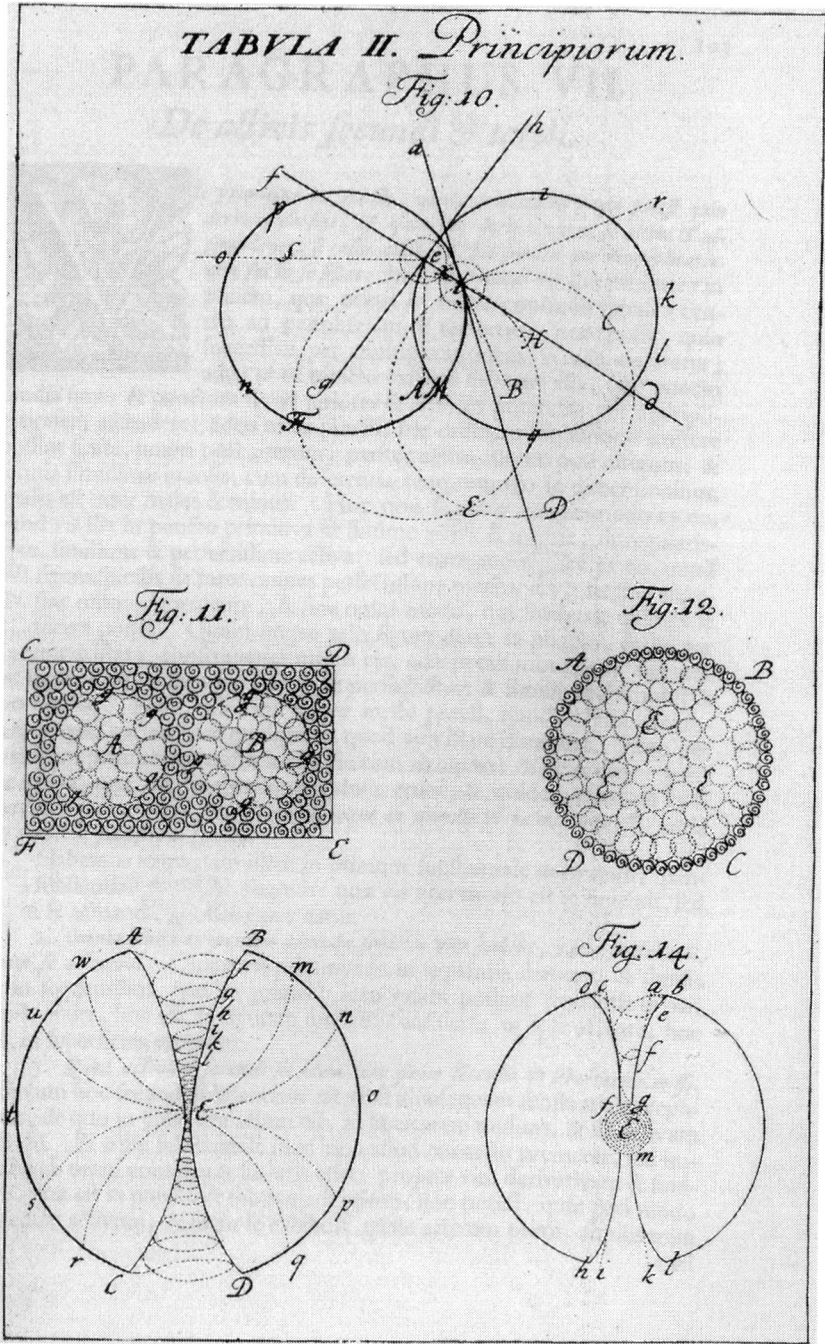

Fig. 9 The first elementary particle. *A* and *B* are actives, surrounded by finites or passives. The actives *E, E, E* constitute the inner core and the passives *A, B, C, D* the outer shell (*12*) (Swedenborg, *Principia rerum naturalium* (1734))

of the micro-macrocosmic world-view of metaphorical thought. From the least, the microcosmic, one can illuminate the greatest, the macrocosmic. Nature is greatest in the least and least in the greatest. Since the world is visible only in large and complex things, one can draw conclusions from the big to the smaller and the smallest, and also from the big to the bigger and the biggest. From a small volume of the smallest particles we can deduce the nature of the firmaments, from a tiny hut to the palace of the heavens, from a grain of dust we can ascend to an immense celestial vault with its boundless multitude of stars and enriched with such beautiful ornament.[112]

Magnetic Effluvia

The magnetic element is the first elementary that stands out visible to the eye.[113] Swedenborg imagines magnetic particles as small bubbles. Experiments, above all from Pieter van Musschenbroek's *Physicæ experimentales et geometricæ de magnete* (1729), are compared with his own theory. If his own theory of elements agrees with every conceivable experiment, no more need be done. All the rest would follow from the theory. Swedenborg applies his spiral theory to magnetic force and tries to explain it as a purely mechanical transfer of the particles' force. One of the most successful parts of the Cartesian vortical theory was that concerning magnetism. Magnetism was one of the stumbling blocks for the mechanistic natural philosopher and had to be explained in mechanical terms, not as an occult force working at a distance. Descartes' own explanation, which is also described in Hoffwenius's *Synopsis physciæ*, was that magnetism consisted of a kind of screw-shaped particles, right- and left-threaded, caused by the vortex around the earth and penetrating into the pores of the magnetic stone through the polar axis.[114]

On the issue of the magnet, the concept of effluvia could be used once again. Magnetism and electricity are combined as one phenomenon, as effluvia, a fine flow of particles which had to be understood in a mechanistic spirit. Before this Gassendi, whom Swedenborg mentions several times in *Principia*, had reckoned with effluvia, that there is an electric and magnetic attraction caused by exhalations from the attracting bodies as a stream of small corpuscles. The concept of liquid became a model for electrical 'fluids' as something flowing in currents. Around the magnetic stone, Swedenborg assumes, there are magnetic effluvia moving constantly and continuously around the active centre of spirals or vortices (Fig. 10). The effluvia form a sphere of vortices around the iron. All magnetic force consists entirely of the regular structure of the parts of the magnet. What clearer evidence can be obtained than from iron filings scattered around a magnet, which follow the lines of the

---

[112]*Principia*, 118, 121f; translation I, 202, 207f.

[113]*Principia*, 125; translation I, 211.

[114]Descartes (1644), IV, § 87–93, 139–187, 146; *Oeuvres* VIII:1, 279–315; Hoffwenius (1698), XVII, 78f.

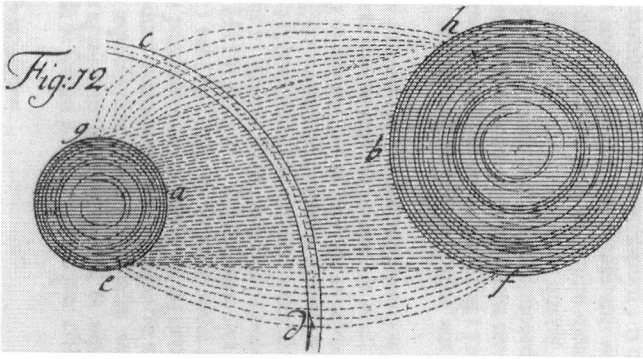

Fig. 10 The magnets, *a* and *b*, with the intervening effluvia (Swedenborg, *Principia rerum naturalium* (1734))

Fig. 11 A magnet hovers above a baroque garden. The magnetic effluvia flow from pole to pole. A queen is shown the sea god Poseidon, who is seated on a dolphin. In the background a man is trying to hit a dog with a stick. The neatly cut hedge extends towards the horizon, the infinite (Swedenborg, *De magnete*, cutting from Valentini's *Museum museorum* (1714))

magnetic force and place themselves in relation to the iron? (Fig. 11). If we could see the magnetic effluvia we would see them lying in a similar way. The magnetic sphere can flow freely through air and ether, but also through water and fire, through hard bodies like wood and stone.[115]

---

[115]*Principia*, 131f, 183, 260, 272; translation I, 223, 312, & II, 74, 93.

Swedenborg declares: 'In eliciting these truths, bringing them under general rules, and reducing them to geometrical principles, philosophy might employ her proportions and analytical computations for many an age, while geometry alone would fill numerous volumes with its illustrations by curves, spaces, and lines.' But whoever wishes to create and systematize principles through geometrical and mechanical rules and then confirm them through experiment should not busy himself by adopting or rejecting other people's opinions and arguments, but only present the causes and demonstrate the links between the first principles and the experiments. If there is no geometrical and mechanical connection between the first principles and the experiments, then the principles are nothing but hallucinations, dreams of the brain. If there really is a connection, then everything will fall into place and all conflicting opinions will be neutralized and silenced. Spending one's time refuting others is just a needless waste of one's time and effort. I therefore do not oppose Musschenbroek, Swedenborg explains. His opinions are largely my own, that is to say, that the effluvia themselves cannot produce any magnetism and would lack it entirely if there were not a circulating and whirling movement. But despite this, Swedenborg rejects all attempts through chemical torture, that is, say chemical decomposition and processing of magnets. It is meaningless, because all its strength and force lies in the regularity of the pores, in the geometrical proportions of its parts, and acquires its ability through its effluvia and mechanics. Such experiments are like trying with the aid of salt and fire to extract the image from a mirror or colours from a prism, like seeking to produce mechanics from the ashes of a machine, or using Archimedes' blood and marrow to try to analyse his mechanical genius.[116]

In the multitude of experiments and facts lie 'all our hopes of solving the problem before us [...], yet posterity may perhaps derive from them far greater profit than ourselves.' Swedenborg has also discovered in others a not infrequent error, that of drawing a universal conclusion from some isolated observations. Yet in his own theory he seems to reason in precisely this way, drawing universal conclusions from a limited number of experiments he had found in books. 'While we are ignorant of the cause of magnetism, we are only like moles burrowing under the earth. We go in search, we investigate, we move onwards in labyrinthine directions, sometimes meeting with soil altogether barren, sometimes with such as slightly rewards our labours.'[117]

The Magnetic Sphere and the Sidereal Heaven

If the least thing is perfectly geometrical, this also applies to the greatest. In the mechanical order of the world, man is between the smallest and the largest, for his senses register things that are equally far from nature's two extremes.[118]

---

[116]*Principia*, 158, 184f, 195; translation I, 276f, 315, 334.

[117]*Principia*, 210, 254; translation I, 360f, & II, 63.

[118]*Principia*, 375f; translation II, 229f.

Man is amazed both by what he sees and by what he does not see. Wherever he turns his eyes he is struck with wonder, one extreme surprise above his senses, the other under him. Since nature remains the same in the greatest and the least, from what we see and feel we can arrive through reason and analysis at a knowledge of what we neither see nor feel.

The magnet and its sphere are an image of heaven, a world system in miniature. In the magnetic sphere there are spiral gyrations or vorticles, and in the same way there are also spiral gyrations and vortices in the sidereal heavens. Around the earth there is a whirl in spiral motion, where the earth is like a core, a child in the arms of its nurse. The spiral motion forms an ecliptic and poles with conical apertures through which elements can flow in and out, like a current from south pole to north pole in spirals. 'Let the mind thus soar into the vast regions of the universe [. . .], enjoy the wonders of the heavens above.' The whole visible firmament is a single huge sphere. Beyond our solar system, the countless stars form vortices as our own sun does. But, Swedenborg adds, the firmament that we see is perhaps only one of an uncountable number of spheres or sidereal heavens in the finite universe, and between these universes there may be connections as between two magnetic spheres; 'and that the whole visible sidereal heaven is perhaps but a point in respect to the universe.' There are probably countless immense spheres or heavens like ours. But all these spheres together are not even a point in comparison with the infinite, just as our visible firmament is merely a point in comparison with the finite universe. We ask ourselves then: What is man?

> Vain-glorious mortal, why so inflated with self-importance? Why deem all the rest of creation beneath thee? Diminutive worm! What makes thee so big, so puffed out with pride, when thou beholdest a creation so multitudinous—so stupendous around thee? Look downward upon thyself, thou puny manikin! behold and see how small a speck thou art in the system of heaven and earth; and in thy contemplations remember this, that if thou wouldest be great, thy greatness must consist in this—in learning to adore Him who is Himself the Greatest and the Infinite.[119]

After this physico-theological exclamation Swedenborg turns to the diversities of worlds: 'how many myriads of heavens may there not be! how many myriads of mundane systems!' Astronomers are totally misled in their calculations of the number of the planets. The earths in the universe may have animal kingdoms like ours, but need not be inhabited by exactly the same kind of living creatures. The perfection of the world consists in its very variation, its mutability, the ability to give rise to increasingly complex things. Descend into Tartarus and Pluto's regions and you will not find one thing like any other. 'Look abroad on the vegetable kingdom; how varied is it! how pleasing, how delightful because of this variation!' Swedenborg exclaims.[120] Walk through groves and woods and see how everything is pregnant. If anything were missing, the world would not be as perfect, a piece would be missing in the order, a link in the chain. The perfection of the world is in

[119] *Principia*, 311, 376f, 380; translation II, 145, 231, 234, 238.
[120] *Principia*, 381–383; translation II, 240–243.

proportion to larger alternations, changes, and temporary properties that collaborate in shaping it. The world is more perfect and beautiful in its composite, connected things than in its simple, separate parts, and in greater and freer motion than in a more limited motion. Perfection, Swedenborg had noted from Wolff's *Ontologia* around 1733–1734, is unity in variation, or the diversity of the fluctuation of differences in unity.[121]

In every world system, however, the principles of geometry and mechanics remain the same. The dissimilarities or variations consist merely of differences in series, degrees, proportions, and figures. But the mechanics can actually differ in different worlds, since the external circumstances diverge, with different proportions, events, and degrees. If air and ether, or something similar, were to be found there, they would not display the same tremulations, and one would therefore see and hear differently. Our sensory organs, which are adapted to our world, would perhaps not even be able to receive these undulations. Machines in a different world would be constructed according to other rules and be applied according to other mechanical forces. The very symbol of inventiveness and the combination of mechanics and mathematics, the ingenious Archimedes who said that he could displace the world, would realize his limitations. Archimedes had a special symbolic value, which Swedenborg, incidentally, assigned to Polhem when he called him 'our Swedish Archimedes' in *Dædalus Hyperboreus*.[122] If the ancient Archimedes were moved to a different world, he would lower his voice at the realization that his genius and skill would be valueless, as if vanished, in a world with proportions and figures totally different from what we find on our globe. The infinite, God, can vary creation in infinitely many ways, and consequently He can also vary geometry and mechanics to infinity.[123]

Swedenborg's comparison between the magnetic sphere and the sidereal heaven rests on metaphor, on the analogy between microcosm and macrocosm. It is not possible to refer to any particular source or a necessary dependence on a reading of other thinkers, such as Aristotle, Paracelsus, or someone else; he is chiefly proceeding from the human ability to think in metaphors. The idea of microcosm and macrocosm is a notion of infinite analogies, metaphors, correspondences between different levels in existence. The classical expression of this agreement between large and small, above and below, could be found by Swedenborg in a book that he owned, *Die gantz neue eröffnete Pforte zu dem chymischen Kleinod* (1728). It is the Emerald Tablet of Hermes Trismegistus: 'that which is below is as that which is above, and that which is above is as that which is below.'[124] The micro-macrocosmic idea is a metaphor that constantly produces and generates new metaphors. Gold below corresponds to the sun above, and the sun corresponds to God the Father, and so on to new metaphors. As in heaven, so on earth.

---

[121] KVA, cod. 88, 325; Wolff (1730), § 503.

[122] *Dædalus Hyperboreus* I, introduction, cf. II, 25; VI, 1.

[123] *Principia*, 384f; translation II, 246f.

[124] Anon. (1728), 17.

Fig. 12 A terella—four
small magnetic globes are
attracted to each other.
Swedenborg cut out the same
picture for his notes about
magnets (Gilbert, *De magnete*
(1600))

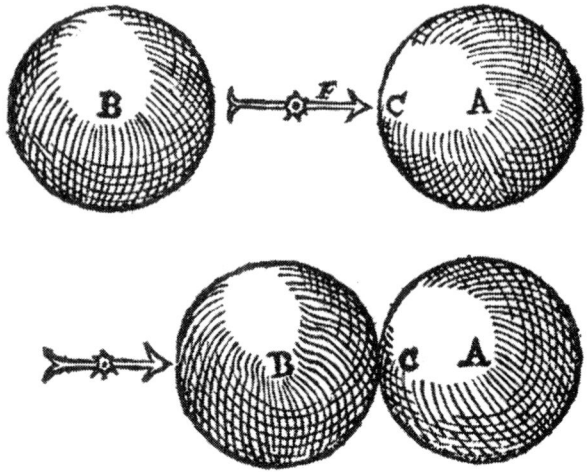

In the analogy between the earth and the magnet, Swedenborg was inspired by William Gilbert's *De magnete* (1600), which he had used frequently in his manuscript with the same name.[125] Gilbert regards the earth as a large magnet, or conversely the magnet as a small earth (Fig. 12). On the sure foundation of geometry, Gilbert says, the ingenious mind can rise above the ether.[126] Since the sphere is the most perfect form, the shape of our own globe, it is also the best form for experiments. Magnetic demonstrations should therefore be performed with a globular magnet. To this round stone we give the name 'Microgē' or 'Terella', that is, a little earth. Gilbert's is a typically analogical mode of thought, alternately understanding the earth and the magnet in terms of analogies.[127] The magnet has the same properties as the globe, such as attraction, polarity, and orbiting. The circular motion of the magnet shows that the whole of our earth is also moving in a daily circular motion. All the movements of the magnet are in harmony with and controlled by geometry and the shape of the earth. Gilbert takes the analogical thinking to extremes when he assumes that the whole world, the globes, all the stars are alive, animate. Why should not the stars have souls, if worms, ants, roaches, plants and morels have? Thales, of whom Aristotle speaks, says that the magnet is alive as a part of the living mother earth and her beloved offspring.[128] Earlier in *De magnete*, Swedenborg too had described the magnet as being animate according to Thales, but in a passage taken from Pliny.[129]

---

[125] *Principia*, 376; translation II, 231.

[126] Gilbert, translation, xlviii.

[127] Gilbert, translation, 23f, 66f, 330–332.

[128] Gilbert, translation, 309–312.

[129] *De magnete*, 237.

The Macrocosmic Vortex

In the beginning was chaos. Light and dark, hard and soft were buried together in a raw, disordered mass. 'The ore of the first world was full of slag and dross / The pools of the abyss were filled with dreadful mist!' wrote Spegel.[130] From this chaotic mass, Swedenborg explains in *Principia*, everything issued like a child from the womb.[131] Thought chooses the least difficult path, as a wanderer in the darkness gropes his way in the direction with the fewest obstacles, and follows the path without seeing it. He touches different objects without knowing what they are, and finally reaches his destination without knowing how he got there. In the same way the ancient philosophers succeeded in reaching the rational stance on chaos. Aristophanes wrote about chaos and the black-winged night laying a wind-egg, Ovid spoke of a confused, formless chaos without life, without sea, without earth or sky.[132] Moses agreed: 'And the earth was without form and void; and darkness was upon the face of the deep. And the spirit of God moved upon the face of the waters.' I was there, says 'Wisdom' in the book of Proverbs, when God set a compass upon the face of the depth.[133]

In our nearest macrocosmic vortex, the solar system, the sun is the world egg, the seed from which the planets are born. It is Burnet's theory of the world egg we glimpse here, but also general metaphorical thought. The egg stands for something living, for vitality, a symbolic beginning, something that develops from a point to full-grown body. It is a seed, an onion, a point. Forsius described the spheres of the universe as an onion, with shell inside shell, as the egg white encloses the yellow 'egg-flower'.[134] Creation takes a different course for Swedenborg. The planets are not extinguished stars, sucked into the vortex of the sun, as in Descartes. Instead the sun gives birth to its satellites, bringing them forth in a successively arising whirling movement. Swedenborg then shows us pictures of the development of the solar system, as he says, 'for a thousand words will not convey the idea which may be derived from a single representation'[135] (Fig. 13). The intellect can form a more distinct idea of an object in proportion to the clarity of the impression it makes on the sensory organs. The sun rests in the centre, surrounded by particles in a vortical movement. On the inside of the vortex the elementary particles are compressed to finites of the fourth order. They increase in number and form a crust around the sun which shuts out the sunlight. This crustaceous matter rotates around the sun, and because of the centrifugal force it moves further and further away from the sun. The vortices of space are like those in our own visible world, as when water whirls

---

[130]Spegel (1685), edition, I:1, 38.

[131]*Principia*, 388–390; translation II, 254–257.

[132]Aristophanes, 691–695; Ovid, *Metamorphoseon*, 1.5–9.

[133]Prov. 8:27; Gen. 1:2.

[134]Forsius, 76, cf. 84.

[135]*Principia*, 394; translation II, 263.

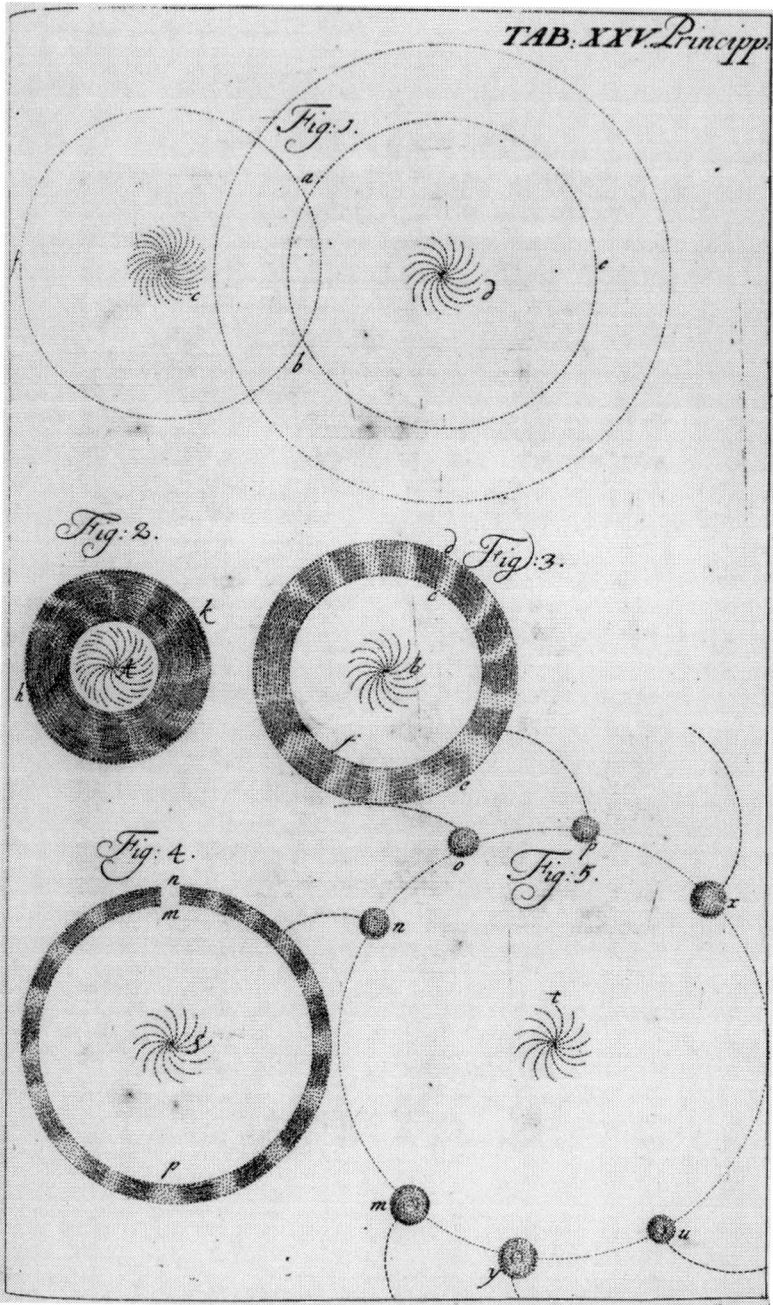

Fig. 13 The solar vortices. The sun, *A*, with its surrounding shell, *kh*, of fourth finites (*2*). The shell, *cdef*, is expanding (*3*). The shell expands further and cracks open at *mn* (*4*). Seven spherical bodies, *mnopruy*, are cast out in spirals from the solar vortex, *t* (*5*) (Swedenborg, *Principia rerum naturalium* (1734))

around a centre and continues to do so even if the first impulse has ended. The crust continues its movement around the sun, not unlike effluvia circulating around a magnet. The huge crust and the enclosed sun are like an elementary particle with an active core and finite particles around it, like images of each other. In its outward journey, in larger and larger circles, as a geometrical consequence the crust becomes thinner and thinner and finally bursts. The debris left by the explosion falls in a belt that revolves around the sun in increasingly large circles, until it thins out and finally explodes once again. The blasted mass of matter sticks together and forms round globes or spheres of matter of the fourth finites, that is to say, the planets and satellites of the solar vortex.

The globular form arises from a constant pressure in every direction, just like all bodies, like liquids brought together in globes, water in air, mercury in water, air, and ether. The planets move away in spirals from the sun until they achieve balance with the solar vortex and enter their final circular orbit. For a long time the sun was covered by darkness, heaven was in the shade, obscured by pitch-black clouds with the rays imprisoned behind the crust. Each planet is like a large finite particle with the same movements. They differ only in degrees and dimensions. The parts resemble the whole and the whole the parts. On a large scale one can see what happens on a small scale, in the visible part of the material world what happens in the invisible part, in the whole machine what happens in the model. If nature 'is invisible you seek her in the visible world, she will never disappoint you, but there present herself as visible before you. Thus will she never elude the eye, nor hide herself within mysterious shrines, but ever be most intimately present, and perpetually about and around both yourself and your senses.'[136]

The earth lies naked, uniform, in an azure, sky-blue tone. No air, no rosy light of dawn, no dew or clouds exist through which Iris with her saffron wings can show her multicoloured rainbows, no forests, no green, violet, red shades over the fields, no shimmer of metals—not a living creature. Swedenborg's product of natural philosophy is combined with mythology. The god Jupiter is the ether and Juno the air. Nature is reborn from herself as the phoenix rising from the ashes. Vignettes and initial letters in *Principia* show the phoenix flapping its wings in the fire. Nature takes place in an eternal circle. From death comes life, from the funeral pyre comes resurrection. The earth itself also forms a vortex like a big magnet, wholly in harmony with the large solar vortex. Under the spiral journey it was at first naked, but was then covered with ether, then air, and finally water. A crust formed over the watery surface of the earth. It was an eternal spring. Years and days were shorter when the earth was closer to the sun and rotating at a greater speed. The inhabitants of Mercury and Venus likewise counted more summers and years than we do. And if the antediluvians were to appear to us they would be astounded at our short springs and long autumns and winters.[137]

---

[136]*Principia*, 397, 403, 412; translation II, 273f, 285, 301.
[137]*Principia*, 410f, 423, 440, 445, 447; translation II, 300, 321, 347, 355, 358f.

The ether, which in Greek is 'always running' through eternity, is the third element (Fig. 14). Movement through the ether creates light. Heat, electricity, the light of phosphorus and meteors, also have their origin in the ether's local motion or conatus to motion. The fifth finites form on the surface of the earth. Air consists of fifth finites on the surface (which nourish the fire) and the first and second elements inside the earth formed as exact spheres which give rise to sound. Fire consists of the actives of the fourth and fifth finites. Water particles resemble air particles, although as highly compressed air particles, and acquire their mobility and fluidity from the ether flowing through them. An ocean without shores embraces the earth. What pleasure nature grants us through this living machinery![138]

The fifth element is the first that can be perceived by vision and touch. It is water vapour, as when one sees from the side how subtle steam rises in perfect round forms from the surface of hot water. This is where the elementary world ends and the visible world begins. We see everything through the elements, but only the effects, not the causes. In the spheres of steam the eye can observe the whole machinery and geometry of the elements. Steam is formed on the surface of water and encloses ether, it is under external pressure from ether and air so that its surface ends up in equilibrium between the forces and retains its spherical form. A large amount of steam can rise from a small quantity of water and can expand with the heat, with such great force that heavy weights can be raised and vessels of iron and brass can be burst. Swedenborg is thinking of the steam engine. Workers at foundries and smelteries can often see how large weights are lifted by steam. The particles of steam differ from water bubbles in that steam contains ether, while bubbles contain both ether and air. A bubble holds all previous stages, from the point to all finite, active, and elementary particles. A bubble contains the whole visible and invisible world. The macrocosm in the microcosm. The world in a particle. The same laws in the large as in the small, the whole in the parts. All the elements are enveloped by something else, and everything is enclosed as in a bubble. It is cosmos and order, seeing the unity in the diversity, the immutable in the change.[139]

The Declination of the Magnetic Needle

'In writing the present work, I have had no aim at the applause of the learned world,' says Swedenborg in a concluding appendix, 'nor at the acquisition of a name or popularity. To me it is a matter of indifference whether I win the favourable opinion of every one or of no one.'[140] The truth can defend itself. Yet it would turn out that he not could dismiss other people's opinions. A certain egocentric need and a pride in his work were expressed when he was passing through Copenhagen in July 1736.

---

[138] *Principia*, 408f, 418f, 429; translation II, 296, 310–312, 331.

[139] *Principia*, 433–436; translation II, 336–341.

[140] *Principia*, 451; translation II, 365.

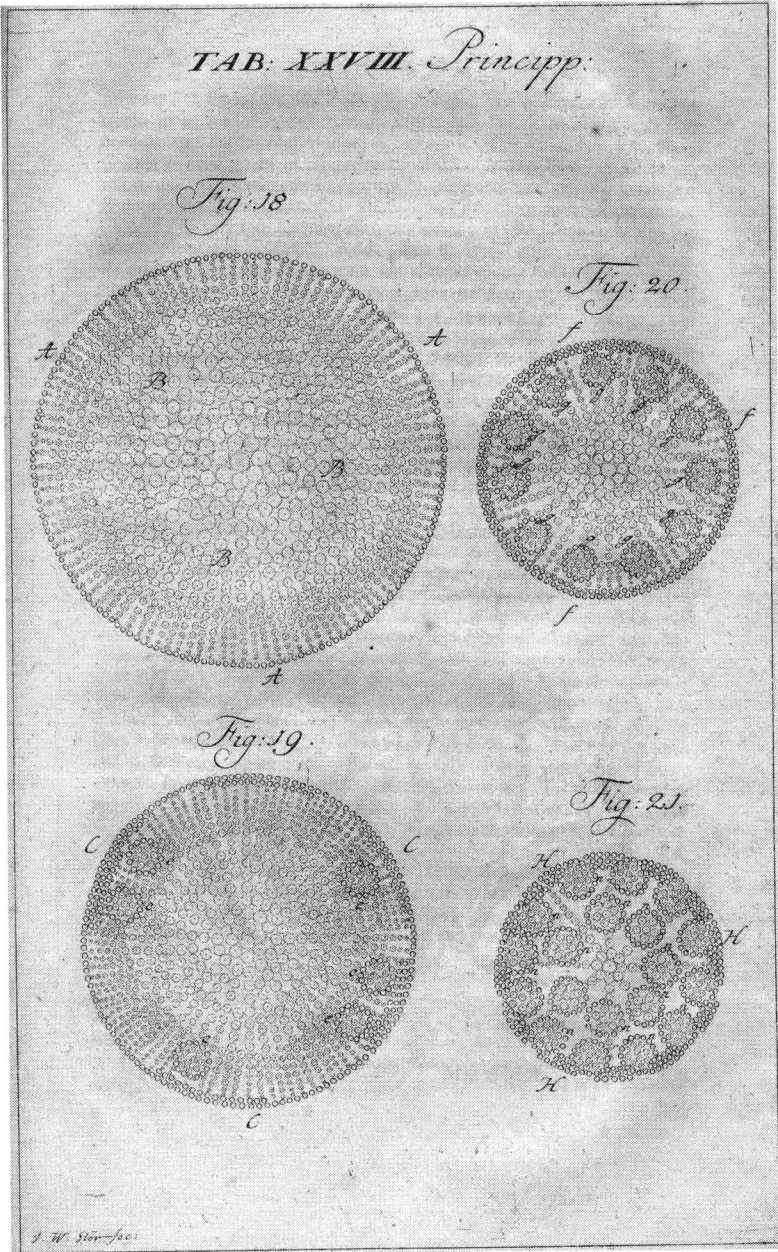

Fig. 14 Ether particles. Swedenborg asks how we can arrive at knowledge of the invisible particle world, which is so infinitely small and can never be observed, other than by reason. An ether particle in its largest extent with a surface of finites, *A, A, A*, and an interior of elementary particles of the first order, *B, B, B* (*18*). After a compression of the ether particle, small spheres are formed, *C, C, C* (*19*). An ether particle compressed even more (*20*). In its greatest degree of compression with small spheres, *n* (*21*). In principle the pictures also illustrate air and water particles, but they are larger (Swedenborg, *Principia rerum naturalium* (1734))

The Swedish ambassador, Anders Skutenhielm, told him that scholars in the city had spoken well of his *Principia*. Two days later, Swedenborg went to the library and looked at his own work, without revealing that he himself was the author.[141] In March 1737 James Theobald presented Swedenborg's *Principia* at a meeting of the Royal Society, describing how Swedenborg in the first chapter explains the path to a true philosophy through experience, geometrical skill, and reason. With philosophy we learn about the world machinery and the three kingdoms of geometry.[142]

Yet the mathematics and physics of the time had overtaken Swedenborg, as is evident from the controversy that flared between him and Anders Celsius and Olof Hiorter. Celsius did not value Swedenborg's *Principia* very highly. From Paris, where he mixed in the circles of the French Academy of Sciences, he wrote to Benzelius in May 1735:

> Mr Swedenborg's *Principia* will probably appeal to those who are still Cartesians and content themselves with particle philosophy. But Englishmen and Frenchmen who follow the manner of Verulam [Francis Bacon] and Newton in physics declare that perhaps 100,000 years from now it will still be too early to seek to determine the figure and size of the elements.[143]

Swedenborg was regarded as an old-fashioned Cartesian and a speculative particle philosopher. In the proceedings of the Swedish Academy of Sciences in 1740 Celsius criticized Swedenborg's explanation in *Principia* of the declination of the magnetic needle.[144] Swedenborg believed that the daily change of the declination, which George Graham had tried to demonstrate, was not a genuine phenomenon but due to erroneous observations. The declination of the magnetic needle, which varies from place to place and changes during the year, had to do with the difficulty of making correct observations because of disturbances and defects in the needle, faulty calculations of the meridian, etc.[145]

Swedenborg tried to defend himself in a letter to the Academy of Sciences in December 1740. Investigations in nature, he says, are grounded on both deductive thought and empirical observation. To explain his abstract thought he uses metaphorical thinking with descriptions of exploration by light and wandering. The metaphor is *to think is to see*:

> The way to discover what is revealed and concealed in nature is twofold, namely a priori, which is also called synthetica, and a posteriori, which is via analytica; both are necessary for the contemplation and the obtainment of one and the same thing, for this requires both light a priori and experience a posteriori, but the ancient scholars followed this light as high and as deep as they could.

---

[141]*Resebeskrifningar* 20 and 22 July 1736, 64f; Jonsson (1967–1968), 35, 40.

[142]Royal Society, London, 9 March 1737; *Green Books* V, no. 612.12.

[143]Celsius to Benzelius, Paris, 6 May 1735. LiSB, Br 10, XIV, 105; Nordenmark (1936), 51.

[144]Celsius (1740a), 296–299; Celsius (1740b), 384–388; *Svenska Vetenskapsakademiens protokoll* I, 284f, 292, & II, 120–125.

[145]*Principia*, 273; translation II, 95; Wilcke, 22.

It will be noticed that Swedenborg uses the terms a priori and a posteriori in an unorthodox way. A priori, that is, thinking independently of experience, is a synthetic approach, he says, while thinking with the aid of experience, a posteriori, is analytical. This was the reverse of what was customarily believed. With Kant people began to speak of a synthetic a priori. The senses give instructions as to what the mind ought to explore, Swedenborg goes on. One should search for the treasure chamber of experience, 'for nature is a bottomless sea'. Exploring nature means wandering and searching:

> Merely seeking and tracking, and feeling and knowing what can ultimately be seen in nature, can be likened to those who fumble and search in a valley between bushes, trees, and stones for the road and the direction to their right abode, as against those who climb high in a tower, from where they discover at a glance the whole wide field with its angles and nooks, and the false steps taken by many.[146]

The declination of the magnetic needle was caused by an 'iron vapour' that can pass through air, through glass, 'indeed, through metals, gold, silver, water, and fire, as has been proved by many.' He goes on to defend his Cartesian vortex theory. Pure ether forms the great vortex around the earth, 'for if one examines this vorticis motum perpetuo spiralem, or endless helical flow, with the correct geometry, one will find that its motion cannot make anything but larger and smaller circles.'[147]

Hiorter demonstrates a great many miscalculations in Swedenborg's reply, which could only make sense with the aid of these mistakes. When Hiorter had checked the calculations, Celsius was able to answer that Swedenborg, after many erroneous assumptions and some 30 faulty calculations, managed to arrive at the result he desired, which Hiorter called, with a certain sarcasm, 'a very odd occurrence'.[148] But Swedenborg, of course, was to have the last word. The miscalculations are not so important, he seems to think, and avoids answering the criticism: 'The other observations are of such little value that they do not merit a response.'[149]

The Membrane Between Body and Soul

In the second part of De infinito, entitled De mechanismo operationis animæ et corporis, Swedenborg tries to ascertain how the body can affect the soul and vice versa, how physical particles can move mental ones in accordance with the laws of mechanics and geometry. This is the classical problem of psychophysical interaction that arose with Descartes' strict dualism of extended substance and thinking substance. To be able to retain the mechanistic theory, Swedenborg advocates the solution that both the body and the soul must actually be extended substances. Like the body, the soul consists of active and passive particles but of a more subtle nature.

---

[146] Uträkning af magnetens declination, edition, 79f.

[147] Uträkning af magnetens declination, edition, 85f.

[148] Celsius to KVA, 23 January 1741. Celsius (1741), 87f, 93.

[149] Swedenborg to KVA, 1 February 1741. Swar uppå magist. Hiorters critiquer, 94.

One step in this requires combining his theory of tremulations with the doctrine of elements in *Principia*. Swedenborg counts four elements from gross to subtle: air, ether, the magnetic element, and the finest of them, the universal element. Larger elements move more slowly, water slower than air, air slower than ether, and so on. Tremulations in air cause undulations in ether, which in turn cause even faster tremulations in the next finer element. Since the smallest elements have the same mechanism as the biggest ones, this can be demonstrated through experiments using balls. For what applies to balls is also true for the smallest particles beyond our senses. If there are membranes that can receive the grosser undulations in the air, why can there not be finer membranes that can receive even finer undulations? Using a microscope to explore the brain, one can find membranes of increasingly fine structure which finally disappear beyond the limits of the microscope. This ever-finer ramification can also be seen in the blood vessels and the nerves. Just because we can no longer see them does not justify us in assuming that they do not exist. Although we cannot proceed any further at present with experience and the external senses, we can go beyond this with the aid of geometry and mechanics. Based on the large and visible we can undoubtedly draw conclusions about the small and invisible. Man is a microcosm analogous to the macrocosm. This is the parallel in metaphorical thought between the microcosm and the macrocosm, between man and the universe, that has many counterparts in other thinkers, as in the books by Kircher that Swedenborg owned.[150]

The membranes adapt to the oscillation and vibrations of the elements, imitating their movements. There is a type of membrane in the form of a spiral which can receive vibrations of different degrees and speeds. He omits the geometrical proof, however, referring this to a future work. A harmonious continuity, the contact between the finer and the grosser, feels pleasing, pure, perfect, and distinct. But if the connection is imperfect and dissimilar, it cannot reach the soul distinctly, purely, harmoniously, or pleasingly, but is obstructed by dissonance and incompatibility. Those who are in constant dissonance with the more perfect world cannot feel any pleasure other than dissonances of the impure that correspond to their own impurity. Swedenborg assumes that there is a kind of membrane in the most subtle parts of the cerebral cortex, where the meninx displays serpentines and meanders, and the cerebellum shows off beautiful trees or bushes with abundant branches and trunks. In these membranes the motor impulse of the physical particles is conveyed mechanically to the mental particles. The great challenge and task is to prove the immortality of the soul to the senses. The theological truth of an immortal soul must be proved through empirical studies and scientific inquiry into the machine that is man.[151]

At the same time, in the manuscript *Psychologica*, Swedenborg elaborates on the idea that the soul is a spiral shell.[152] Since the spiral is a continuous curve

[150]*De infinito*, 238–241, 252f; translation, 205–209, 217; Kircher (1665), book XII, 406.

[151]*De infinito*, 248–260, 268; translation, 213–223, 230.

[152]*Psychologica*, 16f.

that, so to speak, contains all radii, it can also be affected by every kind of tremulation in the different particles. Yet this spiral design has the same potential for individual variation as musical instruments have. There are no two instruments that sound exactly alike, even though they are in harmony with each other, and it is the same with human noises: no person speaks in the same tone as another. The same applies to two different soul spirals. By harmonious proportion he means that the first number is related to the last as the difference between the first two against the difference between the last two, 2, 3, 6 are harmonious, 2:6::3–2:6–3.[153] A harmonious proportion can be seen in the hyperbola; it is arithmetical and geometrical. The spiral of the soul has a hyperbolic curvature.[154] The hyperbolic spiral was, incidentally, described in 1704 by Swedenborg's mentor from his time in Paris, Pierre Varignon. Swedenborg envisages the subtlest membrane enveloped from centre to circumference in spirals with first actives inside, while the membrane itself consists of second finites.

In *Psychologica* Swedenborg has even drawn a sketch of a membrane between body and soul in keeping with his particle theory (Fig. 15). It is also an attempt to reconcile his idea of tremulations with Wolff's metaphysics, that is, to combine physiological and physical principles with purely abstract philosophical ones. The membrane actually consists of seven different levels of membrane, each one containing different elements and filling different functions, such as memory, vision, hearing, and the other senses. These membrane levels transmit the movements in both directions between the body and the soul. An external mechanical sensory movement will thus pass through them to arrive finally at a spiral, shell-like structure—the soul. This is what is illustrated by the chaotic, whirling spiral lines, *RS*, at the bottom of the drawing. Inside the spiral of the soul are actives of the first finite, while the first element is at *T*. A helical coil, *QP*, which is a membrane of third finites, encloses the first and second actives. Adjacent to this is a membrane, *NO*, which encloses the first element. The will is found in the fluids of the second and third membrane. Above is the membrane, *cd*, that contains the particles of the second element, *h, i, k* . . . the magnetic element. Wild animals' memory is found here, while man's memory is in both this and the preceding one. The outermost membrane in the picture, *ab*, contains the third element or the ether that flows along canals. This membrane corresponds to the sense of sight. But there are two more membranes not shown in the picture. One of them consists of a kind of subtle fluid, containing hearing and the other senses, and outside it, finally, is a membrane consisting of the arteries and veins of the blood system. Swedenborg points out self-consciously that future research will verify this theory of the membrane of the soul; if one only had a sufficiently powerful microscope one could see the entire structure of the soul.[155]

---

[153] *Psychologica*, 18f.

[154] *Psychologica*, 20f.

[155] *Photolith.* III, 108; *Psychologica*, 24f, 112f; cf. Jonsson (1969), 73–75.

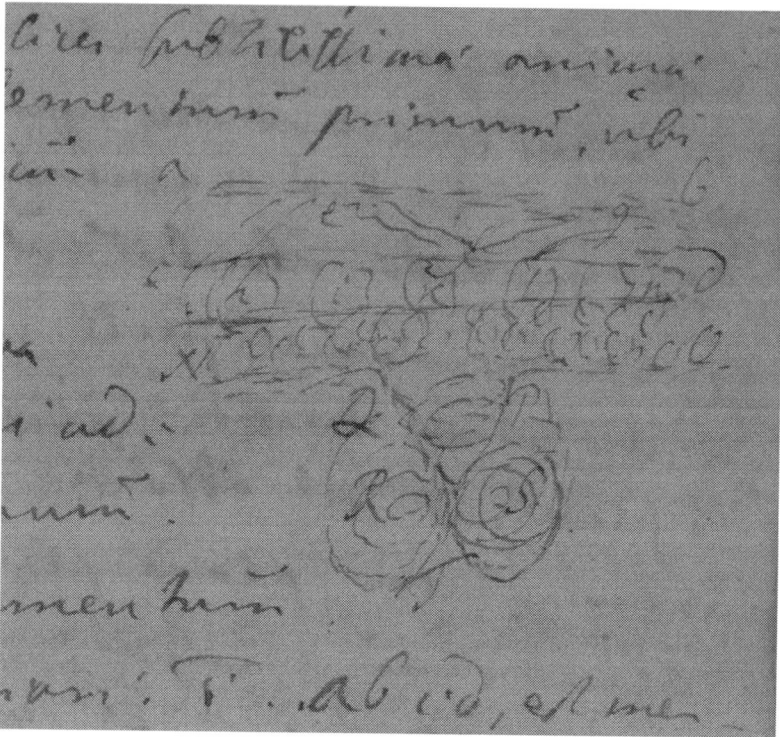

Fig. 15 The membrane between body and soul (Swedenborg, *Psychologica* (1733–1734))

The spiral seemed to Swedenborg like the primeval image of the world, the creative force, the energy in nature. For nature's most perfect thing this is the only appropriate figure of motion. It is the link that joins the microcosm with the macrocosm, a motion that recurs from the inner conatus of the point, via the processes of the particle world, to the vortices of the universe. Swedenborg's natural philosophy is a processual theory in which everything is changed, is transformed, undergoes metamorphoses. In his natural philosophy he seems to have drifted from the core meaning of the spiral to the same figure's associative and subjective meanings or connotations, such as perfection, beauty, universality, and so on. It is in this context especially that cultural and historical explanations are valid. The properties he attributes to the spiral in his mathematical manuscripts can in large part be understood on the basis of his knowledge of the elements of geometry, while an understanding of the associative properties leads into the cognitive ability to think in metaphors and complex cultural-historical explanations in the geometrical ideal of science, the idea of the perfect circle and the geometrical world-view.

Swedenborg's dynamic aesthetic, in which the spiral is—in a double sense—the salient point, is in phase with the rococo ideal. The spiral in Swedenborg's thought, with its wealth of variety, its mutability and gentle curves, is similar to the rococo emphasis on asymmetry and soft, elegant lines. Swedenborg's spiral lacks a specific centre—instead it has several shifting centres. Centre and periphery are everywhere and nowhere. The winding lines and stylized tendrils of the rococo can be seen in the folk painting of Dalarna with its gourd plants and arabesques, and in the interwoven spiral patterns in Swedenborg's books. The cult of the serpentine reached its zenith in the seventeenth century in artists like Bernini and Rubens, and it lived on in the rococo. An aesthetic interest in gentle arches and serpentines around the middle of the eighteenth century is expressed by the otherwise satirical William Hogarth. Like Swedenborg, he assumed the existence of a hierarchy of forms, from the least elegant, the straight line, to the most varied, the waving, meandering serpentine in three dimensions, 'the line of grace'.[156] The serpentine, the most distinguished of all figures, embodied movement, expression, variation, elegance, and beauty.

The perfect spiral recurs in almost all of Swedenborg's works, in the most varied and unexpected contexts, from the motion of the point and the membrane of the soul to evil and good spirits in spiral form, spiral staircases in Babylon and spiral paradisiacal gardens in the spiritual world. There is a constant change proceeding through Swedenborg's system, but also a general validity in which the perfect, dynamic form recurs everywhere from the microcosmic to the macrocosmic world. The finite geometrical world conceals the perfect spiral, the most complete geometrical figure, most beautiful of all forms. This is an extension of the Cartesian vortices of ether, which became a model for all physical movements in both macrocosm and microcosm. The universality of the spiral is significant for Swedenborg's effort to find a harmonious, generally valid, and uniform system that creates order in chaos. The harmony of the universe consists of repetition, the rhythmic replication of the circle, the sphere, and the spiral, where the harmony is crucial for the unity, like the harmony in a work of art. The spiral was one of the repetitive devices for achieving a micro-macrocosmic world-view, a dynamic regularity, in which the solar vortex of the macrocosm is found in the spiral motion of the microcosm, in the figure of the nervous fibre, in the form of ideas, in the harmonious proportions of the soul's membrane. All the poles, equators, ecliptics, and meridians of these geometrical figures are in agreement and harmony with each other. In the same way, there is a correlation between breathing, the spiral movement, and the universe.

---

[156]Hogarth, 55f.

The Infinite

> Having premised so much, let us turn our eyes to the heavens, and contemplate the
> universe. How wonderful to us are the heavens! filled with stars innumerable! enriched
> perhaps with innumerable worlds like our own! their very vastness in our conception,
> sufficiently wonderful! although it is probably little that we, and all of it not a point to the
> Infinite.[1]

Swedenborg, De infinito (1734)

A World That Is Not Even a Point

The stars instil wonder, affording a presentiment of a divinity behind the order.
'The heavens declare the glory of God; and the firmament sheweth his handywork,'
as we read in Psalm 19:1. It was determined that animals should walk prostrate,
looking towards the ground, says Ovid, while it was granted to man to carry his
head high and look up towards the heavens and the stars.[2] 'Let us lift our eyes
up to the heavens and the stars; let us compare their measures, numbers, courses,
harmony, times, tasks, and changes,' Stiernhielm writes in *Archimedes reformatus*,
as we see God's wonderful construction and excellent work.[3] If you take away the
seas, marshes, and deserts, Boethius explains in a thought experiment, not much
remains of the habitable world. Mankind is confined 'in this tightly-enclosed and
tiny point, itself but part of a point.'[4] We are like small, piteous worms on a little
point in the universe, writes Mersenne. A grain of sand could be as large as the whole
earth since it would have as infinitely many parts, for one infinity is not greater than

---

[1] *De infinito*, 53; translation, 44.
[2] Ovid, *Metamorphoseon*, 1.84–86.
[3] Stiernhielm (1644), dedication.
[4] Boethius, 2.7.20–22; cf. Pliny, 2.68.174; Lucretius, 6.648–654; Marcus Aurelius, 12.32.

D. Dunér, *The Natural Philosophy of Emanuel Swedenborg*, Studies in the History
of Philosophy of Mind 11, DOI 10.1007/978-94-007-4560-5_8,
© Springer Science+Business Media Dordrecht 2013

another.[5] Heaven is 'So immensely *large* that earth therewith compared / is like a little *point* drawn on the largest board,' Spegel wrote in a poem.[6] Tyge Brahe believed that there are stars which are as much as 90 times greater than Earth. Our globe, Block calculated, is a cabbage seed in comparison with the rest of the world.[7] And for this little point we fight with fire and sword. Derham, citing Seneca, points out how ridiculous this is, when 'above there are vast spaces, to whose possession the Mind is admitted.'[8]

Euclidean geometry avoided infinity; its geometry was finite and limited. In the classical tradition there was an aversion to mutability, to the ugly and immoral boundlessness. It is the limits and the order that create the beauty. For the Aristotelians infinity was something unfinished, imperfect, and unthinkable.[9] The Aristotelian universe could not be infinite. In the seventeenth century, however, an infinity perspective on reality was opened. Infinity triumphed during the baroque, as the world seemed to extend towards infinity, out into the universe and into microscopic matter. Baroque man sought to exceed the bounds of vision, through the microscope and the telescope, as a way to approach infinity. He constantly lost himself in infinity, constantly examining great and small infinity, its essence and limits. He sought order, a place for the finite in all infinities, in a world where man finds himself between the infinitely large and the infinitely small. The 'two marvellous infinities', the large and the small, Pascal wrote, have been given by nature to 'mankind, not to comprehend, but to admire.'[10] The incomprehensible can exist.

The infinite became present in the finite. It was the precipice of creation, the infinite space, the infinite division and calculation of infinity. People began to assume that the stars were at different distances, thus challenging the old idea of the outermost sphere of the firmament according to the closed Ptolemaic world-view. Things consisted of parts, which in turn consisted of smaller parts of their own. The concepts of space and perspective engendered a feeling for horizons, views, a remote haze. The human gaze and the baroque garden were projected far into the distance. There was a beauty in long perspectives, which astonished Swedenborg when he lifted his gaze towards the stars or down into the abyss of matter, but also when he travelled through the landscape. On his journey down to Germany in May 1733 he passed the stately home of Sturefors in Östergötland. He was struck by the lovely landscape surrounding it, 'which is most delightful, and is calculated to refresh and recreate the mind; since it opens to the eye a long vista of lakes, rivers, meadows, and fields, terminating in a forest.'[11] Apart from the new vision, the geometrization of reality helped to introduce the infinite into the world, to progress from a finite

---

[5]Mersenne (1625), books II & III.

[6]Spegel (1685), edition, I:1, 76f.

[7]Lindroth (1973), 152f.

[8]Derham (1715), 235; Seneca, *Naturales quaestiones*, 1, praef. 8, 11.

[9]Aristotle, *Physikēs*, 3.5.204b5–206a9.

[10]Pascal (1657/58); translation, 443.

[11]*Resebeskrifningar* 15 May 1733, 5f; *Documents* II:1, 7.

Euclidean geometry into the geometry of infinity. An infinity can consist of discrete quantities, as Forsius writes: 'So that, from the addition of a number to another number, one can make it infinite when one continues to add one to one, or point to point, thus: . . . . Belonging to this are years and ages upon ages for all eternity.'[12] The thought of infinity is in itself a form of metaphorical thinking. Infinity is a metaphor in terms of something that has an end, a metaphor based on processes that continue without limit, without stopping, but are conceptualized as having an end.[13]

When Swedenborg regarded the universe he was struck by its beauty, its wonderful harmony and perfect machinery. The finite world is a machine in which everything follows mechanical and geometrical laws. All the parts of the machine are in harmony with each other, from the particular microcosm to the celestial macrocosm. Every part of nature has its set function, and no part can be removed without disturbing the harmony of the whole. How and why was this remarkable world machine created? And how does the finite relate to the infinite? In *Prodromus philosophiæ ratiocinantis de infinito et causa finali creationis* (1734), which was written just after *Principia*, Swedenborg tackles infinity and the ultimate cause of the finite world.[14] He was searching for a solution to the mystery of creation, trying to find the link between the natural finite and the divine infinite. What were the connections between the universe and God, and how is the natural philosopher's insatiable curiosity related to the mystery of the revelation?

The Limits of the Unknowable

Once a person has begun to reason, to wonder about something, he cannot avoid penetrating deeper and deeper, higher and higher.[15] This is the 'law of humanity', writes Swedenborg in *De infinito*. Previously in *Principia* he had speculated about the structure of the world, about the smallest finite particles, the solar system, and the birth of the planets. But his thinking led him ever deeper down into the mystery of reality and he was forced to submit to the law of humanity. Why does the world exist, what is the ultimate cause of this mechanical world? His thoughts on natural philosophy in *Principia* required him to think metaphysically. In *De infinito* he goes into greater depth in metaphysics, ontology, epistemology, and the philosophy of religion. Natural philosophy passes into metaphysics and theology, blended into a synthesis, a two-headed world-view of finite and infinite.

In philosophizing about finite nature, the geometrical world machine, one is ultimately led into the inexplicable and the unknown—the infinite.[16] One reaches

---

[12]Forsius, 48f.

[13]Lakoff and Núñez, 158.

[14]See also Dunér (2003b), 149–164.

[15]*De infinito*, 25; translation, 19.

[16]*De infinito*, 13; translation, 9.

the limits of the unknowable, beyond which reason cannot go. The mind faces insurmountable difficulties when it attempts to explain the ultimate foundations of reality. Reason is presented with a Gordian knot, which rational man cannot untie. But we burn with a passion for the elusive knowledge that is denied us. Our ambitions are whetted by the unknowable, we have an inner desire to philosophize about the unknown. Reason cannot restrain itself in the face of the inexplicable. It does not content itself with stopping there; indeed, the greater the difficulties are, the more deeply the enquiring mind sinks into the task of solving the Gordian knot and questing for the highly coveted knowledge, even the most secret wisdom.

This unknown enigma which reason cannot capture is infinity, the Infinite. But man's curiosity forces him to proceed. He continues indefatigably towards the infinite. He wishes to find the core of the Infinite, its essential properties. This thirst for knowledge, however, leads him to find himself soon enclosed in an esoteric labyrinth. The enigmatic seems increasingly incomprehensible. Since man is a finite being, human thought is finite, capable of knowing only finite things and acquiring all its experience through finite senses. How could this finite mind be able to grasp anything infinite? His answer is that it is not possible for finite man to arrive at the essence of the Infinite through reason. In fact, there is no relation between the infinite and the finite.[17] They are incomparable, incommensurable entities. There is no similarity or dissimilarity whatever, no difference or relation between their proportions. If one thinks of the infinite, Swedenborg reasons, based on the finite with the aid of relations and proportions, one will find that the greatest finite seems like the empty set in comparison with the infinitely large, and the infinitely small will in turn seem like the empty set in comparison with the smallest finite. When Swedenborg draws a sharp line between the world and God, his arithmetic thinking comes into use and his metaphysical reasoning drifts towards arithmetic. In a draft about the infinite, the indefinite, and the finite he says that in the reason and perception of the soul there is nothing infinite except universal divinity. It is called infinite because it is beyond our perception and seems like nothing in comparison with the receiving organs. That is why we confuse the infinite with nothing. We cannot go outside our finite reason. Our wisdom is like nothing in comparison with the wisdom of the Infinite.[18] If we were to attempt to understand it, we would fall into shadows and our reason would be enveloped in darkness. In the investigation of natural things a person can likewise become blind and sunk in dreams of dark nights, 'so that even in finite things there is mere thick darkness, which enfolds, spreads over, and confuses it.'[19]

Since there is a chasm between the infinite and the finite, finite reason can never reach the essence of the Infinite, of God. Trying to arrive at God's essence with the aid of reason will therefore never give any result. Knowledge of God can be obtained only through God's Word, and the Bible's infinite truths can never be proved by

---

[17]*De infinito*, 15; translation, 11; cf. *De cœlo*, n. 273; cf. Cusanus, book 1, ch. 3, p. 3.

[18]*Opera* II, 196; *Treatises*, 96; cf. *Principia*, 376; translation II, 232.

[19]*Photolith.* III, 173; *Treatises*, 122.

finite reason. Reason would consequently do best not to attempt to tackle the truths of the sacred mysteries, since this would be fruitless and could never provide man with more knowledge. It could then be tempting, one might think, to prohibit reason from philosophizing about the divine. But Swedenborg is firmly against this. He follows Wolff in his assertion of the freedom of thought, *libertas philosophandi*, and also considers questions about the boundaries of philosophy that interested Rydelius.[20] Incidentally, Rydelius was also fascinated by the concept of infinity. But the demand for philosophical freedom initially did not include freedom from theological intervention; it had long been aimed particularly at Aristotelian school philosophy. In *Principia* Swedenborg says that, if the philosopher just follows the rules of reason and philosophizes in a truly philosophical way, he need never fear either religion, virtue, or the state. Science can never make any progress unless it is given full freedom to philosophize. It may have been heretical statements like this that made the Catholic Church place Swedenborg's *Principia* on its list of forbidden books in 1739.[21] Swedenborg's *Principia* is thrown on the fire by men in togas, burning on the pope's bonfire in the *Index librorum prohibitorum* (1758) together with Copernicus, Galileo, Descartes, Spinoza, Burnet, Leibniz, Locke, and many others.

If the philosopher merely uses his reason, philosophy can never come into conflict with theology. The error does not consist in using reason; it is that thought is finite and therefore tries to grasp infinity through comparisons and analogies from the finite sphere. But the infinite and the finite are incomparable, with no similarity whatever between them. The efforts of finite reason using analogical thinking are thus fruitless. Swedenborg makes a sharp distinction between infinite and finite, God and man, theology and natural philosophy, faith and knowledge, revelation and reason. In this way he simultaneously defines the limits of natural science. There are truths that reason cannot reach no matter how hard it wishes to. Reason is forced to acknowledge the chasm between the two worlds for wholly rational reasons. Swedenborg thus seeks a solution to the modern conflict between the scientific and the Christian world-view.

The idea of religion and science as two different worlds had been debated since the Cartesian controversies in Uppsala. This particular issue reveals a conflict between the authority of the Bible and the free quest for scientific knowledge, between believing and knowing, reason and revelation. At the same time, the boundary dispute between religion and natural philosophy was not solely an intellectual battle. The state had an interest in maintaining the distinction between theology and philosophy, ensuring that religious matters were not questioned, to avoid discord and disorder in society. In a more general cognitive sense it was a question of man's inclination to establish order and draw boundaries. Swedenborg, like most people, was in both camps. He wanted to forego neither reason nor the

---

[20] *Principia*, 452; translation II, 367; *Treatises*, 97; Rydelius and Almqvist; B. Lindberg (1973–1974), 230; Rydelius and Hiort.

[21] Decree of 13 April 1739. *Index librorum prohibitorum*, 239.

Bible, wishing to retain Christianity without rejecting the gains of philosophy and natural science. Instead he sought a synthesis in which science and theology were one, with different spheres but in interaction. It is at the same time a defence of scientific curiosity, the enormous potential of reason, and the logical necessity of faith.

Swedenborg finds that reason and revelation have ended up in conflict. How can the regularity of mechanics be reconciled with the miracles in the Bible, the immortality of the soul, and God's providence? Reality as mechanism gave rise to new questions about God's relationship to nature. Was He merely a divine watchmaker? In the mechanistic perception of nature, events are subject to a strict causality which distances creation from God and which does not allow any scope for God's intervention. It rests on a belief that every cause has just one possible effect, that there is only one possible world course. Nothing is by chance. Everything occurs with absolute necessity. A mechanistic universe therefore tends to be deterministic and materialistic and hence atheistic, which is counter to human free will and the immortality of the soul. Mechanistic philosophy revived the paradox of how to combine God's providence and advance knowledge of events with mankind's free will. Swedenborg has a similar aspiration to Wolff, reconciling the mechanistic world-view and the view of living beings as machines with a teleological and religious world-view according to which man was created by and for God as the highest goal of creation. Swedenborg's infinity searches for a harmony between nature, man, and God.

The solution is to draw a sharp dividing line between infinite and finite, faith and knowledge, revelation and reason. A similar radical difference between reason and revelation had also been attempted by the French philosopher Pierre Bayle. Knowledge and faith have different paths. Faith is granted by grace, not by logical proof. For Swedenborg, reason has limits that it cannot and should not exceed, since it would be meaningless and foolish. There are thus spheres of knowledge in which reason cannot be used, where we must rely on faith. The only thing that reason can arrive at when it comes to the Infinite is whether the Infinite exists or does not exist. This is how it is often regarded, that mathematics and Christianity are distinct. But, as we find in Swedenborg and in his time, there were many points in common and mutual dependencies between them.[22] Christianity coloured and in some sense influenced the execution and presentation of mathematics. The church needed mathematics to function, for example, the calendars for the ecclesiastical year. Numerology and symbolic meanings in geometrical figures were utilized, such as God's all-encompassing circle and the triangle of the trinity. The creator was a mathematician. Christian doctrine and mathematics were united in an absolute certainty. Not least of all, mathematics rested on God's guarantee of the truth, certainty, and universal applicability of mathematical theory.

Swedenborg's God is an indefinite infinity without modus. The identification of God with infinity can be found in many places in the history of ideas. The advantage

---

[22]Grattan-Guinness, 467–500.

of the concept of infinity is that it makes God unique, leads to a monotheistic religion with a rising hierarchy of power up to the unique highest point, an omnipotent God. Descartes believed in an independent substance, an infinite, absolute, and perfect God, and in man's finiteness and imperfection.[23] Spinoza understood God as an eternal infinite being of infinite attributes, with infinite extent and infinite thought, and like Swedenborg he was adverse to ascribing human properties to him. Everything proceeds from God's infinite nature, with the same necessity as that the sum of the angles of the triangle is eternally equal to two right angles.[24] When we call God infinite, according to Locke, we do so primarily in respect to his duration and ubiquity. Man acquires his idea of infinity through the repetition of adding parts to parts as regards ideas that are assumed to have parts, ideas that—like all others—have their origin in sensory experiences and reflection.[25] Polhem likewise identified God with an infinity that finite man cannot grasp, to be specific, the four infinities of time and space.[26] In a fragment of *De causis rerum* he writes about the cause of life: 'Three original things exist in nature, namely, minimum, maximum, et medium, of which the first two are infinite and therefore by their essence wholly incomprehensible to us, but the latter is so comprehensible that we ourselves are comprehended in it.'[27] Medium consists of matter and movement, minimum solely of movement, and maximum solely of matter or rest. Everything that cannot be demonstrated with mathematics and mechanics is uncertain guesswork.

The Infinite Is the Ultimate Cause of the Finite

When we want to find the essence of a thing, says Swedenborg, we must first ascertain whether the thing exists or not. The question of *existence* comes before the question of *quality*.[28] In other words, before we ask about the essence of the Infinite we must ask the question: does the Infinite exist? The Infinite cannot exist in the finite world, since the Infinite is infinite. In nature, which is finite, then, the existence of the Infinite is impossible, since everything in nature is finite.

Everything has a cause. That is the basic premise on which the whole of Swedenborg's continued reasoning rests. Everyday experience of cause and effect is transferred to the universal, cosmological plane. This is metaphorical thinking that attempts to understand events in the world based on the clear causality of everyday things. Nothing comes from nothing, *nihil ex nihilo*. All finite things must have a cause, they cannot arise without one. Nor can the first natural substance be its

[23]Descartes (1641), III; *Oeuvres* IX, 35f; cf. Mackie, 31, 37f.

[24]Spinoza, edition, 45f, 62.

[25]Locke, book II, ch. XVII, § 1, 6f, 22.

[26]*Polhems skrifter* III, 313, cf. 305, 312, & IV, 138; *Polhems brev*, 43.

[27]KB, Polhem, X 517:1, fol. 176r.

[28]*De infinito*, 28; translation, 22.

own cause since it is finite. Because it is finite it must have bounds, there must be something that limits it, makes it finite. If it did not have a cause outside itself to make it finite, it would be infinite, which would be a contradiction. Everything in the finite sphere thus has a cause that sets limits. The idea of the finite as a limitation of the infinite was embraced by Swedenborg's predecessors Malebranche and Descartes. Nor can finite things arise by chance. If the first simple or first finite arose by chance, one could ask how it comes about that the world is in harmony, that the elements, the planets, and the world possess order. This entire order and harmony cannot arise by chance, since that would mean that no laws can be maintained. The world would be a chaos without order if things and events could arise arbitrarily and entirely without order.

Something finite must thus have a cause, and the birth of the finite must therefore happen at a specific time, a point in time from which it started to exist and which made it finite and limited in time.[29] Everything in nature thus has a temporal cause that limits it. The ultimate cause of the finite cannot be something finite itself, as it would then need a cause that makes it finite, that limits it, which would lead to an infinite regression of causes: the cause of the cause of the cause.... The finite must therefore have a cause outside itself, a cause that is not finite. In other words, the ultimate cause must be infinite, the Infinite.

The Infinite is the ultimate cause of the finite even if we do not or cannot know the properties or essence of the Infinite. That is to say, the Infinite has no specific quality or property and it is independent, cannot be affected by anything else outside itself.[30] Nor can the Infinite exist in time; it is infinite in time as well, thus eternal and timeless. Time exists only in the finite sphere, denoting degrees and series, marking how something follows something. Time is almost identical with movement and is thus mechanical and geometrical. Where there is no substance or nothing finite there is no movement or speed, and where there is no speed there are no degrees or moments, that is to say, no time. Neither past, present, nor future can be ascribed to the Infinite. In the Infinite everything is present. Moreover, the Infinite has no quantity or extent. It thus has no proportions and no similarity to anything finite. Nor can it be limited, since it would then be finite, and if it had a cause outside itself it would be limited and finite. The Infinite is, in other words, its own cause. Swedenborg's conclusion is that the Infinite is God, the ultimate cause of the finite and of nature. Finite reason must thus a priori acknowledge something that is unknown and can never become known, nolens volens, whether it likes it or not.[31] This unknown being, the Infinite, must exist. It is a kind of Euclidean necessity, in which the laws of logic require a rational person to arrive at a conclusion, willy-nilly.

In this train of thought, the basic scientific and quantifiable concepts of time, geometry, and mechanical movement become separate from God and the theological sphere. It is a theology coloured by natural philosophy, in which the finite concepts

---

[29] De infinito, 20; translation, 15f.

[30] De infinito, 32; translation, 26; cf. Principia, 21; translation I, 36.

[31] De infinito, 36; translation, 29.

of science, such as the first natural substance, time, and speed, are used to prove the existence of God a priori. More exactly, it is a *cosmological* argument which proceeds from the fact that the world exists and that something created it, namely, God. The characteristic feature of the cosmological proof of the existence of God, as used by Duns Scotus and Thomas Aquinas, is that it denies an infinite series of movers, causes, or potential beings. The argument also presupposes that if something exists, there must be a reason for it. Here Swedenborg uses a variant that proceeds from cause and effect. The basic premise is that everything has a cause, that nothing comes from nothing, in accordance with the postulation by Leibniz and Wolff of the principle of sufficient reason. The possible must have a cause that explains why and how it is possible. There must be sufficient reason for a thing to exist. If something exists, it is an effect of a cause, and this cause in turn is an effect of another cause, and so on. If this chain of causes is not to continue infinitely, we must assume the existence of a first cause that is its own cause. This ultimate cause we may call God. Swedenborg, however, has a special variant of this argument, in which he proceeds from a distinction between infinite and finite being. Malebranche has a proof of the existence of God based on how the idea of an infinite and perfect being arises from the world of finiteness; the idea of the infinite exists in the soul before we have arrived at a knowledge of the finite.[32] It is more usual instead to proceed from necessary and contingent being. Wolff's cosmological argument for the existence of God proceeds from necessary-contingent, with the contingency of the world and mankind having sufficient reason outside itself, and this reason is a necessary, perfect being which is its own cause. The contingent universe thus has sufficient reason outside itself, while God, the non-contingent being, has sufficient reason in himself. These modal terms have an epistemological-metaphysical significance, whereas the conceptual pair of infinite-finite in Swedenborg leans towards natural philosophy which easily develops into particle physics, mathematics, and anatomy.

The Fantastic Order of the Brain Machine

In *Principia* Swedenborg explored the perfect machinery in the microcosmic particle world and demonstrated the fantastic mechanism of the firmament, perfectly appropriate for its purpose. In *De infinito* he turns to human beings, the fantastic order and appropriate design of the human body. In the cosmological argument for the existence of God he reasoned a priori, but now he uses experience in an argument a posteriori. He investigates human anatomy to show how each part and organ of the body, every nerve thread and blood vessel is necessary and has its special function for the whole, for the total person. Swedenborg asks whether nature herself can cause such a combination of parts. Could such a machine arise

---

[32] Jonsson (1961), 52; translation, 38f.

without an infinite intelligence, without a cause in God?[33] The whole argumentation follows the classical *physico-theological* and *teleological* proofs of the existence of God, that is to say, deducing from nature's order and finality the existence of a divine orderer, a 'divine watchmaker'. Plato, Aristotle, and Thomas Aquinas were among those who used this argument. Incontrovertible, sure principles for physico-theological arguments could of course be found in the Bible.[34] There was a rich physico-theological literature, John Ray in England, Linnaeus in Sweden, Derham's astro-theology and physico-theology, Fabricius's hydro-theology, Lesser's litho-theology, insecto-theology, and testaceo-theology (concerning shellfish), and a dissertation in Åbo on ornitho-theology.[35]

Swedenborg plunges into the fantastic anatomy of the human body. He opens the skull to investigate 'the machinery of the brain' and its inner mechanical parts, the cerebral substances, and the membranes.[36] The description of the brain gives a sense of presence, as if one were witnessing a dissection, in which the scalpel cuts through different parts of the brain, through lobes and nerves, with organs being removed, displayed to medical students, and placed on the dissection table. We follow the nerve threads and blood vessels through the body. Vertebrae and spinal marrow are shown, the knife exposes the machinery of the brain, and the cerebellum is cut in two. It is also a study in physiology, of how nutrition, heat, fluids, and blood flow through the body. And somewhere among all the body's tissues and organs the soul has its domicile. The medulla oblongata seems to be the doorway into the soul.[37] Beyond that, however, the traces vanish into the unknown. The author's investigation of anatomy leads to the frontiers of the unknowable, to the limit where empiricism and anatomical mechanics give way to metaphysics.

The anatomical investigation, which Swedenborg borrowed from the German surgeon Lorenz Heister's *Compendium anatomicum* (1732), continues with the eye.[38] He finds the optic nerve, the iris with the pupil and retina, and describes how light rays are led into and deflected in the eye, how the vibration movements of light are then transmitted by the optic nerve to the inner parts of the brain. The eye's machinery is analogous to and in harmony with the movement of ether particles. There is an admirable harmony, in Swedenborg's opinion, between the ether and the eye. The ether's mechanical movement penetrates through the machinery of the eye, reaches the retina, is conveyed further by the optic nerve, and continues into the medulla oblongata and the brain. There the traces disappear. The 'remaining

---

[33]*De infinito*, 55; translation, 46.

[34]Ps. 19:2–7, Isa. 40:26, Rom. 1:19f.

[35]Frängsmyr (1972), 121f, 148f; Israel, 457.

[36]*De infinito*, 56; translation, 47.

[37]*De infinito*, 61; translation, 52.

[38]Heister, § 266–272, 274, 276; translation, 121–130, 132–137; *The Infinite*, 47, note; Swedenborg also made a translation of C. F. Richter (1722). Swedenborg's manuscript has the title *Observata de corpore humano* (1734). KVA, cod. 88; *Photolith.* III, 142–145; *Treatises*, 147–156.

stages of the progress lie beyond our senses', the author says.[39] The eye itself lacks sensory perception. This is instead to be found in the inner parts, at the other end of the mechanical process, in the soul. And somewhere on the way to the soul, the anatomist loses the trail of the mechanical movement. Swedenborg was concerned with how light from the material world could be conveyed to a perceiving soul, how the material can become spiritual. This seems to have perplexed Linnaeus too:

> What is it that senses within me? I do not see it. The eye is a camera obscura, it depicts objects, yet it is not through the affected nerve that I see. The nerve does not enable me to judge of anything, and although it leads to the brain, it is not there that I see.
>
> There is, then, something which perceives and reasons, but which eludes me. Is it so strange that I should not see God, if I do not see the self within me?[40]

Swedenborg then directs his interest towards the ear, the chief instrument of the soul. With the aid of hearing we can be instructed, with the aid of the ear science and knowledge are imprinted in the memory. The ear is shaped so that it can receive sounds from all directions, whether above or below. Once again he follows a mechanical movement through the body. Vibrations in the air penetrate into the ear and the movement is then conveyed through the auditory nerve to the brain and the medulla oblongata, finally reaching its goal, which is the soul.[41] The entire mechanism of the ear shows how all its different parts lead to the same final goal, that people are able to hear, to perceive the sounds and pleasures of the world, and to give the soul satisfaction and enjoyment from a purer, more subtle world. The ultimate goal is to know God, into whom everything must flow in the end.[42]

The smaller parts, from the smallest substances to the tissues, form increasingly large parts, finally constituting the whole, the human body. God gave man a body in order that he could enjoy worldly pleasures, and to give delight to the soul. But primarily man was given a body for the greatest of all pleasures, the most perfect pleasure to which everything leads, the chief goal of human creation—the Infinite. All this cannot be by chance, this harmony of the parts, how everything fits together in a set pattern. This would not be possible without mechanical and physical causes, an incalculable number of necessary mechanical processes. The whole of this ordered anatomic machine functions harmoniously, a harmony that leads one's thoughts to Wolff's theory of harmony and concept of order. For Wolff the order that exists in and between things consists in everything having a reason, and this reason must be sought outside things. The idea that greater order means more perfection (agreement in order) can be found both in Wolff and in Swedenborg's physico-theological proof of the existence of God and theory of degrees of perfection. With the aid of effect and the senses, through visible or tangible nature, Swedenborg assumes, a person can be led to affirm the infinite cause better and more clearly than with any other spiritual capacity. God gave man the senses and reason precisely so that he could be led by them to acknowledge the divine in the Infinite.

---

[39]*De infinito*, 64; translation, 54.

[40]Linnaeus (1968), 58; translation, 100.

[41]*De infinito*, 67, 75; translation, 57, 64; cf. Duverney, tab. X.

[42]*De infinito*, 78; translation, 67.

When we consider this system built up through the first units, we are filled with admiration and amazement, and the more we wonder about nature, investigate it, find more and more mechanical and geometrical causes, the more we admire the Infinite that created this nature.[43] The more we adore nature, the further back in the causal chain we will go, right down to the simplest units. In our worship of nature we thus become worshippers of God as well. This perfect nature can only come from the Infinite. In the dissection of the human body we find an expression of this nature worship, as the whole examination leads to amazement about how all the parts can cooperate so perfectly with each other, how one part sets another part in motion, how one part holds on to another, how one part sustains and nourishes another. All this leads towards a purpose, to enable man to see and hear, to use his body, to be a rational being, finally arriving at the ultimate goal of creation.[44] In this anatomical physico-theological proof of the existence of God, natural philosophy enters the service of the spirit. Science is given a spiritual mission. Anatomy develops into metaphysics in a distinctive manner, becoming a form of anatomical theology. In the investigation of the fantastic mechanism of the human body, Swedenborg discovers the divine. The purposeful design of the human body, the harmony of the parts, leads to an assurance of the Infinite.

Nature arises from a divine 'infinite fountain', writes Swedenborg, not unlike the Neoplatonic emanations flowing or radiating from the One.[45] Plotinus's idea of the One is the supreme principle, the ultimate foundation of everything. Like Swedenborg's Infinite, the One is without essence; any attempt to assign properties to it is fruitless; neither 'this' or 'not-this' can be assigned to it. But Plotinus took one step further than Swedenborg in that the One can be ascribed neither being nor non-being, since it stands above the difference between existence and non-existence. The question of existence was the only one that reason could answer when it came to the Infinite. Both the Infinite and the One are principles that may be said to symbolize something indivisible, eternal, self-identical, and perfect. The One, like the Infinite, cannot relate to anything but itself and excludes a relation between 'something' and 'something else'. Instead it is a uniform unity. The similarities between Plotinus's and Swedenborg's emanations and levels of different realities does not mean that Swedenborg must be directly influenced by Plotinus. What both do is quite simply to exploit the human ability to think in metaphors. They use a metaphor based on a transfer of the experience of flowing water and sunshine to abstract concepts.

In Pascal's 'wager' a finite stake is pitted against an uncertain, infinite gain of an infinity of infinitely happy life. Placing the wager is the right thing to do. In Swedenborg the perfect order of the body machine leads to the ultimate foundation of the finite, the Infinite, to which all sensory and spiritual pleasures lead. It is precisely in the description of the senses and pleasure that their purposes are hinted

---

[43]*De infinito*, 45; translation, 37f.

[44]*De infinito*, 66f; translation, 56f.

[45]*De infinito*, 86; translation, 74.

at as a *eudaemonistic* argument for the existence of God, that is, human aspiration for pleasure and bliss must lead to the assumption of God's existence. God must exist if man's striving for bliss is to achieve more than just temporary satisfaction. For Swedenborg, divine satisfaction and pleasure were the greatest of feelings, the whole goal of the senses, the body, the soul, and reason. God created man so that he could enjoy the delights of the world. He was to have the earth for himself and he was given wisdom and reason so that he could worship and venerate Him, and also to be able to make the world perfect. God also gave man these properties, Swedenborg assumes, to be able to increase the number of angels in heaven.[46]

The Limits of the Universe

The infinite is the ultimate cause of the finite, and the Infinite is God, the divine constructor of the world. In our soul there is a tacit assurance of both the existence and infinity of God. And it seems as if it is inherent in man to confess to an omnipotent, omnipresent, and all-good God. But the philosophers, Swedenborg reasons, are actually unable to arrive at the properties of this infinite God with the aid of their finite reason. The mistake the philosophers make is in trying to imagine God's properties by using natural terms, analogies based on finite properties. Swedenborg sees a danger in this analogical thinking. By comparing the creator of the world machine and the body's soul machine to a divine constructor, who maintains the machinery so that the necessary processes continue, one makes God finite and ascribes a finite concept to the Infinite, since one compares Him to a secular potentate.[47] Swedenborg opposes this anthropomorphic way of thinking. God is infinite. Everything is present in Him, even what lies in the future. God's providence, that is, His concern for mankind, cannot be separated from His foresight. When one ascribes time and contingent events to God, as with a secular ruler, one makes Him finite. But His foresight and providence are infinite. With the aid of the sharp distinction between infinite and finite, Swedenborg tries to dissolve various 'finite' perceptions of God. By rejecting a number of perceptions of God he simultaneously defines himself through a negative method, and remembering the watchful eye of Lutheran orthodoxy he distanced himself from heretical ideas. It is a natural theology that Wolff used to prove the existence and properties of God. But this natural theology also had an apologetic task. It was to defend the true religion and combat all the false doctrines of every conceivable infidel, such as atheists, deists, fatalists, materialists, naturalists, anthropomorphists, paganists, epicureans, manichaeists, Spinozists, and so forth.

Not even the most learned and wise man can imagine the infinite any more clearly than an ignorant person. For Swedenborg the end of thinking is an apodictic

---

[46]*De infinito*, 56; translation, 47.
[47]*De infinito*, 87; translation, 76.

limit that cannot be transgressed. Thinking has its boundaries and limitations, yet simultaneously he worships reason and rationality. Finite man commits an error, he says, by trying to understand the infinite through the finite. There is no difference between the infinite and the finite, as all differences become the same. There is neither similarity nor dissimilarity between the infinite and the finite. One cannot image the properties of the infinite by analogy with anything finite. This is the mistake made by those who, in their admiration, deify the universe or the atomical world. That leads to idolatry. Swedenborg, in other words, is polemicizing against *deism*, the idea that God created the world but then retired and no longer intervenes in the world, letting it run according to its laws and reason. The metaphor of the world machine easily leads to a deistic stance, that of imagining that the machine can move by itself and of its own accord once it has been set in motion. Deism arose as an attempt to resolve the disharmony of world-views and strove to retain both the mechanical advances of natural science and the words of the Bible in the hope of being able to combine religion with the autonomy of reason. Those who express this idea, in Swedenborg's opinion, believe in a God who made himself finite through the creation of the world.[48] He was then able to withdraw and place Himself outside the world. Leibniz pointed out that, if one assumes that God corrects defects in creation, this entails diminishing his skill as a craftsman.[49]

Those who associate God with infinite spirit, the *panpsychists*, are also mistaken when they suppose that man has a soul that resembles God's spirit but is finite. But there are no similarities between the infinite and the finite, and therefore the infinite divine soul cannot be compared with the finite human soul.[50] God's spirit cannot be understood by comparing it to an 'infinite' human soul. In other words, he once again rejects an anthropomorphic perception of God, his critique being directed against the Stoics. Instead of identifying God with infinite spirit, one can do as the pantheists and the unmentionable Spinoza do, equating Him with the All, believing that the whole finite universe and all finite humans are part of God, that the All and God cannot be separated. Swedenborg opposes this idea, or more specifically a naturalistic variant of it, that is to say, a theory which assumes that nature is what is real and that God is the sum of all that is. But God is infinite whereas the universe is only finite. In other words, the Infinite cannot be identical with something finite. Like Descartes, Leibniz, and Wolff's rationalism, Swedenborg's God is placed outside the world. In a non-mechanistic world-view according to which God intervenes, God is present in space, not in a different space. For Swedenborg there are two worlds, the finite and the infinite. The universe can never be the same as the Infinite since the universe is finite in time and space. As with Aristotle, the only philosopher mentioned by name in *De infinito*, the whole universe is finite, has limits, is immense but not infinite.[51] The reduction of the

---

[48] *De infinito*, 92f; translation, 80–82.

[49] Brooke, 147.

[50] *De infinito*, 95; translation, 82f.

[51] Aristotle, *Physikēs*, 8.9.265a13–265b16.

Infinite to special forms or figures is also rejected by Swedenborg.[52] Comparing the Infinite to a circle without end, or a triangle to a trinity or some other geometrical figure, means imagining the properties of the Infinite based on an analogy with the finite.

Nor can one evade the problem by assuming that the universe is infinite and eternal. The Epicureans, Bruno, Digges, Gilbert, Guericke, and Newton imagined an infinite universe. Since the universe is infinite, according to Nikolaus Krebs (Cusanus), it has neither centre nor circumference. Any point can be its centre, as a mariner on the ocean always sees the horizon at the same distance. The Cambridge Platonist Henry More combined an infinite space with an infinite God by making space an attribute of God. For Leibniz, God created the fullest universe, the most varied and with the greatest degree of simplicity. The universe was therefore infinite and without any vacuum. Descartes' universe was indefinite and God alone infinite. The world, the totality of bodily substances, has no limits in extension. There is always something beyond the boundary. Where we do not see the limit, as in the case of the extension of the universe, the division of things, or the number of the stars, we must regard them not as infinite but as indefinite.[53] Swedenborg draws the limit between God and the world even more firmly. The world is not indefinite but finite. And the boundless universe, he claims, is not the same as God. The universe must have limits, it must be finite. The universe cannot consist of an infinite amount of finite parts. Since all these parts are finite, this must mean that the universe in its entirety is finite. But we know that, if one tries to describe the infinite with the aid of finite terms, in other words, if we make time eternal and the universe boundless, then a difficulty arises. The combination of finite and infinite means that the finite vanishes, one reduces it to nothing in comparison with the infinite and the infinite becomes everything. The greatest finite is like nothing in comparison with the infinite, and the infinite is like nothing in comparison with the smallest finite. Swedenborg's thinking enters a border zone, as theology becomes mathematics. At the same time, he distinguishes mathematical infinity from metaphysical infinity in order that these infinities shall not be confused with each other. A number in arithmetic, he says, can be so indeterminately small that it can almost be compared to nothing, but in fact it is never exactly the same as nothing; in other words, it is nevertheless finite and therefore not infinite. An example: when one multiplies or divides a finite number by 0 the result equals nothing, or if 0 is added to or subtracted from the finite number it remains unchanged. Infinity is a bottomless hole in which everything disappears.

Swedenborg then moves on to more complicated mathematics, to what he calls 'the calculation of indefinites'. The differentials are so small, he says, that they cannot be multiplied by a constant finite; instead they are annihilated by it.[54] Nor are these infinite, since they possess a similarity or analogy to the finite. A reviewer

---

[52]*De infinito*, 96f; translation, 83f.

[53]Descartes (1644), I, § 26f, & II, § 21; *Oeuvres* VIII:1, 14f, 52.

[54]*De infinito*, 101; translation, 87; cf. *Treatises*, 116–118.

in *Nova Acta eruditorum* (1735) pointed out that in this attempt Swedenborg had misunderstood the multiplication of a differential by a constant.[55] It is infinitesimal calculus Swedenborg is referring to, the subject to which he had been introduced in Paris two decades previously. Infinitesimal calculus, the covering term for differential and integral calculus, reckons with infinitesimal, infinitely small entities. The infinitesimal was smaller that any finite entity, but greater than zero. The concept can be imagined as a quantity that decreases indefinitely or without limit yet does not exactly become the same as zero. When conceptualizing derivatives in calculus one uses everyday terms and experiences, such as movement, approaching a limit, and so on. Swedenborg is thinking along similar lines when he says that the boundlessness of indefinitesimal calculus is not the same as infinity since it shows similarity to something finite. The numbers in the calculus are finite, yet they are indefinitely finite even though they seem to be infinitely small.

Infinitesimals say something important about mathematics, namely, that it can be very fruitful in mathematics to ignore certain differences.[56] Infinitely small differences are ignored in calculus. This runs against the whole idea of mathematics as a supremely exact science with precision. Infinitesimals consequently sparked debate.[57] George Berkeley required exact precision of mathematics, which could not be satisfied when calculus ignored infinitesimals. He rejected the existence of the infinite divisibility of anything finite, 'an infinity of an infinity of an infinity *ad infinitum* of parts.'[58] This calculus was also criticized by Fontenelle, who made a distinction between geometrical and metaphysical infinity. Klingenstierna in turn criticized Fontenelle's view that an infinitely small quantity must be of a specific size so small that not even an omnipotent God could make it smaller. In Klingenstierna's opinion, if one inscribes a rectangle in a rhombus, a rhombus can be inscribed in this rectangle and so on ad infinitum.[59]

Infinite division was problematic. Achilles will never catch up with the tortoise in Zeno's paradox. Half of a half must always have a half, and so on without limit, Lucretius wrote: 'Then what difference will there be between the sum of things and the least of things? There will be no difference.'[60] Reason must admit that there are parts which do not have any parts, that is, atoms. For Hobbes the infinite was a non-concept; it is not found in thought, since all our ideas are of finite things. What we call infinite is what we cannot grasp, the limits of which we cannot see. All we have is a concept of our own incapacity. Halley, whom Swedenborg had met in London, instead defended the true existence of infinity.[61] Nature and matter are infinitely divisible, parts into parts, according to Leibniz: 'Each part of matter can be thought

---

[55]*Nova Acta eruditorum*, December (1735), 557; Jonsson (1969), 325f.

[56]Lakoff and Núñez, 224f.

[57]Mancosu, 118–149.

[58]Berkeley, § 130; Lakoff and Núñez, 251f.

[59]Hildebrandsson, 15f.

[60]Lucretius, 1.619–620.

[61]Halley (1692), 556–558; Halley (1720), 22–24.

of as a garden full of plants or as a pond full of fish. But each branch of the plant, each member of the animal, each drop of its humours, is also such a garden or such a pond.'[62] Nature's machines are machines down to their smallest parts ad infinitum. After reading Wolff's *Ontologia*, Swedenborg wrote among his notes on ontology: 'an infinitely infinite number and an infinitely infinite size is impossible.'[63]

The Nexus Between Infinite and Finite

There must be a connection, a relationship, or a nexus between the infinite and the finite.[64] 'Nexus' is yet another term that appears only after Swedenborg's reading of Wolff. In Wolffian philosophy everything has an order, everything is held together in a network. There is an inner coherence in the universe, a *nexus rerum*. Swedenborg embraces a notion of a causal nexus, that there is a link between cause and effect, and he believes that the cause must by natural necessity exert an effect by virtue of this causal nexus. The quest for a nexus enlists the aid of analogy, as David Hume's regularity theory suggests, according to which cause and effect merely mean that, when an event of the same type as the cause occurs, there also occurs an event of the same type as the effect. Swedenborg thus assumes that there is a nexus, a link between the first finite unit and the infinite. Without a nexus the finite cannot exist or arise. It is nexus than connects events in the causal chain.

When it came to the properties of the first and least finite things, natural philosophers could find their way to knowledge, but in their attempts to find the properties of the infinite they were doomed to fail. What can be said is that the Infinite is the immediate, not the mediate, cause of the first finite things. In other words, there is no cause between the Infinite and the finite. This contradicts his assumption in *Principia* about the mathematical point as the nexus mid-way between the Infinite and the finite, through which the finite arose mediately from the Infinite. In *De infinito* he adopts a different stance, according to which the finite arises immediately from the Infinite through nexus.[65] In *Principia*, moreover, the nexus was neither finite nor infinite, whereas in *De infinito* the nexus is infinite. It is difficult to imagine that Swedenborg could have been unaware of this shift in meaning. But the mathematical point and the infinite nexus in *De infinito* are used to explain different things.

If the nexus is infinite it is impossible to ascertain its nature. Reason can never lead man to a knowledge of the essence of the infinite. In other words, there is an unknown and infinite nexus. If this nexus did not exist, the finite world would not be closed, that is to say, the beginning and end of the world, the first and last cause,

---

[62]Leibniz (1714), § 67.

[63]KVA, cod. 88, 306, 308; Wolff (1730), § 800; Scipion Dupleix, 169–179.

[64]*De infinito*, 103; translation, 89.

[65]*De infinito*, 108f; translation, 93f.

would not be linked. There must therefore be a nexus between the first finite unit and
the Infinite as well as between the last caused finite thing and the Infinite. Everything
is related to the ultimate cause, God. The whole universe, the whole finite sphere,
was created by the Infinite, but not for the sake of the infinite, for mankind or souls,
but for itself. God is both the first and the last cause of the world, and these causes are
identical. The creation and the end of the world in God, cause and effect, cannot be
linked without something, they cannot be related to each other without nexus. Since
the nexus is infinite, it is not—as finite things are—something natural, mechanical,
geometrical, or physical, in other words nothing analogous to the properties of the
finite. We cannot know what nexus is, only that it is. Everything is thus related to
the first, ultimate cause through nexus. Everything is created for the Infinite. If 'any
one tells me the same thing that I myself have arrived at,' Swedenborg adds, then 'I
am bound to believe him on the simple ground that I believe myself.'[66]

The only-begotten Son is infinite and God. They are two persons, but in infinity
and divinity they are one and the same. The connection between the Infinite and
the finite takes place through the Son, who is the nexus, and nothing else.[67] The
Son is infinite and the Father is infinite, but there cannot be two infinite beings,
since the Infinite does not acknowledge distinctions or divisions. The Father and
the Son are in other words the same: God. In this divine unity and nexus there is
an echo of the introduction to the Gospel of John.[68] Jesus is identified there with
the divine Word which produces the world in the Old Testament's creation story.
God did not create the finite universe mediately through the Son, but immediately,
Swedenborg continues. It is the same thing if we say the Father instead of the Son
as the immediate cause of creation. But then one can ask whether Swedenborg had
really solved the problem of the chasm between the Infinite and the finite that the
nexus was supposed to bridge. The nexus is itself identical to the Infinite.

The Last Effect of Creation

With the aid of reason we can conclude that there is a God and that He is infinite.
Reason can also establish the existence of a nexus and that it is infinite, but what
the nexus is cannot be said, since it is infinite. Reason cannot tell us any more
about God. Cogitation has its limits; there is a knowledge that lies beyond reason.
The unknown is and remains unknown, but it exists! The finite is the subject of
the first and last effect that follows on the first or ultimate cause, God. The first
effect in the finite sphere is the first created thing, that is, the first and least unit or
the seed, the first simple principle in the world.[69] In *De infinito* Swedenborg seems

---

[66] *De infinito*, 112, 116f; translation, 97, 100.

[67] *De infinito*, 74f; translation, 64f.

[68] John 1:1; Jonsson (1961), 136; translation, 138.

[69] *De infinito*, 121; translation, 104f.

to avoid the terminology of *Principia*; he does not coordinate his particle theory with his metaphysics of infinity. The natural point, actives, elementary particles are not specified; they are absent. Instead he talks in a more general sense, in Wolffian terms, about the first finite, the least created thing, the simplest thing of the world, the smallest natural thing in which all later stages in the chain of development are embedded.[70]

After God, through nexus, had caused the first, smallest finite units, the first effect in nature, there followed a long chain of intervening processes of cause and effect and cause and effect.... Finally we reach the last natural effect in the finite world. The last finite effect finally returns through the nexus to the Infinite. What then 'is the last effect in the natural or finite sphere'? Swedenborg asks. His answer is that on our planet it is man that is the last effect of the finite.[71] On some other planet in the solar system or in some other world there may be something else that is finite or some other being that is the last effect. All the elements, ether, air, fire, water, and earth, are mere instruments for this last effect, solely a stage in the accomplishment of it. The finer elements create a movement in the larger and grosser things, a movement from the simplest things to the more complex.

Almost everything can be found in man. Water, fire, and earth give him form and nourishment. The elements are combined in man as tools or means for the last finite effect, to nourish him and build up his body. The different elements make up his body: blood consists of water, body heat of fire, and through the air we can hear and through the ether we can see. The animal kingdom, like the elements and the vegetable kingdom, cannot be the last effect since the last effect cannot just be a machine, but something more than that. With only a machine the divine purpose cannot be realized, the last effect, through nexus, cannot re-enter the Infinite. The machine is just a means to this goal. Something active is needed over and above the passive thing that professes and ponders on God, something that can grasp and discern the final goal, that professes its faith that the final goal of creation is the Infinite. This final goal of creation, like Leibniz's teleology, is the purpose of the final, ultimate cause of the world. Animals do not have this ability to ponder on the *causa finalis*. The last effect thus cannot only have a body, but must also have a soul and be a union of body and soul. The last effect is thus rational man. This reminds us of the Aristotelian 'ladder of nature', from the lower vegetable soul, via the animal soul to the higher rational human soul. In Swedenborg the world is similarly divided into kingdoms, from the elements, via the vegetable kingdom and the animal kingdom to the kingdom of reason.[72]

Man is thus the last effect in nature. There is something divine in man, a receptivity for the divine. But the divine in man is not the fact that he is an animal with senses through which he can enjoy the pleasures of the world.[73] Nor is it that

---

[70] Jonsson (1969), 50.

[71] *De infinito*, 122; translation, 105.

[72] *De infinito*, 126f; translation, 107–109; cf. *Principia*, 1; translation I, 2.

[73] *De infinito*, 128f; translation, 110f.

he has a soul, his finite soul or his reason, since both body and soul are finite. Swedenborg conceived of reason as a collaboration between body and soul, and thus not something purely spiritual, but also partly as a material mechanical activity. The divine in man is thus not his purposeful mechanism, but that he can profess and confess to God, believe in a God that is infinite and acknowledge His existence. But if he doubts, he does not confess to God and the divine is no longer in him. All worship of God comes from faith and love. The truly divine in man is his faith in God's existence and infinity, and his sense of pleasure in his love of God. This would not be possible without a soul of a purer and more perfect world, without the reason that is the fruit of the collaboration of body and soul.

The Degree of Perfection

Everything in the visible and the invisible world has its origin in the first units. Finite complex things are formed from the first units through a natural, mechanical, geometrical, and physical process. These first units therefore contain the power to produce all things by a natural, mechanical necessity.[74] Swedenborg's system takes on a deterministic bias with an implicit necessary, regular course of cause and effect. In nature there are no contingent causes, he thinks; everything can be derived through necessary causes from the first units. Swedenborg writes time and again in De infinito about the machinery of the world, nature's duty to follow the laws of mechanics, the necessary mechanism of the brain and the soul. In addition, all future events are comprised in the mathematical point. God is infinitely provident— he knows about all events, even those which have not yet happened. But in just one passage is it clear that this does not at all lead to fatalism.[75] Although fatalism and determinism are not identical concepts, one can nevertheless say that he is restraining the deterministic tendencies of mechanics, those which contradict the miracles in the Bible, the immortality of the soul, and God's intervention and providence.

Everything was present in God before creation. He foresaw everything. Since God is infinite, he creates only what is perfect. Nothing imperfect can come immediately out of the Infinite. God is not the cause of imperfection; that arises in the world through the existence of evil, which means that imperfection has its cause in the finite. He thus tries to avoid the theodicy problem through the assumption of an infinity without essence. All-wise, all-good, and all-powerful are finite terms that cannot be predicated on the Infinite. It is therefore an illusory problem. As regards 'evil' and 'imperfection' he is close to Leibniz's idea of 'the best of all possible worlds', and his successor Wolff. God, being perfect, creates only a perfect world,

---

[74]De infinito, 39; translation, 32.

[75]De infinito, 49; translation, 41.

the best possible. In Leibniz and Wolff there is the idea of the necessity of evil for goodness; the harmony of the universe requires both light and shade, day and night.

Perfection in the finite world exists only in the first unit, and it is as perfect as anything finite can be. Being perfect, the first finite units are exactly like each other. The increasingly complex compounds of units that ensue, deriving from the first units, become less and less perfect the further from the first perfect unit they are. Swedenborg puts forward the classical metaphorical idea of series or degrees of perfection in nature, not unlike Leibniz.[76] If everything were perfect, according to Swedenborg, the finite world could not exist, since it would then be the same as the Infinite. In the finite world, therefore, there must be a scale or series of degrees of perfection. Without such a scale no finite would be possible. It would otherwise be infinite, completely perfect. The whole idea of degrees of perfection leads our thoughts to the *gradation* argument for the existence of God, a proof that is found in Anselm of Canterbury and Thomas Aquinas. The argument proceeds from the different degrees of perfection that are found in things. From this it is deduced that there must be something that is completely perfect, that is to say, God, a God who is in the highest degree perfect, true, and good. Swedenborg seeks to prove that the world is imperfect, that it cannot possibly be perfect. If it were perfect it would be the Infinite. But the world is finite and thus imperfect.

The essence of the finite includes having degrees of perfection and a law which says that something follows something else. There are substantial derivations from the most perfect of finite things to increasingly less perfect things. The finite must necessarily follow these laws, whereas the Infinite is the only being to which 'necessity' cannot be applied, precisely because of its infinity. The perfection of all derived things depends on how close to or how far from the first perfect units they are in their derivation. Since a steadily increasing number of natural units become involved in the more complex things, God also increasingly becomes the mediate cause of the thing. Only the first units are an immediate effect of the ultimate cause. The imperfect arises in this way, in other words, nature or the finite and not the Infinite are the cause of imperfection. Nothing imperfect can come from the Infinite, only likeness, perfection, and purity. But imperfection arises through the increasingly complex compounds of the first units. At the same time, Swedenborg says that everything comes from God. Attributing something to a natural cause is the same as attributing it to its ultimate cause, to God. Nature is merely a machine which must follow the will of this ultimate cause. The natural is nothing other than obedience, necessarily heeding God's will.[77] The finite is therefore a tool or instrument that is bound to perform what it does. Swedenborg envisages a determined world, a world where everything of necessity follows God's will. There is nothing natural that is not God's work, nothing natural that is not divine. Consequently, the more we admire the spectacle of nature, the more we admire the Infinite in nature.

---

[76]*De infinito*, 49, 136; translation, 41, 116; Lovejoy, 179f.

[77]*De infinito*, 51; translation, 42.

Escape to the Oracle of Reason

The most finite perfect is impossible for man. The human body cannot be completely perfect, since man is more complex, consisting of combinations of the first perfect units. The whole therefore becomes less perfect because the body is so far in derivation from the perfection of the first units. To counterbalance this, God gave man a perfect and finite soul, which He made rational and capable of governing the body.[78] The totality of man, the soul and the body with all its parts, is all intended to lead to the final divine goal. Man's will is steered by both the soul and the body, or more exactly, the will depends on reason, and reason in turn depends on both the body and the soul. Without this interaction between body and soul the final goal cannot be attained. Despite the distinctly deterministic features in the regular mechanics of the infinite, Swedenborg presupposes that man has a free will.

Since the world is finite it must have both a beginning and an end. There must therefore be a nexus at both the first and the last effect. But because of the fall of man, with the body's dominance over the soul, the link was broken. God therefore provided the world with His infinite, only-begotten Son, who assumed human form, as a link or nexus between cause and effect. The Son thereby became infinite *with* and *in* the finite, and through himself re-established the link or nexus between the Infinite and the finite, that is to say, God in the form of Christ died on the cross for man's sins.[79] The Son is the link connecting the two realities. Through God's only-begotten Son the first finite was connected to the last and both with God, and this itself is infinite.

Those who were and are unaware of the coming of the Messiah, in Swedenborg's opinion, can nevertheless be united with the divine.[80] If they believe in the Infinite they consequently also have a belief in the only-begotten Son and everything else that is found in the Infinite. The purpose of creation exists primarily for the Infinite or the Creator. Everything in the created universe strives towards infinity. There are many secondary goals in the finite sphere, but all lead to the primary final goal, infinity. The human soul, which is bound to the body, receives its information through the senses and can command the body. All this exists for man's aspiration to infinity. Everything adores and worships God, everything strives towards the divine. Swedenborg's perception that all people, whether they know of the Messiah or not, aspire to infinity because they were created by God, echoes the *consensus gentium* put forward by philosophers such as Aristotle and Cicero, based on the idea that all people in principle have a similar belief in a higher being.[81] Pleasure or enjoyment is the force driving man to achieve this final union with the Infinite. Man's pleasure, Swedenborg thinks, leads him to the natural goal of his life, to be master of the

---

[78]*De infinito*, 138f; translation, 118f.

[79]*De infinito*, 144; translation, 124.

[80]*De infinito*, 147; translation, 126.

[81]Jonsson (1961), 135; translation, 136.

planet, to give the body nourishment, to reproduce his species and bring up his children. The source of pleasure is love. Love itself depends on a connection or a harmonious continuity. But there is a purer, finer, and stronger pleasure. This complete delight or satisfaction should be at its peak at the moment of death. This pleasure is man's worship of God.[82]

With the aid of reason man can assure himself of God's existence and infinity, and of the existence of a nexus, the only-begotten Son, between God and man. But reason cannot go further, asking questions about the nature and properties of God, since He is infinite. His infinite properties can never be grasped by man with the aid of his finite reason. The question of God's essence can only be answered through revelation, through the Bible. When reason fails, when man is unable to go further, he must turn towards revelation. When he cannot find what he is looking for or cannot understand what the Bible says, he must enlist the assistance of reason and escape to 'the rational oracle'. Natural theology, as Swedenborg puts it, must extend its hand to revealed theology when the meaning of the revelation is uncertain, and revealed theology must give guidance to rational theology when reason finds itself in difficulty. Revelation and rational theology can never be in opposition, unless true rational theology is tempted to tackle the mysteries of the infinite, but then rational theology is no longer truly rational. Reason must realize its limitations and not go beyond the bounds of the unknowable.[83]

In the conclusion Swedenborg engages in a long discussion of the concepts of natural and divine. The fact that this problem is illuminated in the conclusion and emphasized in this way gives us a clue to what can be interpreted as the original aim of *De infinito*: to reconcile the scientific world-view with that of theology, to show that God in natural philosophy is the same as God in Christianity, that the philosopher's reason is not in opposition to the theologian's revealed truths, that both have their special domains and are mutually dependent. The dividing line between natural and divine, between reason and revelation, is akin to the philosophy of Leibniz and Wolff, who make a distinction between the secular and the divine, where God may be said to stand outside the world. Swedenborg's *De infinito* was written in an attempt to elaborate a system after a personal reading of Wolff's *Ontologia* and *Cosmologia*.[84] The German philosopher Johann Georg Hamann found Swedenborg's *De infinito* entirely in the Wolffian-scholastic style, with a ghostly Latin style and nauseating tautologies: 'I explain this entire miracle to myself as a kind of transcendental epilepsy dissolved in a critical froth.'[85]

Natural or material, in Swedenborg's view, are something that is *affected* in nature, they are instruments and means. Machines and their power are mechanical and natural in that they are instruments for movement through the machine's wheels, levers, and pulleys. In the *elementary* world the rain clouds, storms, the sun's rays

---

[82]*De infinito*, 153f; translation, 131f.

[83]*De infinito*, 156; translation, 134.

[84]Cf. Jonsson (1969), 48.

[85]Hamann to Johann Georg Scheffner, Königsberg, 10 November 1784. Hamann, 256.

that heat and illuminate the world, the ebb and flow of the oceans are natural because they arise rhythmically in periods. In *the mineral kingdom* the fire that melts metals, liquids that boil, steam, and catch fire are substances that form philosophical trees, the magnet that attracts iron and points north are natural. In *the animal kingdom* vision, hearing, touch, movement, the body's growth, age, pain, disease, and death are natural.[86] All that exists, from nature's simplest building blocks to the largest compound, all the parts and their movements are natural. The totality of the finite sphere is natural. At the same time, however, there is nothing natural that is not divine. The first finite was immediately made natural by God, but the other complex things only mediately. Nothing exists that cannot be traced back to its origin and its cause in the Infinite. Everything exists through causes of causes and so on back to the first cause, the Infinite.

The Soul Machine

The soul is bound by mechanical rules and can be explored through mechanics and geometry, Swedenborg writes in a manuscript from 1733 about the mechanism operating body and soul. As a finite thing, the soul must have space and figure. More specifically, the rational soul consists of the actives of the first and second finites which form a small space, and around it are areas consisting of passive second finites. To assume that it is impossible to acquire knowledge of the soul, that it is secret and separate from the senses, is the shortest route to atheism, he says. We see that there is a mechanism in what we see with our eyes, and therefore we can conclude that there is also a mechanism in more subtle things, but we do not understand what kind of mechanism it is, and we remain in the most obscure darkness, believing it to be as imperceptible as the infinite. The deeper we progress into knowledge of nature, the greater the light we enter, as if from darkness itself to a knowledge of God.

> All things concerning the animal body and soul should be proved by a first or metaphysical philosophy, by the analysis of natural things, by geometry and mechanism, by figures and calculus, by experiments, by the anatomy of the human body, by effects, by the passions of the body and the mind, by the Sacred Scripture; in this manner this theory should be investigated.[87]

The reason for our ignorance and our doubt about God is that so many secrets and clouds stand in our way, as when we do not see the sun concealed behind the clouds, and we doubt that there is a sun. But if we follow this plan, the clouds will be dispersed and God's rays will shine into our soul, just like the sun into our eyes.

When we think, our cranium becomes warm, our eyes red, dry, and motionless, and the nerves in the neck stiffen, Swedenborg writes. Thinking wears out the body. With Wolff as a sounding board, he meditates in *Psychologica* about the materiality

---

[86]*De infinito*, 158; translation, 135.
[87]*Photolith.* III, 97; *Treatises*, 138.

of the soul. If the intellect and memory are not mechanical, how then can they be affected mechanically, helped with medicine, enlarged or reduced by disease, revived by smells and touch, decreased or increased through the body's health? Weak sensory impressions become relatively strong in the absence of stronger sensory impressions, as for example when the moon can be seen better in the absence of the sun. This explains why our imagination is stronger in dreams and when one is alone. Ideas are clearer in the dark, when we close our eyes, when other objects that cause disturbance and interference in the other sensory organs are absent. Similarities produce similarities in the senses, arousing associations. When one sees a tree, one immediately sees a garden, one can see men and youths in the garden, and more and more of 'my own thoughts and amusements in the garden.'[88] Swedenborg agrees with Wolff that dreams have their origin in the senses, and are kept alive through a series of thoughts.[89] With cultivation the membranes of the soul can be shaped to receive and adjust to the tremulations. This is what is called the power of imagination. The ability to receive tremulations harmoniously, through the figure and tension of the membranes, and transmit them on to the centre of the soul, could explain many intellectual capacities. Without cultivation and practice there would not be anything in our imagination and our memory beyond what is found in animals.[90]

Swedenborg specifies the particle physics of the soul: the human soul consists of first and second actives. A soul can perceive another soul through an undulation between them, in the same way as two bodies can sense each other at a distance from one star to another. The orientation ability of animals is due the fact that their souls consist of the magnetic element. The angels consist of particles with first actives on the inside and second actives on the surface. The devil's spirits are instead built up of gross substances, namely, fifth finites, the first and second element, which are constantly tormented by fire and movements of air. This explains why they are unable to perceive more subtle movements. The premise for a future particle philosophy is 'That the soul, angels, the body, immortality, life after death can all be demonstrated geometrically.'[91] Finally, God, Christ, and the Holy Spirit, and everything else must be proved with the aid of reason, geometry, anatomy, and experience. It is undeniably the case that our body is mechanical, and our senses are likewise mechanical, as are the intellect, reason, and the soul itself, yet the learned world did not realize this until some time had passed. If in the larger things, why not in the smaller ones?[92]

In the subsequent part of *De infinito*, entitled *De mechanismo operationis animæ et corporis*, Swedenborg examines the relationship between body and soul, and the nature and location of the soul. According to Swedenborg, the laws of geometry

---

[88] *Psychologica*, 62f, cf. 58–61, 76f.

[89] *Psychologica*, 80f; Wolff (1732c), § 123.

[90] *Psychologica*, 96f, 106f.

[91] *Psychologica*, 146f, cf. 140–145.

[92] *Psychologica*, 154f.

and mechanics can also be applied to the soul. The soul can be described with figure and space, motion and power. Like the body, the soul is finite and has extension and hence figure. Everything in the large, visible world has its cause in the more subtle world. They have the same laws. Just because we cannot see the small world, we cannot deny that it has a mechanical nature. With induction and analogy one can reason about the invisible on the basis of what lies within the sphere of the visible. We see cause and effect with our eyes, its mechanics and geometry. With the microscope we can see even smaller parts, allowing us to view the causes of the movements and forms of the larger parts. But when the parts begin to disappear and we can no longer see them either with the naked eye or by microscope, what should we think then?[93] With our senses we see the effect; with our reason we can deduce the causes that we do not see.

The essence of the soul is activity, an ability to affect the body. The passive principle instead receives actions from the body. The chief faculty of the soul is analytical, philosophical reason. That is what distinguishes humans from animals. Nothing can exist in a closed and cohesive entity without being possible to relate to the outermost points or limits of the entity. In the same way, Swedenborg continues, nothing can exist in the body's microcosm that cannot be related to the soul. Nothing can exist in the macrocosm of the world without being related to the Infinite, without Him there is neither subsistence nor existence. He is the source, the cause, and the goal of the whole universe. The soul is a part of the body. It is mechanical and geometrical. 'So,' Swedenborg says, 'that if the reader would speak of the soul by comparison with machines, I have no objection to respond in the same tone; only I cannot allow the comparison with inanimate, but with animate machines.' But merely because the soul is mechanical and geometrical it does not mean that the soul is mortal. It cannot perish without the whole universe being annihilated. The soul machine consists of a mechanism that cannot be destroyed, cannot be damaged by fire, air, or ether. He who believes in God also believes in the immortality of the soul.[94]

Swedenborg promises a detailed study of the mechanism of the soul in the future. A great deal of work remains to be done on this. We are like simple, uneducated inhabitants of a country where people are ignorant of the limits of the earth and do not even know that it has any. Although they see that the country does not end with their own village, they imagine that it must surely end not too far beyond it. And when they travel further after a while and come to a lake, or to the shore of a sea, they exclaim that they have found the end of the earth. But their successors who have learned the art of navigation have found by experience that the end of the earth is not there. Finally they discover that the earth is round or oval, and that they can circumnavigate the whole globe. 'In short the children are more enlightened than the parents, and know the very thing which the parents declared unknowable and gave up accordingly; so they laugh at the simplicity of their forefathers.'[95]

---

[93]*De infinito*, 173; translation, 149.
[94]*De infinito*, 160, 179, 186–191; translation, 137, 157, 161–166; cf. *Psychologica*, 56f.
[95]*De infinito*, 195f; translation, 169; cf. *Psychologica*, 152f; *Treatises*, 134f.

Swedenborg thus sides with the moderns in the battle between the ancients and the moderns, 'la querelle des anciens et des modernes' which the poet Charles Perrault triggered in 1687.[96] Swedenborg believes in progress, that we actually have come further than the ancient writers. The humanist Benzelius took up his stance on the other side when he wrote in the foreword to Jacob Serenius's English-Swedish-Latin dictionary from 1734 that mathematics had not gone beyond where Euclid had left it. The reason for the lack of scientific progress was the abandonment of Greek as the language of science, with the division into a multitude of languages.[97] We have indeed come further, Swedenborg claims, as can be demonstrated in anatomy; we now know, for example, that hearing is caused by undulations of air. The learned world carries on with experiments to reveal nature's secrets, so there is no reason to doubt. One day we shall reach the soul and assess its mechanical operations. Science is finding more and more similarities between finer and grosser substances. 'And why not at the present day? so as to forestall our posterity, and prevent them from laughing at us as we ourselves now laugh at some of the old philosophers.'[98]

We can argue from effects to causes with good certainty using an analytical method, from the known to the knowledge of the unknown that we seek. Nothing happens in the gross world whose principles cannot be found in the subtle world. All connections require contact. This is known in applied mechanics. The invisible world is linked to the visible one through contact. There can be no such thing as an absolute division or breach between the soul and the body. The soul is the centre of all perceptions. Change is motion; we cannot hear or feel anything without some movement. Motion is the cause of sensory perceptions through the external sense organs. Everything has mutual relations, linked from one to the other, from the subtler to the grosser and vice versa. One corner of the earth is connected to the other, as there are connections between the sun, the stars, and the observer's eyes. 'Why not boldly advance by reason to a subtler sphere, and arguing by analogy, infer its possibility and existence?' If we have experiences that confirm reason, 'we ought to go beyond the point where the bare senses begin and end.'[99]

The investigation of the reciprocal action between body and soul in *De mechanismo* is an attempt to provide a solution to the problem of psychophysical interplay. Descartes assumed that the body lives in time and space, has measurable properties and exists objectively. The soul, in contrast, is outside space, immeasurable, and subjective. The question then is how these two essentially different substances can affect each other at all. In Malebranche's Cartesian philosophy there is an attempt to resolve the problem of this interaction. In his occasionalism theory, which was introduced into the Swedish debate by Hoffwenius and Drossander, God is the true efficient cause. Swedenborg borrowed material for his solution from Malebranche's

---

[96] *Principia*, 239; translation II, 34; Perrault, III.
[97] Benzelius (1734), 1–5; B. Lindberg (1984), 91.
[98] *De infinito*, 197; translation, 170.
[99] *De infinito*, 211f, 215f, 222; translation, 182f, 186, 191, 195.

De la recherche de la verité (1674–1675).[100] The body cannot affect the soul, according to the occasionalist, but it is the occasion for God's intervention, since the physical leads God to evoke a psychic state. The same applies in reverse. In other words, Malebranche claims that there are no true causal relations between things; the only real cause is God. Leibniz tried to avoid the problem with the aid of his theory of pre-established harmony, that there is a predetermined harmony established by God. The physical and the psychical thus run parallel, without affecting each other. Swedenborg tries instead to solve the problem by eliminating the essential difference between the two substances and assuming that both are extended and mechanical. Even the soul is spatial, measurable, and can be observed through empirical studies. In this way he seems to evade the problem of interaction, since both the body and the soul are mechanical and spatial and thus can touch and affect each other in either direction. In practice the soul could be studied objectively in an anatomical dissection, and the natural philosopher's mechanical principles could be employed to solve the mystery of the soul.

The soul thus consists of a kind of active and passive particles. Swedenborg therefore tries on rational grounds to find the seat of the mechanical soul in the human body. He dismisses the idea that the soul can be scattered through the whole body. Swedenborg reported in *Psychologica* that Comenius describes the human soul hieroglyphically, like ideas stamped on children's minds with the aid of images (Fig. 1). The scholastics, Swedenborg notes from Wolff, say that the soul is in the whole body and the whole soul is in every part of the body.[101] In *De mechanismo* Swedenborg likewise rejects Descartes' thesis that the way to the soul goes through the pineal gland.[102] The psychical particles that resemble the physical particles, but are of a finer and more subtle quality, must instead be localized in the organs where the bodily membranes are finest, that is to say, somewhere in the cortex of the brain and in the medulla oblongata. The particles of the soul have a mechanical link through the membranes to the particles of the body. Like every connection or nexus in the mechanism, as between body and soul, they also require a form of mechanical contact between the parts, as in machines with levers and cogs. The psychical and physical particles affect and touch each other as in a machine where the membranes have the function of transmitting vibrations between body and soul.

The human machine in its entirety can be compared to a microcosm with its sets limits, all analogous and similar to the macrocosm. In the small 'human world' there is a membranous and elementary nexus between body and soul, just as there is a nexus between the infinite and the finite in the macrocosmic world. The human soul is a part of a purer world, and man belongs to both the more subtle world and the grosser world. Swedenborg goes on to call the human soul a machine, although not a lifeless soul machine but a 'living machine'. Even animals have souls, albeit of a simpler and coarser kind. To be exact, they consist of the second elementary

---

[100]Jonsson (1999), 38; Lindroth (1989), II, 47.

[101]*Psychologica*, 90f; Wolff (1732c), § 152.

[102]Descartes (1649), I, § 31; *Oeuvres* XI, 351f.

Fig. 1 The human soul. 'The soul is the life of the body, one in everything.' Through the geometrical points lacking dimension, Comenius denoted the unity and simplicity of the soul, and through their dispersion over the body its substantiality and union with the body (Comenius, *Orbis sensualium pictus* (1684))

particle, the magnetic one. This explains why animals can orientate themselves and seem to have some sort of built-in compass.

The extension of the human soul, unlike physical matter, cannot be destroyed. The grosser elements cannot annihilate the more subtle particles of the soul. It was in fact one of the purposes of *De mechanismo* to try to demonstrate the indestructibility and immortality of the soul. Believing the soul to be mortal and destructible would lead straight to materialism and atheism. For Leibniz the soul is indestructible, as a mirror of an indestructible universe. What happens at death, Swedenborg speculates in a draft, is that the soul is compressed into a very small ball that is then taken by angels to heaven.[103] There it returns to its original form. Epicurean philosophy has an idea of the atoms of the soul being separated from each other when death occurs. Democritus believed that the air contains a large number of psychical atoms that flow in when a person breathes. Life and death thus depend on inhalation and exhalation. Aristotle criticizes Democritus's idea of indivisible, spherical soul particles whose movement sets the whole body in motion.[104] It is like the comic playwright Philippus's story of Daedalus, who made a wooden image of Aphrodite which could move when he poured mercury into it. According to Lucretius, the

[103] *Photolith.* III, 101; *Treatises*, 145f.

[104] Aristotle, *Peri psychēs*, 1.3.406b16–26.

soul was built up of small round particles scattered sparsely all over the body. Since the senses are mobile, quick in motion, they must consist of very small particles, smooth and round, similar to the larger rolling particles of water but unlike the sluggish, sticky consistency of honey. 'For a kind of thin breath mixed with heat leaves the dying, and the heat, moreover, draws air with it,' writes Lucretius.[105] But Swedenborg's soul particles, unlike those of Epicurus, are impossible to disperse. The soul lives on in heaven after its journey with the angels. The soul as a globe can be found in Origenes, who claimed for a time that the dead are resurrected in the form of a sphere—the form of perfection.[106] The sphere is the self-image of the soul, Marcus Aurelius discovers when considering himself.[107] The soul resembles the perfect, mobile, elusive sphere. Images of tennis in books of spiritual emblems could be used to illustrate how evil powers play with the human soul. The soul is the ball.[108]

It can be envisaged that Swedenborg's mechanical soul particles lead to a form of dualistic materialism. Indeed, some reviewers accused him of materialism. Like the materialist Julien Offray de La Mettrie, he regards man as a machine. In the same way as it may seem that he materializes the soul, the reverse can be true, that he spiritualizes matter. In *De infinito* his thinking is based not just on the soul but also on the body and the interaction between body and soul. In this way matter has a thinking function. A reviewer of *De infinito* in *Nova Acta eruditorum* for 1735 pointed out the similarity between Swedenborg and the German philosopher Andreas Rüdiger as regards the assumption that the soul follows mechanical and mathematical laws.[109] In *Physica divina* (1716) Rüdiger ascribes spatial extension to the soul, as Henry More had done. It is not at all certain that Swedenborg had been inspired by them, as might be assumed. Similar problems and metaphors often lead to similar conclusions. Wolff declared in *Theologia naturalis* that the belief in the spatiality of the soul, which Swedenborg seems to embrace, must necessarily lead to materialism. Above all, Wolff opposes those who assume that the soul is something substantial or resembles mathematical points. But it was not Swedenborg's intention to put forward a materialistic outlook. His hybrid of natural philosophy and metaphysics materializes and spiritualizes the soul. Orthodoxy is watching over him as he writes about the infinite. It is as if Swedenborg will not, cannot, or dare not go the whole way and accept the consequences to which his thoughts lead. With the metaphor of the world machine his thinking is led in a specific direction towards determinism, pantheism, and materialism, which the Lutheran orthodoxy could scarcely excuse.

---

[105] Lucretius, 3.233f, cf. 180–205, 375f.

[106] Brendel, 32.

[107] Marcus Aurelius, 11.12.

[108] Spegel (1966), introduction, 7.

[109] *Nova Acta eruditorum*, December (1735), 558; Rüdiger, 158; Lamm (1915), 41f, 48; translation, 42, 49.

The Philosopher, the Happiest or the Unhappiest of Mortals

Swedenborg's increasingly bold ambitions in research and his curiosity about invisible nature finally led him beyond the limits of the unknowable. In *Principia* he studied the mechanical, finite world after the first movement of the mathematical point; in *De infinito* he continued even deeper into the enigmatic existence of reality and was forced to submit to 'the law of humanity'. He passed the mathematical point and the nexus, crossed the chasm between the finite and the infinite, finally to try to arrive at the essence of the infinite. But there reason, willy-nilly, necessarily had to stop, if reason believes in its own reason. The infinite can never be grasped with finite reason. It was not until ten years later that he took the final step over the chasm and entered into infinity. The nexus of the mathematical point, the gateway to the geometrical field of finity, was thus also the gateway to the spiritual world of infinity.

In his works Swedenborg constantly weaves natural science, metaphysics, and theology together. His double nature is on the borderland of thinking. With the help of anatomy he tried to find the room of the soul and God's purpose with the world. Through the finite numbers of infinitesimal calculus he sought to prove the existence of the infinite. The mechanics of the soul showed that there was no difference between natural science and philosophy. Natural science merged with philosophy to be combined into a form of anatomical metaphysics and mathematical theology. Faith and knowledge are necessary for each other and can never be in conflict. Although reason has proved to be finite and to have its limits which it cannot surpass without losing its rationality, *De infinito* becomes a worship of the finite possibilities of reason. Swedenborg is one in a series of eighteenth-century worshippers of reason. But there is something unreachable, unknown, and unknowable. The quest for knowledge about things ultimately leads to the goal that everything in the finite world strives for—infinity.

Knowledge makes man both happy and unhappy. As if following a theme about Democritus and Heraclitus, the laughing and crying philosophers in emblematics, Swedenborg ends *De mechanismo* with a meditation on the happy or the unhappy philosopher.[110] If the soul in its mortal body is shaped and prepared for its immortal state, then man is the happiest of beings on earth, otherwise he is the unhappiest. The unhappy person is unhappier than the animals whose souls are annihilated and whose lives are erased when their bodies perish. Christians are even more happy or unhappy through their knowledge of faith. The unhappy ones are unhappier than the unhappiest heathen. Priests are even more happy or unhappy, and the unhappy ones unhappier than the most ignorant Christian. The Christian philosopher is even more happy or unhappy, and the unhappy philosopher is unhappier than the one who has acquired his faith solely through revelation. The Christian philosopher can thus become the happiest or unhappiest of mortals. 'For the more knowledge we possess, the more there is to make us happy, and the more to make us unhappy.'

---

[110]*De infinito*, 269f; translation, 231f.

Conclusion

> I observed that their endings or circuits generally coincided with the heavenly
> breathing, which in the ratio to my own is as 3 to 1.[1]

> Swedenborg, Experientiæ spirituales (18 November 1748)

The Convolutions of the Brain

One October day in 1736, Swedenborg was strolling through the Tuileries Garden
in Paris. It was a pleasant walk as he speculated about the forms of atmospherical
particles.[2] It no longer happened so often that he mused about particle physics. His
thoughts were instead turned towards anatomy and physiology. His quest for proof
of the immortality and the seat of the soul had led him on to a different track. He
abandoned his mechanistic explanations and moved to a teleological and organic
world-view. The metaphor *the world is a machine* was replaced by *the world is a
living being*. He began to think about nature in terms of creatures with life, actions,
and aspirations. The universe is alive, animated, a single large organism in which
everything is reflected in everything else.

An attempt in this direction is *De cerebro*, which deals with the brain and its
functions, a manuscript that was presumably completed in Venice in August 1738
before he set off for Padua.[3] The study of the circulation of the blood led him into
the meandering convolutions of the brain. He found that not only the cerebrum and
the blood vessels display serpentine and helical twists, but also the cerebellum, the
heart, and the muscle fibres. The whole universe actually moves this way, from

---

[1] Experientiæ spirituales 18 November 1748, n. 3989; cf. Arcana cœlestia, n. 3884.

[2] *Resebeskrifningar* 4 October 1736, 75; *Documents* II:1, 94.

[3] *Resebeskrifningar* 9 August 1738, 85; *Documents* II:1, 110.

D. Dunér, *The Natural Philosophy of Emanuel Swedenborg*, Studies in the History
of Philosophy of Mind 11, DOI 10.1007/978-94-007-4560-5_9,
© Springer Science+Business Media Dordrecht 2013

the largest to the smallest parts. This form is propelled by itself, it is eternal, spontaneous, the most perfect geometry, and the noblest of all forces.[4] It is the figure of figures. For the most perfect things this is the sole movement. Nature throws herself spontaneously into this movement, since the spiral form does not offer any resistance—it consists of an infinite number of angles and planes, or else a single angle and a single plane. In the eternal and most perfect spiral, the centre is at the circumference and the radius, while the radius and circumference are at the centre. This figure produces all the geometry and mechanics in the world. This infinite Archimedean screw embraces and surpasses all forces in the composite world.

These words of praise are followed by a couple of experiments with twined ropes, with the aid of which Swedenborg renders his spiral idea in concrete form. Like Minerva Lanifica, the divine spinner, who weaves the world with her woollen yarn (and is also the goddess of Roman physicians), the brain twines its fine threads. These spiral and helical cycles in the cerebral tissue—and this 'follows geometrically', he underlines—display poles, equators, ecliptics, and meridians in the same way as the celestial bodies of the universe. Unfortunately, we cannot observe this with the aid of a microscope, but we can draw the conclusion that there is an underlying reality that is even more perfect, a reality beyond the limits of vision. Our conclusions must instead build on reason and geometry. He ends the chapter about the planes, axes, and centres of the brain by stressing that he does not believe that reason suffers from any hallucinations about the geometry of these invisible organs.[5]

Swedenborg's cerebral physiology shows a capacity to think in pictures, to create models of thought, to find patterns in the anatomy and physiology of the brain, which was perhaps facilitated by his familiarity with mechanical instruments, his ability to illustrate and to read technical drawings. Sometimes we see what we want to see, what we expect to be able to see. Among other things, Swedenborg adopted an observation that Malpighi had made of the brain tissue. Malpighi thought he could see round or oval, closely packed small bodies enveloped by blood vessels which he called 'glandulae'. Swedenborg perceived them as small membranes with spheres enclosed in them and called them 'spherulae' or 'cerebellulae', little spheres or brains. In fact this was a chimera. The small balls were optical artefacts that arose in the preparation of the cerebral tissue.[6] Swedenborg's typical method in *De cerebro* and in his other works on anatomy is first to excerpt long passages with 'experientia', experimental evidence obtained by anatomists such as Vieussens, Winsløw, Willis, and others, and then in an 'inductio' or an 'NB' he draws his own conclusions from the assembled material.

---

[4]*Photolith.* IV, 134; *De cerebro*, translation I, n. 409.

[5]*Photolith.* IV, 259f; *De cerebro*, translation I, n. 746.

[6]*De cerebro*, translation II; Bidloo, tab. X, fig. 2; cf. Gordh and Sourander, 103; Wetterberg, 33–41; Gross (1997), 142–147; Gross (1998), 119, 124f.

From Angular to Perpetuo-Spiritual Form

In *Oeconomia regni animalis* (1740–1741) Swedenborg explores the subtlest functions of the human body but does not proceed from laboratory experiments or microscopic studies performed by himself; instead he follows a synthetic approach. He draws conclusions on the basis of excerpts from investigations by well-known microscopists and physiologists, such as Malpighi, Leeuwenhoek, Swammerdam, and Boerhaave. When he comes to the subject of the growth of the chicken in the egg it is clear that he has left Cartesian mechanics behind and is orientated towards an organic view of the world. He now speaks of a shaping power akin to Paracelsus's and van Helmont's creative natural force, 'Archeus'.[7]

In this work, which chiefly deals with the physiology of the human body, with blood and bodily fluids, we find once again his geometrical vision. Swedenborg noticed, for instance, that spiral motion commonly occurs in nature. The spherical form is the most appropriate for forces in nature since it has no angles or protruding parts. It rests in turn on a first principle: the perpetuo-spherical or cubico-spiral form. Nature, according to Swedenborg, displays no more advanced curve than this, when a thing goes from its previous world to its future one, that is, when something changes.[8] The fibres of the heart muscles consist of perfect spirals or helical curves, the forms of which are significant for the expansion and contraction of the heart. The vortices of the smallest substances are in perfect harmony with the solar vortex.

In the chapter about the soul in the second part there is a passage that describes nature as a large circle.[9] This universal circle comprises an unbroken sequence of smaller circles. All points in these circles, irrespective of where they are within the big circle, refer in towards the shared centre. If any point in its orbit around its circle in the big circle does not refer to the shared centre it is thrown out of the big circle because it is false. It is the same with the activities and emotions of the soul and the body. They too rise or fall in relation to the centre of reality—God. These ideas seem Neoplatonic. In Swedenborg's philosophical notes from 1741 there is an excerpt from the pseudo-Aristotelian work known as 'Aristotle's Theology'.[10] In Swedenborg's days it was assumed to be the work of Aristotle but it was probably written by Porphyry the Neoplatonist. This describes the divine essence as a centre in a circle, to which all radii point. Everything is linked to this centre, everything longs to get there, and nothing can exist without it. From this centre emanates the divine, omnipresent light, like the light from the sun. This is an allegory of light created by metaphorical thinking, which is often found in the Neoplatonic tradition,

---

[7] *Oeconomia* I, n. 253; Jonsson (1999), 47; Lamm (1915), 69f; translation, 69f; cf. Fenzl, 20–22.

[8] *Oeconomia* I, n. 101.

[9] *Oeconomia* II, n. 287.

[10] *Note Book*, 241, cf. 508f; Plotinus, *Ennead* IV, 4.2.1–22; Pseudo-Aristotle, book XII, ch. XIX, 1081; cf. Bergquist (1996), 75, 87; Hallengren (1991), 231–238; Hallengren (1997a), 24f; Swedenborg gave Benzelius a copy of Plotinus's *Opera philosophica* (1580) that had once belonged to Stiernhielm. LiSB; cf. Nelson, 43.

as in Stiernhielm or Friedrich Menius. In a similar way, in the spirit of Hermes Trismegistus, as formulated in Upmarck's dissertation on the circle of things, God could be compared to a circle with its centre everywhere and no circumference anywhere.[11] In Comenius's *Orbis sensualium pictus* God is like a triangle in a circle, the trinity in the everywhere and nowhere of the circle.

In Aristotle we find the idea of the soul as a circle, of thinking as a cyclic movement in the circle of reason.[12] The circle, Proclus wrote, is the activity that returns to itself and expresses Nous, while the straight line is the soul. When the soul refers to Nous, then, it moves in a circle.[13] According to the Renaissance humanist Marsilio Ficino, the compasses, the circle, and the sphere were the emblem of melancholy. To attain wisdom and learning, the soul must be attracted inwards, away from outward-moving things, as a movement from the circumference to the centre.[14] Among Swedenborg's philosophical excerpts there are notes from the pseudo-Augustinian text *De spiritu et anima* which consists, among other things, of a compilation of Boethius and other medieval authors.[15] Boethius's *De consolatione philosophiae* (AD 524) describes the relationship between the immobile providence that is at the centre while the mobile destiny orbits around this centre.[16]

Nature as a circle with radii pointing towards a universal centre leads on to a common notion in the history of ideas: the opposition between the curve and the straight line, between the circle and the square, and between the sphere and the cube. The first in each of these pairs represents the spiritual, divine, and celestial. The latter stands for the material, human, and earthly. Aristotelian physics made a distinction between the celestial circular movement and the sublunar rectilinear movement. In a manner akin to this, Plotinus distinguished between the spiritual, circular movement and the physical, rectilinear movement. Cusanus believed that geometrical figures primarily have a symbolic meaning because they extend to the infinite, where the curved line and the straight line join. In his thinking there is a continuous path from the finite figure to the infinite one. The truth relates to the intellect as the circle to the polygon. The intellect approaches the truth in infinity without becoming it, it comes closer to it with more and more angles and then increasingly resembles the circle.[17] Kepler thought in turn that God in the beginning made a distinction between the curve and the straight line, equating the former with Himself, while the straight line refers to His creation.[18] The straight and the curved are opposites which eliminate each other, wrote Harald Vallerius after Bilberg.[19]

---

[11]Upmarck and Fahlenius, 7.

[12]Aristotle, *Peri psychēs*, 1.3; Forsius, 333.

[13]Proclus, 147, cf. 108–109.

[14]Warburg, 645.

[15]Jonsson (1961), 38f; translation, 23; Jonsson (1969), 175.

[16]Boethius, IV, pr. 6.

[17]Cusanus, book 1, ch. 3.

[18]Kepler, 45.

[19]H. Vallerius and Rockman, 35.

Swedenborg made notes about orders and degrees of lines, curves, figures, and numbers, in a long list of geometrical figures from Wolff's *Ontologia*.[20] Wolff mentions Euclid's view of the circle as the noblest figure, followed by rectilinear figures, then triangular, quadrilateral, polygonal, after which come Apollonius's conical sections and Descartes' algebraic equations for distinguishing curves. These were notes in the quest for a hierarchical system of geometrical forms, from the simplest to the most complex, from the least noble to the most perfect. Proclus assumes a hierarchy of figures, starting at the top with the gods themselves and extending down to the lowest beings, the perfect ones of which take precedence over the imperfect ones. At the bottom are the material figures, as well as the intelligible figures of heaven and the soul, and at the top stand the figures of the gods, which are independent, uniform, simple, and contain all perfection within them.[21]

Swedenborg has a similar hierarchy of forms, from the finite, material ones to the increasingly infinite and spiritual forms. He put this idea forward in the posthumously published *De fibra* from 1740, which was intended to be the third part of *Oeconomia regni animalis*.[22] In this theory, the form doctrine, he does not use an Aristotelian concept of form but rather a Wolffian representation of the universe. Swedenborg's form moves gradually from the concrete form of natural objects in physics, via the vital forms in biology, and on through forms of thought and ideas, to spiritual, divine form. It all began when Swedenborg tried to determine the forms of motion of fibres in the human body. He refers to Leeuwenhoek, who assumed that the blood vessels and nerve fibres had a circular or spiral form like shells. Since these fibres could not be seen by microscope, Swedenborg instead assumed them to be derived from a hierarchy of forms.

Behind the forms that the eye cannot distinguish there extends a hierarchy of forms from the simplest angle to the ultimate being, incomprehensible to reason: the perpetuo-spiritual form, the principle of all forms. The forms belong to a series or scale of perfection, rising or falling in relation to the origin of everything in God. This ladder of forms, from the lowest to the highest, has seven steps. Each higher step means that the lower form is 'perpetuated', becoming infinite in several respects, acquiring a higher degree of eternity, infinity, and spontaneity. They are reflected in each other, and the higher up they are in the hierarchy, the more eternal, perpetuated, free, and varied they become. Swedenborg interprets the different forms in the hierarchy as indices, signs with causal connections to something not present in themselves. They point towards properties that exist in forms further down or up the scale.

The lowest form is *angular*, since the angle is bound to straight lines which lack a common centre. Examples of angular forms are tetrahedrons, octagons, and the polygonal forms of salts. These give rise to taste and smell, and they constitute the research objects of geometry, physics, and chemistry. Of higher rank is the *circular*

---

[20] *Geometrica et algebraica*, 278; Wolff (1730), § 246.

[21] Proclus, 138–139, s. 111.

[22] *De fibra*, n. 260–273; cf. Jonsson (1969), 125ff; Jonsson (1999), 88.

or spherical form, which is the form of motion of matter. This form, which is also called perpetuo-angular, has adopted a feature of the lower form in that the radii of the circle consist of straight lines from circumference to centre. The circle constantly has the same angle and plane. The sphere is more perfect than the angle. Its surface resembles an infinite angle, and all radii can be taken back to a single point, its centre. The third form in the hierarchy is the *spiral*, or perpetuo-circular form. The circle, in other words, has been assigned a higher degree of infinity. It can be varied in all kinds of ellipses, cycloids, and curves. But the spiral can be varied in even larger measure, since it does not have a fixed centre but varying centres; its centre consists of spherical surfaces, and all possible curves can be its centre. The spiral is therefore more perfect than the sphere and the circle since it has been granted a degree of eternity and infinity. The spiral form applies to active natural forces. The methods of geometry take us almost all the way to this level, where the spiral challenges its capacity.

The next step is the *vortical* or perpetuo-spiral form. This has been assigned yet another degree of infinity. The whirls orbit in spirals and describe all circles from centre to circumference. This is the form in which the vortex moves and it corresponds to the vortex around the earth. Then comes the fifth form, the *celestial* or perpetuo-vortical form. This is the highest form in nature, and according to Swedenborg it is equivalent to 'the One' in Plato's *Timaeus*, or 'the First' in *Parmenides*, but is also corresponds to Leibniz's 'monad' and Wolff's 'simple substance'.[23] The celestial form is infinite to a higher degree and is also impossible to describe in words or geometry. This form lacks proper shape, extension, size, and weight. Nothing is above or below anything else, just as nothing can be referred to centre, radius, or surface. In addition we have the *spiritual* or perpetuo-celestial form. This too cannot be described in words—it is inaccessible to human reason and can only be described with the aid of an abstract language, like the tongue of the angels. The spiritual form stands for communication between the human soul and the divine. But there is yet another form, finally, the *divine* or perpetuo-spiritual form, which is beyond all natural attributes. It is strictly not form but divine being, the creator of everything, the beginning and the final goal.

This Jacob's ladder of forms, from the lowest earthly ones to the divine, expresses Swedenborg's endeavour to find a universal order, a connection between heaven and earth. It is the long chain of existence in combination with the classical idea of an opposition between the earthly straight line and the divine curve.[24] The doctrine of forms is a Jacob's ladder leading up to God, an echo of the Platonic idea that geometry leads the mind up towards supersensual, divine forms. All the forms on this ladder aspire towards the supercelestial form, and from this form the divine flows down to the natural forms, as in the Neoplatonic emanation. From the divine comes the spiritual, from which is created the celestial, from which arises

---

[23]*De fibra*, n. 266; Jonsson (1961), 231; translation, 250; Jonsson (1969), 127.
[24]*Oeconomia* I, n. 580, 583, 586.

the vortical form, and so on down to the lowest angular form.[25] The geometrical-mechanical construction of the world in *Principia* has given way to an organic and teleological system. Nature is a chain of forms that all lead towards their final goal, the divine form.

A Blind Man Who Can See, and the Form of Ideas

In July 1743 Swedenborg presented to the Royal Academy of Sciences a new research plan which was intended to result in a work consisting of 17 parts about the whole of human anatomy, physiology, and psychology.[26] Only three parts about the kingdom of the soul, *Regnum animale* (1744–1745), appeared during his lifetime. In the chapter about the stomach in the first part, which contains a recapitulation of the doctrine of forms, he writes that a perpetuo-circular or spiral form of motion can be found in the stomach, the intestines, the brain, and all over the body. It is the essential form of motion of organic substances. The spiral can be observed through its convolutions and tracks, in the fibres of the muscle membranes, the nerve fibres, the arteries and the veins, the ligaments, the glands, and in the form of the cochlea, which corresponds to the spiral motion of air. These spiral movements are exactly synchronous with the respiration of the lungs, where the spiral describes a revolution during each breath. All the parts of the organic body follow the law that larger, composite, and visible forms are dependent on smaller, simpler, and more perfect ones. Ultimately they are dependent on the smallest, invisible forms which are so perfect and universal that they contain an idea that represents the whole universe.[27]

A manuscript about the five senses, *De sensibus* (1744), which was supposed to be the fourth part of *Regnum animale*, gives an account of the spiral form of the cochlea and the movements of the eye, which can describe not only circles but also spirals.[28] Vision was of special interest to Swedenborg. He had personally experienced remarkable light phenomena or photisms in Amsterdam in 1739.[29] His philosophical notes from the 1740s show his particular interest in biblical passages with metaphors of light and dark. He also made excerpts from Augustine about analogies between the intellect and the sense of sight.[30] Among his optical notes in the manuscript *Experimenta physica et optica* (1744), and later in *Regnum animale*,

---

[25]*De fibra*, n. 268, 272.

[26]*Photolith.* III, 141; Swedenborg became a member of the Royal Academy of Sciences, proposed by Linnaeus on 26 November 1740, elected on 10 December, and introduced on 8 January 1741. *Svenska Vetenskapsakademiens protokoll* I, 271, 272, 275, 282; Jonsson (1969), 79f, 333f.

[27]*Regnum animale* I, n. 97, 100.

[28]*De sensibus*, 76–82, 105f; translation, n. 230–261, 395.

[29]Jonsson (1967–1968), 48–51.

[30]*Note Book*, 32, 337ff; *De anima*, 51f; translation, n. 96; Jonsson (1969), 185.

Swedenborg brings up the controversial 'Molyneux problem'.[31] This problem was formulated by the Irish philosopher William Molyneux in a letter to Locke in 1688: 'Suppose a Man born blind, and now adult, and taught by his touch to distinguish between a Cube and a Sphere of the same metal, and nighly of the same bigness, so as to tell, when he felt one and t'other, which is the Cube, which the Sphere. Suppose then the Cube and Sphere placed on a Table, and the Blind Man to be made to see. Quære, Whether by his sight, before he touch'd them, he could now distinguish, and tell, which is the Globe, which the Cube.'[32] No, replied Locke. He would lack experience of which form affects his touch and simultaneously affects his sight. Rationalist thinkers tended to answer in the affirmative, while empiricists replied in the negative. *Philosophical Transactions* for 1728 published an article by the surgeon William Cheselden which was to raise the temperature of the debate. He had succeeded in performing a cataract operation on a young man who had been born blind.[33] Cheselden describes what the world was like for the man who had had his eyes opened, how he was amazed at the colours, found it difficult to judge distances, and at first could not determine what was a dog or a cat without lifting it up. The persons he liked most were those he found beautiful, and things that were agreeable to his sight also pleased his taste.

It is Cheselden's thread that Swedenborg picks up. If you want to get to know things properly, Swedenborg says in *Regnum animale*, touch must instruct sight and sight must instruct touch. Touch and sight are like a married couple, sight being the wife and touch the husband. An idea that is born of sight and touch together is a legitimate child, but a child born of either sight or touch alone, such as a person who is blind at birth, is a bastard. Sight has to learn from touch about what lies in things beyond its external and contingent form. To distinguish between art and nature, touch acts like a scout sent out by sight. Touch is instructed by sight about quantities, such as the extent of forests, lakes, cities, palaces, houses, or such things as are so small that they can scarcely be discerned in a microscope. The truth of this is clearly demonstrated, Swedenborg argues, by the case of people who are born without the gift of light and only acquire it by operation. They scarcely rely on sight if they do not enlist the aid of touch. Swedenborg's answer to the question is therefore likewise 'no'. In support he refers to Locke: the man who regained his sight by operation could not distinguish sizes, distances, or shapes by contrasting light with shade unless he used touch to assist in his interpretation of what he saw.[34]

De anima (1742), the seventh, posthumously published part of *Regnum animale*, which is in part an investigation of the pure intellect, leads up to an attempt at a universal mathematics with the aid of which one could calculate all scientific

---

[31] *Experimenta physica et optica*. KVA, cod. 58; *Photolith.* VI; *Regnum animale* II, n. 557.

[32] Locke, book II, ch. IX, § 8; The problem was discussed by Berkeley, Voltaire, Condillac, Diderot, and others; Degenaar, 17.

[33] Cheselden (1728), 447–450.

[34] Locke, book II, ch. IX, § 8; Cheselden (1741), book IV, ch. IV.

propositions.[35] The pure intellect, which in Swedenborg's hierarchy of form corresponds to nature's highest form, the celestial, comprises the body's innermost nature, all knowledge and the fundamental principles of all sciences. If we knew these basic scientific principles, we could use a universal mathematics to work out all simple ideas. This is a rationalistic idea in Swedenborg, according to which man has an innate capacity for all knowledge. All we have to do is to identify these ideas in the pure intellect. We know that ideas are changes of state in the brain cells. If one can describe these changes geometrically as circular and spiral forms, it should also be possible to use a calculus to perform computations with these ideas. We would thus arrive at a universal mathematics.

Swedenborg's Euphoria

Swedenborg was still a busy assessor in the Board of Mines. Life mostly consisted of meetings of the Board, anatomical studies, and travels. Swedenborg appears humane in his skilled management of a case in the Board of Mines in 1741 concerning the negligence of a clerk named Duseen. He goes through the testimony with a forensic and logical eye, to answer the question whether the clerk's 'main weaknesses result from drunkenness, or his drunkenness from his main weaknesses'.[36] Perhaps he recognized himself in Duseen's experiences of religious brooding, severe anxiety of the heart, and strange voices within him. But Duseen had been in a fight where he had lost his wig. The barber-surgeon Petter Daumont testified that Duseen had 'had a strong bout of melancholy, since he had raged, taken off his shirt, run naked; and that his illness derives from an Orgasmo Sangvinis or fermentation in the blood, which comes from movements of the senses—and from drunkenness.'[37] Swedenborg drew the conclusion that the madness was not caused by drunkenness, but that the drunkenness was caused by madness, and he suggested that the man's pay should not be entirely withheld, as this would plunge him into an even more wretched state. He should instead be freed from his duties and retain at least half of his salary.

Swedenborg himself requested leave of absence two years later in order to have the first two parts of *Regnum animale* printed in the Hague. In July 1743 he left Sweden and set off for Holland. Early on the morning of 10 April 1744 Swedenborg woke up:

> This night as I was sleeping quite tranquilly, between 3:00 and 4:00 o'clock in the morning, I wakened and lay awake but as in a vision; I could look up and be awake, when I chose, and so I was not otherwise than waking; yet in the spirit there was an inward and sensible

---

[35]*De anima*, 255–258; translation, n. 562–567; *Oeconomia* II, n. 206, 211; Jonsson (1969), 119, 129f, 342.

[36]RA, Bergskollegii arkiv, AI:96, II, 1499 (17 December 1741).

[37]RA, Bergskollegii arkiv, AI:96, II, 1502.

gladness shed over the whole body; seemed as if it were shown in a consummate manner how it all issued and ended. It flew up, in a manner, and hid itself in an infinitude, as a center. There was love itself. And it seems as though it extended around therefrom, and then down again; thus, by an incomprehensible circle, from the center, which was love, around, and so thither again.

He was seized by a vertiginous euphoria, an infinite expanding, circling movement from the very centre of love and back again, unfathomable circular movements from centre to periphery and back to the centre. This infinite circular movement expanding from the centre and returning towards the same centre is like the infinite spiral motion of the mathematical point that cannot be grasped by reason. The striving of the mathematical point to move becomes an inner state of supersensual spiral motion. In the very next sentence Swedenborg gives his own simile for this euphoric circular movement: 'This love, in a mortal body, whereof I then was full, was like the joy that a chaste man has at the very time when he is in actual love and in the very act with his mate.'[38] The spiral-like movement in Swedenborg's euphoria is in agreement with a distinctive feature in *Drömboken* ('Journal of Dreams', 1743–1744), namely, the erotic undertones of the dreams.

The whole of the preceding day, Swedenborg had been sunk in prayer, singing hymns of praise to the Lord, reading the Bible, and fasting. This was in the middle of the most eventful period during Swedenborg's dream crisis. His journal from this month is full of accounts of dreams, how he was grappling with the new emergent direction his life was taking, combating his own conceit, his scientific arrogance, and his ambition. He became increasingly aware of his spiritual mission, that he must leave the scientific course to devote himself entirely to the spiritual. Keeping a journal became a form of solipsism shaped according to the printed word, in which he scrutinized his own world, his thoughts, emotions, and visions in dreams. The obscure messages of the dreams and their interpretation became palpable as never before. He had probably had lively dreams before this April month in 1744. 'The word "dreaming",' he explains in his dissertation, 'sometimes seems to denote immense desires,' and he cites Erasmus of Rotterdam's *Adagia*: 'Even the sleeper dreams.'[39] What we notice is not just the content of the dreams, but also, and in particular, his interpretation of them. The difference between the dreams of the dream crisis and those of the visionary is the changed interpretation of the inner sensory experiences. The dreams in his *Journal* are the visionary's dreams in naked form, without an established interpretative framework. One can regard the journal of dreams as an archaic variant of the spiritual diary, before he had found his method of interpretation. The actual inner sensory experience may have been similar for the visionary and for the author of the journal of dreams.

A couple of days later he made a note in his *Journal of Dreams* which reveals how natural science inspires thinking in metaphors and colours the Latinized prose, when he compares thinking to the breathing of the lungs:

---

[38] *Swedenborgs drömmar*, 9–10 April 1744, 19f; KB, Swedenborg, I, p. 57; translation, 36.

[39] *Selectæ sententiæ*, 12; translation, 9f; Erasmus, II, 799.

that the will influences the understanding most in inspiration [breathing in]. The thoughts then fly out of the body inward, and in expiration are as it were driven out, or carried straight forth; showing that the very thoughts have their alternate play like the respiration of the lungs; because inspiration belongs to the will, expiration to nature. Thus the thoughts have their play in every act of respiration; therefore when evil thoughts entered, the only thing to do was to draw to oneself the breath; so the evil thoughts vanished.[40]

With his background in science he tries to understand his situation, his experiences, to give them words and concepts. Swedenborg had a special breathing technique which had arisen involuntarily in his childhood. Respiration ceased, his gaze was directed upwards, and he sank into meditation.[41] He discovered that breathing affected his thinking and was related to the heart. In *Oeconomia regni animalis* he writes about the relationship of mental activity to breathing, with slow breathing when his thoughts were calm and heavy breaths when he was anger.[42] Linnaeus referred in his lectures to Swedenborg's opinion that cerebral movements were dependent on breathing.[43] The movement of the brain is synchronous with that of the lungs, not the movement of the heart.

Swedenborg did not know whether he should continue with his scientific studies or devote himself to the spiritual. On the night between 28 and 29 April 1744 he dreamt: 'It seemed that I passed my water; a woman in the bed looked at me meanwhile: she was fat and red; I took her afterwards by the bosom; she withdrew herself somewhat; she showed me her secret parts and her obscenity; I declined to have any dealings with her.' Science entices him with its secrets. He ought to use his time for higher things, not for worldly things. 'God be so gracious and enlighten me further about what my duty is; for I am still in some darkness as to whither I should turn.'[44] Underlying this is metaphorical thinking. He understands and describes his situation in terms of the metaphor that *a life with a purpose is a journey*. A life without a purpose means that one is lost, lacking direction. Purposes or goals in life are destinations, actions are movements, and life plans are itineraries. He himself is a traveller. A life with a purpose requires planning since the journey can involve difficulties, a plan that states what goals are to be aspired to, and in what order. This is the situation in which he finds himself. But he is in the dark, lost, standing listlessly having lost his itinerary: 'It seemed that I went wandering astray in the darkness and did not go out with the others. I felt my way along the walls and came at last out into a beautiful house in which there were people who were puzzled as to how I could come this way.'[45] A new destination, a new itinerary gives him a direction in life.

---

[40] *Swedenborgs drömmar*, 12–13 April 1744, 25; translation, 45.

[41] *Experientiae spirituales*, n. 3320; Lamm (1915), 63; translation, 63; Lagerborg, 14ff.

[42] *Oeconomia* II, n. 10; *Diarium spirituale*, n. 111, 3464, 11215.

[43] Broberg (1975), 116; Bergquist (1998), 1–22.

[44] *Swedenborgs drömmar*, 28–29 April 1744, 39; translation, 70.

[45] *Swedenborgs drömmar*, 21–22 April 1744, 35; translation, 62.

Swedenborg's natural science led to a dead end. There were several different reasons why Swedenborg's scientific theories had not had a greater impact. His mistake, it could be said, was that he adopted the wrong style, not conforming to the prevailing scientific ideal. He had too much of the culture of the Ancients according to positivist science, and he lacked strict, personally acquired empirical evidence and mathematical stringency. His aims in his later works were also different, with theological and spiritual intentions. Another reason concerns his social circumstances. He had no disciples, was not attached to any university, had a marginal position in relation to the scientific academies, and carried on relatively limited scholarly correspondence.

Spiral Dances in Paradise

Swedenborg gave up the gigantic project of describing the whole human body. It was not by anatomy that he would find the soul. Instead he had a religious work printed in 1745, *De cultu et amore Dei*. It is a poetic paraphrase of the Bible's creation narrative, from the birth of the solar system out of the cosmic egg to the marriage of the first humans, a creation story in symbolic paintings, emblematic images, the symbols of a world theatre. The cosmogonic element, in line with his *Principia*, include the birth of the planets from the sun, how they move from the sun like the clinging of vines, following a spiral path from the hot centre. In the wonderfully beautiful grove of paradise, the rivers meandered with playful courses in constant rounds and convolutions. Snails carried their houses on their backs, shells that gleamed like precious jewels and were twisted in eternal circles or spirals in the same way as the vaulting of heaven.[46] In a book that was part of Swedenborg's library, *Biblia naturae* by the microscopist Swammerdam, there is a section which shows that all spiral shells can be derived from one and the same form, the tube. In a notebook from 1743 Swedenborg comments on Swammerdam's description of the eternal circular forms of shells and muscles, with thoughts about eternal spirals that can be seen in the shells of snails, and combines this with his hierarchy of forms.[47]

The sixth day, Swedenborg says, saw the celebration of the last day of creation and the first day of humanity, with a special kind of dance or ring game, 'the paradise game'. This consists of spheres and labyrinthine spirals, with the heavenly bodies rotating in a ring and bending from the circumference in towards the centre, in a circular and continuous bend until all are united in the centre. The paradise game became a counterpart of the spiral or perpetuo-circular movement of the mathematical point and the vortical motion of the planetary system. A similar dance recurs a little later, when Adam finds himself surrounded by naked young girls who

---

[46]*De cultu*, n. 11, 21, 28.

[47]KVA, cod. 53, 80; *Photolith*. VI, 177–264; Swammerdam, I, 141, 151; Swedenborg gave it to A. J. von Höpken. *Opera* I, 338; *Letters* II, 528.

start a winding ring dance. They moved in towards the centre in coils and spirals and upwards to the heights, ever forward. The girls radiated like stars, Swedenborg writes, casting their beams from centre to periphery and thereby forming a shining circuit around the spiral ring. These beings represented goodnesses, which means that the dance was the game of love and goodness.[48] The figurative spiral dance in *De cultu* has its counterpart in the geometry of the ballets, the serpentine lines of the minuet, the S-shaped movements of folk dances, which John Milton, for example, admired and which, according to William Hogarth, contrasted with the wild skipping in 'the dances of barbarians' with their convulsive shrugs and distorted gestures.[49]

The dance in *De cultu* becomes a language of correspondences. The spiral dance is an expression of progress, circling goodnesses that give rise to truth. The visual image of the spiral leads to a philosophical description of the relationship between goodness and truth. A distinct feature of Swedenborg is this close affinity between truth, goodness, and beauty, which can be found in many places in the history of philosophy, as in Plotinus and Anthony Ashley Cooper, the Earl of Shaftesbury.[50] The concepts proceed from a shared foundation: God, the All-wise and All-good, creates only what is perfectly true, good, and beautiful. Something that is true is thus also something good, even if they are regarded as separate. In Swedenborg's case the geometrical imagery expresses the relationship between the true and the good. Goodness is the centre, from which the truths spread to the circumferences.[51] All truths are referred to goodnesses as the first and last object.

The hierarchy of forms in *De fibra* returns when the creation story turns to Adam and Eve and the innermost anatomy of their bodies. An angel performs a dissection of himself for Eve in order to demonstrate the order of the forms. The angel opens a nerve to expose a fibre rolled in spiral coils. Its membrane is then pulled away to reveal a little 'brain' with small spheres ordered in a subcelestial form. Eve then opens one of these small spheres and inside it she sees countless new small vortices twisted in celestial form in infinite curves and circles, the form of intellectual reason. When one of these vortices is opened, the highest of all forms emerges, the supercelestial, the innermost heaven or the soul's holy of holies. One has finally reached the supercelestial form, which is not really a form but a being: 'there is nothing but what is Perpetual, Infinite, Eternal, Incomprehensible, the Order, Law, Idea of the universe, and the Essence of all essences.'[52] The whole series of forms, from the angular to the supercelestial, enclose each other and one comes further and further in towards the innermost core of creation: God. The whole of reality becomes a single universal spiral, a circling, falling, and rising movement from the supercelestial form, God or the Infinite. All meaning radiates from the centre and

---

[48]*De cultu*, n. 57.

[49]Milton, edition, 159f; cf. Hogarth, 157, 160.

[50]Plotinus, *Ennead*, 1.6.5–6.

[51]*De cultu*, ch. 2, section 2, note q; cf. *De Nova Hierosolyma*, n. 13.

[52]*De cultu*, n. 93.

can be led back to it. The universal spiral stands out as a teleological *imago mundi*, in which God is the ultimate cause of causes, the foundation of and precondition for geometry. Without Him 'not even a point or a line can be made except from Him and by Him'.[53]

The Primary Metaphors of Correspondences

Swedenborg felt that his scientific efforts were unsuccessful and fruitless.[54] His many years spent studying the movements of the muscles, the lungs, and the nerves had taken him nowhere. To be sure, he published an essay about an inlaid marble table in the proceedings of the Academy of Sciences in 1763, but otherwise it was the spiritual world that captivated his systematic interest.[55] In his now constant companion, the Bible in Latin in Sebastian Schmidt's edition, he had underlined some words in Ecclesiastes: 'to seek and search out by wisdom concerning all things that are done under heaven.'[56] The verse continues: 'this sore travail hath God given to the sons of man to be exercised therewith.' He now turned himself inwards, towards the spiritual. In his night's sleep, but also in a slumbering state in the daytime, he could travel away in his dreams to the world of the spirits and the angels. When he woke up he took a cup of coffee with a lot of sugar, ate a bun dipped in milk, and then made notes in his spiritual diary, *Diarium spirituale* (1747–1765), also called *Experientiæ spirituales*, of everything that he had seen and heard in the spiritual world. His first theological work was a commentary on the books of Genesis and Exodus, *Arcana cœlestia* (1749–1756) in eight volumes. With 'arcana' or 'secret' Swedenborg signals that he possesses knowledge that no one else has. Concealed in this secret is his chosen status, but simultaneously a way to demonstrate the significance of his discovery. People are enticed by and always curious about arcane secrets.

One feature that distinguishes Swedenborg's theology, apart from a belief in the existence of a spiritual world, is the doctrine of correspondences. This means that each physical thing corresponds to an intellectual or moral meaning and a theological or divine truth. This allows a different way of reading the Bible. For him the words of the Bible contain an *inner* meaning, while the literal meaning of the words is adapted to those who are poor in spirit.[57] In Great Tartary, far away in the east, the angels tell him in *Apocalypsis revelata* (1766), this original

---

[53]*De cultu*, n. 113.

[54]*Experientiae spirituales*, n. 4010.

[55]*Beskrifningar huru inläggningar ske uti marmorskifwor*, 107–113; cf. Zenzén, 90–94.

[56]Swedenborg, *Schmidius Biblia sacra*, 465; Eccles. 1:13.

[57]*De equo albo*, n. 13.

revelation, a Word solely consisting of correspondences, is supposed to live on.[58] Swedenborg undertakes to expose and explain the inner meaning of the Bible. He sought in particular to explain the Pentateuch, the prophets, and the book of Revelation. This was surely not by chance. These are the parts of the Bible which are hardest to reconcile with reason and a literal reading. The creation story in Genesis is not consistent with the latest findings of natural science, the wrathful God of the prophets contrasted with the message of love in the New Testament, and the meaning of the Apocalypse was shrouded in mystery. With the doctrine of correspondences he could make these problems vanish. For example, the creation story, according to Swedenborg, is not about the creation of the world but about man's own rebirth and maturity, with the six days of creation corresponding to six spiritual states.[59] The less problematic books of the Bible are the ones that he says lack an inner meaning.[60]

The idea that there are several levels of meaning in the Bible went back a long way in Christian exegesis, as in Philo of Alexandria, Origen, and Augustine. Apart from the literal historical meaning, the existence of three spiritual meanings was assumed: allegorical, anagogical, and tropological.[61] Many attempts have been made to link Swedenborg to an allegorical tradition, especially to Neoplatonism and the Kabbalah. But from a cognitive perspective one could instead claim that all these currents proceed from the human cognitive ability to think in metaphors. It need not be assumed that Swedenborg was immediately influenced by any specific thinkers or schools of thought when he developed his doctrine of correspondences. The doctrine of spirits and the correspondences is a fantastically constructed system of metaphors. He was very close to seeing, from a completely different starting point, that metaphorical thinking and petrified metaphors can be found in language. In a note in his spiritual diary from around 1757 Swedenborg says that the language of the spirits is in many cases to be found in human language: when one says 'seeing' (*videre*) instead of 'understanding' (*intelligere*), 'smelling' (*odorari*) instead of 'perceiving' (*percipere*), 'to hear' instead of 'to obey', and so on.[62] These modes of speech have their origin in the spiritual world, Swedenborg says. He has thus found typical metaphors in human thought. To put in extreme terms, it could be said that he did not invent but discovered the correspondences between the natural and spiritual meanings of words. But naturally this rested on a completely different foundation from the modern cognitive theory of metaphor. There is an obvious difference between them. For Swedenborg, the metaphors or the correspondences

---

[58]*Apocalypsis revelata*, n. 11; cf. Swedberg (1941), 617; Jonsson (1969), 249; Jonsson (1988), 201; Hallengren (1997b), 28ff; Hallengren (1998), 20–23.

[59]Bergquist (1999), 289–291; translation, 248f; Bergquist (2001), 86.

[60]Ruth, Chron., Ezra, Neh., Esther, Job, Prov., Eccles., Song of S., Acts, and all the epistles; *Arcana cœlestia*, n. 10,325; Jonsson (1969), 245.

[61]Jonsson (1969), 254f.

[62]*Experientiæ spirituales*, n. 5595; Jonsson (1969), 232.

were not located in the human brain or the mind but in the angels, and they were constituted by God.

Swedenborg's correspondences can be understood as links binding reality together, with the meaning of one word transferred to a different context. The doctrine of correspondences is a pattern or a structure that corresponds to a structure of ideas. The spiritual world and the doctrine of correspondences is built up through a highly developed metaphorical system based on primary metaphors, that is to say, fundamental metaphors that exist in language and human thought.[63] We therefore cannot point in particular to any other thinkers or their texts as the direct sources from which the doctrine of correspondences originated. Ordinary spatial metaphors are those based on the direction up-down. God, paradise, goodness, virtue, the sun, and light are up, while man, hell, evil, sin, the earth, and darkness are down. Swedenborg proceeds with the fundamental metaphors in human thought and develops them into a system.

There are a number of primary metaphors in human thought that Swedenborg uses, for example: *devotion is heat* (God's love warms), *bad is stinking* (evil spirits smell), *likeness is nearness* (similar spirits live near each other in the same community), *states are places* (spiritual states have different places in the spiritual world), *organization is physical structure* (God's order is the spiritual world in the form of a human being), *purposes are destinations* (the spirits travel towards heaven or hell), *causes are physical forces* (there are tributary flows, outflows, inflows, through-flows of the divine). Other primary metaphors can be *important is big*, *linear scales are paths*, and *time is movement*. A common metaphor is *change is motion*, for example, the way that changes in a sense mean a kind of *emotion* that takes the spirit from one place to another, or the way a room is transformed according to spiritual variations. This happened with Philipp Melanchthon, the German reformer whom Swedenborg saw sitting in his study. Melanchthon had been assigned a room when he came to the spiritual world, with a desk, a cupboard, and a small collection of books. As soon as he arrived there he sat down at his desk as if he had just woken out of his sleep and continued to write about justification through faith alone. After a few weeks 'the objects he was using in the room began to grow dim and finally to fade away, until at length nothing was left there but a table, paper and an ink-pot. Moreover, the walls of his room appeared to be covered with whitewash and the floor was paved with yellow brick; he himself wore coarser clothes.'[64] He was amazed and asked how this came about, to which he received the answer that the reason was that he had driven charity from the church. But he continued to write and finally ended up in a workhouse.

The spiritual world as a whole builds on the metaphor *the spiritual is a world*, and *the soul is a body*. The soul is like a little independent person, a 'homunculus', which can move in the spiritual world. A very common metaphor, which he also used frequently in his earlier natural philosophy, is *to know is to see*. When he says

---

[63]Lakoff and Johnson (1999), 50–54; Dunér (2008a), 53–65.

[64]*Vera christiana religio*, n. 797.

'I have seen and heard this many times in the spiritual world', it means a certainty, a knowledge, as if what is seen were more certain than what is thought. He writes, for instance, about the celestial library that 'There appeared places, or repositories, more and more bright, for interior Libraries—but to me and to them, in a dimmer light, because we were incapable of penetrating those depths of wisdom which are there'.[65]

Memorabilia from Earthly Life

The understanding of geography is one of the ways in which human cognition models space and organizes its thoughts. If the world comes from the Creator, there is a secret message in the spatial structure. In Christian thought, especially in the Middle Ages, a movement in geographical space was a movement on a vertical scale of religious and moral values. Heaven was at the top of the scale and hell at the bottom. Moral values and locations merged, as in Swedenborg. Places acquired moral meanings, while moral values, conversely, acquired locations. Geography became ethics. The righteous or sinful person was on a journey from home to the monastery or to the den of sin, to the holy land or the pagan land, from this earth to paradise or hell. In the *Divina Commedia* Dante Alighieri describes a spatial axis of up and down with symbolic meanings. The weight of the sin corresponds to the depth at which the sinner is. He distinguished between the good, rectilinear movement upwards and the sinful, horizontal circular movement. Sinners run constantly in closed circles, and Dante's own journey becomes a rising spiral which ends up plumb vertical.[66] This is the metaphorical thinking on which Swedenborg builds his spiritual world, that is, the fundamental human ability to conceptualize the spiritual on the basis of geographical and spatial experiences.

Swedenborg's doctrine of spirits is no free fantasy; there is nothing arbitrary about it. He is not a poet, although he does employ what poetry and thought have in common: metaphors. What he uses is the metaphorical quality of language, his acquired experience and knowledge. Swedenborg's spiritual world is constructed of memory, or as Augustine writes about his mind, his memory, 'For there have I in a readiness the heaven, the earth, the sea, and whatever I could perceive in them, besides those which I have forgotten.'[67] Swedenborg's doctrine of spirits appeared at a time when the spatial order was being challenged. The earth was no longer at the centre and the universe was expanding towards infinity. The Christian vision of the spatial reality of heaven and hell was threatening to fall outside spatiality.

---

[65] *Diarium spirituale*, n. 5999.

[66] Lotman (1990), 171f, 175, 180–182.

[67] Augustine, 10.8.

Swedenborg restores this space by treating his immaterial world as if it were spatial.[68] In this world he could move freely. It was an inner room.

The vision of the eye reaches the sun and the stars, but the inner vision reaches even further.[69] There is an 'interior sight', according to Swedenborg. 'If interior sight does not exist, the eye cannot possibly see.' In the spiritual life even the blind man can see. 'This also is why when someone is asleep he sees in his dreams just as clearly as when awake. With my internal sight I have been allowed to see the things that exist in the next life more clearly than I see those which exist in the world.'[70] There are three kingdoms in the spiritual world—heaven, the spiritual world, and hell—which correspond to three parts in man: the soul with the heavenly warmth, goodness, and the heavenly light, truth; reason with will and understanding; and the body with heart and lungs. The inhabitants of the spiritual world see it like a valley between high mountains with a narrow path up to heaven in a haze of white clouds, with openings in the cliffs that lead down to the black holes and stinking swamps of hell, where flames fly up as from a furnace when someone falls with a horrifying scream into the abyss.[71]

The spiritual world that Swedenborg explores with his inner vision lacks time and space, however. Therefore there is no geometry as on earth. Geometry is in fact something purely earthly, something that links humans with matter. Time and space exist only in earthly life, and these concepts confine man's ideas and make them natural.[72] The more we devote ourselves to heaven and to God, however, the more we distance ourselves from time and space, and vice versa, the more we distance ourselves from the idea of heaven, the more our thoughts are linked to time and space and the more we therefore distance ourselves from the idea of infinity and eternity. By visualizing the spiritual world and making it comprehensible with the aid of geometry and spatial concepts, Swedenborg concretized his spiritual message. By linking the spiritual message with objects and spatial events, one could more easily understand and learn it. Swedenborg's 'spiritual experiences' can also be regarded as part of a rhetorical genre. The concrete narratives from the spiritual world are 'exempla', examples in the sense of 'patterns', rhetorical evidence to reinforce his abstract spiritual message.

Everything is created in analogy, in connection with God, as God's image in a mirror. Space and the compass points stand in relation to the Lord. All the heavens face the sun of heaven, the Lord, who corresponds to the sun in our world. With the metaphor *God is the sun* Swedenborg is able to generate a multitude of new metaphors. In the spiritual world gravitation acts on the front of the body and the angels are turned to the east where God shows himself as the sun. God is like

---

[68] Miłosz, 143f.

[69] *De cælo*, n. 85.

[70] *Arcana cælestia*, n. 994.

[71] *De cælo*, n. 583, 585, 429; Lamm (1915), 295f; translation, 291f.

[72] *Diarium spirituale*, n. 4609m, 5625; *De telluribus*, n. 127f; *De divino amore*, n. 7, 51, 69, 81; *Vera christiana religio*, n. 29; cf. Pendleton, 5, 64.

the sun, light is divine truth, heat is divine goodness.[73] The solar metaphysics, as in Plato, Plotinus, Augustine, and the introduction to John's gospel, the idea that the sun represented the highest, God or the king, can be found scattered in many cultures.[74] The Arabian astrologer Ja'far ibn Muhammad Abû Ma'shar al-Balhî, whose dream book was translated into Swedish in 1701, writes about how Indians, Persians, and Egyptians identify the sun in dreams with the king.[75] Among Rydelius's 'necessary exercises in reason', Swedenborg had read about different sources of errors committed by the senses, such as the mistake of proceeding from similarities between things.[76] In the same part Rydelius asks why 'so many philosophers in all times have believed light to be something spiritual and living, indeed, even an offspring or efflux of God's own being?' Many still persist in this notion, he says, and it comes of the fact 'that the properties of light have a particular figural or symbolic likeness to a living spirit, and especially to GOD himself, with regard to his wisdom, truth, and revealed glory. The light reveals everything that can be seen, and we see everything in the light. No truth or revelation of truth can therefore occur without GOD.'[77] In *Oeconomia regni animalis* Swedenborg compares the sun of nature with the sun of life: 'We are not forbidden to approach the divine sanctuary by the path of comparison [. . .]. Therefore let us go on in the path of comparison, remembering always that although comparison illustrates, yet it does not teach the nature of that with which the comparison is made.'[78] This is a clear example of metaphorical thought, which need not necessarily have originated in other thinkers or writings; it can come from human thinking itself.

The great world machine was pushed aside by the great world man. Swedenborg's idea is that the whole universe represents a single macrocosmic human, a Maximus Homo, the greatest man.[79] The idea that man is a microcosm, a miniature of the macrocosm, whose body and its parts make up a symbol of unity, can be found in many places in the history of ideas, for instance in Plato, Aristotle, Cicero, in the epistles of Paul, and in Kircher.[80] Man is an image of the universe, a part of the body of Christ. The greatest man in Swedenborg is like an anamorphosis, a popular mirror effect in the optics of the day, by which a seemingly distorted picture appeared

[73]*Diarium spirituale*, n. 944; *De cœlo*, n. 116–124, 141–143, 154–160; *De divino amore*, n. 56, 151–157, 244–246, 291; *De commercio*, n. 6, 18; *Clavis hieroglyphica*, ex. 7–8; translation, 11–13;*Note Book*, 246; Jonsson (1969), 273f.

[74]Plato, *Politeia*, 6.508; Lamm (1915), 74, 252; translation, 74f, 248f.

[75]Abû Ma'shar al-Balhî, 167, 170.

[76]*Note Book*, 113f; Rydelius (1737), 176ff.

[77]Rydelius (1737), 277f.

[78]*Oeconomia* II, n. 254, cf. 251; Jonsson (1961), 116–118; translation, 115–117.

[79]*Note Book*, 21, 137, 462ff, 492; *Diarium spirituale*, n. 1362f, 3221f, 3968, 4098, 4424, 4581m, 4686; *Arcana cœlestia*, n. 4041f, 4302; *Vera christiana religio*, n. 119; Jonsson (1961), 243f; translation, 265.

[80]Rom. 8:29, 12:4f, 1 Cor. 6:15, 10:17, 12:12–14, 27, Eph. 4:12, 15f; Col. 1:18; Cicero, 1.18; Swedberg (1941), 491f; *Note Book*, 266.

regular when reflected in a cylinder.[81] The greatest man is a metaphorical idea in which language's concepts for spatial orientation proceed from man's intuition of the body. Cognition seeks to understand an organic whole based on the image and the organization of a human body. The objective world becomes understandable through an analogy with the human body.

In Swedenborg's world-view the Maximus Homo is connected to the doctrine of correspondences. The parts of the human body correspond to heaven; this is because God the Messiah is a human being and fills the universe. Heaven is therefore God the Messiah himself, and all the spheres and vortices of heaven correspond to Him.[82] Each part of the body in the microcosmic man corresponds to a similar body part of the greatest man, and also to different spiritual meanings. During his travels in the spiritual world Swedenborg visited the different regions of the greatest man, which meant that he also travelled to the most intimate regions around the pelvis. He also lists the correspondences to the greatest man's skin, hair, and bones. Among the spirits in the greatest man's skin are those which correspond to inner states. This means that material things are in accord with spiritual things. Swedenborg writes that one finds in them a beautiful azure pattern shaped in spiral rolls and wonderfully interwoven like lace embroidery that beggars all description.[83] On the skin of a person who has been saved, however, there are even finer and more ornate forms than those just described. In contrast to these, the exterior of a treacherous and deceitful spirit looks like clumps of serpents. Yet even worse is the appearance of magicians, resembling rotten entrails.

Since the spirits can see, there are seemingly spatial and temporal relations in the spiritual world. Progress in time and space in the spiritual world actually consists of changes of internal state, and these changes are dependent on the observer's spiritual state. Spatial and temporal states in spiritual reality are therefore subjective and not, as in material reality, objective. It is like when the road seems to differ in length depending on how much one longs to reach the destination. 'I have often seen this, much to my surprise.'[84] Those who are close to each other spiritually are thus also close to each other in the illusory space of the spiritual world. When a spirit thinks of another spirit, the latter will immediately be beside the former. Everything that the spirits see, all the illusory objects in the spiritual world, are nothing but interior counterparts, mere reflections of inner emotions and thoughts. The things in heaven thus become representations of angels' thoughts and the illusory objects, the spiral spirits, the steps and the gardens, become projections of spiritual states, of their thoughts and feelings. All things in the material world are also found in the spiritual world, but there they are of a more perfect kind.

---

[81]*Diarium spirituale*, n. 2164; *Arcana cælestia*, n. 1871; Kircher (1646); Schott (1657), I, 163; Bergquist (1999), 304f; translation, 260f; Bergquist (2001), 65ff.

[82]*Diarium spirituale*, 29 November 1747, n. 279.

[83]*Arcana cælestia*, n. 5559, cf. 5050–5060; *Diarium spirituale*, n. 4082–4084.

[84]*De cælo*, n. 195.

Swedenborg's travels in the spiritual world are mental changes. In the spiritual world there are cities as on earth, pastoral landscapes, gardens, and snow-covered Nordic winter scenes with people on their way to a rural church.[85] His delight is given visual expression in the form of beautiful countryside, gardens, cities, and palaces. Evil appears in the guise of places with stinking swamps, filled with filth and dirt. He sees cities being swallowed up by the earth in a rotating vortex, like water around a hole.[86] At the turn of the year 1756–1757 Swedenborg came to the Babylon of the spiritual world, which represents human greed and depravity.[87] The city is surrounded by high walls but has no gate. To be able to enter Babylon one must therefore first climb a high mountain outside the city. On the top of this mountain there is a shaft down which you make your way. You then follow passages and continuous wide spiral staircases that lead up towards the city. The spiral passages in Swedenborg's Babylon are a kind of inverted Tower of Babel: here the people go upwards through underground spiral staircases. The avaricious inhabitants of Babylon acquire their beloved wealth under the city. These treasure chambers can be reached if you first make your way down through a shaft, then follow winding passages and descend vertiginous spiral staircases. Since Swedenborg, the former mining assessor, had seen the Babylon of the spiritual world, he was also able to draw a sketch of its subterranean shafts and galleries (Fig. 1). In mining, as Agricola mentions, screw-shaped shafts were sometimes used, with steps carved out of the rock.[88]

The spiritual world is a kind of inner landscape where Swedenborg is on a voyage of discovery into himself. Memories and experiences from his life are processed and recur as mental images. His descriptions of the people he meets are in many ways more about himself. They are self-reflecting, a self-image, a way of coming to terms with himself and the emotions that flared up when he encountered the outside world. He walks along Stora Nygatan in the Stockholm of the spiritual world and meets his old colleague Johan Bergenstierna, a hypocrite on his way to hell.[89] Eric Benzelius's haughtiness, like that of Wolff and Lars and Gustaf Benzelstierna, made Swedenborg's anus itch.[90] He wins all the debates. Even Newton acknowledges to him that the idea of a vacuum is 'a totally destructive notion'.[91] Newton shuddered at this destructive idea and he issued a stern recommendation to 'beware the notion of nothing'. Swedenborg is perfectly right.

Charles XII was in the deepest inferno. According to Swedenborg, this king was the most obstinate and stubborn of mortals on this earth. He had good intentions for his country, as he stated himself, but refused to admit that the outcome was

---

[85] *Vera christiana religio*, n. 185.

[86] *Diarium spirituale*, n. 4992, 5057; *Vera christiana religio*, n. 79.

[87] *Diarium spirituale*, n. 5280–5304; Gen. 11:3–9

[88] Agricola, translation, 214.

[89] *Diarium spirituale*, n. 5711, cf. 4351, 4396, 5132.

[90] *Diarium spirituale*, n. 4851, cf. 4749, 4757, 4787, 5074, 5148, 5702, 5722, 5885, 6016.

[91] *De divino amore*, n. 82; *Diarium spirituale*, n. 6064; cf. Jonsson (1969), 30.

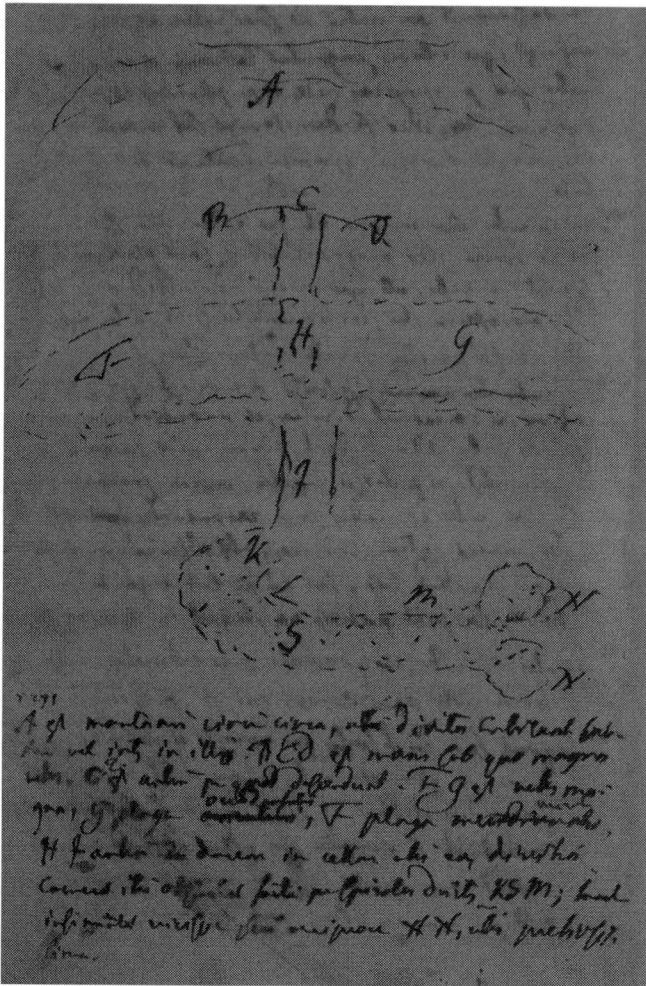

Fig. 1 The subterranean passages of Babylon. At the mountain, *BCD*, there is an entrance to the city, *FG*, through shaft *CE*. Shaft *HI* leads down to the cellars where the inhabitants of the city store their wealth. Through winding corridors, *KSM*, one comes to the cellar, *NN*, where they keep the most precious things (Swedenborg, *Diarium spirituale* (1756–1757))

otherwise. He is insane, Swedenborg writes laconically. The king was a clear example of someone who is all self-love on the inside but polite and virtuous on the surface. He regarded himself as a God but preferred the Mohammedan religion to Christianity. Although he could think better than most people and was very astute, in his actions he was violent and cruel, totally lacking respect for human life. He aspired to become the devil himself and lord of hell, declared war on the Lord, whom he hated, and maligned God's name through atheistic doctrines. He wanted to take power over the whole universe. Bloodthirsty men like Charles XII, who desire to murder everyone and are arrogant beyond belief, but lack compassion, end up copulating with pigs or other bestial creatures in hell. In the hell of hells they claim to know everything, that things are this way and not that way. Using only his powers of persuasion, Charles XII was able to destroy others. One could write a whole book about all the details in his life, Swedenborg says.[92]

The author of the big biography of Charles XII, Jöran Nordberg, who had departed this earthly life in 1744, is also in Swedenborg's deepest hell.[93] He stood on the back of Swedenborg's head and did not hesitate even to injure or murder someone in order to obtain gold, totally lacking a sense of guilt. Another time he appeared in the form of a large green serpent. He admitted loud and clear that he did not believe in the Father, the Son, and the Holy Spirit, even though he had been a priest. Like the subject of his biography, he had strong powers of persuasion which could blind the people to whom he posed questions, so that some were left without an answer. Nordberg was flung deep down into hell. Swedenborg had a profound aversion, above all, to the dishonest, hypocritical, flattering, corrupt, and faithless society in which he lived and of which he himself had been a part, where people concealed their selfish, ambitious, power-hungry interior behind a façade of affected friendliness, virtue, and piety. It was both self-criticism and a critique of the brutal despotism and warmongering of Sweden's time as a great power, when liberty, justice, and truth were swept aside.

Swedenborg's ability to re-establish contact with dead philosophers, royals, and other celebrities naturally caused a great sensation. The young Swiss theologian Johann Caspar Lavater tried to test Swedenborg by sending a code which he expected Swedenborg to be able to decipher if he did actually have this remarkable gift of divination.[94] The mining councillor Daniel Tilas tells how Swedenborg had visited the dead architect Carl Hårleman to obtain the plan of a building. Tilas himself felt some concern about his late wife, Hedvig Reuterholm. 'I am all in a flutter before conversing with him and hearing whom' she has married, he wrote to the inspector of mines Axel Fredrik Cronstedt. 'I should not like it, forsooth, if she had become sultaness. All this he reports without a screw seeming to be loose in the clock-work'.[95] The politician Anders Johan von Höpken, who was a good friend of Swedenborg, writes to Linnaeus to tell that he had heard that the naturalist and freethinker Anders Celsius 'was in hell'.[96] Did the information come from Swedenborg, who was in a position to know?

---

[92]*Diarium spirituale*, n. 4741f, 4748, 4750–4752, 4763, 4764, 4857, 4884, 4934, 6015, 6028, 6034; *De ultimo judicio*, n. 236–238; cf. Westerlund, 152–160.

[93]*Diarium spirituale*, n. 4543–4544, 4811–4812.

[94]Lavater to Swedenborg, Zurich, 24 September 1769. *Letters* II, 687f.

[95]Tilas to A. F. Cronstedt, 16 March 1760. KB, Ep. T 14; *Documents* II:1, 396.

[96]Höpken to Linnaeus, 28 March 1763. Linnaeus (1917), 160.

The Geometry of the Spiritual World

In Swedenborg's natural philosophy, geometry stands for order, clarity, reason, and perfection. If we find an elevated view of geometry in the earlier scientific works, the subject has a much less agreeable position in the doctrine of spirits. There the geometricians' endeavour to achieve clarity and order is an expression of human vanity, finiteness, and simplicity. Geometry is an obstacle to the understanding of universal concepts because it only accepts what is geometrical and mechanical. Consequently, geometry cannot extend beyond worldly and bodily things. In Swedenborg's spiritual world Polhem became an example of the delusions of atheistic mechanics. When Polhem was buried in 1751, Swedenborg followed the coffin at the funeral. He said that Polhem himself had walked beside him 'asking why they were burying him when he is alive? Then also why the priest said that he would be raised up at the last judgment, when yet he was already raised up.'[97] What was wrong with Polhem was, quite simply, that he could only think of material things, was interested only in mechanics and physics, indifferent to the spiritual, to all that is independent of time, space, and people. Swedenborg saw him sitting in the spiritual world, designing mechanical birds, mice, cats, and infants, and when he invented a way to communicate with evil spirits, he was cast down into the darkest regions of hell and stripped of his powers of invention.[98]

Geometry has its limits. Even the lowest human forms, such as the intestines, surpass the forms that can be described with geometrical ideas. The spiral form of the intestines is in turn surpassed by the forms of their activity. Swedenborg goes on to assert in his spiritual diary that the most subtle forms in this activity cannot be conceived with the aid of geometry or infinitesimal calculus, since they infinitely surpass such calculations.[99] How could geometry describe the forms that receive life, forms that go so enormously far beyond the organic and defy vision, Swedenborg wonders. Thus does human reason relate to the spiritual, the celestial, and the divine: since reason cannot even show how excrement is separated from the intestine, Swedenborg draws the conclusion that the mind thereby reasons on the basis of the worst secretions of this defecation, the most disgusting and polluted of all things. It is not a pretty picture of human reason that Swedenborg displays; it is totally unlike the enormous, undreamed-of possibilities of reason during his mechanistic period. He now claims that geometricians, above all, deny the idea that nature is God. They think that nothing can be explained beyond their science. But not even with the aid of the widest range of geometrical methods, not even with its chief tool, infinitesimal calculus, can geometricians determine the basest and coarsest processes in the evacuation of faeces or the form of the intestine.

---

[97]*Diarium spirituale*, n. 4752 m [4773], 5837; *Arcana cœlestia*, n. 4527, 4622; Ehrenheim, 355, cf. 368f.

[98]*Diarium spirituale*, n. 4722, 6049, 6071.

[99]*Diarium spirituale* 5 October 1748, n. 3482f.

Despite this, geometrical associations and metaphors, combined with an underlying fascination with the abstract beauty of geometry, seem to live on in Swedenborg's doctrine of sprits. It becomes a transcendent, supersensual geometry, elevated high above the reality of the senses and of reason. On one occasion God showed him forms that went beyond anything that the geometricians can imagine. This is a passage that echoes the geometrical hierarchy of his hierarchy of forms. The geometricians' science ends in the circle or in the curves which are earthly and do not comprise even the lowest forms of air and water. These lowest or earthly forms, which are confined in time and space, are obtained by removing imperfections, such as gravitation, rest, cold, and so on. But emerging from these forms there are forms that are freer, and then others that are even more free. We arrive at forms in which nothing can be imagined except centres in each point. These forms consist solely of centres which refer to all circles and circumferences, where every point represents centres and these have similar centres. Swedenborg then finds forms that almost entirely lack limits and relation to time and space. But all these forms are nevertheless finite, since one can imagine an idea of them by abstracting from things that are finite. In other words, they are within nature's sphere and are without spiritual life. As long as reason dwells on such natural forms it will not succeed with the forms of life. This statement reflects the course of his own life. He had previously been busily occupied with nature's forms in his quest for the soul, but during the dream crisis he increasingly realized how hopeless that project was, and he was instead granted access to the higher forms of the spiritual world.

All things below and above the forms of life, however, originally come from God. For Swedenborg there are forms in nature that human reason can never understand: 'One can never, therefore, by some kind of removal, have any conception of the forms within the earthly ones, as I now realize, having written about forms on this page that within the most subtle ones of nature there are spiritual forms, completely beyond understanding.'[100] The forms in the higher spheres, in the spiritual world, are therefore even more incomprehensible to man, as we see in a memorandum from 7 October 1748. Swedenborg saw how some spirits formed a circular spiral and God's life poured like an infinite, perpetuo-spiral form, 'a form known to no one but the Lord.'[101]

The Helical Motions of Emotions

There is a correspondence between the motions in the space of the spiritual world and our emotions. They are motions of the senses, and thus a kind of changes of state, and as movement they consequently mean that they also describe space, extension, and form. In the spiritual world the thoughts and speech of the spirits

---

[100]*Diarium spirituale*, n. 3484.

[101]*Diarium spirituale*, n. 3495.

circulate almost in accordance with the helical convolutions that are found in the human brain, convolutions that Swedenborg had previously noticed during his work with *De cerebro*. There are, as Swedenborg writes in his spiritual diary, 'wonderful winding bending, flowing in, returning, which are beyond all comprehension because they follow the patterns of turnings of the world of spirits.'[102] In heaven there are even more wonderful helical motions in the form of ideas, thoughts, speech, and representations, which follow twists that correspond to the celestial form, a form that we cannot possibly grasp. Emotions produce these forms. The circulating movements are so wonderful that no one is able ever to grasp them even in their most general form. All this is dependent on God. Without Him there would not be any circling convolutions, ideas, forms, laws, differences, genera, species, or order.

Swedenborg explains what these forms must be like in the thoughts of the evil and of the good. In *Sapientia angelica de divino amore et de divina sapientia* (1763) the spirituality of an evil person is closed, like a nerve fibre that contracts when touched or like a compressed spiral spring. The change in state of the spiritual mind takes on the metaphysical form of the spiral. Before reformation or salvation it is a spiral that twists down towards hell. After reformation it is instead a spiral that twists upwards towards heaven. This metaphysical spiral thus describes the movement and return to the centre of its being. Moreover, he envisages the earthly mind and the spiritual mind as two intertwined spirals. The former is an image of the world and it is the foundation of evil and falsity, while the latter is an image of heaven. The natural spiral moves from right to left, anti-clockwise and down towards hell, while the spiritual spiral moves from left to right, clockwise and up towards heaven. The movement from left to right is based on the general idea that a clockwise motion is of higher rank than an anti-clockwise motion, which implies adversity. Swedenborg was able to observe in the spiritual world that this is the case. Evil spirits can only move their bodies from right to left, whereas good spirits prefer movement from left to right. In other words, the outward movement of the spirits is a direct counterpart to their inner emotions.[103]

In *Sapientia angelica de divina providentia* (1764) Swedenborg also explains the spirals in evil and good. Evil spirals go backwards and are turned towards hell, while good ones go backwards and are turned towards the Lord. The spiral form is based on Swedenborg's description of thoughts and feelings as a kind of movement. To be more specific, emotions and thoughts in a person are changes and variations in the state and form of his mind's organic substances. How and what these changes and variations are can be understood by considering the heart and the lungs. In these organs there are alternating expansions and compressions, such as the pumping of the heart and the respirations of the lung. Such movements are also found in the other bodily organs and in their small parts, where the blood and the bodily

---

[102] *Diarium spirituale* 3 August 1748, n. 2728–2731.

[103] *De divino amore*, n. 254, 263, 270; cf. *De amore conjugiali*, n. 203; *Vera christiana religio*, n. 258.

fluids are carried back and forth. The variations and changes in the organic forms of the mind, that is to say, in human feelings and thoughts, are similar. But the expansions and contractions of feelings and thoughts are of much greater perfection than purely physical ones. Swedenborg underlines that these changes cannot be described with natural language, only with spiritual language. But what form of motion do these changes of state in thoughts and emotions correspond to? It must be the spiral movement, 'like whirlpools spiraling in and out like endless twisted coils joined together in forms that are wonderfully receptive of life.'[104] It is the archaic metaphorical idea of the spiral as the loom of life, the sign of birth, growth, and death. Swedenborg himself felt these movements in his own thinking: 'When I allowed my imagination to wander freely, its thoughts were led around in a spiral path from left to right, coming toward the center.'[105]

The Spiral Forest at Adramandoni

> In the eastern quarter there appeared to me a wood of palm-trees and laurels arranged in spiralling curves. I approached and went in, walking though several curving paths, until at the end of the paths I saw a garden, which occupied the central position in the wood. There was a small bridge which divided off the garden, with a gate on the side of the wood and another on the garden side. When I approached, the gates were opened by the guard. I asked him what the garden was called. 'Adramandoni', he said, 'which means the delights of conjugal love.'[106]

Swedenborg entered Adramandoni, saw olive trees, vines, and flowering bushes. In the middle of the garden, in a circle of grass, sat young men and women and two angels. They were conversing about conjugal love. Adramandoni is the enclosed garden, *hortus conclusus*, like Solomon's garden in the Song of Solomon 4:12–15, the closed room of virginity. *Delitiæ sapientiæ de amore conjugiali* (1768) portrays spiral heavenly gardens. The spiral is an image of the innocent paradise, a winding labyrinth with long passages, like the testing road of life, leading in towards its inner court. In another garden, Swedenborg says, the spirits at first could not see the garden. All they could see was a single tree with golden fruit and leaves of silver. But they were in fact already in the middle of the garden, in front of the tree of life, the symbol of rebirth, the axis between heaven and earth (Fig. 2). The angels then said to them: 'But go on closer and your eyes will be opened, and you will see the garden.' The spirits did so and were able to see fruit trees with clinging vines arranged in endless circles and infinite spirals. The trees formed a perfect spiral, with each species following another depending on the nobility of the fruit. The innermost trees were the most outstanding, with the finest fruit, and they were

---

[104] *De divina providentia*, n. 319.

[105] *Diarium spirituale* 13 June 1748, n. 2318, cf. 2846.

[106] *De amore conjugiali*, n. 183.

L E X I C O N
GRAECOLATINVM.

SEV,

EPITOME THESAVRI GRÆCÆ LINGVÆ
ab HENRICO STEPHANO conſtructi, quæ hactenus ſub nomine
IOH.SCAPVLÆ prodiit:Lexicon ſanè vltra præcedentes edi-
tiones, innumeris dictionibus, è probatis autori-
bus petitis, locupletatum :

Duplici Methodo conſtans:vna Naturali,eáque ditiſſimis Græcæ linguæ theſauris bre-
ui faciléque comparandis aptiſſima, quæ ex Primitiuorum & Simplicium fonti-
bus Dériuata atque Compoſita dilucidè deducuntur:altera merè Alphabetica ad-
ſcriptis numeris,qua,tanquam Indice,minùs exercitatis oſtenditur, vbi in illa Na-
turali methodo ſingulæ voces inueniantur:vt Epiſtola ad Lectorem declarat.

Accesſerunt opuſcula perquàm neceſſaria,de dialectis de inueſti-
gatione thematum,& alia.

Accesſit & INDEX earum vocum quibus hæc editio vltra
præcedentes locupletata fuit.

Ex Typis SOCIETATIS HELV.CALDORIANÆ

M· DC· XXIII·

Fig. 2 'The tree of knowledge' is written by hand under the tree. Raindrops, small drops of
knowledge, fall into a round pool enclosing the tree. On 14 September 1700 Swedenborg became
the owner of this Greek-Latin dictionary, which had belonged to his father, Jesper Swedberg. The
tree of knowledge leads to the tree of life, the visionary wrote (Scapula and Estienne, *Lexicon
Graecolatinum* (1623))

called the trees of paradise, trees of a kind never seen before, which cannot grow on
earth. They were followed by olive trees and vines, then aromatic trees, and finally
trees that are useful for making objects. There were also gates through the spiral
row of trees. If you entered one of them you came to luxuriant flower gardens which

in turn led to meadows and vegetable gardens. When the spirits saw all this they exclaimed: 'Here is heaven made visible! Whichever way we turn our gaze, there is an impression made on us of heavenly paradise, beyond description.'[107] All gardens in heaven, the angels replied, are visible representations of heaven's blessed origin. The garden is a metaphor for paradise.

In Swedenborg's own garden, in the Mullvaden block in the district of Södermalm in Stockholm, there were likewise winding paths. He had built 'a labyrinth of boards, which is so designed that, if a stranger enters it he cannot then find the way out without help.'[108] His friends' children used to play in the maze. Such mazes were popular in the baroque garden. The paths in the spiral paradisiacal garden in the spiritual world are also like labyrinths. One such labyrinthine heavenly garden in *De amore conjugiali* forced the spirits to circle hither and thither during their walk towards the centre. They just wandered around and around, unable to find the way out, and came deeper and deeper into the garden. In their despair an angel said to them: 'This garden-maze is in fact an entrance to heaven. I know the way and will take you out.'[109] True heavenly bliss does not lie in bodily and earthly things but inside a person, in the soul, in the union of love and wisdom in utility. In the history of art we find not infrequently that the spiral and the labyrinth are closely related figures. Both describe the longest way in towards a centre. Often the labyrinth, like the spiral, came to symbolize the hazardous journey of life through all the difficulties of the world, a serious test for the soul, rebirth, a pilgrimage to the Holy Land, as in John Bunyan's *Pilgrim's Progress* or the 40-year journey through the desert in Exodus. The labyrinth also has the character of initiation, and sometimes one can find paradise in the very centre of the central room.

Like the gates of the heavenly garden through which the spirits were able to enter, his own garden had similar entrances affording access to another garden. There was a door, and if one opened it one saw another door with a mirror in which could be seen a reflection of the garden looking towards a green hedge and a birdcage. Behind that, if one managed to open this blind door, one could find an even more beautiful garden, Swedenborg used to say.[110] He took a keen interest in gardening, with a particular fondness for the strict geometry of the baroque garden, its geometrical ground plan and the distant visual points with symbols, canals, fountains, allegorical sculptures, ornaments, and *parterres de broderie*. The geometry of the French style was also applied in his own artfully clipped garden in Södermalm. It was not wild, chaotic nature that he admired, but the ordered, arranged, inhabited cultural landscape. The fertility of cultivated land, which has its parallel in the useful trees of the heavenly garden, appealed to him. In some notes on a journey from August 1743 in *The Dream Book*, before it takes on its dreamlike character, he is struck by the beauty of the cultural landscape: 'The 17th, travelled from Hamburg, over

---

[107] *De amore conjugiali*, n. 13.

[108] *Beskrifning på afledne assessorens* . . . ; Lindh (1992), 25; B. Sahlin, 28–32.

[109] *De amore conjugiali*, n. 8.

[110] Robsahm, 31f.

the river to Buxtehude, where, for the space of a mile I saw the prettiest country I had seen in Germany; the route lay through a continuous garden of apples, pears, plums, walnuts, chestnut trees, limes and elms.'[111] The series of useful trees echoes the tree-lined path of the paradisiacal spiral from the most wonderful fruit trees to the trees that are useful for their timber.

Epilogue on a Garden

Swedenborg used to sit writing in his summerhouse, his rallying point in the earthly world. Thoughts ran through his brain as he lifted the pen from the paper and looked out into the garden. In his thoughts he was in the inner world. In the outer world, in the three dimensions of space, on the other side of the hand-blown window panes, lay the garden with its paths, spaces, and angles. The light from the wax candle spread in waves out into the darkness, as his thoughts became clear in the obscurity of the unknown. The air vibrated with small sounds, the buzz of the city, occasional verses from the aviary with its birds big and small. He could feel the garden with the movement of his body, with his senses. The scent particles from the avenue of lime trees caused a pleasant tickling sensation in his nose, the taste particles of the beetroot rolled nicely on his tongue. This very movement, the action, the interaction in the physical environment, affected his thinking. The world is structured with the body's spatial orientation. Above and below, in front and behind, inside and outside, all proceed from the body. The garden was, in a sense, divided into three worlds: the natural world at the manure heap, the spiritual world where the flower seeds were germinating, and the divine world where the summerhouse stood, the starting point and the beginning.[112] The roof of the summerhouse ended in a sphere and a star.[113] Along the central path in front of the summerhouse were ornamentally clipped figures in box bushes, symbolizing living birds and pyramids, which he had ordered together with flower bulbs from Amsterdam.[114] The geometrically dug vegetable beds were arranged in rectangles, the onions set in straight lines to give a surveyable, secluded island in nature's chaos. To the right of the summerhouse he stored his books, which afforded roads out into the world and into thought. Just around the corner, to the right, was the board maze with its winding spiral paths, where steps and thoughts could easily get lost. To the left of the central path lay the mirror pavilion, the counterpoised doors of which reflected out into infinity, and 'when all the three doors are opened, and a mirror is placed in front of the fourth wall, which is along the board-fence, three gardens are seen reflected in it, in which every thing is represented in the same order as in the original garden.'[115]

---

[111]*Swedenborgs drömmar*, 2; translation, 4.

[112]Lindh (1992), 27f.

[113]Alm, 48.

[114]Joachim Wretman to Swedenborg, Amsterdam, 27 September 1760. *Letters* II, 535f, cf. 512f.

[115]Lindh (1992), 24; *Documents* I, 392.

In year 5754 after the creation, Swedenborg noted in his diary what seeds he had sown in the garden. It was the spring of 1752. Two eclipses of the sun would occur during the year, over South America and the Bay of Bengal. A new volume of *Arcana cælestia* was with the printer in London. Swedenborg sowed lemon pips, dwarf peas, parsley root, larkspur, sweet william, linseed, sunflower....[116] The lame gardener, the former guardsman Nils Ahlstedt, toiled away. His wife, Maria Norman, stood in the kitchen, and the carpenter's hand Jakob Spångberg was hammering at an extension to the house. Beetroot, spinach, and carrots were set in the soil. Globe artichokes, white scented roses, and hollyhocks would appear. In order to find and see what would come up, he set the seeds in a rectangular geometrical orientation with divisions into the large quarter, top and bottom, right and left. On 23 April he sowed American seeds and mulberry seeds: 'in the fourth box nearest the yard 1 pea tree from America is sown, three peas, the side towards the garden 2 in the middle plane or beech, 3 at the far end towards the tree nursery cornus Americana.'[117] How did it all grow that summer?

Swedenborg thought in his own way. He 'does not understand himself,' wrote the university librarian Johan Hinric Lidén.[118] People laugh at his printed follies, he said, and call him, quite simply, 'the illuminated Apocalyptic Historiographer, Father Fantastic and the Old Man.[119] Others who met Swedenborg in his garden in Hornsgatan found an amiable, tranquil, smiling person, always friendly and unexpectedly sober and sensible. This book about *The World Machine* has sought to understand *how* Swedenborg thought, not just what he thought. Swedenborg is comprehensible. His thought uses cognitive skills that are universally human. Experience of space affected his thoughts, and he imagined the unknown and the invisible with the aid of metaphors from the known and visible; he saw reality with a gaze for geometry, and he interpreted texts in terms of his own needs and purposes. What he read was reshaped into thoughts of his own. In his natural philosophy he employed metaphors such as *the world is a machine, the world is geometry, thinking is working with mathematics, to live is to tremble, micromechanics is macromechanics* and many more. The small particles, the solar vortex, the human body, thoughts, the soul became machines. Geometry was transcendental, the sure method, the ideal objectivity. His thinking was influenced by emotions, by war, by political and economic conditions. The waves of the sea, music, circulations, steam, all this gave him models for nature's invisible reality. Waves washed in towards land, through the air, the ether, and the nervous fluids. Thoughts could travel through the air from one person to another. Mines and engineering gave clues to the mechanics of small machines. The world could be understood in analogies, proportions, and numerical relations. He searched for order, harmony, and wholeness in the chaos

---

[116]KB, Swedenborg, I, p. 56, 4, 6; ed., *Swedenborgs almanacka*, 9f, 12f; *Arcana cælestia*, n. 4700–5993.

[117]KB, Swedenborg, I, p. 56, 25; ed., *Swedenborgs almanacka*, 27.

[118]Lidén to Anders Schönberg, 18 September 1769. UUB, G 151 t; Rydberg, 357.

[119]Lidén to Samuel Älf, 17 May 1769, Cited in Afzelius, 15.

of phenomena. The natural philosopher was the spider in the web, walked in light and darkness in nature's labyrinth. Everything was created from small points that the prime mover had set in motion. The microcosm and the macrocosm reflected each other, whirled in spirals. The soul could be seen under a magnifying glass. But knowledge had its limits. Finite reason can never understand the Infinite. As a visionary he built up a metaphorical system, a doctrine of correspondences and a spiritual world, all predicated upon metaphorical thought. Swedenborg created his own work of art, into which he finally stepped, to disappear out of the prison of reality. The work of art was confused with reality, or perhaps rather became his reality, the artist's dream of becoming one with his work. The world exists in thought.

Bibliography

Unpublished Sources
The Military Archives, Stockholm (Krigsarkivet, KrA)
Artilleribok, Artilleriet. Läro- och handböcker. XVI:47.
Benzelstierna, Jesper Albrecht, *En dehl Hrr volontairers af Fortificationen examen pro anno 1737,* Fortifikationen, Chefsexpeditionen, Examenshandlingar 1737, F2:1.
Fyra stycken projecter öfwer sluyser af trää wedh Trollhättan. Kungsboken 16:1.
Journaler öfwer arbetet på dockan. ifrån den 2: ianuary 1717. till den 1: october 1720. då entreprenaden begynttes, Militieräkningar 1717:1.
Rappe, Niklas, *Åtta böcker om artilleriet, uti den moskovitiska fångenskapen sammandragna och till slut bragta, av generalmajor Niklas Rappe* (1714), Artilleriet. Läro- och handböcker, XVI:18a–b.

The Royal Library, Stockholm (Kungliga Biblioteket, KB)
Bromell, Magnus von, *Doctoris Magni Bromelii prælectiones privatæ in regnum minerale Upsaliæ habito in martio etc anno 1713,* copy by J. Troilius, X 601.
Buschenfelt, Samuel, *Den äldre fadren Buschenfelts marchscheider Relation 1694. tilhörige Ritningar,* L 70:54:2.
Nordberg, Jöran, *Anecdotes, eller Noter till kyrckoherdens doctor Jöran Norbergs Historia, om konung Carl den XIIte, glorwyrdigst i åminnelse, wid censureringen uteslutne,* D 809.
Nordberg, Jöran, *Anecdotes, eller Noter till kyrkioherdens doctor Jöran Nordbergs Historia om konung Carl den XIIte, hwilka wid censureringen blifwit uteslutne,* part one, D 812.
Nordberg, Jöran, *Kyrckoherden doctor Jöran Nordbergs Anedoter til des Historia om konung Carl den XII. glorwordigst i åminnelse, hwilcka blifwit uteslutne wid censurerandet,* D 814.
Polhem, Christopher, *Anteckningar och utkast rörande ett af honom uppfunnet 'Universalspråk',* . . . , N 60.
Polhem, Christopher, *Filosofiska uppsatser,* P 20:1–2.
Polhem, Christopher, *Mindre uppsatser och fragment i praktisk mekanik,* X 267:1.
Polhem, Christopher, *Uppsatser i allmänt naturvetenskapliga ämnen,* X 517:1.
Polhem, Christopher, *Ordatecken på naturens materialer och dess egenskaper,* X 519.
Polhem, Christopher, *Ordatecken på naturens materialer och dess egenskaper,* copy by J. Troilius, X 521.
Polhem, Christopher, *Matematiska uppsatser, fragmenter och utkast,* X 705:1–2.
Polhem, Christopher, *Uppsatser om mått, mål och vigt,* X 706.
Stiernhielm, Georg, *Arithmetica mnemonica universalis,* Wasúla 1642, Fd 15.
Stiernhielm, Georg, *Linea Carolina,* copy, Uppsala, 11 December 1705, X 727.

D. Dunér, *The Natural Philosophy of Emanuel Swedenborg,* Studies in the History
of Philosophy of Mind 11, DOI 10.1007/978-94-007-4560-5,
© Springer Science+Business Media Dordrecht 2013

Swab, Anton von, *En kort berättelse om Afwesta crono bruk, så till thess belägenhet och forna, som för tiden warande tillstånd*, 1723, Rålamb, fol. 143.

Swedenborg, Emanuel, *Almanach för skott-året, efter wår Frälsares JEsu Christi födelse, 1752. Til Stockholms horizont*, ..., Stockholm [1751], Almanacksant:r 1752, I s 56.

Swedenborg, Emanuel, *Drömmar 1744*, I s 57.

Swedenborg, Emanuel, *En ny räkenkonst som omvexlas wid 8 i stelle then wahnliga wid thalet 10, hwarigenom all ting angående mynt, wicht, mål och mått, monga resor lettare än effter wahnligheten uträknas*, Karls grav 1718, X 722.

Tilas, Daniel, *Bref från Daniel Tilas till A. F. Cronstedt 1745–63 m.fl.*, Ep. T 14.

Vallerius, Johan, lecture notes by Gabriel Gyllengrijp, X 219.

The Royal Swedish Academy of Sciences, Stockholm (Kungliga Vetenskapsakademien, KVA)
Swedenborg, Emanuel, *Index variorum philosophicorum*, cod. 37 (90).

Swedenborg, Emanuel, *Anatomica et physiologica*, cod. 53 (114).

Swedenborg, Emanuel, *Physiologica et metaphysica*, cod. 54 (113).

Swedenborg, Emanuel, *Anatomica et physiologica*, cod. 55 (105).

Swedenborg, Emanuel, *Riksdagsskrifter*, cod. 56.

Swedenborg, Emanuel, *Varia anatomica, physiologica et philosophica*, cod. 57 (104).

Swedenborg, Emanuel, *Anatomica, physiologica et physica*, cod. 58 (119).

Swedenborg, Emanuel, *Anatomica, physiologica et philosophica*, cod. 65 (98).

Swedenborg, Emanuel, *De magnete et diversis ejus qualitatibus*, cod. 81 (74).

Swedenborg, Emanuel, *De sulphure et pyrite*, cod. 82 (83:2).

Swedenborg, Emanuel, *De sale communi; h. e. de sale fossili vel gemmeo, marino et fontano*, cod. 83 (83:4).

Swedenborg, Emanuel, *De secretione argenti a cupro, quae seger-arbete vocatur*, cod. 84 (83:1).

Swedenborg, Emanuel, *De victriolo, deque modis victriolum elixandi, etc*, cod. 85 (83:3).

Swedenborg, Emanuel, *Geometrica et algebraica*, cod. 86 (53).

Swedenborg, Emanuel, *Varia philosophica, anatomica et itineraria*, cod. 88 (93).

Linköping Diocesan Library, Linköping (Linköpings Stiftsbibliotek, LiSB)
Benzelius, Eric, the Younger, *Erici Benzelii jun: egenhänd: diarium sine barn tillägnat*, B 53.

Benzelius, Eric, the Younger, *Erici Benzelii utriusque ... epistolæ etc.*, B 56.

Benzelius, Eric, the Younger, *Gustaf Benz. bref til sin broder Eric Benzelius*, Bf 11.

Benzelius, Eric, the Younger, *Bref til ärkebiskop Eric Benzelius d.y.*, Br 10.

Benzelius, Eric, the Younger, *Collectanea physica-mathematica, såsom ock åtskilligt rörande Sveriges natural historia, hvaraf gjordes bruk vid Societatis literariae inrättande i Upsala*, N 14a.

Benzelius, Eric, the Younger, *Bref til Benzelius 1709–1710*, S III, no. 93.

Lund University Library, Lund (Lunds universitetsbibliotek, LUB)
Polhem, Christopher, *Anmärckningar om barometrerns stigande och fallande på följande luftens förändringar* (copy), Naturvet. Handskrift.

Rationarium Bibliothecæ Carolinæ, Utlåningsjournal, 4 February 1699–6 November 1728, Lunds universitetsbiblioteks arkiv, FIa 1.

Swedenborg, Emanuel, *Rent swar på falskia satser. Dedicerat till R: O: B. Renhorn wid Riksdagen 1761*, Dela Gardies arkiv, Riksdagshandlingar cod. VII:b) 19, [copy no. 7].

The National Archives, Stockholm (Riksarkivet, RA)
Bergenstierna, Johan & Emanuel Swedenborg, *Commisions förrättning angående bergs frälse skogar vid Stora Kopparberget. År 1730*, Bergskollegii arkiv, Bergverksrelationer, Stora Kopparberg 1730, Huvudarkivet, EIIc:8.

Bergskollegii arkiv, kungl. brev 1717–1721, fol. 150 1/2. Letter from Ulrica Eleonora to the Board of Mines, Stockholm, 14 November 1719.

Bergskollegii arkiv, Huvudarkivet, Protokoll 1725, AI:71.
Bergskollegii arkiv, Huvudarkivet, Protokoll 1731, vol. 1, AI:77.
Bergskollegii arkiv, Huvudarkivet, Protokoll 1741, AI:96.
Bergskollegii arkiv, Huvudarkivet, Protokoll 1747, vol. I, AI:107.
Bergskollegii arkiv, Huvudarkivet, Bref och suppliker, 1725, vol. I, EIV:169–70.
Collegium medicum, Protokoll 1706–1720, vol. 2, A1A.
Collegium medicum, Register till protokoll, 1690–1732, vol. 1, A1B.
Ekman, Natanael, *Kort relation om smälte-wärcket wijd Stora Kåppar Berget uppsatt åhr 1704*, Bergskollegii arkiv, Bergverksrelationer, Stora Kopparberg 1695–1710, Huvudarkivet, EIIc:3.
Harmens, Lars, *Berättelse om Afwesta cronobruk*, 21 December 1723, Bergskollegii arkiv, Bergverksrelationer, Stora Kopparberg 1723–1727, Huvudarkivet, EIIc:6.
Kommerskollegium, Gruvkartor, markområden, utländska.
Polhem, Christopher, *Kort berettelse och anmerckning om slysswerckens inrettning wid Trollhettan, Gullspång etc.* (11 June 1717), Swedenborg's copy, Kommunikationsväsendet, Kanaler, IVb, Trollhättekanaler.
Skrivelser till Kansliämbetsmän. Till ombudsrådet C. Feif. från myndigheter och enskilde. 15:38. L–Ö.
Swedenborg, Emanuel, *Beskrivning öfwer swenska masugnar* ... (2 November 1719), Bergskollegii arkiv, Bergverksrelationer, Huvudarkivet, EIIa:9.

The Russian Academy of Sciences, Saint Petersburg (Rossiiskaya akademiya nauk, RAN)
Ausgehende Briefe 1734–1735, fond 1, opis 3, no. 19, razryad V, opis 1–C, no. 6.
Tatishchev, Vasily Nikitich, *Tetradi Tatishcheva k ego radotash po geometrii*, razryad II, opis 1, no. 211.

Skara Diocesan and County Library, Skara (Stifts- och landsbiblioteket i Skara, SLBS)
Carlmark 34:18, letter to E. Benzelius the Younger from Johan Moraeus concerning the finding in 1705 of 'ossa gigantis', 21 November 1705.

Stockholm City Library, Stockholm (Stockholms stadsarkiv, SSA)
Jakob och Johannes, Stockholm, dopbok 1680–1689, CIa:10.

Swedenborg Library, Bryn Athyn PA (SLBA)
Academy collection of Swedenborg documents I–X, 'Green books'.
Review of Swedenborg's De ferro, Academy of Sciences, St. Petersburg, S2 Ac12.

The National Museum of Science and Technology, Stockholm (Tekniska museet, TM)
Cronstedt, Carl, *Machiner, som till största dehlen äro uti wärket stelte [av Polhem] och af Ehrensverd och mig afritade åhr 1729: tillika med andra tilökningar som iag sielf giort tid effter annan*, Ms 7405.

Uppsala University Library, Uppsala (Uppsala universitetsbibliotek, UUB)
Benzelius, Eric, the Younger, [Letters from D. Tiselius], G 19:6–8.
Benzelstierna, Gustaf, *Bref till och från G. Benzelstjerna*, G 20 a, no. 75.
Lidén, Johan Hinric, *Samling af bref, till Johan Hinric Lidén. 1767 och 1768*, vol. III, G 151:b.
Lidén, Johan Hinric, [Letter from J. H. Lidén], G 151 t.
Roberg, Lars, *Collegium Chymicum*, Uppsala 1705, 1713, 1713, D 1426–1428.
Swedenborg, Emanuel, 'Appendix till "Om jordenes ... gång och stånd".' (1718), *Diverse fragment m.m.*, D 27.
Utlåningsjournal 1694–1727, vol. I, Bibliotekets arkiv, G 1.
Vallerius, Göran, *G. Vallerii Brefvexling med De la Hire och Polhem*, G 321.

Vallerius, Harald, *Matematik, optik m.m.*, A 518.
Vallerius, Johan, 'Observatio eclipseos solaris quæ Upsaliæ contigit totalis anno 1715 d 22 apr. st: v. horis antemeridianis', *Geometri, optik, astronomi*, A 519:6.

Published Sources

Abû Ma'shar al-Balhî, Ja'far ibn Muhammad, *Ny Apomasaris dröm-book, thet är en upsats på allehanda slags drömmar hwilka man icke allenast effter the indianers, persers och aegyptiers lära, har befunnit sanne, . . .*, translation G. P. Lillieblad, Stockholm 1701.
Acta eruditorum, Leipzig 1684, 1697, 1706, 1722–1723*, 1726.
Acta Germanica: Or the Literary Memoirs of Germany, &c. I, London 1742.
Acta literaria et scientiarum Sveciæ III–IV, Uppsala 1730–1739.
Acta literaria Sveciæ I–II, Uppsala & Stockholm 1720–1729.
Agner, Eric, *Kort och ny method til at extrahera radices quantitatum, per tabulam logarithmorum*, Stockholm 1710.
Agner, Eric, 'Eric Agners brev till Eric Benzelius den 23 juni 1711', ed. A. Liljencrantz, *Lychnos* 1940.
Agricola, Georgius, *De re metallica*, Basel 1657*; translation H. C. Hoover & L. H. Hoover, *De re metallica*, New York NY 1950.
Anon., *En kort methode och genwäg, om interessens uträknande på heela och brutne tahl i diverse tijder, med heela och brutne pro cento; . . .*, Stockholm 1710.
Anon., *Vträkning, som wijsar hwad kopparm:t giör uthi caroliner, efter den nya valvationen*, Stockholm 1716.
Anon., *Die gantz neue eröffnete Pforte zu dem chymischen Kleinod oder einige vornehmste chymische Arcana, . . .*, Nürnberg 1728.*
Archimedes, *The Works of Archimedes*, ed. T. L. Heath, Cambridge 1897.
Archimedes, *Peri elikōn*; ed. J. L. Heiberg, *Opera omnia cum commentariis Eutocii* II, Leipzig 1908.
Archimedes, *Psammitēs*; ed. J. L. Heiberg, *Opera omnia* II.
Aristophanes, *Ornithes*; ed. B. B. Rogers, *The Peace. The Birds. The Frogs*, Cambridge MA & London 1989.
Aristotle, *Analytikōn ysterōn*; ed. H. Tredennick & E. S. Forster, *Posterior Analytics. Topica*, Cambridge MA & London 1966.
Aristotle, *Mēchanika*; ed. W. S. Hett, *Minor Works*, Cambridge MA & London 1980.
Aristotle, *Meteōrologikōn*; ed. H. D. P. Lee, *Meteorologica*, Cambridge MA & London 1987.
Aristotle, *Peri ouranou*; ed. W. K. C. Guthrie, *On the Heavens*, Cambridge MA & London 1986.
Aristotle, *Peri poiētikēs*; ed. S. Halliwell, *Poetics. Longinus on the Sublime. Demetrius on Style*, Cambridge MA & London 1995.
Aristotle, *Peri psychēs*; ed. W. S. Hett, *On the Soul. Parva Naturalia. On Breath*, Cambridge MA & London 1986.
Aristotle, *Physikēs*; ed. P. H. Wicksteed & F. M. Cornford, *The Physics* I–II, Cambridge MA & London 1980.
Aristotle, *Tōn peri ta zōia istoriōn*; ed. D. M. Balme, *History of Animals: Books VII–X*, Cambridge MA & London 1991.
Aristotle, Pseudo-, 'De secretiore parte divinæ sapientiæ, secvndvm Ægyptios', *Aristotelis opervm* II, ed. G. Du Val, Paris 1629.
Augustine, Aurelius, *Confessiones*; ed. W. Watts, *Confessions* I–II, Cambridge MA & London 1977–1979.
Bacon, Francis, *Novum organum . . .*, London 1620; ed. L. Jardine & M. Silverthorne, *The New Organon*, Cambridge 2000.
Bacon, Francis, *New Atlantis*, London 1627; new ed., in *Three Modern Utopias*, ed. S. Bruce, Oxford 1999.
Bacon, Francis, *De augmentis scientiarum lib. IX*, Leiden 1652.*
Baglivi, Giorgio, *Opera omnia medico-practica, et anatomica*, new ed., Antwerpen 1715.

Baker, Thomas, *Reflections upon Learning, Wherein is Shewn the Insufficiency Thereof, in its Several Particulars: In Order to Evince the Usefulness and Necessity of Revelation*, 5th ed., London 1714; 4th ed. 1708.

Barchusen, Johann Conrad, *Elementa chemiæ, quibus subjuncta est confesctura lapidis philosophici imaginibus repræsentata*, Leiden 1718.*

Basilius Valentinus, *Chymische Schriften*, Hamburg 1694; *Chymische Schriften alle, so viel derer verhanden*, ... I–II, 3rd ed., Hamburg 1700.

Bauer, Georg, see Agricola.

Bazin, Aîné, *Traité sur l'acier d'Alsace*, Strasbourg 1737; translation C. H. König, *Tractat om stålhtilwärkning i Alsas: Eller konsten, at af tackjärn tilwärka ståhl*, Stockholm 1753.

Becher, Johann Joachim, *Physica subterranea profundam subterraneorum genesin, è principiis hucusque ignotis, ostendens*, new ed., G. E. Stahl, Leipzig 1703.

Becher, Johann Joachim, *Chymischer Glücks-Hafen, oder Grosse chymische Concordantz und Collection, von funffzehn hundert chymischen Processen, ... oder Bedencken von der Gold-Macherey, Herrn Georg Ernst Stahls, ...*, new ed., Halle 1726.*

Becke, David von der, *Mindani, Experimenta et meditationes circa naturalium rerum principia. Quibus quæ circa fixi & alcalisati salis, ante calcinationem in misto præexistentiam, ac causas volatilisationis, obscura aut dubia esse poterant, clarè solvuntur*, Hamburg 1674.*

Bellman, Johan Arent (pres.), *De antiquae & medii ævi musica*, resp. Göran Vallerius, Uppsala 1706.

Below, Jacob Fredrik (pres.), *De natura, arte et remediis in morborum cura necessariis*, resp. Mathias Ribe, Uppsala 1695.*

Benson, William, *A Letter to Sir J- B- [Jacob Bancks], by Birth a Swede, but Naturaliz'd, and a M-r of the Present P-t, Concerning the Late Minehead Doctrine, which was Establish'd by a Certain Free Parliament of Sweden, to the Utter Enslaving of that Kingdom*, 2nd ed., London 1711.

Benzelius, Eric, the Elder, *Breviarium historiæ ecclesiasticæ, veteris et novi testamenti, ...*, Strängnäs 1695.

Benzelius, Eric, the Younger, 'Praefatio Erici Benzelii, Episcopi Lincopensis', in Jacob Serenius, *Dictionarium Anglo-Suethico-Latinum ...*, Hamburg 1734.

Benzelius, Eric, the Younger, *Utkast till swenska folkets historia, ifrån desz första uprinnelse, til och med konung Gustaf den förstes tid*, ed. C. J. Benzelius, Lund 1762.

Benzelius, Eric, the Younger, *Brefwäxling imellan ärke-biskop Eric Benzelius den yngre och dess broder, censor librorum Gustaf Benzelstierna*, ed. J. H. Lidén, Linköping 1791.

Benzelius, Eric, the Younger, *Anecdota Benzeliana. Eric Benzelius d. y:s anteckningar i svensk historia*, ed. H. Lundgren, Stockholm 1914.

Benzelius, Eric, the Younger, *Letters to Erik Benzelius the Younger from Learned Foreigners* I–II, ed. A. Erikson, Göteborg 1979.

Benzelius, Eric, the Younger, *Erik Benzelius' Letters to his Learned Friends*, ed. A. Erikson & E. Nilsson Nylander, Göteborg 1983.

Benzelius, Henric, 'Henric Benzelius' brev till Eric Benzelius d.y. från Lapplandsresan 1711', ed. C.-O. von Sydow, *Lychnos* 1962.

Berkeley, George, *A Treatise Concerning the Principles of Human Knowledge*, Dublin 1710; new ed., G. J. Warnock, Glasgow 1981.

Bernoulli, Johann, *Essay d'une nouvelle théorie de la manoeuvre des vaisseaux: avec quelques lettres sur même sujet*, Basel 1714.*

Beskrifning på afledne assessorens herr Emanuel Svedenborgs gård på Södermalm, Stockholm 1772.

Biancani, Guiseppe, *Sphæra mvndi sev Cosmographia demonstratiua, ac facili methodo tradita ... III. Echometria, id est geometrica traditio de echo*, Bologna 1620.

Biblia sacra sive Testamentum Vetus et Novum ex linguis originalibus in linguam latinam, translation S. Schmidt, Strasbourg 1696.*

Biblia, thet är then Heliga Skrift på swensko; efter konung Carl then tolftes befalning ..., Stockholm 1703; facsimile, Stockholm 1978.

Bidloo, Govard, *Anatomia hvmani corporis,* ..., Amsterdam 1685.

Bilberg, Johan, *Elementa geometriæ planæ ac solidæ, una cum sphæricorum doctrina atq; praxi trigonometrica* ..., Stockholm 1691.*

Bilberg, Johan, *Refractio solis inoccidui, in Septemtrionalibus Oris / Midnats solens rätta och synlige rum uti Norrlanden,* ..., Stockholm 1695.

Bilberg, Johan (pres.), *De magnetismis rerum,* resp. Erik Odhelius, Stockholm 1683.

Bilberg, Johan (pres.), *De coloribus,* resp. Pehr G. Elvius, Uppsala 1685.

Blake, William, *The Complete Poems,* ed. W. H. Stevenson, 2nd ed., London 1989.

Block, Magnus Gabriel von, 'Blocks och Leibniz' brevväxling 1698–1700', ed. J. Nordström, *Lychnos* 1965–1966.

Boerhaave, Hermann, *Institutiones medicæ, in usus annuæ exercitationis domesticos,* Leiden 1708; new ed., Paris 1735.*

Boerhaave, Hermann, *Aphorismi de cognoscendis et curandis morbis in usum doctrinæ domesticæ digesti,* Leiden 1709; new ed., Leiden 1737.*

Boerhaave, Hermann, *Elementa chemiae, qvae anniversario labore docuit, in publicis, privatisque scholis* I, Leipzig 1732.

Boethius, Anicius Manlius Severinus, *De consolatione philosophiae libri quinque* (524); translation S. J. Tester, *The Theological Tractates: The Consolation of Philosophy,* Cambridge MA & London 1973.

Bokwetts Gillets protokoll (1719–1731), ed. H. Schück, Uppsala 1918.

Bonde, Gustaf, *Clavicula hermeticæ scientiæ ab hyperboreo quodam horis subsecivis calamo consignata,* Marburg 1746.

Borelli, Giovanni Alfonso, *De motu animalium,* 2nd ed., Leiden 1685.

Boyle, Robert, *Certain Physiological Essays and Other Tracts* ..., 2nd ed., London 1669.

Boyle, Robert, 'Of the Excellency and Grounds of the Corpuscular or Mechanical Philosophy' (1674), in Maria Boas Hall, *Robert Boyle on Natural Philosophy: An Essay with Selections from his Writings,* Bloomington IN 1965.

Boyle, Robert, *Opera varia, quorum posthac exstat catalogus,* Genève 1680.*

Brandt, Georg, *En grundelig anledning til mathesin universalem och algebram, efter herr And. Gabr. Duhres håldne prælectioner sammanskrifwen af Georg Brandt,* Stockholm 1718.

Brandt, Georg, 'Acta laboratorii chymici', *Kongl. swenska wetenskaps academiens handlingar* II, 1741.

Brasser, Jacob R., *Regula cos, of algebra, Zijnde de alder-konstrijcksten Regel om het onbekende bekent te maken. Ofte Een korte Onderwijsinge, waer in geleert werdt het Upttrecken der Wortelen, soo bezre men begeeren mach,* Amsterdam 1663.

Bredberg, Sven, *Greifswald–Wittenberg–Leiden–London. Västgötamagistern Sven Bredbergs resedagbok 1708–1710,* ed. H. Sandblad, Skara 1982.

Bromell, Magnus von, *Lithographiæ svecanæ* I–II, Uppsala 1726–1727.*

Bromell, Magnus von, *Mineralogia, eller inledning til nödig kundskap at igenkiänna och upfinna allahanda berg-arter, mineralier, metaller samt fossilier, och huru de måge til sin rätta nytta anwändas,* 2nd ed., Stockholm 1739.*

Brossard, Sebastien de (James Grassineau), *A Musical Dictionary:* ..., London 1740.

Browallius, Johan, *Känningar af Guds försyn vid nyttiga vetenskapers främjande; i et tal hållit för Kongl. Svenska Vetenskaps Academien,* Stockholm 1762.

Brückmann, Franz Ernst, *Magnalia Dei in locis svbterraneis oder unterirdische Schatz-Kammer Königreiche und Länder, in ausführlicher Beschreibung aller, mehr als MDC. Bergwerke durch alle vier Welt-Theile, welche* ..., Braunschweig 1727.*

Bunyan, John, *The Pilgrim's Progress from this World, to That Which Is to Come Delivered under the Similitude of a Dream, Wherein is Discovered the Manner of His Setting Out, His Dangerous Journey, and Safe Arrival at the Desired Country,* London 1678–1684; translation M. Lagerström, *En christens resa til den saliga ewigheten,* ..., Stockholm 1727.

Burman, Eric, 'Observatio circa lumen boreale d. 20. sept. ao. 1717. prope Upsal.', *Acta literaria Sveciæ* 1724.

Burman, Eric (pres.), *De minere ferreæ per magnetem investigatione*, resp. Jonas El. Ask, Uppsala 1728.

Burnet, Thomas, *Telluris theoria sacra: orbis nostri originem & mutationes generales, quas aut jam subiit, aut olim subiturus est, complectens, libri duo priores de diluvio & paradiso*, London 1681; new ed. Frankfurt 1691*; new ed., Amsterdam 1694.

Burnet, Thomas, *The Theory of the Earth: Containing an Account of the Original of the Earth, and of all the General Changes which it Hath Already Undergone, or Is to Undergo Till the Consummation of all Things*, 3rd ed., London 1697.

Campanella, Tommaso, *La città del sole. Dialogo poetico* (1623); translation D. J. Donno, *The City of the Sun: A Poetical Dialogue*, Berkeley CA 1981.

Campani, Giovanni, see Ripa.

Carlberg, Bengt Wilhelm, 'Sannfärdig berättelse af några få omständigheter, som sig tildrogo den natten då … Carl XII … olyckeligen blef skuten', in Bring 1920.

Catalogus lectionum publicarum, …, Uppsala 1709, 1714–1715.

Catalogus prælectionum publicarum, …, Uppsala 1708; translation E. S. Price, 'The Curricula in Swedenborg's Student Years', *The New Philosophy* 1935:1.

Celsius, Anders, 'Anmärkningar öfwer magnet-nålens stundeliga förändringar uti des misswisning', *Kongl. swenska wetenskaps academiens handlingar* 1740a.

Celsius, Anders, 'Magnet-nålens misswisning eller afwikande från norr-streket, observerad i Upsala', *Kongl. swenska wetenskaps academiens handlingar* 1740b.

Celsius, Anders, 'Prof. And. Celsii anmärkning vid. H. Assess. Swedenborgs uträkning av magnetens declination till Upsala meridian' (1741), ed. Nordenmark 1933.

Celsius, Anders, *Samtal emellan en herre och en fru, om geometriens nytta för unga studerande*, Stockholm 1743.

Cherubini, Le père d'Orléans, *La dioptrique oculaire ou la théorique, la positive, et la méchanique, de l'oculaire dioptrique en toutes ses espèces*, Paris 1671.*

Cheselden, William, 'An Account of Some Observations Made by a Young Gentleman, Who Was Born Blind, or Lost His Sight So Early, That He Had No Remembrance of Ever Having Seen, and Was Couch'd between 13 and 14 Years of Age', *Philosophical Transactions* 1728:402.

Cheselden, William, *The Anatomy of the Human Body*, 6th ed., London 1741.

Chydenius, Anders, *Rikets hjelp genom en naturlig finance-system*, Stockholm 1766.

Cicero, Marcus Tullius, *De natura deorum academica*, ed. H. Rackham, Cambridge MA & London 1979.

Clauberg, Johann, *Physica, quibus rerum corporearum vis & natura, mentis ad corpus relatæ proprietates*, …, Amsterdam 1664.

Collegium curiosorums protokoll (1711), ed. N. C. Dunér 1910.

Columbus, Samuel, *Odae Sveticae. Thet är, Någre werlds-betrachtelser, sång-wijs författade*, Stockholm 1674; ed. B. Olsson & B. Nilsson, *Samlade dikter* I, Stockholm 1994.

Comenius, John Amos, *Orbis sensualium pictus. Hoc est: Omnium fundamentalium in mundo rerum & in vita actionem, pictura & nomenclatura. Editio trilinguis auctior et emendatior*; …, Riga 1683; new ed., Åbo 1684; 4th ed., Stockholm 1716.

Constitutiones nationis Dalekarlo-Vestmannice Upsaliæ die X maji MDCC datæ jämte några anteckningar om Emanuel Swedenborgs studenttid i Uppsala 1699–1709, ed. H. Ruuth, Uppsala 1910.

Coster, Johann, *Affectuum totius corporis humani præcipuorum theoria et praxis, tabulis exhibitæ.* …, Frankfurt 1663.*

Cusanus, Nicolaus, 'De docta ignorantia', *Opera*, …, Basel 1565.

Dahlbergh, Erik, *Svecia antiqua et hodierna*, Stockholm 1716; new ed., Stockholm 1924.

Dalin, Olof von, *Sagan om hästen*, Stockholm 1740; facsimile, ed. A. Blanck & F. Askeberg, Stockholm 1953.

Dalin, Olof von, *Åminnelsetal öfver kongl. vetenskaps academiens medlem ärcke-biskopen herr doct. Eric Benzelius*, Stockholm 1744.

Dalin, Olof von, *Svea rikes historia ifrån dess begynnelse til wåra tider* I, Stockholm 1747.

Dante Alighieri, *Divina Commedia* (*c.* 1320); translation R. M. Durling, *The Divine Comedy*, Oxford 1996–2003.

Dee, John, *The Mathematicall Praeface to the Elements of Geometrie of Euclid of Megara (1570)*, ed. A. G. Debus, New York NY 1975.

Dela Gardiska archivet, eller handlingar ur grefl. Dela-Gardiska bibliotheket på Löberöd, IX, ed. P. Wieselgren, Lund 1837.

Derham, William, 'Farther Observations and Remarks on the Same Subject', *Philosophical Transactions* 1704–1705:303.

Derham, William, *Physico-theology: Or, a Demonstration of the Being and Attributes of God, from his Works of Creation. Being the Substance of XVI Sermons …*, London 1713; translation A. Nicander, *Physico-theologie eller: Til Gud ledande naturkunnighet, medelst jord-klotets och de däruppå befintelige kreaturens upmärksama betraktelse; til ögonskenligt bewis, at en Gud är til, och at Han det högsta goda, samt et alsmägtigt och allwist wäsende är*, 2nd ed., Stockholm 1760.

Derham, William, *Astro-theology or a Demonstration of the Being and Attributes of God, from a Survey of the Heavens*, 2nd ed., London 1715; translation A. Nicander, *Astro-theologie eller Himmelska kropparnes betracktelse. Til ögonskenligit bewis at en Gud är til, och at han det aldrabästa, aldrawisaste och alsmäktigste wäsende är med kopparstycke författad*, Stockholm 1735.

Descartes, René, *Discours de la méthode pour bien conduire sa raison et chercher la vérité dans les sciences …*, Leiden 1637; ed. G. Heffernan, *Discours de la Methode pour bien conduire sa raison et chercher la verité dans les sciences. Discourse on the method of conducting one's reason well and of seeking the truth in the sciences. A bilingual edition and an interpretation of René Descartes' philosophy of method*, Notre Dame IN 1994.

Descartes, René, *Meditationes de prima philosophia …*, Paris 1641; translation F. E. Sutcliffe, 'Meditations on the First Philosophy in which the Existence of God and the Real Distinction between the Soul and the Body of Man Are Demonstrated', *Discourse on Method and the Meditations*, new ed., London 1988.

Descartes, René, *Principia philosophiæ*, Amsterdam 1644.

Descartes, René, *Les passions de l'âme*, Paris 1649.

Descartes, René, *Le monde ou le traité de la lumiere*, Paris 1664a.

Descartes, René, *L'homme …*, Paris 1664b.

Descartes, René, *Oeuvres de Descartes* II–XI, ed. C. Adam & P. Tannery, Paris 1898–1909.

Diogenes Laertius, *Biōn kai gnōmōn tōn en philosophiai eydokimēsantōn tōn eis deka to ekton*; ed. R. D. Hicks, *Lives of Eminent Philosophers* II, Cambridge MA & London 1979.

Ditton, Humphry, *An Institution of Fluxions: Containing the First Principles, the Operations, with Some of the Uses and Applications of That Admirable Method*, London 1706.

Dress, Otto, 'Beskrifningh om deth som ståhl och en extraordinarie godh järntillwärkningh … angåår … sampt … om metallgrufwor och mallms påfinnande …' (1687), ed. H. Carlborg, *Blad för bergshandteringens vänner* XVIII, 1925.

Duhre, Anders Gabriel, *Första delen af en grundad geometria, bewijst uti de föreläsningar, som äro håldne på swänska språket uppå kongl. fortifications contoiret i Stockholm*, Stockholm 1721.

Duhre, Anders Gabriel, *Wälmenta tanckar, angående … et laboratorium mathematico-oeconomicum …*, Stockholm 1722.*

Dupleix, Scipion, *Corps de philosophie contenant la logique, la physique, la metaphysique, et l'etique*, Genève 1636.

Duræus, Samuel, *Utkast til föreläsningar öfver naturkunnigheten*, Uppsala 1759.*

Duverney, Joseph G., *Tractatus de organo auditus oder Abhandlung vom Gehör …*, German translation from French, J. A. Mischel, Berlin 1732.*

Ehrenheim, Fredrik Wilhelm von, *Tessin och Tessiniana. Biographie med anecdoter och reflexioner, samlade utur framledne riks-rådet m.m. grefve C. G. Tessins egenhändiga manuscripter*, Stockholm 1819.

Elvius, Pehr, the Elder, *Tabula compendiosa logarithmorum sinuum ad quadrantis gradus, eorumq; partes decimas, nec non numerorum absolutorum ab unitate ad 1 000*, Uppsala 1698.

Elvius, Pehr, the Elder, 'Observatio circa eclipsin solis totalem, quæ in diem 22 aprilis st. jul. anni 1715 incidit, habita Upsaliæ', *Acta literaria Sveciæ* 1722.

Elvius, Pehr, the Elder (pres.), *De re metallica Sueo-Gothorum schediasma*, resp. Lars Benzelius, Uppsala 1703.

Elvius, Pehr, the Elder (pres.), *Historiam astronomiæ ellipticæ*, resp. Eric Frondin, Uppsala 1703.

Elvius, Pehr, the Elder (pres.), *De perspecillis*, resp. A. Walingius, Uppsala 1704.

Elvius, Pehr, the Elder (pres.), *Astronomicum argumentum de asterismis*, resp. Bengt M. Bredberg, Uppsala 1707.

Elvius, Pehr, the Elder (pres.), *De longitudine geographica dissertatio*, resp. Andreas Duræus, Uppsala 1710.

Elvius, Pehr, the Elder (pres.), *De proportione harmonica;* ... I, resp. & author Eric Burman, Uppsala 1715.

Elvius, Pehr, the Elder (pres.), *De fatis philosophiae naturalis* II, resp. Carl Bechstadius, Uppsala 1716.

Elvius, Pehr, the Elder (pres.), *De planeta Venere*, resp. Birger Jonas Wassenius, Uppsala 1717.

Erasmus of Rotterdam, *Opera omnia* II, Leiden 1703.

Euclid, *De sex första jämte ellofte och tolfte böckerna af Euclidis Elementa, eller grundeliga inledning til geometrien*, ed. M. Strömer, Uppsala 1784.

Faggot, Jacob, *Rön af mätekonsten til utletande af hwarjehanda kärils innehåll uti swenskt mått och mål*, Stockholm 1739.

Faggot, Jacob, 'Om tijotälning, eller decimalers häfd i bokhålleri och räkning, som rörer mått, mål, wigt och mynt, utan rubning i de wanlige inrättningar', *Kongl. swenska wetenskaps academiens handlingar* III, 1742.

Ferner, Bengt, *Tvisten om vattu-minskningen, föreställd uti et tal, för kongl. vetenskaps academien*, Stockholm 1765.

Fontana, Niccolò, see Tartaglia.

Fontenelle, Bernard Le Bovier de, *Entretiens sur la pluralité des mondes*, Amsterdam 1686; ed. A. Calame, Paris 1966; translation J. Glanvill, *A Plurality of Worlds*, London 1702.

Forelius, Hemming (pres.), *De symbolis pythagoricis*, resp. J. M. Gardmann, Uppsala 1697.

Forsius, Sigfridus Aronus, *Physica (Cod. Holm. D 76)*, ed. J. Nordström, Uppsala universitets årsskrift 1952:10, Uppsala 1952.

Förtekning på afl. wälborne herr assessor Swedenborgs efterlämnade wackra boksamling, i åtskilliga språk och wetenskaper, som kommer at försäljas på bok-auctions-kammaren i Stockholm d 28 nov. 1772, Stockholm 1772; facsimile, ed. A. H. Stroh, *Catalogus bibliothecae Emanuelis Swedenborgii*, Stockholm 1907.

Frese, Jacob, *Echo, å Sweriges allmänne frögdeqwäden uttryckt, på hans kongl. maj.ts wår allernådigste konungs konung Carl den XII:tes Sweriges, Giötes, och Wändes konungs, &c. högsthugnelige namns-dag,* ..., Stockholm 1715.

Frese, Jacob, *Andelige och werldslige dikter,* ..., Stockholm 1726.

Frölich, David, *Bibliotheca, seu Cynosura peregrinantium, hoc est, Viatorium, omnium hactenus editionum absolutissmum,* ... II:1, Ulm 1644.

Frondin, Elias (pres.), *De Schedvia Westergothiæ urbe, antiqua S. Helenæ sede* I, resp. Andreas H. Forssenius, Uppsala 1734.

Galilei, Galileo, *Istoria e dimonstrazioni intorno alle macchie solari* ..., Roma 1613.

Galilei, Galileo, *Dialogo* ... *sopra i due massimi sistemi del mondo* ..., Firenze 1632; translation G. de Santillana, *Dialogue on the Great World Systems*, Chicago IL 1953.

Geber, *Curieuse vollständige chymische Schriffte* ... *an Tag gegeben von Phileletha*, Frankfurt & Leipzig 1710.*

The Gentlemen's Magazine, September 1754.

Gepriesenes Andencken von Erfindung der Buchdruckerey wie solches in Leipzig beym Schluss des dritten Jahrhunderts von den gesammten Buchdruckern daselbst gefeyert worden, Leipzig 1740.

Gestrinius, Martinus Erici, *In geometriam Euclidis demonstrationum libri sex* ..., Uppsala 1637.

Gestrinius, Martinus Erici (pres.), *De primo ac simplicissimo geometriæ principio puncto*, resp. Petrus A. Schomerus, Uppsala 1627.

Gilbert, William, *De magnete, magneticisqve corporibvs, et de magno magnete tellure; Physiologis noua, plurimis & argumentis, & experimentis demonstrata*, London 1600*; translation P. Fleury Mottelay, *De magnete*, new ed., Mineola NY 1958.

Glauber, Johann Rudolf, *Furni novi philosophici, sive descriptio artis destillatoriæ novæ;* ..., Amsterdam 1651.*

Granberg, Lars Bengtson, *Curieuse och snälle uthräkningar* [*c.* 1700].

Grassineau, James, see Brossard.

Gregory, David, *Astronomiæ physicæ & geometricæ elementa* I, 2nd ed., Genève 1726.*

Grimaldi, Francesco Maria, *Physico-mathesis de lumine, coloribus, et iride, aliisque adnexis libri duo* ..., Bologna 1665.

Grönwall, Anders (pres.), *De ferro svecano osmund*, resp. Petrus Saxholm, Uppsala 1725.

Grönwall, Anders (pres.), *De territorio Cuprimontano, et Sæterensi, nec non Jarnberia occidentali*, resp. Samuel Resenberg, Uppsala 1734.

Grundell, Daniel, *Nödig underrättelse om artilleriet till lands och siös, så wäl till theoriam, som praxin beskrifwen, och med nödige kopparstycken förklarad*, Stockholm 1705.*

Gudhemius, Petrus, *Undersöknings-skrift om jordbäfningar: deras naturliga orsaker, forna wärkan, de länder hwilka af dem mäst hemsökas, och teknen som dem föregå*, translation C. P. Ekblad, Stockholm 1756 (1702).

Guglielmini, Domenico, *De sanguinis natura & constitutione exercitatio physico-medica. Accedit ejusdem pro theoria medica adversus empiricam sectam prælectio*, 2nd ed., Utrecht 1704; Venezia 1701.*

Halley, Edmond, 'An Account of the Several Species of Infinite Quantity, and of the Proportions they Bear One to the Other, as it was Read before the Royal Society', *Philosophical Transactions* 1692:195.

Halley, Edmond, 'An Account of the Circulation of the Watry Vapours of the Sea, and of the Cause of Springs', *Philosophical Transactions* 1694:192.

Halley, Edmond, 'Of the Infinity of the Sphere of Fix'd Stars', *Philosophical Transactions* 1720:364.

Hamann, Johann Georg, *Briefwechsel. Fünfter Band 1783–1785*, ed. A. Henkel, Frankfurt am Main 1965.

Hammarström, Eric, *Äldre och nyare märkwärdigheter wid St. Kopparberget* I, Falun 1789.

Happel, Eberhard Werner, *Denna werldennes största tänckwärdigheeter eller dhe så kallade relationes curiosæ* ..., Stockholm 1682.

Harvey, William, *Exercitatio anatomica de motu cordis et sanguinis in animalibus*, Frankfurt 1628; translation R. Willis, *On the Motion of the Heart and Blood in Animals*, Buffalo NY 1993.

Hauksbee, Francis, *Physico-mechanical Experiments on Various Subjects:* ..., London 1709.

Heister, Lorenz, *Compendium anatomicum* ... , Nürnberg & Altdorf 1732*; translation of 4th Latin ed., G. F. Claudern, *Compendivm anatomicvm d. i. kurtzer Begriff derjenigen Kunst, welche von denen Theilen des menschlichen Cörpers, und anderer Thiere, nebst deroselben künstlichen Zerlegung handelt*, Nürnberg 1736.

Hellwig, Christoph von (pseud. Valentin Kräutermann), *Der accurate Scheider und künstliche Probierer,* ..., Frankfurt & Leipzig 1717.*

Helvetius, see Schweitzer.

Henckel, Johann Friedrich, *Pyritologia, oder Kieß-Historie, als des vornehmsten Minerals, nach dessen Nahmen, Arten, Lagerstätten, Ursprung, Eisen, Kupffer, unmetallischer Erde, Schwesel, Arsenic, Silber, Gold, einfachen Theilgen, Vitriol und Schmeltz-Nutzung,* ..., Leipzig 1725.*

Henckel, Johann Friedrich, *Idea generalis de lapidum origine per observationes experimenta & consectaria succincte adumbrata*, Dresden & Leipzig 1734.

Henrion, Denis, see Mangin.

Heraclitus, *Peri physeos*; ed. & translation T. M. Robinson, *Fragments*, Toronto 1987.

Hermansson, Johan, *Memoria vitæ et mortis viri quondam amplissimi & celeberrimi, mag. Johannis Vallerii,* ..., Uppsala 1718.

Hermansson, Johan (pres.), *De virgula divinatoria,* resp. Johannes Maehlin, Uppsala 1718.

Hermansson, Johan (pres.), *De existentia mentis,* resp. Anders Celsius, Uppsala 1728.

Herodotus, *Historiai;* ed. A. D. Godley, *Herodotus* I, Books I and II, Cambridge MA & London 1975.

Hesiod, *Theogonia; Erga kai hēmerai;* ed. G. W. Most, *Theogony, Works and Days, Testimonia,* Cambridge MA & London 2006.

Hesselius, Andreas, 'Magr Andreæ Hesselii probstens i Gagnef anmärckningar, öfwer några besynnerliga occurrencer, som honom bemött under hans resa till America, under hans wistande thär, och under hans hemresa til Sverige, ifrån åhr 1711. till åhr 1724 inclusive', ed. N. Jacobsson, *Svenska Linnésällskapets årsskrift* 1938.

Hiärne, Urban, *Den korta anledningen, til åthskillige malm och bergarters, mineraliers och jordeslags &c. efterspörjande och angifwande, beswarad och förklarad, jämte deras natur, födelse och i jorden tilwerckande, samt vplösning och anatomie, i giörligaste måtto beskrifwen,* Stockholm 1702.*

Hiärne, Urban, 'Memorabilia nonulla lacus Vetteri', *Philosophical Transactions* 1704–1705:298.

Hiärne, Urban, *Den beswarade och förklarade anledningens andra flock, om jorden och landskap i gemen,* Stockholm 1706.*

Hiärne, Urban, *Actorum chymicorum Holmensium parasceue id est præparatio ad tentamina, in regio laboratorio Holmensi peracta,* ..., Stockholm 1712.*

Hiärne, Urban, *Ortographia svecana, eller den retta svenska bookstafveringen stelt i ett samtal emellan Neophilum och Eustathium,* Stockholm 1716/17.

Hiärne, Urban, *Samlade dikter,* ed. B. Olsson & B. Nilsson, Stockholm 1995.

Hippocrates, *Peri chymon;* ed. W. H. S. Jones, *Hippocrates* IV, Cambridge MA & London 1979.

Historie der Gelehrsamkeit unserer Zeiten, Leipzig 1722.

Hobbes, Thomas, *Leviathan, or the Matter, Forme, & Power of a Common-wealth, Ecclesiasticall and Civill,* London 1651; new ed., C. B. Macpherson, London 1985.

Hoffmann, Friedrich, *Observationvm physicochymicarvm selectiorvm libri III in qvibis mvlta cvriosa experimenta et lectissimæ virtvtis medicamenta exhibentvr ad solidam et rationalem chymiam stabiliendam præmissi,* Halle 1722.*

Hoffmann, Friedrich, *Opera omnia physico-medica* ... I, Genève 1748.

Hoffwenius, Petrus, *Synopsis physica, disputationibus aliquot academicis comprehensa, quas in gratiam alumnorum regiorum publice instituit, ac postea ad requisitionem & desiderium suorum auditorum, proprijs sumptibus prælo subjecit,* Stockholm 1678; 2nd ed., ed. H. Vallerius, Stockholm 1698.

Hogarth, William, *The Analysis of Beauty, Written with a View of Fixing the Fluctuating Ideas of Taste,* London 1753; new ed., Oxford 1955.

Homer, *Ilias;* ed. A. T. Murray, *The Iliad,* Cambridge MA & London 1999.

Homer, *Odyssea;* ed. A. T. Murray, *The Odyssey* I, Cambridge MA & London 1995.

Hooke, Robert, *Micrographia or Some Physiological Descriptions of Minute Bodies Made by Magnifying Glasses with Observations and Inquiries Thereupon,* London 1665; facsimile, New York NY 1961.

Höpken, Anders Johan von, *Riksrådet grefve A. J. von Höpkens skrifter* I, ed. C. Silfverstolpe, Stockholm 1890.

Horatius Flaccus, Quintus, *Carmina;* ed. C. E. Bennett, *The Odes and Epodes,* Cambridge MA & London 1978.

Hugo of Saint Victor, *Didascalicon;* translation J. Taylor, 'Classification of the Sciences', *A Source Book in Medieval Science,* ed. E. Grant, Cambridge MA 1974.

Hultman, Johan, *Annotationer öfver konung Carl XII:s hjeltebedrifter,* Stockholm 1986.

Hume, James, *Traicté de l'algebre, d'vne methode novvelle et facile.* ..., Paris 1635.

Humerus, Bonde (pres.), *De camera obscura,* resp. Johannes Gane, Lund 1706.

Huygens, Christiaan, *Horologivm oscillatorivm sive de motv pendvlorvm ad horologia aptato demonstrationes geometricæ,* Paris 1673; facsimile, Brussel 1966.

Huygens, Christiaan, *Traite de la lvmiere, Où sont expliquées les causes de ce qui luy arrive dans la reflexion, & dans la refraction. Et particulierement dans l'etrange refraction dv cristal d'Islande...*, Leiden 1690; facsimile, Brussel 1967.

Index librorum prohibitorum ..., Roma 1758.

Jobert, Louis, *La science des médailles pour l'instruction de ceux qui s'appliquent à la connaissance des médailles antiques et modernes*, Amsterdam 1693.

Kant, Immanuel, *Träume eines Geistersehers, erläutert durch Träume der Metaphysik*, Riga & Mietau 1766; translation G. R. Johnson & G. A. Magee, *Kant on Swedenborg: Dreams of a Spirit-seer and Other Writings*, ed. G. R. Johnson, West Chester PA 2002.

Kellner, David, *Kurtz abgefastes sehr nutz- und erhauliches Berg- und Saltzwercks-Buch* ..., Frankfurt & Leipzig 1702.*

Kemble, John M. (ed.), *State Papers and Correspondence Illustrative of the Social and Political State of Europe from the Revolution to the Accession of the House of Hanover*, London 1857.

Kepler, Johannes, *Prodromus dissertationvm cosmographicarvm, continens mysterivm cosmographicvm de admirabili proportione orbium coelestium* ..., Frankfurt 1621; new ed. F. Hammer & W. von Dyck, *Gesammelte Werke* VIII, München 1963.

Kertzenmacher, Peter, *Alchimia, das ist alle Farben, Wasser, olea, salia und alumina, ...*, 1720.*

Kircher, Athanasius, *Magnes siue de arte magnetica opvs tripartitvm, ...*, Köln 1643.*

Kircher, Athanasius, *Ars magna lvcis et vmbrae, in decem libros digesta. ...*, Roma 1646; new ed., Amsterdam 1671.

Kircher, Athanasius, *Mvsvrgia vniversalis sive ars magna consoni et dissoni in X. libros digesta* ..., Roma 1650.

Kircher, Athanasius, *Mundus subterraneus, in XII libros digestus;* ..., Amsterdam 1665.

Kircher, Athanasius, *Magneticum naturæ regnum sive disceptatio physiologica de triplici in natura rerum magnete, juxta triplicem ejusdem naturæ gradum digesto inanimato animato sensitivo* ..., Amsterdam 1667.*

Kirchmaier, Georg Caspar, *Institutiones metallicæ, das ist wahr- und klarer Vnterricht vom edlen Bergwerck, ...*, Wittenberg 1687.

Kirchmaier, Georg Caspar, *Hoffnung besserer Zeiten durch das edle Bergwerck, von Grund, und aus der Erden zuerwarten; Nebenst Vorbericht vom Bergwerck selbst, dessen Rechten und Freyheiten, besserer Schmeltz-Scheide- und Seygerkunst, ...*, Wittenberg 1698.*

Klingenstierna, Samuel (pres.), *De usu algebræ* II, resp. Lars Julius Kullin, Uppsala 1743.

Komenský, Jan Amos, see Comenius.

Kongl. maj:ts förnyade förordning, angående måhl, mått och wigt, Stockholm, 29 May 1739.

Kongl. maj:ts förnyade förordning, angående mått, måhl och wigt, Stockholm, 27 May 1737.

Kongl. maj:ts nådige öpne bref, och privilegium, angående ett saltsiuderij-wercks inrättande, Lund, 26 June 1717, Stockholm 1717.

Kongl. swenska wetenskaps academiens handlingar I–III, XXIV, Stockholm 1740–1742, 1763.

König, Emanuel, *Regnum minerale* ..., Basel 1686.*

Kräutermann, Valentin, see Hellwig.

Krebs, Nikolaj, see Cusanus.

Kunckel von Löwenstern, Johann, *Ars vitraria experimentalis, oder vollkommene Glasmacher-Kunst, ...*, Frankfurt & Leipzig 1679.*

Kunckel von Löwenstern, Johann, *Collegium physico-chymicum experimentale, oder Laboratorium Chymicum, ...*, Hamburg & Leipzig 1716.*

Küster von Rosenberg, Johannes, see Coster.

Lagerbring, Sven, *Sammandrag af Swea-rikes historia, ifrån äldsta til de nyaste tider* IV:3, 2nd ed., Stockholm 1779.

Lagerholm, Johannes, *Uthräknings-book uthi trenne delar fördelt. ...*, Stockholm 1704.

Lagerlööf, Petrus (pres.), *De siphonibus marinis*, resp. Andreas Högwall, Uppsala 1697.

La Motraye, Aubry de, *A. de Motraye's Travels through Europe, Asia, and into Part of Africa; ...*, London 1723.

Leeuwenhoek, Antonie van, *Arcana naturæ detecta*, Delft 1695; Leiden 1722.*

Leibniz, Gottfried Wilhelm von, 'Nova methodvs pro maximis et minimis, itemque tangentibus, quæ nec fractas, nec irrationales quantitates moratur, & singulare pro illis calculi genus', *Acta eruditorum*, October 1684.

Leibniz, Gottfried Wilhelm von, 'Système nouveau de la nature et de la communication des substances, aussi bien que de l'union qu'il y a entre l'âme et le corps', *Journal des savants*, 27 June 1695; translation L. E. Loemker, 'A new system of the nature and the communication of substances, as well as the union between the soul and the body', *Philosophical Papers and Letters*, vol. II, Chicago IL 1956.

Leibniz, Gottfried Wilhelm von, *Monadologie* (1714); ed. & translation L. E. Loemker, 'The Monadology', *Philosophical Papers and Letters*, vol. II, Chicago IL 1956.

Leibniz, Gottfried Wilhelm von, *Epistolae ad diversos*, ... I, ed. C. Kortholt, Leipzig 1734.

Lémery, Nicolas, *Cours de chymie contenant la maniere de faire les operations qui sont en usage dans la medicine*, Paris 1675; Paris 1683*; German translation, *Cours de chymie, oder: Der vollkommene Chymist*, ..., Dresden 1713; Dresden 1726.*

Leupold, Jacob, *Theatrum arithmetico-geometricum, das ist: Schau-Platz der Rechen- und Meß-Kunst*, ..., Leipzig 1727.

Leutmann, Joh. Georg, *Vvlcanvs famvlans oder sonderbahre Feuer-Nutzung welche durch gute Einrichtung der ... mit wenigem Holtze starcke Wärme und grosse Hitze gemachet auch das Rauchen in Stuben verhindert werden ...*, 2nd ed., Wittenberg 1723.*

L'Hospital, Guillaume-François-Antoine de, *Analyse des infiniment petits, pour l'intelligence des lignes courbes*, 2nd ed., Paris 1715.

Linder, Johan (Lindestolpe), *De venenis in genere, & in specie exercitatio, videlicet eorum natura, & in corpus agendi modo: atque eadem, pro morbi acuti vel chronici ex iisdem oborientis indole, curandi; & in esculentis, potulentisque indagandi ratione, juxta veterum quorumdam & recentiorum dogmata, ad solidorum & fluidorum corporis organici leges mechanicas, deducta & explicata*, Leiden 1707.

Linnaeus, Carl, 'Qværenda', *Iter Lapponicum. Lappländska resan 1732* II, ed. I. Fries, S. Fries & R. Jacobsson, Umeå 2003.

Linnaeus, Carl, *Diæta naturalis 1733*, ed. A. Hj. Uggla, Uppsala 1958.

Linnaeus, Carl, *Vulcanus Docimasticus Fahlun 1734*, ed. G. A. Granström, Uppsala 1925.

Linnaeus, Carl, *Linnés Dalaresa (Iter Dalekarlicum) jämte Utlandsresan (Iter ad exteros) och Bergslagsresan (Iter ad fodinas)* (1734), ed. A. Hj. Uggla, Stockholm 1953; translation, *The Dalarna Journey Together with Journeys to the Mines and Works*, Örebro 2007.

Linnaeus, Carl, *Tal, om märkwärdigheter uti insecterne*, Stockholm 1739.

Linnaeus, Carl, 'Minne af Johannis Moræi ... Lefnad' (1742), ed. T. Fredbärj, *Svenska Linnésällskapets årsskrift* 1962.

Linnaeus, Carl, 'Minne af Johannis Moræi ... Lefnad', *Kongl. swenska wetenskaps academiens handlingar* III, 1742.

Linnaeus, Carl, 'Extract af Prof. Linnaei brev till Secret. Elvius', *Lärda tidningar* 1746:7.

Linnaeus, Carl, 'Kinne-kulle aftagen i profil och beskrifven af volontairen vid Kongl. Fortification Herr Johan Svenson Lidholm', *Kongl. swenska wetenskaps academiens handlingar* VIII, 1747.

Linnaeus, Carl, *Systema naturæ sistens regna tria naturæ, in classes et ordines genera et species*, 6th ed., Stockholm 1748.

Linnaeus, Carl, *Pluto Svecicus och Beskrifning öfwer stenriket*, ed. C. Benedicks, Uppsala 1907.

Linnaeus, Carl, *Bref och skrifvelser af och till Carl von Linné* I:7, ed. Th. M. Fries & J. M. Hulth, Stockholm 1917.

Linnaeus, Carl, *Nemesis divina*, ed. E. Malmeström & T. Fredbärj, Stockholm 1968; ed. & translation M. J. Petry, *Nemesis Divina*, Dordrecht 2001.

Linnaeus, Carl (pres.), *Oeconomia naturæ*, resp. Isac J. Biberg, Uppsala 1749.

Linnaeus, Carl (pres.), *Generatio ambigena*, resp. Christian Ludvig Ramström, Uppsala 1759.

Linnaeus, Carl (pres.), *De politia naturae*, resp. Christ. Daniel Wilcke, Uppsala 1760.

Lister, Martin, 'De fontibus medicatis Angliæ exercitatio nova & prior', *Acta eruditorum* 1684.

Locke, John, *An Essay Concerning Human Understanding*, London 1689; new ed., P. H. Nidditch, Oxford 1975.

Löhneyß, Georg Engelhard von, *Gründlicher und aussführlicher Bericht von Bergwercken*, ..., Stockholm & Hamburg 1690.*

Lucretius Carus, Titus, *De rerum natura*; ed. W. H. D. Rouse, Cambridge MA & London 1982.

Maigret, Philippe, 'Brev den 23 december 1723 om Karl XII:s död', in Bring 1920.

Malebranche, Nicolas, *De la recherche de la verité: Où l'on traitte de la nature de l'esprit de l'homme, & de l'usage qu'il en doit faire pour éviter l'erreur dans les sciences* I–II, 4th ed., Amsterdam 1688; *De inquirenda veritate* ..., Genève 1689.*

Mandeville, Bernard de, *The Fable of the Bees: Or Private Vices, Publick Benefits.* ..., 5th ed., London 1728–1729.

Mangin, Clément Cyriaque de (pseud. D. Henrion), *Traicté d'algebre*, Paris 1620.

Marcus Aurelius, *The Meditations of Marcus Aurelius Antonius*, translation A. S. L. Farquharson, Oxford 1989.

Martin, Pehr, 'Then salige herrens, professorens och medicinæ doctorens Herr Assessor Petri Martins bref, anmärkningar och påminnelser wid Wätters beskrifningen' (18 April 1723), in D. Tiselius 1730.

Materialy dlya istory Imperatorskoy akademy nauk, vol. 2 (1731–1735), St. Petersburg 1886.

Melle, Jacob à, *De lapidibvs figvratis agri litorisque Lvbecensis*, Lübeck 1720.*

Mersenne, Marin, *La vérité des sciences contre les sceptiques ou pyrrhoniens*, Paris 1625.

Mersenne, Marin, *Harmonie universelle*, Paris 1636–1637; facsimile, Paris 1963.

Micrander, Julius (pres.), *Demophili similitudines seu vitæ curatio ex Pythagoreis. Ejusdemque sententiæ Pythagoricæ. Cum versione & scholiis L. Holstenii. Quas observationibus moralibus illustravit* ..., resp. Jesper Swedberg, Stockholm 1682.

Milliet Dechales, Claude François, *Cursus seu mundus mathematicus* III, 2nd ed., Lyon 1690.

Milton, John, *Paradise Lost: A Poem Written in Ten Books*, London 1667; new ed., ill. W. Blake, Liverpool 1906.

More, Thomas, *Utopia*, London 1551; new ed., in *Three Modern Utopias*, ed. S. Bruce, Oxford 1999.

Morhof, Daniel Georg, *Epistola de scypheo vitreo per certum humanæ vocis sonum rupto*, Kiel 1672.

Musgrave, W., 'Some experiments and observations concerning sounds', *Philosophical Transactions* 1698:247.

Musschenbroek, Pieter van, *Physicæ experimentales et geometricæ de magnete, tuborum capillarium vitreorum* ..., Leiden 1729.*

Musschenbroek, Pieter van, *Tentamina experimentorum naturalium captorum in Academia del Cimento* ..., Leiden 1731.*

Mylius, Gottlieb Friedrich, *Memorabilium Saxoniæ subterraneæ* I–II, Leipzig 1709–1718.*

Napier, John, *Rabdologiae, seu numerationis per virgulas libri duo*, Edinburgh 1617.

Neue Zeitungen von gelehrten Sachen, Leipzig 1719, 1721–1723.

Newton, Isaac, *Philosophiæ naturalis principia mathematica*, London 1687*; facsimile, Brussel 1965.

Newton, Isaac, *Optice: sive de reflexionibus, refractionibus, inflexionibus & coloribus lucis libri tres*, London 1706; *Opticks: Or, a Treatise of the Reflections, Refractions, Inflections and Colours of Light*, 3rd ed., London 1721.

Newton, Isaac, *Analysis per quantitatum series fluxiones, ac differentias: cum enumeratione linearum tertii ordinis*, London 1711.

Norcopensis, Andreas (pres.), *De sono*, resp. & author Harald Vallerius, Stockholm 1674.

Nordberg, Jöran Andersson, *Konung Carl XII. historia, Senare delen ifrån slutet af junii månad år 1709, til hans May:ts. död och begrafning*, Stockholm 1740; translation C. G. Warmholtz, *Histoire de Charles XII. roi de Suéde* III–IV, Den Haag 1748.

Nordberg, Jöran Andersson, *Anmärkningar, wid hög'stsalig i åminnelse Konung Carl den XII:tes historia, som auctoren apart sina wänner meddelt, och under sina pagina och §. §. komma at intagas*, Stockholm 1767.

Nova Acta eruditorum, Leipzig 1735.

Odel, Anders, *De makalöse högstsalige konungarnas, konung Gustav Adolphs och konung Carl den tolftes rop, ifrån de dödas rike til hieltinnan Swea,* ..., Stockholm 1741.

Odhelius, Erik, *Observationes chemico-metallurgicas circa ortum & effluvia metallorum,* Brussel 1687.

Ordinaire Stockholmiske post tidender 1716:2, 17 & 36.

Ovid, *Metamorphoseon*; ed. F. J. Miller, *Metamorphoses* I–II, Cambridge MA & London 1984; translation J. Dryden et al., ed. S. Garth, *Metamorphoses*, Ware 1998.

Ovid, *Ars amatoria*; ed. J. H. Mozley, *The Art of Love, and Other Poems*, Cambridge MA & London 1985.

Ovid, *Fasti*; ed. J. G. Frazer, Cambridge MA & London 1989.

Palmroot, Johannes (pres.), *De consummatione mundi*, resp. Andreas Amberni Unge, Uppsala 1710.

Pascal, Blaise, *L'esprit de la géométrie: De l'art de persuader* (1657/58); new ed., Paris 1986; translation O. W. Wright, 'Of the Geometrical Spirit', *Minor Works*, Vol. XLVIII, Part 2. The Harvard Classics, ed. C. W. Eliot, New York NY 1909–1914.

Pascal, Blaise, *Pensées sur la religion et sur quelques autres sujets, qui ont esté trouvées après sa mort parmy ses papiers*, Paris 1670; ed. L. Lafuma, Paris 1963; translation H. Levi, *Pensées and Other Writings*, Oxford 1999.

Peringskiöld, Johan, the Elder, *En book af menniskiones slächt, och Jesu Christi börd; eller Bibliskt slächt-register, från Adam, til Iesu Christi heliga moder jungfru Maria, och hennes trolåfwade man Ioseph;* ..., Stockholm 1713.

Perrault, Charles, *Parallèle des anciens et modernes en ce qui regarde les arts et les sciences*, Paris 1688–1697.

Philo of Alexandria, *Peri tēs kata Mōysea kosmopoiias*; ed. F. H. Colson, *Philo* I, Cambridge MA & London 1981.

Philosophical Transactions, London 1692, 1694, 1698–1699, 1704–1705, 1720, 1728.

Picinelli, Filippo, *Mundus symbolicus, in emblematum universitate formatus, explicatus, et tam sacris, quàm profanis eruditionibus ac sententiis illustratus*, Köln 1687.

Plato, *Timaios*; ed. R. G. Bury, *Timaeus. Critias. Cleitophon. Menexenus. Epistles*, Cambridge MA & London 1989.

Plato, *Parmenides*; ed. H. N. Fowler, Cambridge MA & London 1926.

Plato, *Politeia*; ed. P. Shorey, *The Republic II, Books VI–X*, Cambridge MA & London 1987.

Pliny the Elder, *Naturalis historia*; ed. H. Rackham, *Natural History* I–III, Cambridge MA & London 1979–1983.

Plotinus, *Plotini Platonicorum facile coryphæi opervm philosophicorvm omnivm libri LIV. in sex enneades distribvti* ... *cum latina Marsilii Ficini interpretatione & commentatione*, Basel 1580.*

Plotinus, *Ennead*; ed. A. H. Armstrong, *Plotinus* I & IV, Cambridge MA & London 1984–1989.

Poleni, Giovanni, *De motv aqvae mixto libri dvo. Quibus multa nova pertinentia ad aestuaria, ad portus, atque ad flumina continentur*, Padua 1717.*

Polhem, Christopher, *Wishetens andra grundwahl til vngdoms prydnad mandoms nytto och ålderdoms nöje; lempadt för vngdomen efter theras tiltagande åhr, vti dagliga lexor fördelt. Första boken innehållande en liten försmak af thet, som widare följandes warder*, Uppsala 1716.

Polhem, Christopher, 'Index experimentorum, quæ in montibus vallibusque Laponiæ ut instituerentur, digna judicavit Chri[s]toph. Polhem Reg. Coll. Commerc. Consiliarius. d. 15 April. 1711', *Acta literaria Sveciæ* 1722.

Polhem, Christopher, *Kort berättelse om de förnämsta mechaniska inventioner som tid efter annan af commercierådet Christopher Polhem blifwit påfundne och til publici goda nytta och tienst inrättade, sampt om det öde, som en del af dem hafft genom tidernas oblida förändringar.* ..., Stockholm 1729.

Polhem, Christopher, 'Epistola ad Andream Celsium ... de nova sua ignis theoria', *Acta literaria et scientarum Sveciae* III, Uppsala 1731.

Polhem, Christopher, 'Tanckar om mechaniquen', *Kongl. swenska wetenskaps academiens handlingar* I, 1740.

Polhem, Christopher, *Samtal emellan en swär-moder och son-hustru, om allehanda hus-hålds förrättningar*, Stockholm 1745; facsimile, Stockholm 1987.

Polhem, Christopher, *Christopher Polhems Patriotiska testamente, eller underrättelse om järn, stål, koppar, mässing, tenn och bly för dem, som wilja begynna manufacturer i dessa ämnen. Jemte en förtekning på alla dess mechaniska inventioner*, Stockholm 1761.*

Polhem, Christopher, *Christopher Polhems brev*, ed. A. Liljencrantz, Uppsala 1941–1946.

Polhem, Christopher, *Christopher Polhems efterlämnade skrifter* I–IV, ed. H. Sandblad, G. Lindeberg, A. Liljencrantz & B. Löw, Uppsala 1947–1954.

Polhem, Gabriel, *Tal om mathematiska vetenskapernes nytta uti åtskilliga bygnaders varaktiga sammansätningar, . . .*, Stockholm 1745.

Porta, Giambattista della, *Magiæ naturalis libri 20. Ab ipso authore expurgati, . . .*, Napoli 1589.

Proclus, *A Commentary on the First Book of Euclid's Elements*, translation G. R. Morrow, Princeton NJ 1970.

Prosperin, Erik, *Tal, om kongliga vetenskaps societeten i Upsala; hållet för kongl. vetenskaps academien, vid præsidii nedläggande, den 18 november 1789*, Stockholm 1791.

Pufendorf, Samuel von, *De officio hominis et civis juxta legem naturalem libri duo*, Lund 1673.

Quensel, Conrad, 'Conrad Qvensels math. prof. Lundensis amicum judicium de methodo dn. Swedenborgii, pro invenienda longitudine locorum per lunam', *Acta literaria Sveciæ* 1722.

Rålamb, Åke, *Adelig öfnings första tom*, Stockholm 1690.

Réaumur, René Antoine Ferchault de, *L'art de convertir le fer forgé en acier, et l'art d'adoucir le fer fondu, ou de faire des ouvrages de fer fondu aussi finis que de fer forgé*, Paris 1722.*

Regnard, Jean-François, *Voyage de Flandres, d'Hollande, Suède, Danemark, la Laponie, la Pologne et l'Allemagne*, Paris 1731; translation, 'A Journey through Flanders, Holland, &c.', in John Pinkerton, *A General Collection of the Best and Most Interesting Voyages and Travels in All Parts of the World* I, London 1808.

Reyneau, Charles René, *Analyse demontrée, ou la methode de resoudre les problêmes des mathematiques, et d'apprendre facilement ces sciences* I, new edition, Paris 1708.

Reyneau, Charles René, *Usage de l'analyse, ou la maniere de l'appliquer à découvrir les proprietés de figures de la geometrie simple & composée, à resoudre les problêmes de ces sciences & les problêmes des sciences physico-mathematiques, en employant le calcul ordinaire de l'algebre, le calcul differentiel & le calcul integral . . .* II, Paris 1708.*

Rhyzelius, Andreas Olai, *Brontologia theologico-historica, thet är enfaldig lära och sanferdig berettelse, om åske-dunder, blixt och skott; . . .*, Stockholm 1721.

Rhyzelius, Andreas Olai, *Biskop A. O. Rhyzelii anteckningar om sitt lefverne*, ed. J. Helander, Uppsala 1901.

Riccioli, Giovanni Battista, *Almagestum novum astronomiam veterem novamque complectens . . .*, Bologna 1651.

Richter, Christian Friedrich, *Die höchst-nöthige Erkenntniß des Menschen, sonderlich nach dem Leibe und natürlichem Leben, oder ein deutlicher Unterricht, von der Gesundheit und deren Erhaltung: . . . Haus- Reise- und Feld-Apothecken . . .*, Leipzig 1722.*

Riddermarck, Andreas (pres.), *De intendendis sonis per tubos acusticos, in usum imprimis maritimum, ex suffragio*, resp. Samuel Lychovius, København 1693.

Rinman, Sven, 'Beskrifning på vals- och skär-verk, med förbättringar', *Kongl. swenska wetenskaps academiens handlingar* XXXIII, 1772.

Ripa, Cesare, *Iconologia ouero descrittione dell'imagini vniuersali cauate dall'antichita et da altri luoghi*, Roma 1593; facsimile, Hildesheim 1970.

Roberg, Lars, *Lijkrevnings tavlor, déd ær tydelig utreedning óm een människje-cropp til dés förnæmre fasta deelars kjænningar, om-égor och vérkan*, Uppsala 1718.

Roberg, Lars, *Tal, holne för publique promotioner, vid Upsala academie*, ed. C. Linnaeus, Stockholm 1747.

Roberg, Lars, *Orationes, . . .*, Stockholm 1748.

Roberg, Lars (pres.), *De vitriolo*, resp. Johan Moræus, Uppsala 1703.

Roberg, Lars (pres.), *De metallo Dannemorensi*, resp. Magnus Håkan Sunborg, Uppsala 1716.
Roberg, Lars (pres.), *De usu methodi mechanicae in medicina*, resp. & author Nils Rosén, Uppsala 1728.
Robsahm, Carl, *Anteckningar om Swedenborg* (1782), Sollentuna 1989.
Rohr, Julius Bernard von, *Physikalische Bibliothek*, Leipzig 1754.*
Rosén von Rosenstein, Nils, *Academiskt prof, om the halliska läkedomars tilredelse samt theras sanna och inskränkta nytto*, ... , Stockholm 1744.
Rößler, Balthasar, *Speculum metallurgiæ politissimum: Oder hell-polierter Berg-Bau-Spiegel*, Dresden 1700.*
Rudbeck, Olof, the Elder, *De circulatione sangvinis*, pres. Olof Stenius, Västerås 1652.
Rudbeck, Olof, the Elder, *Nova exercitatio anatomica, exhibens ductus hepaticos aquosos, & vasa glandularum serosa*, Västerås 1653.
Rudbeck, Olof, the Elder, *Sorg- och klage-sång till Axel Oxenstiernas begravning* (1654); ed. C.-A. Moberg & J. O. Rudén, *Drei Vokalwerke der schwedischen Grossmachtepoche / Three Vocal Compositions from the Swedish Great Power Period*, Stockholm 1968.
Rudbeck, Olof, the Elder, *Atland eller Manheim* ... I, Uppsala 1679; ed. A. Nelson, *Atlantica I*, Uppsala 1937.
Rudbeck, Olof, the Elder, *Bref af Olof Rudbeck d.ä. rörande Upsala universitet* IV, ed. C. Annerstedt, Uppsala 1905.
Rudbeck, Olof, the Elder, *Taflor till Olaus Rudbecks Atlantica*, ed. A. Nelson, Uppsala 1938.
Rudbeck, Olof, the Younger, 'Rudbeck d.y:s dagbok från Lapplandsresan 1695. Med inledning och anmärkningar. 1', ed. C.-O. von Sydow, *Svenska Linnésällskapets årsskrift* 1968–1969.
Rudbeck, Olof, the Younger, *Iter Lapponicum. Skissboken från resan till Lappland 1695*, Stockholm 1987.
Rudbeck, Olof, the Younger, 'Thet svarta Nordens sorgemoln fördrifvit ...' (1697), *Samlade vitterhetsarbeten af svenska författare från Stjernhjelm till Dalin* XII, ed. P. Hanselli, Uppsala 1869.
Rudbeck, Olof, the Younger, *Nora Samolad sive Laponia illustrata* ... , Uppsala 1701.
Rudbeck, Olof, the Younger, 'Index plantarum praecipuarum, quas in Itinere Laponico anno 1695 observavit', *Acta literaria Sveciæ* 1720.
Rudbeckius, Johannes, *Enchiridion eller then swenska Psalm-Boken, sampt andra wanligha handböcker: flitigt öffuersedt och corrigerat, såsom ock medh många nödhtorfftigha stycker förbättrat och förmerat*, Västerås 1627.
Rüdiger, Andreas, *Physica Divina, recta via, eademque inter superstitionem et atheismum media, ad utramque hominis felicitatem, naturalem atque moralem*, ... , Frankfurt 1716.*
Runcrantz, Carl, *Minne af kongl. maj:ts trotjenare, lectorn i historien och moralen* ... *Nils Gustaf Sandberg*, Kalmar 1798.
Ruysch, Frederik, *Observationum anatomico-chirurgicarum centuria* ... , Amsterdam 1691.*
Rydelius, Andreas, *Nödiga förnufftz öfningar* ... I–V, Linköping 1718–1722; 2nd ed., *Nödiga förnufts-öfningar, at lära kenna thet sundas wägar och thet osundas felsteg*, Linköping 1737.
Rydelius, Andreas (pres.), *De infinito*, resp. Adam Lars Hiort, Lund 1726.
Rydelius, Andreas (pres.), *De moderamine libertatis philosophicae*, resp. Sven Almqvist, Lund 1726.
Sandels, Samuel, *Åminnelse-tal, öfver kongl. vetenskaps-academiens framledne ledamot* ... *Emanuel Swedenborg*, ... , Stockholm 1772.
Savery, Thomas, 'An Account of Mr. Tho Savery's Engine for Raising Water by the Help of Fire', *Philosophical Transactions* 1699:253.
Scapula, Johann, *Lexicon graecolatinvm sev, epitome thesavri græcæ lingvæ*, Yverdon 1623.*
Schefferus, Johannes, *De natura & constitutione philosophiæ Italicæ seu pythagoricæ liber singvlaris*, Uppsala 1664.
Schefferus, Johannes, *Lapponia, id est regionis Lapponum et gentis nova et verissima descriptio. In qua multa de origine, superstitione, sacris magicis, victu, cultu, negotiis Lapponum, item animalium, metallorumque indole, quæ in terris eorum proveniunt, hactenus incognita* ... , Frankfurt 1673; French translation, *Histoire de la Laponie, sa description, l'origine, les moeurs,*

la maniere de vivre de ses habitans, leur religion, leur magie, & les choses rares du pais, Paris 1678; English translation, *The History of Lapland Containing a Geographical Description, and a Natural History of That Country; With an Account of the Inhabitants, their Original, Religion, Customs, Habits, Marriages, Conjurations, Employments, &c*, 2nd ed., London 1704.

Schott, Gaspar, *Magia universalis naturæ et artis, sive Recondita naturalium & artificialium rerum scientia*, …, Würzburg 1657.

Schott, Gaspar, *Technica curiosa, sive mirabilia artis, libris XII. comprehensa;* …, Würzburg & Nürnberg 1664.

Schultze, Lars G., 'Kort berättelse, om myrugnar eller såkallade bläster-wärk, uti östra och wästra Dahle-orterne brukelige' (1732), *Jern-kontorets annaler* 1845.

Schweitzer, Johann Friedrich (Helvetius), *Vitulus aureus, quem mundus adorat & orat, in quo tractatur de rarissimo naturæ miraculo transmutandi metalla,* …, Amsterdam 1667.*

Seneca, Lucius Annaeus, *Ad Lucilium epistulae morales* III, ed. R. M. Gummere, Cambridge MA & London 1971.

Seneca, Lucius Annaeus, *Naturales quaestiones* I, ed. T. H. Corcoran, Cambridge MA & London 1971.

Sohren, Peter, *Musicalischer Vorschmack der jauchzenden Seelen im ewigen Leben. Das ist neuausgefärtigtes … Lutheranisches Gesangbuch* …, Hamburg 1683.*

Spegel, Haquin, *Gudz werk och hwila, thet är hela werldenes undersamma skapelse, uti sex dagar af then altzmäktige Gud fullbordad:* …, Stockholm 1685; ed. B. Olsson & B. Nilsson, *Samlade skrifter av Haquin Spegel* I:1–2, Stockholm 1998.

Spegel, Haquin, *Emblemata*, ed. B. Olsson, Stockholm 1966.

Spencer, John, *Sinnrike betänkiande om allehanda widunder och betydande tekn, hwaruti theras förebådelses fäfängia bewises och förkastas, och theras egentelige art och egenskap utmärkes*, translation M. G. Block, Stockholm 1709.

Spinoza, Baruch, *Ethica ordine geometrico demonstrata*, Amsterdam 1677; ed. C. Gebhardt, *Opera* II, Heidelberg 1925.

Spole, Anders, *Andreas Spole: Hans självbiografiska anteckningar*, ed. P. Wilstadius, Stockholm 1946.

Spole, Anders (pres.), *De myopia*, resp. O. Kiilberg, Uppsala 1679.

Spole, Anders (pres.), *De radio*, resp. Nicolaus J. Petriin, Uppsala 1679.

Spole, Anders (pres.), *Hydro-pneumatica de elaterio & phoenomenis globulorum aereorum sub aqua* …, resp. B. Humerus, Stockholm 1686.

Spole, Anders(pres.), *De trigonometria*, resp. Johannes L. Lindelius, Uppsala 1687.

Spole, Anders (pres.), *De telescopiis*, resp. Pehr Elvius, Uppsala 1688.

Spole, Anders (pres.), *De sagacitate canvm*, resp. Lars Odhelius, Uppsala 1692.

Spole, Anders (pres.), *De usu et necessitate matheseos in philosophia disputatio*, resp. Gothofredus J. Alinus, Uppsala 1693.

Stahl, George Ernst, *Anweisung zur Metallurgie, oder der metallischen Schmeltz- und Probier-Kunst. Nebst dessen Einleitung zur Grund-Mixtion derer unterirrdischen mineralischen und metallischen Cörper.* …, Leipzig 1720.*

Steuchius, Matthias (pres.), *De vacuo*, resp. Harald Vallerius, Stockholm 1678.

Stiernhielm, Georg, *Archimedes reformatus*, Stockholm 1644; facsimile, Borås 1987.

Stiernhielm, Georg, *Hercules*, Stockholm 1668a.

Stiernhielm, Georg, *Parnassus triumphans* …, Stockholm 1668b.

Stiernhielm, Georg, *Samlade skrifter av Georg Stiernhielm* I–III, ed. J. Nordström, Stockholm 1924–1990.

Stiernman, Anders Anton von, *Anonymorum centuria prima ex scriptoribus gentis Sviogothicæ quorum auctores in lucem publicam protraxit*, Stockholm 1724.

Stockholmiske kundgiörelser 1717:14, 15.

Strömer, Mårten, *Åminnelse-tal, öfver kongl. maj:ts troman, stats-secreteraren … Samuel Klingenstjerna, på kongl. vetensk. academiens vägnar, hållit, den 27 julii 1768*, Stockholm 1783.

Sturm, Johann Christopher, *Collegium experimentale, sive curiosum* ... I–II, Nürnberg 1676–1685; Nürnberg 1701.*

Sturm, Johann Christopher, *Mathesis juvenilis tomus prior. Accessit consilium de mathesi in scholarum trivialium & gymnasiorum classes omnes, etiam puerorum legere discentium infimas, ingenti juventutis commodo postliminiò introducenda*, Nürnberg 1699.

Suetonius Tranquillus, Gaius, *De vita Caesarum*; ed. J. C. Rolfe, *Suetonius* I, Cambridge MA & London 1979.

Svenska Vetenskapsakademiens protokoll för åren 1739, 1740 och 1741 I–II, ed. E. W. Dahlgren, Stockholm 1918.

Swab, Anton von, *Anton von Swabs berättelse om Avesta kronobruk 1723*, ed. S. Högberg, Stockholm 1983.

Swammerdam, Jan, *Bybel der natuure* ... */ Biblia naturae* ... , Leiden 1737–1738.*

Swedberg, Jesper, *Herre, ho tror vår predikan? I ett gudeligit bref, tå the fast sorgeliga tidender om the swenskas nederlag i Ukranien, i augusti månad innewarande åhr i wisshet förspordes; Skara stiffts församlingar förestält*, Stockholm 1709a.

Swedberg, Jesper, *Vngdoms regel och ålderdoms spegel, af Salomos Predik. XII kapitel, förestelt i ene wisa, med thess förklaring, i twå predikningar, hålna til afsked i Vpsala åhr 1703*, Skara 1709b.

Swedberg, Jesper, *Betenckiande om Sweriges olycko, hwadan then härkommer: och huru wida then, igenom stora bönedagars firande, står at lindras eller botas. Af Gudz heliga ord vpsatt, och hwart redeligit swenskt barn, som sig och sitt fädernesland wel wil, at med Skrifften och samwetet jemnföra, welment lemnat*, Skara 1710.

Swedberg, Jesper, *Gudz barnas heliga sabbatsro; Vti christeliga predikningar öfwer söndags och högtidzdags Evangelierna, med Kongl. Maij:tz nådiga frihet, gudeliga förestellt*, Skara 1710–1712.

Swedberg, Jesper, *Gudelige dödstanckar, them en christen altid, helst i thessa dödeliga krigs- och pestilens tider, bör hafwa*, Skara 1711.

Swedberg, Jesper, *Schibboleth: Swenska språkets rycht och richtighet*, Skara 1716.

Swedberg, Jesper, *Biskopens i Skara; D. Jesper Swedbergs Rettmätiga heders förswar, emot V. præsidentens och archiaterns, D. Vrban Hiernes obetenckta skrifft emot thess Schibboleth, angående swenska språkets retta skrifart,* ... , Skara 1719.

Swedberg, Jesper, *Gudeliga betenckande om then owahnliga wäderleken, och thet stadiga regnandet, som synes wilja förderwa säden och gräset på jorden. Och huru then wrede guden skal blidkas*, Skara 1725.

Swedberg, Jesper, *America illuminata, skrifwen och vtgifwen af thes biskop, doct. Jesper Swedberg, åhr 1732*, Skara 1732*; new ed., Stockholm 1985.

Swedberg, Jesper, *Jesper Swedbergs Lefwernes beskrifning*, ed. G. Wetterberg, Lund 1941.

Swedenborg, Emanuel, 'Jesperi Swedbergii, Doct. et Episcopi Scarensis, parentis optimi, canticum svecicum, Vngdoms regel och ålderdoms spegel, ex Ecclesiast: c. XII, latino carmine exhibitum ab Emanuele Swedbergio, filio', in Swedberg 1709b.

Swedenborg, Emanuel, *L. Annæi Senecæ & Pub. Syri Mimi forsan & aliorum selectæ sententiæ cum annotationibus Erasmi & græca versione Jos. Scaligeri*, pres. Fabian Törner, Uppsala 1709; translation A. Acton, *Selected Sentences from L. Annaeus Seneca and Publius Syrus the Mime*, Bryn Athyn PA 1967. [*Selectæ sententiæ*]

Swedenborg, Emanuel, 'Pl. rev. studiis, meritis atqve pietate clarissimo viro, dn. Andreæ Amb. Unge, poës. et eloq. apud Scarenses lectori, secundum tenorem constitut. reg. pro sparta ibid. theologica, Upsalis disputaturo', in Palmroot & Unge 1710.

Swedenborg, Emanuel, *Festivus applausus in Caroli XII in Pomeraniam suam adventum*, Greifswald 1714/15; ed. & translation H. Helander, *Festivus applausus in Caroli XII in Pomeraniam suam adventum*, Uppsala 1985.

Swedenborg, Emanuel, *Ludus Heliconius, sive carmina miscellanea, quæ variis in loco cecinit Eman. Swedberg*, Greifswald 1714/15*; ed. & translation H. Helander, *Ludus Heliconius and Other Latin Poems*, Uppsala 1995.

Swedenborg, Emanuel, *Camena Borea cum heroum et heroidum factis ludens: sive fabellæ Ovidianis similes sub variis nominibus scriptæ ab E. S. Sveco*, Greifswald 1715; ed. & translation H. Helander, *Camena Borea*, Uppsala 1988.

Swedenborg, Emanuel, *Cantus Sapphicus in charissimi parentis, Doct. Jesperi Swedbergii, Episcopi Scarensis reverendissimi, diem natalem, d. xxiix Augusti, ann: 1716. Ætatis 63. sive anni climacterici magni*, Skara 1716.

Swedenborg, Emanuel, *Machine att flyga i wädret* (1716), facsimile, Stockholm 1960.

Swedenborg, Emanuel, *Experiment om echo* (1716); facsimile, *Photolith.* I; ed. & translation E. E. Sandstrom, 'Experiment on Echo', *The New Philosophy*, January–June 2009.

Swedenborg, Emanuel, *Dædalus Hyperboreus: eller några nya mathematiska och physicaliska försök och anmerkningar: som wälborne herr assessor Polhammar och andre sinrike i Swerige hafwa giordt och nu tijd efter annan til almen nytta lemna* I–VI, Uppsala & Skara 1716–1718*; facsimile, N. C. Dunér 1910.

Swedenborg, Emanuel, *En ny theorie om jordens afstannande* (1717); ed., *Opera* III; translation H. Lj. Odhner, 'A New Theory about the Retardation of the Earth', *The New Philosophy*, April 1950.

Swedenborg, Emanuel, *Underrättelse, om thet förtenta Stiernesunds arbete, thess bruk, och förtening*, Stockholm 1717; facsimile, C. Sahlin 1923.

Swedenborg, Emanuel, *Regel-konsten författad i tijo böcker*, Uppsala 1718.*

Swedenborg, Emanuel, *Försök at finna östra och westra lengden igen igenom månan, som til the lärdas omprøfwande framstelles*, Uppsala 1718.

Swedenborg, Emanuel, *En ny räkenkonst som omvexlas wid 8 i stelle then wahnliga wid tahlet 10, hwarigenom all ting angående mynt, wicht, mål och mått, monga resor lettare än effter wahnligheten uträknas* (1718); ed. C. W. Oseen, 'Ett manuskript av Emanuel Swedenborg', *Lychnos* 1937; translation A. Acton, *A New System of Reckoning Which Turns at 8 instead of the Usual Turning at the Number 10 Whereby Everything Respecting Coinage, Weights, Dimensions, and Measures, Can Be Reckoned Many Times More Easily Than in the Ordinary Way*, Philadelphia PA 1941.

Swedenborg, Emanuel, *Beskrifning öfver swenska masvgnar och theras blåsningar* (1719); ed., *Noraskogs arkiv. Bergshistoriska samlingar och anteckningar* IV, Stockholm 1901–1903.

Swedenborg, Emanuel, *Underrettelse om docken, slysswercken och saltwercket*, Stockholm 1719.

Swedenborg, Emanuel, *Om jordenes och planeternas gång och stånd: thet är några bewisliga skiäl at jorden aftager i sitt lopp och nu går longsammare än tillförene; giörande winter och sommar, dagar och nätter lengre i anseende til tiden nu än förr*, Skara 1719.*

Swedenborg, Emanuel, *Om watnens högd och förra werldens starcka ebb och flod. Bewjs vtur Swergie*, Stockholm 1719.

Swedenborg, Emanuel, *Förslag til wårt mynts och måls indelning, så at rekningen kan lettas och alt bråk afskaffas*, Stockholm 1719; new ed., *Förslag til vårt mynts och måls indelning, så at räkningen kan lättas och alt bråk afskaffas*, Stockholm 1795.

Swedenborg, Emanuel, *Anatomi af wår aldrafinaste natur, wisande att wårt rörande och lefwande wäsende består af contremiscentier* (1720); facsimile, *Photolith.* I; ed. D. Dunér, *Om darrningar*, Lund 2007; translation C. T. Odhner, *On Tremulation*, 2nd ed., Bryn Athyn PA 1976.

Swedenborg, Emanuel, 'Epistola nobiliss. Emanuelis Svedenborgii ad vir. celeberr. Jacobum à Melle', *Acta literaria Sveciæ* 1721; translation, 'Some Indications of the Deluge in Sweden', *Acta Germanica: Or the Literary Memoirs of Germany, &c.* I, London 1742; translation C. E. Strutt, 'Letter to Jacob a Melle', *Miscellaneous*.

Swedenborg, Emanuel, *Prodromus principiorum rerum naturalium sive novorum tentaminum chymiam et physicam experimentalem geometrice explicandi*, Amsterdam 1721*; ed., *Opera* III; translation C. E. Strutt, *Some Specimens of a Work on the Principles of Chemistry, with Other Treatises*, new ed., Bryn Athyn PA 1976.

Swedenborg, Emanuel, *Nova observata et inventa circa ferrum et ignem, et præcipue circa naturam ignis elementarem, una cum nova camini inventione*, Amsterdam 1721; ed., *Opera* III; translation C. E. Strutt, 'Observations on Iron and Fire', *Principles of Chemistry*.

Swedenborg, Emanuel, *Methodus nova inveniendi longitudines locorum terra marique op lunæ*, Amsterdam 1721; ed., *Opera* III; translation C. E. Strutt, 'A New Method of Finding the Longitude of Places', *Principles of Chemistry*.

Swedenborg, Emanuel, 'Novæ regulæ de caloris conservatione in conclavibus', *Acta literaria Sveciæ* 1722; translation, 'New Rules for the Conservation of Heat in Rooms', *Acta Germanica: Or the Literary Memoirs of Germany, &c.* I, London 1742; translation C. E. Strutt, 'New Rules for Maintaining Heat in Rooms', *Miscellaneous*.

Swedenborg, Emanuel, 'Expositio legis hydrostaticæ, qua demonstrari potest effectus & vis aquæ diluvianæ altissimæ in saxa & materias fundi sui', *Acta literaria Sveciæ* 1722; translation C. E. Strutt, 'An Elucidation of a Law of Hydrostatics, Demonstrating the Power of the Deepest Waters of the Deluge, and their Action on the Rocks and Other Substances at the Bottom of Their Bed', *Miscellaneous*.

Swedenborg, Emanuel, *Miscellanea observata circa res naturales & præsertim circa mineralia, ignem & montium strata* I–III, Leipzig 1722*; *Pars quarta miscellanearum observationum circa res naturales & præcipue circa mineralia, ferrum & stallactitas in cavernis Baumannianis etc.*, Schiffbeck bey Hamburg 1722; translation C. E. Strutt, *Miscellaneous Observations Connected with the Physical Sciences*, new ed., Bryn Athyn PA 1976.

Swedenborg, Emanuel, 'Amicum responsum ad objectionem factam a celeberr. dn. profess. C. Quensel contra nobiliss. dn. assessor. E. Svedenborgii novam methodum longitudinis inveniendæ (vid. Act. Liter. Svec. p. 270.) datum in absentia auctoris ab amico', *Acta literaria Sveciæ* 1722.

Swedenborg, Emanuel, *Oförgripelige tanckar om svenska myntetz förnedring och förhögning*, Stockholm 1722; new ed., *Oförgripelige tankar om myntets uphöjande och nedsättjande*, Uppsala 1771.

Swedenborg, Emanuel, *Qvia typis vulgandus est liber, De genuina metallorum tractatione* …, 1722.

Swedenborg, Emanuel, *De sale communi hoc est de sale fossili vel gemmeo marino et fontano* (*c.* 1725–1728), ed. A. Acton, Philadelphia PA 1910; translation M. David, 'On Common Salt', *The New Philosophy* 1983–1992.

Swedenborg, Emanuel, *Principia rerum naturalium ab experimentis et geometria sive ex posteriori et priori educta* (*c.* 1729–1731); ed., *Opera* II; translation I. Tansley, *The Minor Principia*, London 1913.

Swedenborg, Emanuel, *Psychologica: Being Notes and Observations on Christian Wolf's Psychologia empirica* (1733–1734), ed. & translation A. Acton, Philadelphia PA 1923.

Swedenborg, Emanuel, *Opera philosophica et mineralia* I–III, Dresden & Leipzig 1734.

Swedenborg, Emanuel, *Principia rerum naturalium sive novorum tentaminum phænomena mundi elementaris philosophice explicandi*, Dresden & Leipzig 1734; translation A. Clissold, *The principia; Or, the First Principles of Natural Things, Being New Attempts toward a Philosophical Explanation of the Elementary World* I–II, 3rd ed., Bryn Athyn PA 1988.

Swedenborg, Emanuel, *Regnum subterraneum sive minerale: De ferro deque modis liquationum ferri per Europam passim in usum receptis; Deque conversatione ferri crudi in chalybem; De vena ferri et probatione ejus; Pariter de chymicis præparatis et cum ferro et victriolo ejus factis experimentis &c. &c.*, Dresden & Leipzig 1734.

Swedenborg, Emanuel, *Regnum subterraneum sive minerale: De cupro et orichalco, deque modis liquationum cupri per Europam passim in usum receptis; De secretione ejus ab argento; De conversione in orichalcum, inque metalla diversi generis; De lapide calaminari; De zinco; De vena cupri et probatione ejus; Pariter de chymicis præparatis, et cum cupro factis experimentis &c. &c*, Dresden & Leipzig 1734*; translation A. H. Searle, *Treatise on Copper* I–III, London 1938.

Swedenborg, Emanuel, *Ex Principiis rerum naturalium meis* (1734); ed., *Opera* II; translation A. H. Stroh, *Summary of the Principia*, Bryn Athyn PA 1904.

Swedenborg, Emanuel, *Prodromus philosophiæ ratiocinantis de infinito et causa finali creationis: deque mechanismo operationis animæ et corporis*, Dresden & Leipzig 1734; translation J. J. G. Wilkinson, *Forerunner of a Reasoned Philosophy Concerning the Infinite, the Final Cause of Creation Also the Mechanism of the Operation of the Soul and Body*, new ed., London 1992. [*De infinito*]

Swedenborg, Emanuel, *Transactionum de cerebro fragmenta* (1738–1740); facsimile, *Photolith.* IV; translation A. Acton, *Three Transactions on the Cerebrum* I–II, new ed., Bryn Athyn PA 1976. [*De cerebro*]

Swedenborg, Emanuel, *Uträkning af magnetens declination till Upsala meridian; och hwad nytta sådane uträkningar till östra och westra lengdens påfinnande kunna medbringa* (1740); ed. Nordenmark 1933.

Swedenborg, Emanuel, *Swar uppå magist. Hiorters critiquer öfwer min communicerade uträkning om magnetens declination i Upsala åhr 1740*; ed. Nordenmark 1933.

Swedenborg, Emanuel, *Oeconomia regni animalis in transactiones divisa* I–II, London & Amsterdam 1740–1741; translation A. Clissold, *The Economy of the Animal Kingdom, Considered Anatomically, Physically, and Philosophically* I–II, New York NY 1955.

Swedenborg, Emanuel, *Oeconomia regni animalis in transactiones divisa* III (1740), ed. J. J. G. Wilkinson, London 1847; translation A. Acton, *The Economy of the Animal Kingdom Considered Anatomically, Physically and Philosophically. Transaction III. The Medullary Fibre of the Brain and the Nerve Fibre of the Body. The Arachnoid Tunic. Diseases of the Fibre*, new ed., Bryn Athyn PA 1976. [*De fibra*]

Swedenborg, Emanuel, *Varia philosophica et theologica* (1741–1744); translation A. Acton, *A Philosopher's Note Book, Excerpts from Philosophical Writers and from the Sacred Scriptures on a Variety of Philosophical Subjects; Together with Some Reflections, and Sundry Notes and Memoranda*, 2nd ed., Bryn Athyn PA 1976. [*Note book*]

Swedenborg, Emanuel, *Clavis hieroglyphica arcanorum naturalium & spiritualium, per viam repræsentationum et correspondentiarum* (1742), ed. R. Hindmarsh, London 1784; translation A. Acton, 'A Hieroglyphic Key to Spiritual and Natural Arcana by Way of Representations and Correspondences', *Psychological Transactions*.

Swedenborg, Emanuel, *Ontologia* (1742); facsimile, *Photolith.* VI; translation A. Acton, *Ontology, or the Signification of Philosophical Terms*, Boston MA 1901.

Swedenborg, Emanuel, *Regnum animale anatomice, physice et philosophice perlustratum, cujus pars septima de anima agit* (1742), ed. I. Tafel, Tübingen & London 1849; translation N. H. Rogers & A. Acton, *Rational Psychology*, new ed., Bryn Athyn PA 2001. [*De anima*]

Swedenborg, Emanuel, *Swedenborgs drömmar 1744 jemte andra hans anteckningar* (1743–1744), ed. G. E. Klemming, Stockholm 1860; translation J. J. G. Wilkinson, ed. W. R. Woofenden, *Swedenborg's Journal of Dreams, 1743–1744*, 2nd ed., Bryn Athyn, PA 1989.

Swedenborg, Emanuel, *Regnum animale anatomice, physice et philosophice perlustratum, cujus pars quarta de caro tidibus, de sensu olfactus, auditus et visus, de sensatione et affectione in genere, ac de intellectu et ejus operatione agit* (1744), ed. I. Tafel, Tübingen & London 1848; translation E. S. Price, *The Five Senses. The Animal Kingdom* III, new ed., Bryn Athyn PA 1976. [*De sensibus*]

Swedenborg, Emanuel, *Regnum animale anatomice, physice et philosophice perlustratum*, I–II Den Haag 1744, III London 1745; translation J. J. G. Wilkinson, *The Animal Kingdom, Considered Anatomically, Physically, and Philosophically* I–II, London 1843–1844.

Swedenborg, Emanuel, *De cultu et amore Dei*, London 1745; translation A. H. Stroh & F. Sewall, *The Worship and Love of God*, Boston 1914.

Swedenborg, Emanuel, *De Messia venturo in mundum* (1745); *Photolith.* VIII; translation A. Acton, *Concerning the Messiah about to Come and Concerning the Kingdom of God and the Last Judgment*, Bryn Athyn PA 1949.

Swedenborg, Emanuel, *Diarium spirituale* I–VII (1745–1765), ed. I. Tafel, Tübingen & London 1843–1846; translation G. Bush, J. H. Smithson & J. F. Buss, *The Spiritual Diary of Emanuel Swedenborg* I–V, London 1883–1902; *Experientiae spirituales* I–VI, ed. J. D. Odhner, Bryn Athyn PA 1983–1997; translation J. D. Odhner, *Emanuel Swedenborg's Diary, Recounting Spiritual Experiences during the Years 1745 to 1765* I–III, Bryn Athyn PA 1998–2002.

Swedenborg, Emanuel, *Schmidius Biblia sacra* (1747), facsimile, ed. R. L. Tafel, Stockholm 1872; translation E. E. Iungerich, *The Schmidius Marginalia Together with the Expository Material of the Index Biblicus*, Bryn Athyn PA 1917.

Swedenborg, Emanuel, *Arcana coelestia quæ in Scriptura Sacra seu Verbo Domini sunt detecta* I– VIII, London 1749–1756; translation John Elliott, *Arcana Caelestia: Principally a Revelation of the Inner or Spiritual Meaning of Genesis and Exodus*, London 1983–1999.

Swedenborg, Emanuel, *Anteckningar i Swedenborgs almanacka för år 1752, förvarad å Kungl. Biblioteket i Stockholm* (1752), ed. A. H. Stroh, Stockholm 1903.

Swedenborg, Emanuel, 'Utdrag af någre herr assessor Suedenborgs anmärkningar, om ståhl' [from *De ferro*], in Bazin 1753.

Swedenborg, Emanuel, *De Nova Hierosolyma et ejus doctrina cælesti*, London 1758.

Swedenborg, Emanuel, *De telluribus in mundo nostri solari, quæ vocantur planetæ*, London 1758; translation J. Chadwick, *The Worlds in Space*, London 1997.

Swedenborg, Emanuel, *De coelo et ejus mirabilibus, et de inferno*, London 1758; translation G. F. Dole, *Heaven and its Wonders and Hell, Drawn from Things Heard & Seen*, West Chester PA 2000.

Swedenborg, Emanuel, *De equo albo de quo in Apocalypsi, cap: XIX. Et dein de verbo et ejus sensu spirituali seu interno, ex Arcanis coelestibus*, London 1758.

Swedenborg, Emanuel, *Ödmjukt memorial* (1761); ed. A. Kahl, *Nya Kyrkan och dess inflytande på theologiens studium i Swerige. Ett bidrag till sednare tidens svenska kyrkohistoria* II, Lund 1849.

Swedenborg, Emanuel, 'Traité du fer' [from *De ferro*], in Gaspard Le Compasseur Créqui Montfort Marquis de Courtivron & Etienne-Jean Bouchu, *Art des forges et fourneaux à fer*, Paris 1762; translation J. H. G. von Justi, *Abhandlung von den Eisenhammern und hohen Oefen*, Berlin, Stettin & Leipzig 1763–1764.

Swedenborg, Emanuel, *De ultimo judicio* (1762); facsimile, *Photolith.* VIII; translation J. Whitehead, 'The Last Judgment (Posthumous), . . .', *Posthumous Theological Works of Emanuel Swedenborg* I, New York NY 1928.

Swedenborg, Emanuel, 'Beskrifningar huru inläggningar ske uti marmorskifwor, til bord eller annan huszirat', *Kongl. swenska wetenskaps academiens handlingar* XXIV, 1763.

Swedenborg, Emanuel, *Sapientia angelica de divino amore et de divina sapientia*, Amsterdam 1763; translation G. F. Dole, *Angelic Wisdom about Divine Love and about Divine Wisdom*, West Chester PA 2003. [*De divino amore*]

Swedenborg, Emanuel, *Sapientia angelica de divina providentia*, Amsterdam 1764; translation G. F. Dole, *Angelic Wisdom about Divine Providence*, West Chester PA 2003. [*De divina providentia*]

Swedenborg, Emanuel, *Apocalypsis revelata in qua deteguntur arcana quæ ibi prædicta sunt, et hactenus recondita latuerunt*, Amsterdam 1766.

Swedenborg, Emanuel, *Delitiæ sapientiæ de amore conjugiali; post quæ sequuntur voluptates insaniæ de amore scortatorio*, Amsterdam 1768; translation J. Chadwick, *The Delights of Wisdom on the Subject of Conjugial Love, Followed by the Gross Pleasures of Folly on the Subject of Scortatory Love*, London 1996. [*De amore conjugiali*]

Swedenborg, Emanuel, *De commercio animæ et corporis, quod creditur fieri vel per influxum physicum, vel per influxum spiritualem, vel per harmoniam præstabilitam*, London 1769; translation J. Whitehead, 'Interaction of Soul and Body, Which is Believed to be Either by Physical Influx, or by Spiritual Influx, or by Preestablished Harmony', in *Miscellaneous Theological Works*, West Chester PA 1996.

Swedenborg, Emanuel, *Vera christiana religio continens universam theologiam Novæ ecclesiæ*, Amsterdam 1771; translation J. Chadwick, *The True Christian Religion, Containing the Complete Theology of the New Church as Foretold by the Lord in Daniel 7:13, 14 and in Revelation 21:2, 3*, London 1988.

Swedenborg, Emanuel, *Em. Svedenborgii autographa ed: photolith* I–VIII, ed. R. L. Tafel, Stockholm 1869–1870. [*Photolith.*]

Swedenborg, Emanuel, *Documents Concerning the Life and Character of Emanuel Swedenborg* I–II:2, ed. R. L. Tafel, London 1875–1877.

Swedenborg, Emanuel, *Opera quædam aut inedita aut obsoleta de rebus naturalibus* I–III, ed. A. H. Stroh, Stockholm 1907–1911. [*Opera*]

Swedenborg, Emanuel, *Opera poetica*, Uppsala 1910.

Swedenborg, Emanuel, *Resebeskrifningar af Emanuel Swedenborg under åren 1710–1739*, Uppsala 1911.

Swedenborg, Emanuel, *The Mechanical Inventions of Emanuel Swedenborg,* translation A. Acton, Philadelphia PA 1939.

Swedenborg, Emanuel, *The Letters and Memorials of Emanuel Swedenborg* I–II, ed. A. Acton, Bryn Athyn PA 1948–1955. [*Letters*]

Swedenborg, Emanuel, *Psychological Transactions and Other Posthumous Tracts 1734–1744*, translation A. Acton, new ed., Bryn Athyn PA 1984.

Swedenborg, Emanuel, *Scientific and Philosophical Treatises (1716–1740)*, ed. W. R. Woofenden, Bryn Athyn PA 1992. [*Treatises*]

Swift, Jonathan, 'An Account of a Battel between the Ancient and Modern Books in St. James's Library', in *A Tale of a Tub: Written for the Universal Improvement of Mankind*, London 1704; new ed. H. J. Real, Berlin 1978.

Tartaglia, Niccolò, *Nova scientia*, Venezia 1537; translation S. Drake, *Mechanics in Sixteenth-century Italy: Selections from Tartaglia, Benedetti, Guido Ubaldo, & Galileo*, Madison WI 1969.

Tatishchev, Vasily Nikitich, 'Generosiss. Dn. Basilii Tatischow Epistola ad D. Ericum Benzelium de Mamontowa Kost, id est, de ossibus bestiæ Russis Mamont dictæ', *Acta literaria Sveciæ* 1725.

Then swenska psalm-boken med the stycker som ther til höra, och på föliande blad opteknade finnas, ..., Stockholm 1694.*

Then swenska psalm boken. Medh the stycker som ther til höra, och på följande bladh vpteknade finnas, ..., Stockholm 1697.

Tiselius, Daniel, *Uthförlig beskrifning öfwer den stora Swea och Giötha siön, Wätter, til des belägenhet, storlek och märkwärdiga egenskaper; samt anmärkningar och berättelser, om några uthi och omkring siön belägna öjar, nääs, berg och strömar; med mera, som wärdt är at i ackt taga*, Uppsala 1723.

Tiselius, Daniel, *Ytterligare försök och siö-profwer vthi Wättern i anledning af lärde mäns anmärkningar och påminnelser, wijsandes wattnets nogare beskaffenhet, af de grundsatser som i Wätters beskrifningen äro anförde*, Stockholm 1730.

Törner, Fabian (pres.), *De Alexandria Ægypti*, resp. Henric Benzelius, Uppsala 1711.

Triewald, Mårten, *Kort beskrifning, om eld- och luftmachin wid Dannemora grufwor*, Stockholm 1734; facsimile, Stockholm 1985.

Triewald, Mårten, *Mårten Triewalds år 1728 och 1729 håldne föreläsningar, på Riddarehuset i Stockholm, öfwer nya naturkunnigheten, hwarutinnan densamma genom rön och försök är worden förklarad och stadfästad. Til allmänn tienst utgifne* I–II, Stockholm 1735–1736.

Tschirnhaus, Ehrenfried Walther von, 'De magnis lentibus seu vitris causticis, quorum diameter trium quatuorve pedum, nec non eorundem usu & effectu plene & perspicue indicato', *Acta eruditorum* 1697.

Upmarck, Johan (pres.), *De circulo rerum*, resp. Jonas Fahlenius, Uppsala 1705.

Upmarck, Johan (pres.), *Certamen naturæ et artis*, resp. Eric Ström, Uppsala 1710.

Valentini, Michael Bernhard, *Museum museorum oder der vollständigen SchauBühne frembder Naturalien ...* III, Frankfurt 1714.*

Vallemont, Pierre Le Lorrain de, *La physique occulte, ou traité de la baguette divinatoire et de son utilité pour la découverte des sources d'eau, des minières, des tresors cachez, des voleurs & des meurtriers fugitifs. ...*, new ed., Paris 1696*; Amsterdam 1693.

Vallerius, Harald, the Elder (pres.), *De certitudine sensuum*, resp. Nicolaus Schultin, Uppsala 1681.

Vallerius, Harald, the Elder (pres.), *De centro terræ*, resp. Jacob Moell, Uppsala 1693.

Vallerius, Harald, the Elder (pres.), *De puncto mathematico*, resp. Johannes E. Mozelius, Uppsala 1693.

Vallerius, Harald, the Elder (pres.), *Adumbrationem Alpium, quas habet Jemtia, perbrevem*, resp. Daniel Touscher, Uppsala 1694.

Vallerius, Harald, the Elder (pres.), *Theorema de matheseos incrementis*, resp. Jonas J. Dryander, Uppsala 1694.

Vallerius, Harald, the Elder (pres.), *De habitu terræ tempore diluvii*, resp. Johan [Joh. Fil.] Vallerius, Stockholm 1696.

Vallerius, Harald, the Elder (pres.), *De angulo*, resp. Andreas O. Rockman, Uppsala 1698.

Vallerius, Harald, the Elder (pres.), *De tactu*, resp. Olof Retzelius, Uppsala 1698.

Vallerius, Harald, the Elder (pres.), *De camera obscura*, resp. & author Harald Vallerius the Younger, Uppsala 1700.

Vallerius, Harald, the Elder (pres.), *De logarithmis* ..., resp. Sven Laurel, Uppsala 1700.

Vallerius, Harald, the Elder (pres.), *De superficie corporum naturalium*, resp. Magnus Melander, Uppsala 1701.

Vallerius, Harald, the Elder (pres.), *De parallelismo*, resp. Olof Swebilius, Uppsala 1702.

Vallerius, Harald, the Elder (pres.), *De tarantula*, resp. Göran Vallerius, Uppsala 1702.

Vallerius, Harald, the Elder (pres.), *De fallaciis visionis ratione objectorum*, resp. Johannes Rosell, Uppsala 1703.

Vallerius, Harald, the Elder (pres.), *De linea perpendiculari positiones*, resp. Petrus Ternerus, Uppsala 1703.

Vallerius, Harald, the Elder (pres.), *De parallelogrammis geometricam* ..., resp. Lars Rimmius, Uppsala 1703.

Vallerius, Harald, the Elder (pres.), *De aquarum motu per circulum in globo terraqueo*, resp. Matthias Ramzelius, Uppsala 1704.

Vallerius, Harald, the Elder (pres.), *De statione fluminum*, resp. Andreas Wennerwall, Uppsala 1704.

Vallerius, Harald, the Elder (pres.), *De æquilibrio corporum naturalium*, resp. & author Pehr Martin, Uppsala 1705.

Vallerius, Harald, the Elder (pres.), *De fallaciis sensuum indultu*, resp. Christian Bredenberg, Uppsala 1705.

Vallerius, Harald, the Elder (pres.), *De progressione arithmetica et geometrica*, resp. Petrus Erichsson, Uppsala 1705.

Vallerius, Harald, the Elder (pres.), *De figura*, resp. Erengislus Bohm, Uppsala 1707.

Vallerius, Harald, the Elder (pres.), *De vallibus*, resp. Eric Dahl, Uppsala 1708.

Vallerius, Harald, the Elder (pres.), *Disputatio philosophica parallelismum microcosmi et macrocosmi breviter delineans*, resp. & author Andreas Sam. Pijl, Uppsala 1711.

Vallerius, Johan (pres.), *De proportione harmonica,* ... II, resp. & author Eric Burman, Uppsala 1716.

Vallerius, Johan (pres.), *Exercitium academicum instrumenta musica leviter delinean*, resp. Olof O. Bergrot, Uppsala 1717.

Vauban, Sébastien le Prestre de, *Véritable manière de bien fortifier, avec un traité préliminaire des principes de géometrie. Le tout expliqué d'une manière nouvelle* ..., Amsterdam 1692.

Vico, Giambattista, *Principi di una scienza nuova intorno alla natura delle nazioni*, Napoli 1725; 3rd ed. (1744); translation T. G. Bergin & M. H. Fisch, *The New Science*, Ithaca NY 1968.

Vieussens, Raymond, *Nevrographia universalis. Hoc est, omnium corporis humani nervorum, simul & cerebri, medullæque spinalis descriptio anatomica;* ..., 2nd ed., Frankfurt 1690.

Virgil, Publius Vergilius Maro, *Georgica*; ed. H. R. Fairclough, *Eclogues. Georgics. Aeneid I–VI*, Cambridge MA & London 1999.

Virgil, Publius Vergilius Maro, *Aeneid*; ed. H. R. Fairclough, *Aeneid VII–XII. Appendix Vergiliana*, Cambridge MA & London 2000.

Vitali, Girolamo, *Lexicon mathematicvm astronomicvm geometricvm, hoc est Rerum omnium ad utramque immò & ad omnem ferè mathesim quomodocumque spectantium, collectio, & explicatio*, Paris 1668.*

Vogler, Johann Hermann, 'Bericht eines Augenzeugen über Leibnizens Tod und Begräbnis' (1716), *Zeitschrift des historischen Vereins für Niedersachsen* 81, 1916.

Voltaire, François Marie Arouet de, *Histoire de Charles XII: Roi de Suède* II, Basel 1731; new ed., Dresden 1749.

Voltaire, François Marie Arouet de, *Élémens de la philosophie de Neuton*, London 1738.*

Voltaire, François Marie Arouet de, *Œuvres complètes de Voltaire* XXXVI, Paris 1880.

Wallerius, Johan Gottschalk (pres.), *Kort afhandling, om malmgångars upsökande*, resp. Jacob Leonhard Roman, Uppsala 1757.

Wallerius, Johan Gottschalk (pres.), *Kort afhandling om malmförande bergs egenskaper*, resp. Claes Fredric Scheffel, Uppsala 1759.

Wallerius, Nils, 'De ascensu vaporum in vacuo, demonstratio', *Acta literaria et scientiarum Sveciæ* 1738.

Wallin, Georg, *Gothländske samlingar* II, Göteborg 1776.

Wargentin, Pehr Wilhelm, [Wargentin to Mallet, Stockholm, 18 September 1766], ed. Nordenmark 1944–1945.

Warnmark, Peter Ol., *Then swänske Ulysses eller En nyttig rese-book: ...*, Göteborg 1709.

Wassenius, Birger, 'Cogitationes de incremento & decremento Lacus Vener', *Acta literaria et scientiarum Sveciæ* 1730.

Weidler, Johann Friedrich, *Institvtiones mathematicæ decem sex pvræ mixtæqve matheseos disciplinas complexæ. ...*, Wittenberg 1718; translation J. Mört, *En klar och tydelig genstig eller anledning til geometrien och trigonometrien, innehållandes dess utan några nödiga capitel af arithmetica, som til geometriens och trigonometriens klarhet tiena, såsom om proportionerne, extractione radicum, och logarithmernes grund och egenkaper, såsom och ett uttåg af logarithmiska taflor. ...*, Stockholm 1727.

Wilcke, Johan Carl, *Tal, om magneten*, Stockholm 1764.

Wilkins, John, *An Essay towards a Real Character and a Philosophical Language*, London 1668.

Wilkins, John, *The Mathematical and Philosophical Works ...*, London 1708.

Willis, Thomas, *Opera omnia ...*, Genève 1676.*

Wolff, Christian von, *Der Anfangs-Gründe aller mathematischen Wissenschaften*, Halle 1710; 4th ed., Frankfurt & Leipzig 1732a.

Wolff, Christian von, *Elementa matheseos universæ*, Halle 1715.

Wolff, Christian von, *Mathematisches Lexicon*, Leipzig 1716.

Wolff, Christian von, *Allerhand nützliche Versuche, dadurch zu genauer Erkäntniß der Natur und Kunst der Weg gebähnet wird* I–III, Halle 1721–1723*; new ed., Halle 1737–1738.

Wolff, Christian von, *Philosophia prima sive ontologia methodo scientifica pertractata, qua omnis cognitionis humanæ principia continentur*, Frankfurt & Leipzig 1730*; 2nd ed., Frankfurt & Leipzig 1736.

Wolff, Christian von, *Cosmologia generalis, methodo scientifica pertractata, qva ad solidam, imprimis Dei atqve natvræ, cognitionem via sternitvr*, Frankfurt & Leipzig 1731*; new ed., Frankfurt & Leipzig 1737.

Wolff, Christian von, *Elementa matheseos universæ* I, 2nd ed., Genève 1732b.

Wolff, Christian von, *Psychologia empirica methodo scientifica pertractata, qua ea, quæ de anima humana indubia experientiæ fide constant, continentur ...*, Frankfurt & Leipzig 1732c.

Wolff, Christian von, *Psychologia rationalis, methodo scientifica pertractata*, Frankfurt & Leipzig 1734.

Wolff, Christian von, *Theologia naturalis, methodo scientifica pertractata*, Frankfurt & Leipzig 1736–1737.

Zahn, Johann, *Oculus artificialis teledioptricus sive telescopium, ...*, Nürnberg 1702.*

Secondary literature

Afzelius, Nils, 'Brev från J. H. Lidén om Swedenborg', *Svenskarnas syn på Swedenborg under 1800-talet*, ed. G. Appelgren & B. Ekengren, Swedenborgiana 2000/2001, Stockholm [2002].

Ahnlund, Nils, 'Källorna', *Sanning och sägen om Karl XII:s död*, Stockholm 1941.

Alm, Henrik, 'Huset och trädgården', Hjern, Eriksson & Hallengren 1992.

Almquist, Karl Gustaf, *Andreas Rydelius' etiska åskådning*, Lund 1955.

Almqvist, Daniel, 'Några karolinska kanalprojekt', *Karolinska förbundets årsbok* 1935.

Althin, Torsten, 'Kring Dædalus Hyperboreus 1716–1717', *Dædalus* 1958.

Åmark, Karl, *Sveriges statsfinanser 1719–1809*, Stockholm 1961.

Andersen, Hanne, Peter Barker & Xiang Chen, *The Cognitive Structure of Scientific Revolutions*, Cambridge 2006.

Andriesse, Cornelis Dirk, *Huygens: The Man Behind the Principle*, Cambridge 2005.

Arrhenius, Svante, 'Emanuel Swedenborg as a Cosmologist', *Emanuel Swedenborg as a Scientist*, ed. A. H. Stroh, Stockholm 1908.

Aste, Tomaso & Denis Weaire, *The Pursuit of Perfect Packing*, Philadelphia PA 2000.

Atran, Scott, *Cognitive Foundations of Natural History: Towards an Anthropology of Science*, Cambridge 1990.

Atran, Scott & Douglas L. Medin, *The Native Mind and the Cultural Construction of Nature*, Cambridge MA 2008.

Aurivillius, Carl Wilhelm Samuel, *Der Wal Svedenborg's (Balæna Svedenborgii Lilljeborg), nach einem Funde im Diluvium Schwedens*, Stockholm 1888.

Báez-Rivera, Emilio, 'Swedenborg and Borges: From the Mystic of the North to the Mystic *in puribus*', in McNeilly 2004.

Baigrie, Brian S., 'Descartes's Scientific Illustrations and "la grande mécanique de la nature"', *Picturing Knowledge: Historical and Philosophical Problems Concerning the Use of Art in Science*, ed. B. S. Baigrie, Toronto 1996.

Baldwin, Martha, 'Alchemy and the Society of Jesus in the Seventeenth Century: Strange Bedfellows?', *Ambix* 40:2, 1993.

Beckman, Anna, 'Två svenska experimentalfysiker på 1700-talet: Mårten Triewald och Nils Wallerius', *Lychnos* 1967–1968.

Bengtsson, Frans G., *Karl XII:s levnad: Från Altranstädt till Fredrikshall*, new ed., Stockholm 1954; translation N. Walford, *The Life of Charles XII, King of Sweden 1697–1718*, Stockholm 1960.

Bense, Max, *Konturen einer Geistesgeschichte der Mathematik* II, Hamburg 1949.

Benz, Ernst, *Emanuel Swedenborg: Naturforscher und Seher*, München 1948.

Berg, Fredrik, *Bidrag till oftalmologiens äldre historia i Sverige*, Uppsala 1958.

Berggren, Maria, 'Det vetenskapliga latinet i Emanuel Swedenborgs medicinska verk', *Sjuttonhundratal* 2005.

Berggrén, Per Gust., 'Karl XII:s galärtransport från Strömstad till Idefjorden och striderna därstädes år 1718', *Karolinska förbundets årsbok* 1920.

Bergquist, Lars, *Swedenborgs drömbok: Glädjen och det stora kvalet*, Stockholm 1988; translation A. Hallengren, *Swedenborg's Dream Diary*, West Chester, PA 2001.

Bergquist, Lars, *Biblioteket i lusthuset: Tio uppsatser om Swedenborg*, Stockholm 1996.

Bergquist, Lars, 'Linné och Swedenborg: spegelbilder i svenskt 1700-tal', *Världarnas möte: Nya Kyrkans Tidning* 1998:1–2.

Bergquist, Lars, *Swedenborgs hemlighet: Om Ordets betydelse, änglarnas liv och tjänsten hos Gud: En biografi*, Stockholm 1999; translation, *Swedenborg's Secret: The Meaning and Significance of the World of God, the Life of the Angels, and Service to God: A Biography*, London 2005.

Bergquist, Lars, 'Mystik: om att "vara i anden"', in Tysk 2000a.

Bergquist, Lars, 'Swedenborgs galenskap', *Mänskliga gränsområden: Om extas, psykos och galenskap*, ed. J. Cullberg, K. Johannisson, O. Wikström, new ed., Stockholm 2000b.

Bergquist, Lars, *Ansiktets ängel och Den stora människan: Emanuel Swedenborg om livet och lyckan – en sammanfattning*, Stockholm 2001.

Blay, Michel, *Reasoning with the Infinite: From the Closed World to the Mathematical Universe*, translation M. B. DeBevoise, Chicago IL 1998.

Borges, Jorge Luis, 'El otro, el mismo', *Obras completas: Obra poética 1923–1964*, Buenos Aires 1964; *Obras completas* II, 1952–1972, Barcelona 1989.

Borges, Jorge Luis, 'Prólogo', *Swedenborg, testigo de lo invisible*, ed. S. Synnestvedt, Buenos Aires 1982; 'Testimony to the Invisible', translation C. Rodriguez-Nieto et al., *Testimony to the Invisible: Essays on Swedenborg*, ed. J. F. Lawrence, West Chester, PA 1995.

Bos, Henk J. M., *Redefining Geometrical Exactness: Descartes' Transformation of the Early Modern Concept of Construction*, New York NY 2001.

Boyd, Brian, *On the Origin of Stories: Evolution, Cognition, and Fiction*, Cambridge MA 2009.

Brendel, Otto J., *Symbolism of the Sphere: A Contribution to the History of Earlier Greek Philosophy*, Leiden 1977.

Brewster, David, *Memoirs of the Life, Writings, and Discoveries of Sir Isaac Newton* II, 2nd ed., Edinburgh 1860.

Brinck, Ingar, 'Situated Cognition, Dynamic Systems, and Art: On Artistic Creativity and Aesthetic Experience', *Janus Head* 9(2), 2007.

Bring, Samuel E., 'Bidrag till Christopher Polhems lefnadsteckning', *Christopher Polhem*, ed. S. E. Bring, Stockholm 1911a.

Bring, Samuel E., 'Några bref från Casten Feif till Christopher Polhem', *Karolinska förbundets årsbok* 1911b.

Bring, Samuel E., 'Bidrag till frågan om Karl XII:s död', *Karolinska förbundets årsbok* 1920.

Broberg, Gunnar, *Homo sapiens L.: Studier i Carl von Linnés naturuppfattning och människolära*, Stockholm 1975.

Broberg, Gunnar, 'Naturvetenskapsmännen om sin egen historia: halvseklet före Nordström', *Lychnos* 1983.

Broberg, Gunnar, 'Swedenborg och Uppsala', *Världarnas möte: Nya Kyrkans Tidning* 1990:4.

Brock, Erland J. (ed.), *Swedenborg and His Influence*, Bryn Athyn PA 1988.

Brooke, John Hedley, *Science and Religion: Some Historical Perspectives*, Cambridge 1996.

Brulin, Herman, 'Handskriftsmaterial till Voltaires Charles XII', *Karolinska förbundets årsbok* 1940.

Bruzelli, Birger & Håkan Carlestam, *Svensk mått-, mål- och vikthistoria 1605–1889*, Nora 1999.

Buhl, Hans, *Sfærernes harmoni – en videnskabshistorie om forholdet mellem musik og fysik*, Århus 2000.

Calatrello, Robert Lawrence, *The Basic Philosophy of Emanuel Swedenborg with Implications for Western Education*, Los Angeles CA 1966.

Calvo, Paco & Toni Gomila (eds.), *Handbook of Cognitive Science: An Embodied Approach*, Oxford 2008.

Carlborg, Harald, 'Om gruvkompasser, malmletning och kompassgångare', *Med hammare och fackla* 23, 1963.

Carruthers, Peter, Stephen Stich & Michael Siegal (eds.), *The Cognitive Basis of Science*, Cambridge 2002.

Cassirer, Ernst, *Philosophie der symbolischen Formen* I–II, Berlin 1923–1925.

Cassirer, Ernst, *An Essay on Man: An Introduction to a Philosophy of Human Culture*, new ed., Garden City NY 1953.

Cassirer, Ernst, *The Platonic Renaissance in England*, translation J. P. Pettegrove, New York NY 1970.

Chadwick, John, *On the Translator and the Latin Text: Essays on Swedenborg*, London 2001.

Chadwick, John & Jonathan S. Rose (eds.), *A Lexicon to the Latin Text of the Theological Writings of Emanuel Swedenborg (1688–1772)*, London 2008.

Chartier, Roger, *L'ordre des livres: Lecteurs, auteurs, bibliothèques en Europe entre XIVe et XVIIIe siècle*, Aix-en-Provence 1992; translation L. G. Cochrane, *The Order of Books: Readers, Authors, and Libraries in Europe between the Fourteenth and Eighteenth Centuries*, Oxford 1994.

Choluj, Bozena & Jan C. Joerden, *Von der wissenschaftlichen Tatsache zur Wissensproduktion: Ludwik Fleck und seine Bedeutung für die Wissenschaft und Praxis*, Frankfurt am Main 2007.

Clark, Andy, *Being There: Putting Brain, Body, and World Together Again*, Cambridge MA 1997.

Clark, Andy & David Chalmers, 'The Extended Mind', *Analysis* 58.1, 1998.

Clark, Stuart, *Vanities of the Eye: Vision in Early Modern European Culture*, Oxford 2007.

Clericuzio, Antonio, *Elements, Principles and Corpuscles: A Study of Atomism and Chemistry in the Seventeent Century*, Dordrecht 2000.

Cook, Theodore Andrea, *The Curves of Life Being an Account of Spiral Formations and Their Application to Growth in Nature, to Science and to Art with Special Reference to the Manuscripts of Leonardo da Vinci*, new ed., New York NY 1979.

Crasta, Francesca Maria, *La filosofia della natura di Emanuel Swedenborg*, Milano 1999.

Crasta, Francesca Maria, 'Metaphysics and Biology: Thoughts on the Interaction of the Soul and Body in Emanuel Swedenborg', in McNeilly 2002.

Crombie, Alistair C., *Styles of Scientific Thinking in the European Tradition: The History of Argument and Explanation Especially in the Mathematical and Biomedical Sciences and Arts* II, London 1994.

Cromwell, Peter R., *Polyhedra*, Cambridge 1999.

Dahl, Per, *Svensk ingenjörskonst under stormaktstiden: Olof Rudbecks tekniska undervisning och praktiska verksamhet*, Uppsala 1995.

Dahlin, Ernst Mauritz, *Bidrag till de matematiska vetenskapernas historia i Sverige före 1679*, Uppsala 1875.

Dal, Björn, 'Om den graverade frontespisen till Acta literaria Sveciae', *Den idéhistoriska bilden*, ed. G. Broberg & J. Christensson, Lund 1995.

Dal, Björn, *Sveriges zoologiska litteratur: En berättande översikt om svenska zoologer och deras tryckta verk 1483–1920*, Kjuge 1996.

Danesi, Marcel, 'The Dimensionality of Metaphor', *Sign Systems Studies* 27, Tartu 1999.

Dear, Peter, *Discipline and Experience: The Mathematical Way in the Scientific Revolution*, Chicago IL 1995.

Dear, Peter, *The Intelligibility of Nature: How Science Makes Sense of the World*, Chicago IL 2006.

Debus, Allen G., *The Chemical Philosophy: Paracelsian Science and Medicine in the Sixteenth and Seventeenth Centuries* I, New York NY 1977.

Degenaar, Marjolein, *Molyneux's Problem: Three Centuries of Discussion on the Perception of Forms*, Dordrecht 1996.

Des Chene, Dennis, *Spirits and Clocks: Machine and Organism in Descartes*, Ithaca NY 2001.

Dictionary of Scientific Biography I & VII, ed. C. C. Gillispie, New York NY 1970–1990.

Dijksterhuis, Fokko Jan, *Lenses and Waves: Christiaan Huygens and the Mathematical Science of Optics in the Seventeenth Century*, Dordrecht 2005.

Dintler, Åke, *Lars Roberg: Akademiska sjukhusets grundare*, Uppsala 1958.

Djurberg, Vilhelm, *Läkaren Johan von Hoorn: Förlossningskonstens grundläggare i Sverige*, Uppsala 1942.

Donald, Merlin W., 'A View from Cognitive Science', *Was ist der Mensch?*, ed. D. Ganten et al., Berlin & New York NY 2008.

Duchesneau, François, 'Epistemological Problems of Iatromechanism', *Reason, Experiment, and Mysticism in the Scientific Revolution*, ed. M. L. Righini Bonelli & W. R. Shea, New York NY 1975.

Dunbar, Kevin, N., 'Understanding the Role of Cognition in Science: The *Science as Category* Framework', in Carruthers, Stich & Siegal 2002.

Dunér, David, 'Swedenborgs spiral', *Lychnos* 1999; 2nd ed., *Swedenborgs spiral*, Swedenborgiana 1999, Grödinge & Stockholm [2000]; translation, 'Swedenborg's Spiral', *Studia Swedenborgiana* 2002:4.

Dunér, David, 'Sextiofyra och åtta istället för tio: Karl XII, Swedenborg och konsten att räkna', *Scandia* 67:2, 2001.

Dunér, David, 'Bubblor, kanonkulor och en tunna ärtor: Polhem och Swedenborg om materiens struktur', *Polhem: Tidskrift för teknikhistoria* 2000/2001 [2002].

Dunér, David, 'Sinnen som darrar', *Vidgade sinnen*, ed. L. G. Andersson & E. Mansén, Nora 2003a.

Dunér, David, 'Världsmaskinen och oändligheten: Emanuel Swedenborg vid det ovetbaras gräns', *Personhistorisk Tidskrift* 2003b:2.

Dunér, David, 'Polhems huvudvärk', *Sjuttonhundratal* 2005a.

446   Bibliography

Dunér, David, 'Daedalus flykt', *Polhem: Teknikhistorisk årsbok* 2005b.

Dunér, David, 'Q. E. D. (Euklides): Om oemotsägliga bevis', *Filosofiska citat: Festskrift till Svante Nordin*, ed. G. Broberg, J. Hansson & E. Mansén, Stockholm 2006a.

Dunér, David, 'Om darrningar: Emanuel Swedenborgs iatromekanik', *Svensk medicinhistorisk tidskrift* 2005 [2006b].

Dunér, David, 'Språket i universum: Polhem och alfabetskonsten', *Lychnos* 2007.

Dunér, David, 'Om kunskap och metaforer', *Kunskapens kretsar: Essäer om kunskap, bildning och vetenskap genom tiderna*, ed. C. Christensen-Nugues, G. Broberg & S. Nordin, Stockholm 2008a.

Dunér, David, 'Helvetet på jorden: Resor till Stora Kopparbergs gruva', *Förmoderna livshållningar: Dygder, värden och kunskapsvägar från antiken till upplysningen*, ed. M. Lindstedt-Cronberg & C. Stenqvist, Lund 2008b.

Dunér, David, 'Maskinen Människa: Iatromekaniken i Sverige', *Til at stwdera läkedom: Tio studier i svensk medicinhistoria*, ed. G. Broberg, Lund 2009a.

Dunér, David, 'Modeller av verkligheten: Modellbyggaren Polhem, seendet och det spatiala tänkandet', *Vetenskapssocieteten i Lund Årsbok* 2009b.

Dunér, Nils C., *Kungliga vetenskaps societetens i Upsala tvåhundraårsminne*, Uppsala 1910.

Dutton, Denis, *The Art Instinct: Beauty, Pleasure, and Human Evolution*, Oxford 2009.

Eby, S. C., *The Story of the Swedenborg Manuscripts*, New York NY 1926.

Eco, Umberto, *Semiotics and the Philosophy of Language*, Bloomington IN 1986.

Eco, Umberto, *The Search for the Perfect Language*, translation J. Fentress, new ed., Oxford 1997.

Edenborg, Carl-Michael, *Gull och mull: Den monstruöse Gustaf Bonde, upplysningens fiende i frihetstidens Sverige. Historien om hans exkrementalkemi, hans krets och värld.* ..., Lund 1997.

Edenborg, Carl-Michael, *Alkemins skam: Den alkemiska traditionens utstötning ur offentligheten*, Stockholm 2002.

Edgerton, Samuel Y., 'The Renaissance Artist as Quantifier', *The Perception of Pictures* I, ed. M. Hagen, New York NY 1980.

Edgerton, Samuel Y. *The Mirror, the Window, and the Telescope: How Renaissance Linear Perspective Changed Our Vision of the Universe*, Ithaca NY 2009.

Edwards, Jess, *Writing, Geometry and Space in Seventeenth-Century England and America: Circles in the Sand*, London 2005.

Ekenvall, Asta, 'Eric Benzelius d.y. och de utländska lärda tidskrifterna', *Lychnos* 1950–1951.

Ekenvall, Asta, 'Eric Benzelius d.y. och G. W. Leibniz', *Linköpings biblioteks handlingar*, ny serie 4:3, Linköping 1953.

Ellenius, Allan, *De Arte Pingendi: Latin Art Literature in Seventeenth-century Sweden and its International Background*, Uppsala 1960.

Eneström, Gustaf, *Meddelande om Swedenborgs matematiska arbeten*, Öfversigt af Kongl. Vetenskaps-Akademiens Förhandlingar 8, Stockholm 1889.

Eneström, Gustaf, *Emanuel Swedenborg såsom matematiker*, Bihang till K. svenska vet.-akad. handlingar, vol. 15:1, no. 12, Stockholm 1890.

Engelfriet, Peter M., *Euclid in China: The Genesis of the First Chinese Translation of Euclid's Elements Books I–VI (Jihe Yuanben; 1607) and its Reception up to 1723*, Leiden 1998.

Erici, Einar & R. Axel Unnerbäck, *Orgelinventarium: Bevarade klassiska kyrkorglar i Sverige*, Stockholm 1988.

Ericsson, Peter, *Stora nordiska kriget förklarat: Karl XII och det ideologiska tilltalet*, Uppsala 2002.

Eriksson, Gunnar, 'Framstegstanken i de cartesianska stridernas Uppsala: Kring debatten om naturens konstans och vetenskapernas tillväxt', *Lychnos* 1967–1968.

Eriksson, Gunnar, 'Epikuros i Uppsala', *Vetenskapens träd: Idéhistoriska studier tillägnade Sten Lindroth*, ed. G. Eriksson, T. Frängsmyr & M. von Platen, Stockholm 1974.

Eriksson, Gunnar, *The Atlantic Vision: Olaus Rudbeck and Baroque Science*, Canton, MA 1994.

Eriksson, Gunnar, *Rudbeck 1630–1702: Liv, lärdom, dröm i barockens Sverige*, Stockholm 2002.

Falkman, Ludvig B., *Om mått och vigt i Sverige; Historisk framställning* II, Stockholm 1885.

Fenzl, Annelise, *'De Cerebro' von Emanuel Swedenborg*, München 1960.

Ferguson, Eugene S., 'The Mind's Eye: Nonverbal Thought in Technology', *Science* 197:4306, 1977.

Fleck, Ludwik, *Entstehung und Entwicklung einer wissenschaftlichen Tatsache: Einführung in die Lehre vom Denkstil und Denkkollektiv*, Basel 1935; translation F. Bradley & T. J. Trenn, *Genesis and Development of a Scientific Fact*, Chicago 1979.

Florschütz, Gottlieb, *Swedenborgs verborgene Wirkung auf Kant: Swedenborg und die okkulten Phänomene aus der Sicht von Kant und Schopenhauer*, Würzburg 1992.

Fors, Hjalmar, 'Occult Traditions and Enlightened Science: The Swedish Board of Mines as an Intellectual Environment 1680–1760', *Chymists and Chymistry: Studies in the History of Alchemy and Early Modern Chemistry*, ed. L. M. Principe, Sagamore Beach MA 2007.

Forssell, H. L., 'Minne af erkebiskopen doktor Erik Benzelius den yngre', *Svenska akademiens handlingar ifrån år 1796* LVIII, Stockholm 1883.

Foucault, Michel, *Les mots et les choses: une archéologie des sciences humaines*, Paris 1966.

Frängsmyr, Tore, *Geologi och skapelsetro: Föreställningar om jordens historia från Hierne till Bergman*, Uppsala 1969.

Frängsmyr, Tore, *Wolffianismens genombrott i Uppsala: Frihetstida universitetsfilosofi till 1700-talets mitt*, Uppsala 1972.

Frängsmyr, Tore, John L. Heilbron & Robin E. Rider (eds.), *The Quantifying Spirit in the 18th Century*, Berkeley CA 1990.

Frank, Robert G., *Harvey and the Oxford Physiologists: Scientific Ideas and Social Interaction*, Berkeley CA 1980.

Fredbärj, Telemak, 'De offentliga anatomierna i Sverige under 1600- och 1700-talen', *Medicinhistorisk årsbok* 1958.

Gabay, Alfred J., *Covert Enlightenment: Eighteenth-century Counterculture and its Aftermath*, West Chester PA 2005.

Gandt, François de, *Force and Geometry in Newton's Principia*, translation C. Wilson, Princeton NJ 1995.

Gärdenfors, Peter, *Blotta tanken*, Nora 1992.

Gärdenfors, Peter, *Conceptual Spaces: The Geometry of Thought*, Cambridge MA 2000a.

Gärdenfors, Peter, *Hur Homo blev sapiens: Om tänkandets evolution*, Nora 2000b; translation, *How Homo Became Sapiens: On the Evolution of Thinking*, Oxford 2006.

Gärdenfors, Peter, 'Cognitive Science: From Computers to Ant Hills as Models of Human Thought', in Gärdenfors & Wallin 2008a.

Gärdenfors, Peter, 'The Evolution of Thought', in Gärdenfors & Wallin 2008b.

Gärdenfors, Peter, 'Concept Learning', in Gärdenfors & Wallin 2008c.

Gärdenfors, Peter & Annika Wallin (eds.), *A Smorgasbord of Cognitive Science*, Nora 2008d.

Gardiner, Perry F., 'Cassirer, Swedenborg, and the Problem of Meaning', *Studia Swedenborgiana* 1978:1.

Garrett, Clarke, 'Swedenborg and the Mystical Enlightenment in Late Eighteenth-Century England', *Journal of the History of Ideas* 1984:45.

Gaukroger, Stephen, *Descartes' System of Natural Philosophy*, Cambridge 2002.

Gentner, Dedre & Michael Jeziorski, 'Historical Shifts in the Use of Analogy in Science', *Psychology of Science: Contributions to Metascience*, ed. B. Gholson et al., Cambridge 1989.

Giere, Ronald, 'Scientific Cognition as Distributed Cognition', in Carruthers, Stich & Siegal 2002.

Giere, Ronald N. & Barton Moffatt, 'Distributed Cognition: Where the Cognitive and the Social Merge', *Social Studies of Science* 33/2, April 2003.

Ginzburg, Carlo, 'Clues: Morelli, Freud, and Sherlock Holmes', *The Sign of Three: Dupin, Holmes, Peirce*, ed. U. Eco & T. A. Sebeok, Bloomington IN 1979.

Ginzburg, Carlo, *Miti, emblemi, spie: Morfologia e storia*, Torino 1986; translation J. & A. C. Tedeschi, *Clues, Myths, and the Historical Method*, Baltimore MD 1989.

Gooding, David, 'Cognitive History of Science: The Roles of Diagrammatic Representations in Discovery and Modeling Discovery', *Theory and Application of Diagrams*, Berlin & Heidelberg 2000.

Gordh J:r, Torsten & Patrick Sourander, 'Swedenborg, Linné och hjärnforskningen', *Nordisk medicinhistorisk årsbok* 1990.

Granström, G. A., 'Swedenborg och "De Ferro"', *Jernkontorets annaler* 1926.

Grattan-Guinness, Ivor, 'Christianity and Mathematics: Kinds of Link, and the Rare Occurrences after 1750', *Physis: Rivista internazionale di storia della scienza* XXXVII, 2000.

Gross, Charles G., 'Emanuel Swedenborg: A Neuroscientist before his Time', *The Neuroscientist*, 1997:3.

Gross, Charles G., *Brain, Vision, Memory: Tales in the History of Neuroscience*, Cambridge MA 1998.

Hag, Torgny, 'Karoliner och behemoter: 1700-talets svenska diskussion om mammuten', *Svenska Linnésällskapets årsskrift* 1979–1981.

Hag, Torgny, *Vatten, Vättern och Vetenskapssocieteten: Kring Daniel Tiselius' beskrivningar av sjön Vättern*, Skara 1983.

Häll, Jan, *I Swedenborgs labyrint: Studier i de gustavianska swedenborgarnas liv och tänkande*, Stockholm 1995.

Hallengren, Anders, *Universum som hieroglyfisk text: Swedenborg, Emerson, Whitman och det adamitiska språket*, Stockholm 1989.

Hallengren, Anders, 'Kristendomens Plotinos: Till frågan om swedenborgianismens nyplatonska ursprung', *Lychnos* 1991.

Hallengren, Anders, *The Code of Concord: Emerson's Search for Universal Laws*, Stockholm 1994.

Hallengren, Anders, *Tingens tydning: Swedenborgstudier*, Stockholm 1997a.

Hallengren, Anders, *Öarna under vinden: Färder i Swedenborgvärlden*, Stockholm 1997b.

Hallengren, Anders, *Gallery of Mirrors: Reflections of Swedenborgian Thought*, West Chester PA 1998.

Hallyn, Fernand, *The Poetic Structure of the World: Copernicus and Kepler*, New York NY 1990.

Hanegraaff, Wouter J., *New Age Religion and Western Culture: Esotericism in the Mirror of Secular Thought*, Leiden 1996.

Hanegraaff, Wouter J., *Swedenborg, Oetinger, Kant: Three Perspectives on the Secrets of Heaven*, West Chester PA 2007.

Hannaway, Owen, *The Chemists and the Word: The Didactic Origins of Chemistry*, Baltimore MD 1975.

Hård, Mikael, '*Mechanica och Mathesis:* Några tankar kring Christopher Polhems fysikaliska och vetenskapsteoretiska föreställningar', *Lychnos* 1986.

Hatfield, Gary, 'Descartes' Physiology and its Relation to his Psychology', *The Cambridge Companion to Descartes*, ed. J. Cottingham, Cambridge 1992.

Hatton, Ragnhild Marie, *Charles XII of Sweden*, London 1968.

Heilbron, John L., 'Introductory Essay', in Frängsmyr, Heilbron & Rider 1990.

Heintz, Christophe, 'Introduction: Why There Should Be a Cognitive Anthropology of Science', *Journal of Cognition and Culture* 2004:3.

Helander, Hans, 'Om Swedenborgs latin', *Kyrkohistorisk årsskrift* 1988.

Helander, Hans, *Neo-Latin Literature in Sweden in the Period 1620–1720: Stylistics, Vocabulary and Characteristic Ideas*, Uppsala 2004.

Helander, Hans, see also Swedenborg.

Henschen, Folke, *Emanuel Swedenborg's Cranium: A Critical Analysis*, Uppsala 1960.

Hildebrand, Bengt, *Kungl. svenska vetenskapsakademien: Förhistoria, grundläggning och första organisation* I–II, Stockholm 1939.

Hildebrand, Karl-Gustaf, 'Swedenborg och Karl XII', *SvD* 30 November 1947.

Hildebrandsson, H. Hildebrand, *Samuel Klingenstiernas levnad och verk: I. Levnadsteckning*, Stockholm 1919.

Hjern, Olle, *Swedenborg och hans vänner i Göteborg*, Stockholm 1991.

Hjern, Olle, Arne Eriksson & Anders Hallengren (eds.), *Swedenborg som Söderbo*, Stockholm 1992.

Hoffmeyer, Jesper, *En snegl på vejen: Betydningens naturhistorie*, København 1993; translation B. J. Haveland, *Signs of Meaning in the Universe*, Bloomington IN 1996.

Högbom, Arvid Gustaf, *Nivåförändringarna i Norden: Ett kapitel ur den svenska naturforskningens historia*, Göteborg 1920.

Högnäs, Sten, *Människans nöjen och elände: Gyllenborg och upplysningen*, Lund 1988.

Holm, Nils F., 'Jöran Nordbergs Konung Carl XII:s historia: Ett 200-årsminne', *Karolinska förbundets årsbok* 1940.

Holmer, Birger, 'Emanuel Swedenborgs manuskript om en flygmaskin: En tolkning av texten med tekniska kommentarer', *Dædalus* 1988.

Holmquist, Hjalmar, 'Från Swedenborgs ungdom och första stora verksamhetsperiod', *Bibelforskaren: Tidskrift för skrifttolkning och praktisk kristendom*, Uppsala 1909a.

Holmquist, Hjalmar, 'Ur Swedenborgsforskningens historia', *Kyrklig tidskrift*, Uppsala 1909b.

Holmquist, Hjalmar, 'Från Swedenborgs naturvetenskapliga och naturfilosofiska period', *Finska kyrkohistoriska samfundets årsbok* III, Helsingfors 1913.

Hoppe, Hans, 'Die Kosmogonie Emanuel Swedenborgs und die Kantsche und Laplacesche Theorie', *Emanuel Swedenborg 1688–1772: Naturforscher und Kundiger der Überwelt*, ed. H. Bergmann & E. Zwink, Stuttgart 1988.

Horn, Friedemann, *Schelling und Swedenborg: Ein Beitrag zur Problemgeschichte des deutschen Idealismus und zur Geschichte Swedenborgs in Deutschland*, Lörrach-Stetten 1954; translation G. F. Dole, *Schelling and Swedenborg: Mysticism and German Idealism*, West Chester, PA 1997.

Hult, Olof, 'Några anteckningar om Olof och Magnus Bromelius', *Svenska Linnésällskapets årsskrift* 1926.

Hultkrantz, Johan Vilhelm, *The Mortal Remains of Emanuel Swedenborg*, Uppsala 1910–1912.

Hultman, Frans W., 'Svenska aritmetikens historia', *Tidskrift för matematik och fysik, tillegnad den svenska elementar-undervisningen*, vol. 3, Uppsala 1870.

Hyde, James, *A Bibliography of the Works of Emanuel Swedenborg Original and Translated*, London 1906.

Israel, Jonathan I., *Radical Enlightenment: Philosophy and the Making of Modernity 1650–1750*, Oxford 2001.

Janiak, Andrew, *Newton as Philosopher*, Cambridge 2008.

Jansson, Sam Owen, *Måttordboken*, Stockholm 1995.

Johns, Adrian, *The Nature of the Book: Print and Knowledge in the Making*, Chicago IL 1998.

Johnson, Mark, *The Body in the Mind: The Bodily Basis of Meaning, Imagination, and Reason*, Chicago IL 1987.

Jonasson, Gustaf, 'Karl XII', *Den svenska historien* VIII, Stockholm 1993.

Jonsson, Inge, *Swedenborgs skapelsedrama De Cultu et Amore Dei: En studie av motiv och intellektuell miljö*, Stockholm 1961; translation M. McCarthy, *A Drama of Creation: Sources and Influences in Swedenborg's Worship and Love of God*, West Chester, PA 2004.

Jonsson, Inge, 'Köpenhamn–Amsterdam–Paris: Swedenborgs resa 1736–1738', *Lychnos* 1967–1968.

Jonsson, Inge, *Swedenborgs korrespondenslära*, Stockholm 1969.

Jonsson, Inge, *Emanuel Swedenborg*, translation C. Djurklou, New York NY 1971.

Jonsson, Inge, *I symbolens hus: Nio kapitel litterär begreppshistoria*, Stockholm 1983a.

Jonsson, Inge, 'Swedenborg i Tyskland – Resor, reflexer, reception', *Bland böcker och människor: Bok- och personhistoriska studier till Wilhelm Odelberg den 1 juli 1983*, ed. K.-I. Hillerud, E. Ljungdahl & M. von Platen, Stockholm 1983b.

Jonsson, Inge, *Humanistiskt credo*, ed. B. Ståhle Sjönell & C. Wijnbladh Bergin, Stockholm 1988.

Jonsson, Inge, 'Swedenborg and Italian Science', *Sidereus Nuncius & Stella Polaris: The Scientific Relations between Italy and Sweden in Early Modern History*, ed. M. Beretta & T. Frängsmyr, Canton MA 1997.

Jonsson, Inge, *Visionary Scientist: The Effects of Science and Philosophy on Swedenborg's Cosmology*, West Chester PA 1999.

Jonsson, Inge, 'Swedenborg som vetenskapsman', in *Tysk* 2000.

Jonsson, Inge, [review of Crasta 1999], *Isis* 2002:2.

Jonsson, Inge, 'Die Swedenborgforschung: ein persönlicher Überblick', *Kant and Swedenborg / Kant und Swedenborg: Approaches to a Controversial Relationship / Zugänge zu einem umstrittenen Verhältnis*, ed. F. Stengel, Berlin & New York 2008.

Jonsson, Inge & Olle Hjern, *Swedenborg: Sökaren i naturens och andens världar. Hans verk och efterföljd*, Stockholm 1976.

Jukht, Aleksandr Isayevich, *Gosudarstvennaya deyatelnost V. N. Tatishcheva v 20-ch – nachale 30-ch godov XVIII v.*, Moskva 1985.

Jukht, Aleksandr Isayevich (ed.), *Nauchnoye nasledstvo*. Vol. 14. *Vasily Nikitich Tatishchev. Zapiski. Pisma 1717–1750 gg.*, Moskva 1990.

Jütte, Robert, *A History of the Senses: From Antiquity to Cyberspace*, Cambridge 2005.

Kargon, Robert Hugh, *Atomism in England from Hariot to Newton*, Oxford 1966.

Kemp, Martin, 'Temples of the Body and Temples of the Cosmos: Vision and Visualization in the Vesalian and Copernican Revolutions', *Picturing Knowledge: Historical and Philosophical Problems Concerning the Use of Art in Science*, ed. B. S. Baigrie, Toronto 1996.

Kenny, Neil, 'The Metaphorical Collecting of Curiosities in Early Modern France and Germany', *Curiosity and Wonder from the Renaissance to the Enlightenment*, ed. R. J. W. Evans & A. Marr, Aldershot 2006.

Kirven, Robert H., *Emanuel Swedenborg and the Revolt against Deism*, Waltheim MA 1965.

Kirven, Robert H. & Robin Larsen, 'Emanuel Swedenborg: A Pictorial Biography', *Emanuel Swedenborg: A Continuing Vision. A Pictorial Biography & Anthology of Essays & Poetry*, ed. R. Larsen, New York NY 1988.

Kjellberg, Erik & Jan Ling, *Klingande Sverige: Musikens vägar genom historien*, Göteborg 1991.

Kleen, Emil A. G., *Swedenborg: En lefnadsskildring* I–II, Stockholm 1917–1920.

Kragh, Helge S., *Conceptions of Cosmos: From Myths to the Accelerating Universe: A History of Cosmology*, Oxford 2007.

Krois, John, et al. (eds.), *Embodiment in Cognition and Culture*, Amsterdam 2007.

Kutik, Ilja, *Hieroglyphs of Another World: On Poetry, Swedenborg, and Other Matters*, Evanston IL 2000.

Küttner, Juri, 'V. N. Tatiščevs mission i Sverige 1724–1726', *Lychnos* 1990.

Lachièze-Rey, Marc & Jean-Pierre Luminet, *Celestial Treasury: From the Music of the Spheres to the Conquest of Space*, Cambridge 2001.

Lagerborg, Rolf, *Fallet Swedenborg i belysning av nyare undersökningar*, Stockholm 1924.

Lagercrantz, Olof, *Dikten om livet på den andra sidan*, Stockholm 1996; translation A. Hallengren, *Epic of the Afterlife: A Literary Approach to Swedenborg*, West Chester PA 2002.

Lakoff, George, *Women, Fire, and Dangerous Things: What Categories Reveal about the Mind*, Chicago IL 1990.

Lakoff, George & Mark Johnson, *Metaphors We Live By*, Chicago IL 1980.

Lakoff, George & Mark Johnson, *Philosophy in the Flesh: The Embodied Mind and its Challenge to Western Thought*, New York NY 1999.

Lakoff, George & Rafael E. Núñez, *Where Mathematics Comes From: How the Embodied Mind Brings Mathematics into Being*, New York NY 2000.

Lamm, Martin, *Swedenborg: En studie öfver hans utveckling till mystiker och andeskådare*, Stockholm 1915; translation T. Spiers & A. Hallengren, *Emanuel Swedenborg: The Development of his Thought*, West Chester PA 2000.

Lamm, Martin, *Upplysningstidens romantik: Den mystiskt sentimentala strömningen i svensk litteratur* I–II, Stockholm 1918–1920.

Lawson, E. Thomas, 'Counterintuitive Notions and the Problem of Transmission: The Relevance of Cognitive Science for the Study of History', *Historical Reflections/Réflexions Historique* 20(3), 1994.

Lawson, E. Thomas, 'The wedding of psychology, ethnography, and history: methodological bigamy or tripartite free love?', *Theorizing Religions Past: Archaeology, History, and Cognition*, ed. H. Whitehouse & L. H. Martin, Walnut Creek CA 2004.

Lenhammar, Harry, *Tolerans och bekännelsetvång: Studier i den svenska swedenborgianismen 1765–1795*, Uppsala 1966.
Lenhammar, Harry, 'Swedenborgsbilden', *Kyrkohistorisk årsskrift* 1988.
Lenhammar, Harry, *Siarens återkomst: Swedenborg i Uppsala domkyrka*, Stockholm 2010.
Lesch, John E., 'Systematics and the geometrical spirit', in Frängsmyr, Heilbron & Rider 1990.
Liljegren, Bengt, *Karl XII i Lund: När Sverige styrdes från Skåne*, Lund 1999.
Liljegren, Bengt, *Karl XII: En biografi*, Lund 2000.
Liljencrantz, Axel, 'Polhem och grundandet av Sveriges första naturvetenskapliga samfund jämte andra anteckningar rörande Collegium Curiosorum' I–II, *Lychnos* 1939 & 1940.
Liljencrantz, Axel, 'Eric Benzelius d.y:s naturvetenskapliga studier och biblioteksverksamhet', *Corona amicorum: Studier tillägnade Tönnes Kleberg*, Uppsala 1968.
Lindberg, Bo, 'Den eklektiska filosofien och "libertas philosophandi": Svensk universitetsfilosofi under 1700-talets första decennier', *Lychnos* 1973–1974.
Lindberg, Bo, *De lärdes modersmål: Latin, humanism och vetenskap i 1700-talets Sverige*, Göteborg 1984.
Lindberg, Bo, *Stoicism och stat: Justus Lipsius och den politiska humanismen*, Stockholm 2001.
Lindberg, David C., *Theories of Vision from Al-Kindi to Kepler*, Chicago IL 1976.
Lindborg, Rolf, *Descartes i Uppsala: Striderna om 'nya filosofien' 1663–1689*, Uppsala 1965.
Lindborg, Rolf, 'Om Stiernhielm och cartesianismen', in Ohlsson & Olsson 2000.
Lindgren, Michael, 'Den Kongliga Modellkammaren – en trädimensionell upplevelse', *Polhem* 1992:4a.
Lindh, Frans G., 'Swedenborgs ekonomi', *Nya kyrkans tidning* 1927–1929.
Lindh, Frans G., 'Söderbon', in Hjern, Eriksson & Hallengren 1992.
Lindqvist, Svante, *Technology on Trial: The Introduction of Steam Power Technology into Sweden, 1715–1736*, Uppsala 1984.
Lindqvist, Svante, 'Trä, vatten och muskelkraft: 1720–1815', *Svensk teknikhistoria*, ed. S. Rydberg, Hedemora 1989.
Lindroth, Sten, *Christopher Polhem och Stora Kopparberget: Ett bidrag till bergsmekanikens historia*, Uppsala 1951.
Lindroth, Sten, *Gruvbrytning och kopparhantering vid Stora Kopparberget intill 1800-talets början* I–II, Uppsala 1955.
Lindroth, Sten, *Kungl. svenska vetenskapsakademiens historia 1739–1818*, Stockholm 1967.
Lindroth, Sten, *Magnus Gabriel von Block*, Stockholm 1973.
Lindroth, Sten, *Svensk lärdomshistoria* II–III, 2nd ed., Stockholm 1989.
Linton, Olof, 'Skapelsens år, månad och dag: Till den kristna tideräkningens historia', *Lychnos* 1937.
Löfkvist, Hans-Eric, 'Några kommentarer till Emanuel Swedenborgs flygplansprojekt', *Dædalus* 1988.
Lotman, Yuri Mikhailovich, 'K probleme tipologii kultury', *Trudy po znakovym sisteman* III, Tartu 1967.
Lotman, Yuri Mikhailovich, *Universe of the Mind: A Semiotic Theory of Culture*, translation A. Shukman, London 1990.
Lotman, Yuri Mikhailovich, 'Semiotiken och historievetenskapen', *Den inre teatern: Filosofiska dialoger 1986–1996*, ed. M. Florin & B. Göranzon, Stockholm 1996.
Lovejoy, Arthur O., *The Great Chain of Being: A Study of the History of an Idea*, new ed., Cambridge MA 1964.
Mackie, John Leslie, *The Miracle of Theism: Arguments for and against the Existence of God*, new ed., Oxford 1990.
Mahoney, Michael Sean, 'The Mathematical Realm of Nature', *The Cambridge History of Seventeenth-century Philosophy* I, ed. D. Garber & M. Ayers, Cambridge 1998.
Mancosu, Paolo, *Philosophy of Mathematics and Mathematical Practice in the Seventeenth Century*, Oxford 1996.
Mansén, Elisabeth, *Ett paradis på jorden: Om den svenska kurortskulturen 1680–1880*, Stockholm 2001.

Marshall, David L., *Vico and the Transformation of Rhetoric in Early Modern Europe*, Cambridge 2010.

Martin, Luther H. & Jesper Sørensen, *Past Minds: Studies in Cognitive Historiography*, London 2010.

Mazzeo, Joseph Anthony, *Renaissance and Seventeenth-century Studies*, New York NY 1964.

McAllister, James W., *Beauty and Revolution in Science*, Ithaca NY 1996.

McMullin, Ernan, 'Conceptions of Science in the Scientific Revolution', *Reappraisals of the Scientific Revolution*, ed. D. C. Lindberg & R. S. Westman, Cambridge 1990.

McNeilly, Stephen (ed.), *On the True Philosopher and the True Philosophy: Essays on Swedenborg*, London 2002.

McNeilly, Stephen (ed.), *In Search of the Absolute: Essays on Swedenborg and Literature*, London 2004.

McNeilly, Stephen (ed.), *Between Method and Madness*, London 2005.

McNeilly, Stephen (ed.), *The Arms of Morpheus: Essays on Swedenborg and Mysticism*, London 2007.

Meheus, Joke, 'Analogical Reasoning in Creative Problem Solving Processes: Logico-philosophical Perspectives', *Metaphor and Analogy in the Sciences*, ed. F. Hallyn, Boston MA 2000.

Miłosz, Czesław, *The Land of Ulro*, translation L. Iribarne, 3rd ed., New York NY 2000.

Miłosz, Czesław, 'Swedenborg the Mystic', in McNeilly 2007.

Mithen, Steven, *The Prehistory of the Mind: The Cognitive Origins of Art, Religion, and Science*, London 1996.

Moberg, Carl-Allan, 'Olof Rudbeck d. ä. och musiken', *Rudbecksstudier: Festskrift vid Uppsala universitets minnesfest till högtidlighållande av 300-årsminnet av Olof Rudbeck d. ä:s födelse*, Uppsala 1930.

Nathorst, A. G., 'Emanuel Swedenborg såsom geolog', *Geologiska föreningens förhandlingar* 28:243, Stockholm 1906.

Nelson, Axel, 'Eric Benzelius' anteckningar till J. Schefferus' Svecia literata', *Linköpings biblioteks handlingar*, ny serie 3, Linköping 1940.

Nersessian, Nancy J., 'How do Scientists Think? Capturing the Dynamics of Conceptual Change in Science', *Cognitive Models of Science*, ed. R. N. Giere, Minneapolis MN 1992.

Nersessian, Nancy J. 'Opening the Black Box: Cognitive Science and History of Science', *Osiris*, 2nd Series, vol. 10, 1995.

Nersessian, Nancy J. 'Interpreting Scientific and Engineering Practices: Integrating the Cognitive, Social, and Cultural Dimensions', *Scientific and Technological Thinking*, ed. M. E. Gorman et al., Mahwah NJ 2005.

Netz, Reviel, *The Shaping of Deduction in Greek Mathematics: A Study in Cognitive History*, Cambridge 1999.

Netz, Reviel, 'The Aesthetics of Mathematics: A Study', *Visualization, Explanation and Reasoning Styles in Mathematics*, ed. P. Mancosu, K. F. Jørgensen & S. A. Pedersen, Dordrecht 2005.

Neumeyer, Friedrich, 'Christopher Polhem och hydrodynamiken', *Arkiv för matematik, astronomi och fysik* 28A:15, Stockholm 1942.

The New Philosophy, Bryn Athyn PA 2003.

Newton, Norman, *The Listening Threads: The Formal Cosmology of Emanuel Swedenborg*, Bryn Athyn PA 1999.

Nilsson, Stig, *Terminologi och nomenklatur: Studier över begrepp och deras uttryck inom matematik, naturvetenskap och teknik* I, Lund 1974.

Nilsson, Stig, 'Materieuppfattningens termer i svenskan fram till år 1800', *Lychnos* 1975–1976.

Nilsson, Stig, *Yta – djupare sett: Ett begrepp och dess benämningar i historiskt perspektiv*, Kalmar 1992.

Nisser, Marie, 'Fortifikationsofficerarnas utbildning under 1600- och 1700-talen', in Runnberg 1986.

Nordenmark, N. V. E., *Swedenborg som astronom*, Arkiv för matematik, astronomi och fysik, vol. 23 A, no. 13, Stockholm 1933.

Nordenmark, N. V. E., *Anders Celsius: Professor i Uppsala 1701–1744*, Uppsala 1936.

Nordenmark, N. V. E., 'Swedenborg och longitudproblemet: Med anledning av ett nyfunnet brev från Wargentin', *Lychnos* 1944–1945.

Nordenmark, N. V. E., *Astronomiens historia i Sverige intill år 1800*, Uppsala 1959.

Nordin, Svante, *Från tradition till apokalyps: Historieskrivning och civilisationskritik i det moderna Europa*, Stockholm, Lund & Stehag 1989.

Nordin, Svante & Jonas Hansson (eds.), *Att skriva filosofihistoria*, Lund 1998.

Nordström, Johan, 'Inledning', *Samlade skrifter av Georg Stiernhielm* II:1, Stockholm 1924.

Norlind, Tobias, *Svensk musikhistoria*, 2nd ed., Stockholm 1918.

Núñez, Rafael E., 'Mathematics, the Ultimate Challenge to Embodiment: Truth and the Grounding of Axiomatic Systems', in Calvo & Gomila 2008.

Núñez, Rafael E. & George Lakoff, 'The Cognitive Foundations of Mathematics: The Role of Conceptual Metaphor', *Handbook of Mathematical Cognition*, ed. J. I. D. Campbell, New York NY 2005.

Nystedt, Lars, 'Metersystemet – en pigg tvåhundraåring', *Tid, längd och vikt*, ed. I. Elmqvist & J. Florén, Stockholm 1999.

Nyström, Anton, *Karl XII och sammansvärjningen mot hans envälde och lif*, Stockholm 1900.

Odelberg, Wilhelm, *Lärdom för livet*, Stockholm 1999.

Ohlon, Rolf, *Från Stiernhielms Carl-Staf till metern*, Stockholm & Borås 1989.

Ohlon, Rolf, 'Stiernhielm som metrolog och skapare av det svenska måttsystemet', in Ohlsson & Olsson 2000.

Ohlsson, Stig Örjan & Bernt Olsson (eds.), *Stiernhielm 400 år: Föredrag vid internationellt symposium i Tartu 1998*, Tartu 2000.

Olofsson, Rune Pär, *Georg Stiernhielm – diktare, domare, duellant: En levnadsteckning*, Hedemora 1998.

Olsén, Jan Eric, 'Molyneuxs problem', in Nordin & Hansson 1998.

Olson, David R., *The World on Paper: The Conceptual and Cognitive Implications of Writing and Reading*, Cambridge 1996.

Ong, Walter J., *Orality and Literacy: The Technologizing of the Word*, London 1982.

Önnerfors, Andreas, *Svenska Pommern: Kulturmöten och identifikation 1720–1815*, Lund 2003.

Oredsson, Sverker, 'Karl XII och det svenska stormaktsväldets fall i historieskrivning och tradition', *Tsar Peter och kung Karl: Två härskare och deras folk*, ed. S. Oredsson, Stockholm 1998.

Örneholm, Urban, *Four Eighteenth-century Medical Dissertations under the Presidency of Nils Rosén*, Uppsala 2003.

Ornstein [Bronfenbrenner], Martha, *The Role of Scientific Societies in the Seventeenth Century*, New York NY 1975.

Pagel, Walter, '"Circulatio" – Its Unusual Connotations and William Harvey's Philosophy', *Studies on William Harvey*, ed. I. B. Cohen, New York NY 1981.

Panofsky, Erwin, *Die Perspektive als 'symbolische Form'*, Leipzig 1927; translation C. S. Wood, *Perspective as Symbolic Form*, New York NY 1991.

Park, David, *The Fire within the Eye: A Historical Essay on the Nature and Meaning of Light*, Princeton NJ 1997.

Pedersen, Olaf, *Niels Stensens videnskabelige liv*, 2nd ed., Århus 1994.

Pehrsson, Anna-Lena, 'Nils Rosén von Rosenstein och iatromekaniken', *Svenska Linnésällskapets årsskrift* 1965.

Pendleton, Charles Rittenhouse, *Space and Extense in the Spiritual World*, Bryn Athyn PA 1962.

Pesic, Peter, 'Secrets, Symbols, and Systems: Parallels between Cryptanalysis and Algebra 1580–1700', *Isis* 1997:4.

Piotrowska, Ewa, 'Elementy filozofii matematyki w twórczości Emanuela Swedenborga', *Zeszyty naukowe Wydziału Humanistycznego. Studia Scandinavica* 10, Gdańsk 1988.

Pipping, Gunnar, 'Några drag ur det svenska mått- och justeringsväsendets historia', *Dædalus* 1968.

Pipping, Gunnar, *The Chamber of Physics: Instruments in the History of Sciences Collections of the Royal Swedish Academy of Sciences, Stockholm*, 2nd ed., Stockholm 1991.

Pipping, Gunnar, 'Georg Stiernhielm and his System of Weights and Measures', in Ohlsson & Olsson 2000.

Platen, Magnus von, *Skandalen på Operakällaren och andra essayer*, Stockholm 1996.

Pochat, Götz, 'Janus hos Swedenborg och i nyplatonsk och emblematisk tradition', *Tidskrift för litteraturvetenskap* 1975:2.

Pomian, Krzysztof, 'Vision and Cognition', *Picturing Science, Producing Art*, ed. C. A. Jones & P. Galison, New York NY 1998.

Portis-Winner, Irene, 'The Dynamics of Semiotics of Culture; Its Pertinence to Anthropology', *Sign Systems Studies* 27, Tartu 1999.

Potts, John Faulkner, *The Swedenborg Concordance: A Complete Work of Reference to the Theological Writings of Emanuel Swedenborg* I–VI, London 1888–1902.

Powers, John C., '"Ars sine arte": Nicholas Lemery and the End of Alchemy in Eighteenth-century France', *Ambix* 45:3, 1998.

Principe, Lawrence M., *The Aspiring Adept: Robert Boyle and his Alchemical Quest*, 2nd ed., Princeton NJ 2000.

Principe, Lawrence M., 'Wilhelm Homberg: Chymical Corpuscularianism and Chrysopoeia in the Early Eighteenth Century', *Late Medieval and Early Modern Corpuscular Matter Theories*, ed. C. Lüthy et al., Leiden 2001.

Purrington, Robert D., *The First Professional Scientist: Robert Hooke and the Royal Society of London*, Basel 2009.

Quennerstedt, August Wilhelm, 'Karl XII i Lund', *Karolinska förbundets årsbok* 1912.

Quennerstedt, August Wilhelm, *Ur Carl XII:s lefnad*, Stockholm 1916.

Räf, Per-Ola, *Organister och orglar i Skara stift t o m 1857*, Skara 2002.

Ramström, Martin, *Om Emanuel Swedenborg som naturforskare och i synnerhet hjärnanatom*, Uppsala 1910.

Ratcliff, Marc J., *The Quest for the Invisible: Microscopy in the Enlightenment*, Farnham 2009.

Reisberg, Daniel, *Cognition: Exploring the Science of the Mind*, New York NY 1997.

Renfrew, Colin, Chris Frith & Lambros Malafouris (eds.), *The Sapient Mind: Archaeology Meets Neuroscience*, Oxford 2009.

Rescher, Nicholas, *Leibniz's Metaphysics of Nature*, Dordrecht 1981.

Retzius, Gustaf, *Emanuel Swedenborg as an Anatomist and Physiologist*, Bryn Athyn PA 1903.

Richardson, Alan & Francis F. Steen, 'Literature and the Cognitive Revolution: An Introduction', *Poetics Today* 23:1, Spring 2002.

Richter, Herman, *Geografiens historia i Sverige intill år 1800*, Uppsala 1959.

Ridderstad, Per Soldan, *Konsten att sätta punkt: Anteckningar om stenstilens historia 1400–1765*, Stockholm 1975.

Rix, Robert, *William Blake and the Cultures of Radical Christianity*, Aldershot 2007.

Rodhe, Staffan, *Matematikens utveckling i Sverige fram till 1731*, Uppsala 2002.

Roos, Anna Marie, 'Luminaries in Medicine: Richard Mead, James Gibbs, and Solar and Lunar Effects on the Human Body in Early Modern England', *Bulletin of the History of Medicine*, 2000:74.

Rorty, Richard, 'The Historiography of Philosophy: Four Genres', *Philosophy in History*, ed. R. Rorty, J. B. Schneewind & Q. Skinner, new ed., Cambridge 1993.

Rosch, Eleanor, 'Principles of Categorization', *Cognition and Categorization*, ed. E. Rosch & B. B. Lloyd, Hillsdale NJ 1978.

Rose, Jonathan S., 'Holding On and Letting Go', *Logos*, spring 2004.

Rose, Jonathan S. (ed.), *Emanuel Swedenborg: Essays for The New Century Edition on his Life, Work, and Impact*, West Chester PA 2005.

Rosengren, Cecilia, 'Form, materia och barocka figurer i Anne Conways naturfilosofi', *Lychnos* 2003.

Rossi, Paolo, *Logic and the Art of Memory: The Quest for a Universal Language*, translation, Chicago IL 2000.

Runnberg, Bertil (ed.), *Fortifikationen 350 år: 1635–1985*, Stockholm 1986.

Rydberg, Sven, *Svenska studieresor till England under frihetstiden*, Uppsala 1951.

Ryman, Björn, *Benzelius d. y.: En frihetstida politiker*, Stockholm & Lund 1978.

Ryman, Björn, 'Eric Benzelius den yngre, de mångas patronus', *Klient och patron: Befordringsvägar och ståndscirkulation i det gamla Sverige*, ed. M. von Platen, Stockholm 1988.

Sahlin, Björn, 'Swedenborgs trädgård', *Parnass* 1999:1.

Sahlin, Carl, *Vår järnindustris äldsta reklamtryck*, Örebro 1923.

Sahlin, Carl, *Svenskt stål före de stora götstålprocessernas införande*, Stockholm 1931.

Sahlin, Carl, *Valsverk inom den svenska metallurgiska industrien intill början av 1870-talet*, Stockholm 1934.

Sandblad, Henrik, 'Det copernikanska världssystemet i Sverige: II. Cartesianismen och genombrottet', *Lychnos* 1944–1945.

Sandstedt, Erik, *Studier rörande Jöran Nordbergs Konung Carl XII:s historia*, Lund 1972.

Sanner, Inga, *Den segrande Eros: Kärleksföreställningar från Emanuel Swedenborg till Poul Bjerre*, Nora 2003.

Sawday, Jonathan, *Engines of the Imagination: Renaissance Culture and the Rise of the Machine*, London 2007.

Schlieper, Hans, *Emanuel Swedenborgs System der Naturphilosophie besonders in seiner Beziehung zu Goethe-Herderschen Anschauungen*, Berlin 1901.

Scholz, Bernhard F., 'Alchemy, metallurgy and emblematics in the works of the seventeenth-century Dutch "Bergmeester" Goossen van Vreeswijck (1626–after 1689)', *Emblems and Alchemy*, ed. A. Adams & S. J. Linden, Glasgow 1998.

Schubert, Gotthilf Heinrich von, *Symbolik des Traumes*, 3rd ed., Leipzig 1840.

Schuster, John A., '"Waterworld": Descartes' Vortical Celestial Mechanics—A Gambit in the Natural Philosophical Contest of the Early Seventeenth Century', *The Science of Nature in the Seventeenth Century: Patterns of Change in Early Modern Natural Philosophy*, ed. P. R. Anstey & J. A. Schuster, Dordrecht 2005.

Sellberg, Erland, *Filosofin och nyttan: I. Petrus Ramus och ramismen*, Göteborg 1979.

Sellberg, Erland, 'Stiernhielms världsbild', in Ohlsson & Olsson 2000.

Sellberg, Erland, 'Bilberg i den akademiska katedern', *Mellan teologins krav och filosofins fria bruk: Johan Bilberg som filosof och kyrkoman*, ed. S.-O. Lindeberg, Örebro 2002.

Seth, Ivar, *Universitetet i Greifswald och dess ställning i svensk kulturpolitik 1637–1815*, Uppsala 1952.

Shank, John Bennett, *The Newton Wars and the Beginning of the French Enlightenment*, Chicago IL 2008.

Sigstedt, Cyril Odhner, *The Swedenborg Epic: The Life and Works of Emanuel Swedenborg*, New York NY 1952.

Siljestrand, Karl K:son, *Karl XII såsom filosof*, Linköping 1891.

Sjödén, Karl-Erik, *Swedenborg en France*, Stockholm 1985.

Sjögren, Hjalmar, 'Några ord om Swedenborgs manuskript: "Nya anledningar til grufwors igenfinnande" etc.', *Geologiska föreningens i Stockholm förhandlingar* 29:7, Stockholm 1907.

Sjögren, Otto, *Karl den tolfte och hans män: Lifsbilder från vår sjunkande storhetstid*, Stockholm 1899.

Skinner, Quentin, 'Meaning and Understanding in the History of Ideas', *History and Theory* 8:1, Middletown CT 1969.

Snickare, Mårten, 'Landskrona', *Karolinska förbundets årsbok* 1999.

Söderberg, Henry, '"En machine att flyga i wädret": Emanuel Swedenborgs förslag till en flygmaskin år 1714', *Dædalus* 1988a.

Söderberg, Henry, *Swedenborg's 1714 Airplane: A Machine to Fly in the Air*, New York NY 1988b.

Söderberg, Henry, 'Swedenborgdokumenten i Linköpings Stiftsbibliotek', *Linköpings biblioteks handlingar*, ny serie 12, Linköping 1989.

Spaak, George & Torsten Althin, 'Enkelmikroskop, som möjligen tillhört Emanuel Swedenborg', *Dædalus* 1950.

Spear, William, *Emanuel Swedenborg: The Spiritual Columbus*, London 1876.

Spengler, Oswald, *Der Untergang des Abendlandes: Umrisse einer Morphologie der Welt-geschichte: Erster Band. Gestalt und Wirklichkeit*, 2nd ed., Wien & Leipzig 1919; translation C. F. Atkinson, *The Decline of the West. Volume 1. Form and Actuality*, new ed., New York 1983.

Spranzi, Marta, 'Galileo and the Mountains of the Moon: Analogical Reasoning, Models and Metaphors in Scientific Discovery', *Journal of Cognition and Culture* 2004:3.

Stengel, Friedemann (ed.), *Kant und Swedenborg: Zugänge zu einem umstrittenen Verhältnis*, Tübingen 2008.

Stroh, Alfred H., *The Cartesian Controversy at Upsala, 1663–1689, and Its Connection with Swedenborg's Nebular Hypothesis*, Heidelberg 1908.

Stroh, Alfred H., *A Series of Reports Concerning Investigations and Proceedings in Sweden from 1902 to 1912*, Stockholm 1912.

Stroh, Alfred H., *Investigations in Sweden 1902 to 1918*, Stockholm 1918.

Stroh, Alfred H. & Greta Ekelöf, *Kronologisk förteckning öfver Emanuel Swedenborgs skrifter 1700–1772*, Uppsala 1910.

Suzuki, Daisetz Teitaro, *Swedenborg: Buddha of the North*, translation A. Bernstein, West Chester PA 1996.

Tafel, Rudolf Leonard, *Results of an Investigation into the Manuscripts of Swedenborg*, Edinburgh 1869.

Taton, René & Curtis Wilson (eds.), *Planetary Astronomy from the Renaissance to the Rise of Astrophysics. Part A: Tycho Brahe to Newton*, Cambridge 1989.

Taylor, John R., *Linguistic Categorization*, 3rd ed., Oxford 2003.

Thanner, Lennart, *Revolutionen i Sverige efter Karl XII:s död: Den inrepolitiska maktkampen under tidigare delen av Ulrika Eleonora d.y:s regering*, Uppsala 1953.

Thomas, Keith, *Man and the Natural World: Changing Attitudes in England 1500–1800*, new ed., Harmondsworth 1984.

Thomson, Ann, *Bodies of Thought: Science, Religion, and the Soul in the Early Enlightenment*, Oxford 2008.

Tingström, Bertel, 'Stormaktstidens penningväsen', *Myntningen i Sverige 995–1995*, ed. K. Jonsson, U. Nordlind & I. Wiséhn, Stockholm 1995.

Tingström, Bertel, 'Görtz' caroliner', *Karolinska förbundets årsbok* 1997.

Tiselius, Elias, *Daniel Tiselius: En kulturbild från tidigt 1700-tal*, Uppsala 1951.

Toksvig, Signe, *Emanuel Swedenborg: Scientist and Mystic*, New Haven CT 1948.

Tomasello, Michael, *The Cultural Origins of Human Cognition*, Cambridge MA 1999.

Tomasello, Michael, 'Uniquely Human Cognition Is a Product of Human Culture', *Evolution and Culture: A Fryssen Foundation Symposium*, ed. S. C. Levinson & P. Jaisson, Cambridge MA 2005.

Transactions of the International Swedenborg Congress 1910, London 1912.

Turner, Mark, 'The Cognitive Study of Art, Language, and Literature', *Poetics Today* 23:1, Spring 2002.

Tweney, Ryan D., 'Scientific Thinking: A Cognitive-Historical Approach', *Designing for Science: Implications from Everyday, Classroom, and Professional Settings*, ed. K. Crowley, C. D. Schunn & T. Okada, Mahwah NJ 2001.

Tysk, Karl-Erik (ed.), *Vetenskap, mystik och religion: Den mångdimensionelle Emanuel Sweden-borg*, Skara 2000.

Ullmann, Dieter, 'Athanasius Kircher und die Akustik der Zeit um 1650', *Internationale Zeitschrift für Geschichte und Ethik der Naturwissenschaften, Technik und Medizin*, 2002:2.

Varela, Francisco J., Evan Thompson & Eleanor Rosch, *The Embodied Mind: Cognitive Science and Human Experience*, Cambridge MA 1991.

Wainscot, A. S., *A Bibliography of the Works of Emanuel Swedenborg Original and Translated by the Rev. James Hyde. Lists of Additions to the Bibliography since its Publication in 1906*, London 1967.

Wald, Melanie, *Welterkenntnis aus Musik: Anthanasius Kirchers 'Musurgia universalis' und die Universalwissenschaft im 17. Jahrhundert*, Kassel 2006.

Warburg, Aby, *The Renewal of Pagan Antiquity: Contributions to the Cultural History of the European Renaissance*, translation D. Britt, Los Angeles CA 1999.

Wardhaugh, Benjamin, *Music, Experiment and Mathematics in England, 1653–1705*, Farnham 2008.

Weimarck, Torsten, *Akademi och anatomi: Några aspekter på människokroppens historia i nya tidens konstnärsutbildning och ateljépraktik*, . . . , Stockholm & Stehag 1996.

Westerlund, Olov, *Karl XII i svensk litteratur från Dahlstierna till Tegnér*, Lund 1951.

Westfall, Richard S., *The Construction of Modern Science: Mechanisms and Mechanics*, Cambridge 1977.

Wetterberg, Lennart, 'Swedenborgs syn på hjärnan', *Värld och vetande* 1993:2–3.

Whitehouse, Harvey, 'Cognitive Historiography: When Science Meets Art', *Historical reflections/Réflexions historiques* 2005:2.

Widegren, Ragnar, 'Fortifikationsofficeren som väg- och vattenbyggare', in Runnberg 1986.

Wilkinsons, Lynn R., *Dream of an Absolute Language: Emanuel Swedenborg and French Literary Culture*, Albany NY 1996.

Williams-Hogan, Jane, *A New Church in a Disenchanted World: A Study of the Formation and Development of the General Conference of the New Church in Great Britain*, Philadelphia PA 1985.

Williams-Hogan, Jane, 'Swedenborg Studies: "On the Shoulders of Giants"', *The New Philosophy* 2002:1–2.

Williams-Hogan, Jane, 'The Place of Emanuel Swedenborg in the Spiritual Saga of Scandinavia', *Western Esotericism: Based on Papers Read at the Symposium on Western Esotericism Held at Åbo, Finland on 15–17 August 2007*, ed. T. Ahlbäck, Åbo 2008.

Wilson, Catherine, *The Invisible World: Early Modern Philosophy and the Invention of the Microscope*, Princeton NJ 1995.

Woofenden, William Ross, *Swedenborg's Philosophy of Causality*, St. Louis MO 1970.

Woofenden, William Ross, *Swedenborg Explorer's Guidebook: A Research Manual for Inquiring New Readers, Seekers of Spiritual Ideas, and Writers of Swedenborgian Treatises*, West Chester PA 2002.

Yates, Frances A., *The Art of Memory*, new ed., London 2001.

Yeo, Richard, 'Classifying the Sciences', *The Cambridge History of Sciences: Volume 4: Eighteenth-Century Science*, Cambridge 2003.

Zenzén, Nils, 'Om den s. k. Swedenborgsstammen och det swedenborgska marmorbordet', *Svenska Linnésällskapets årsskrift* 1931.

*Works listed in *Förtekning på afl. wälborne herr assessor Swedenborgs efterlämnade wackra boksamling* (1772), or known to have been in Swedenborg's possession.

Index

A

Abraham (18th century BC), Jewish patriarch, 4

Abû Ma'shar al-Balhî, Ja'far ibn Muhammad (c. 786/787–886), Arabian astrologer, 403

Acton, Alfred (1867–1956), American Swedenborgian, doctor, 10

Agner, Eric Nilsson (c. 1642–1727), Swedish mathematician, surveyor, 62, 116

Agricola, Georgius (Georg Bauer) (1494–1555), German mineralogist, metallurgist, 4, 195, 208, 250–254, 258–262, 275, 405

Ahlstedt, Nils Larsson (1700–1761), Swedish former guardsman in Svea Livgarde, Swedenborg's gardener, married to Maria Norman, 415

Åhrman, Johan (died c. 1714), Swedish organ builder, 162

Älf, Samuel (1727–1799), Swedish priest, poet, 415

Alstrin, Eric (1683–1762), Swedish professor of theoretical philosophy in Uppsala, bishop of Växjö, Strängnäs, 7

Anselm of Canterbury (1033–1109), English philosopher, theologian, saint, 373

Antram, Joseph (died 1723), English clockmaker in London, 209

Apelles (c. 370–c. 300 BC), Greek painter, 9

Apollonius of Perga (c. 262–c. 190 BC), Greek mathematician, astronomer, 159

Apomasar. See Abû Ma'shar al-Balhî

Archimedes (c. 287–212 BC), Greek mathematician, 3, 147, 156, 219, 304, 337, 339

Archytas of Tarentum (c. 430–345 BC), Greek philosopher, Pythagorean, 146

Aristarchus of Samos (c. 310–c. 230 BC), Greek astronomer, mathematician, 3

Aristippus of Cyrene, the Elder (c. 435–c. 355 BC), Greek philosopher, 69

Aristophanes (c. 445–c. 385 BC), Greek comic playwright, 341

Aristotle (384–322 BC), Greek philosopher, natural scientist, 21, 120, 122, 148, 149, 194, 195, 225, 230, 249, 252, 283, 286, 309, 313, 339, 340, 354, 362, 366, 374, 381, 387, 388, 403

Ask, Jonas Elias (18th century), Swedish student in Uppsala, 262

Atterbom, Per Daniel Amadeus (1790–1855), Swedish poet, 9

Augustine, Aurelius (354–430), Roman church father, 284, 309, 391, 399, 401, 403

B

Bach, Johann Jakob (1682–1722), German oboist, court musician, brother of Johann Sebastian Bach, 161

Bach, Johann Sebastian (1685–1750), German composer, brother of Johann Jakob Bach, 161

Bacon, Sir Francis (1561–1626), English philosopher, statesman, Baron Verulam, 24, 46, 75, 156, 222, 282, 296, 346

Bacon, Roger (c. 1220–c. 1292), English philosopher, natural scientist, 209

Baglivi, Giorgio (1668–1707), Italian anatomist, physician, professor in Rome, 182, 183, 187

Baker, Thomas (1656–1740), English author, antiquarian, 122

Balzac, Honoré de (1799–1850), French author, 12

Barchusen, Johann Conrad (1666–1723), German apothecary, physician, chemist, 194, 258, 267

Bartholin, Rasmus (1625–1698), Danish physician, mathematician, physicist, 240

Basilius Valentinus (born 1394), German chemist, physician, Benedictine monk, 252, 253, 262, 263, 270

Baudelaire, Charles (1821–1867), French author, critic, 12

Bauer, Georg. *See* Agricola

Bayer, Gottlieb Siegfried (1694–1738), German-Russian orientalist, member of the Academy in St Petersburg, 84

Bayle, Pierre (1647–1706), French philosopher, 358

Becher, Johann Joachim (1635–1682), German alchemist, 177, 237, 252, 258, 268, 313

Beeckman, Isaac (1588–1637), Dutch physicist, 148

Behm, Sara (1666–1696), Swedish, married to Jesper Swedberg, mother of Emanuel Swedenborg, 6

Bellini, Lorenzo (1643–1704), Italian physiologist, medical scholar, 187

Bellman, Johan Arent (1664–1709), Swedish professor of eloquence, 42, 167

Benzelius, Eric, the Elder (1632–1709), Swedish archbishop, 111, 277

Benzelius, Eric, the Younger (1675–1743), Swedish university librarian, professor of theology in Uppsala, archbishop, son of Eric Benzelius the Elder, 5, 115

Benzelius, Erik Eriksson, ennobled as Benzelstierna (1705–1767), Swedish inspector of mines, councillor of mines, son of Eric Benzelius the Younger, 64

Benzelius, Henric (1689–1758), Swedish professor of Oriental and Greek languages in Uppsala, archbishop, son of Eric Benzelius the Elder, 151, 152

Benzelstierna, Gustaf (1687–1746), librarian, son of Eric Benzelius the Elder, 82, 108, 405

Benzelstierna, Jesper Albrecht (1716–1743), Swedish fortification officer, son of Lars Benzelstierna, 217

Benzelstierna, Lars (1680–1755), Swedish assessor in the Board of Mines, councillor of mines, son of Eric Benzelius the Elder, married to Hedvig Swedenborg, 139, 250, 405

Bergenstierna, Johan, before ennoblement Frondberg (1668–1748), Swedish assessor in the Board of Mines, councillor of mines, 208, 405

Bergquist, Lars (born 1930), Swedish author, diplomat, 12, 17, 87

Berkeley, George (1685–1753), Irish philosopher, theologian, bishop, 75, 242, 368, 392

Bernini, Giovanni Lorenzo (Gianlorenzo) (1598–1680), Italian sculptor, architect, painter, 351

Bernoulli, Jakob (Jacques) (1654–1705), Swiss mathematician, 304, 306

Bernoulli, Johann (Jean) (1667–1748), Swiss mathematician, 67, 135

Biancani, Giuseppe (1566–1624), Italian astronomer, 156

Bidloo, Govard (1649–1713), Dutch anatomist, 188, 386

Bignon, Abbé Jean Paul (1662–1743), French member of the Académie des Sciences, editor of *Journal des Savants*, 42, 58, 237

Bilberg, Johan (1646–1717), Swedish professor of mathematics in Uppsala, bishop of Strängnäs, 150, 181, 388

Bilmarck, Johan (1687–1750), Swedish lecturer in theology in Uppsala, dean, 184

Björk, Mathias Andreæ (Björkstadius) (1604–1651), Swedish mathematician, 104

Blake, William (1757–1827), British poet, artist, 5, 12

Block, Magnus Gabriel von (1669–1722), Swedish physician, assessor in Collegium Medicum, 111, 112, 139, 203, 244, 257, 354

Blondel, François (1618–1686), French mathematician, architect, 216

Boë, François de le (Franciscus Sylvius) (1614–1672), Dutch physician, 187

Boerhaave, Hermann (1668–1738), Dutch chemist, physician, professor of medicine, botany, and chemistry in Leiden, 187, 188, 190, 192, 256, 259, 268, 387

Boethius, Anicius Manlius Severinus (*c.* 480–524), Roman philosopher, musical theorist, 353, 388

Bonde, Gustaf (1682–1764), Swedish count, politician, president of the Board of Mines, Councillor of the Realm, 5, 38, 267

Borelli, Giovanni Alfonso (1608–1679), Italian professor of mathematics in Messina and Pisa, 51, 183, 187, 200

Borges, Jorge Luis (1899–1986), Argentinian author, 9

Boström, Christopher Jacob (1797–1866), Swedish philosopher, professor of practical philosophy in Uppsala, 12

Boyle, Robert (1627–1691), English natural scientist, chemist, 5, 70, 139, 145, 172, 179, 226, 232, 237, 238, 240, 242, 253, 258, 272

Brahe, Tyge (Tycho) (1546–1601), Danish astronomer, 354

Brandt, Georg (1694–1768), Swedish chemist, doctor of medicine, extraordinary councillor of mines, 65, 258

Brask, Hans (1464–1538), Swedish bishop of Linköping, 134

Brasser, Jacob R. (17th century), Dutch mathematician, 83

Bredberg, Bengt (1686–1740), Swedish vicar of Stenstorp, Västergötland, brother of Sven Bredberg, 311

Bredberg, Sven (1681–1721), Swedish master of arts, vicar of Fågelås, Västergötland, brother of Bengt Bredberg, 7

Brenner, Elias (1647–1717), Swedish copperplate engraver, numismatist, 310

Briggs, Henry (1561–1630), English mathematician, 93

Bromell, Magnus von, before ennoblement Bromelius (1679–1731), Swedish physician, professor of anatomy in Stockholm, assessor in Collegium Medicum and the Board of Mines, 15, 49, 145, 184, 188, 191, 249, 251, 253, 275, 308

Bruno, Giordano (1548–1600), Italian philosopher, 288, 367

Buddha, Shakyamuni (Siddhartha Gautama) (died 544/43 BC), Indian teacher of wisdom, 9

Buffon, Georges Louis Leclerc de (1707–1788), French count, natural scientist, 178

Bunyan, John (1628–1688), English author, 413

Bure, Anders (1571–1646), Swedish cartographer, 104

Burman, Eric (1692–1729), Swedish professor of astronomy in Uppsala, musician, 54, 55, 64, 135, 167, 168, 184, 262, 269

Burnet, Thomas (1635–1715), English theologian, geologist, 319, 320, 357

Buschenfelt, Samuel, the Elder (1666–1706), Swedish surveyor in Falun, 166, 260, 261

C

Caesar, Gaius Julius (100–44 BC), Roman general, statesman, 99

Cahman, Hans Heinrich (c. 1640–1699), German-Swedish organ builder, 162

Cahman, Johan Niclas (c. 1679–1737), German-Swedish organ builder, 162

Campanella, Tommaso (1568–1639), Italian philosopher, 20

Campani, Giovanni (c. 1560–c. 1623), Italian iconographer, 313

Canal, Giovanni Antonio 'Canaletto' (1697–1768), Italian artist, 38

Cardano, Gerolamo (1501–1576), Italian physician, mathematician, astrologer, 250

Carlberg, Bengt Wilhelm (1696–1778), Swedish fortification lieutenant, borough engineer, 87

Cartesius. See Descartes

Cassini, Giovanni Domenico (Jean Dominique) (1625–1712), Italian-French astronomer, 51

Cassirer, Ernst (1874–1945), German philosopher, 19, 20, 72, 73, 75

Catherine I., Yekaterina Alekseyevna (1684–1727), Russian Empress, married to Peter I, 273

Cato, Marcus Porcius, the Younger (95–46 BC), Roman senator, 99

Cavalieri, Bonaventura (1598–1647), Italian mathematician, 289

Cederhielm, Germund, the Younger (1661–1741), Swedish baron, county governor of Södermanland, 102

Cederholm, Bernhard (1678–1750), Swedish official, secretary of the purchasing deputation, 48, 82, 83

Celsius, Anders (1701–1744), Swedish professor of astronomy in Uppsala, son of Nils Celsius, 69, 74, 151, 290, 293, 346, 347, 407

Celsius, Nils (1658–1724), Swedish professor of astronomy in Uppsala, brother of Olof Celsius, 258

Celsius, Olof (1670–1756), Swedish professor of Greek, Oriental languages, theology in Uppsala, botanist, cathedral dean, brother of Nils Celsius, 42

Chapman, Fredric Henric af (1721–1808), Swedish master shipbuilder, vice-admiral, 135

Charles X Gustavus (1622–1660), Swedish king, 95

Charles XI (1655–1697), Swedish king, 101, 112, 150, 166

Charles XII (1682–1718), Swedish king, 7, 11, 34, 43, 64, 77–79, 82–88, 90, 91, 93, 99–101, 103, 105, 106, 109–119, 127, 152, 155, 160, 161, 166, 170, 317, 405–407

Cherkasov, Ivan Antonovich (1692–1757), Russian cabinet secretary, 105

Cherubini, Le père d'Orléans (1613–1697), French Capuchin monk, 39

Cheselden, William (1688–1752), British surgeon, anatomist, 392

Christina (1626–1689), Swedish queen, 195, 312

Chrysippus (c. 280–c. 205 BC), Greek philosopher, Stoic, 284

Chydenius, Anders (1729–1803), Finnish-Swedish political economist, 107, 108

Cicero, Marcus Tullius (106–43 BC), Roman orator, author, politician, 7, 374, 403

Clauberg, Johann (1622–1665), German philosopher, professor in Herborn and Duisburg, 45, 288

Clavius, Christoph (1537–1612), German mathematician, astronomer, 50, 69

Coleridge, Samuel Taylor (1772–1834), British poet, critic, 12

Columbus, Christopher (1451–1506), Italian navigator, 9

Columbus, Samuel (1642–1679), Swedish poet, 56, 133, 196

Comenius, John Amos (Jan Amos Komenský) (1592–1670), Czech theologian, educator, 49, 222, 380, 381

Condillac, Etienne Bonnot de (1714–1780), French philosopher, abbot, 392

Copernicus, Nicolaus (Mikołaj Kopernik) (1473–1543), Polish astronomer, 70, 314, 357

Cortesi, Giovanni Battista (1554–1636), Italian physician, professor of anatomy in Messina, 183

Coster, Johann. See Küster von Rosenberg

Crasta, Francesca Maria, Italian historian of philosophy, 12

Croll, Oswald (c. 1560–1609), German medical scholar, chemist, 167

Cronstedt, Axel Fredrik (1722–1765), Swedish chemist, mineralogist, inspector of mines, 122, 407

Cusanus, Nicolaus (Nikolaj Krebs) (1401–1464), German theologian, philosopher, mathematician, 356, 367, 388

D

Dahl, Eric (1679–1720), Swedish student in Uppsala, vicar of Dala, 134, 149, 150, 218

Dahlbergh, Erik Jönsson (1625–1703), Swedish count, fortification officer, draughtsman, 29

Dalbeck (18th century), Swedish widow of a bookbinder in Stockholm, 102

Dalgarno, George (c. 1626–1687), English educator, 122

Dalin, Olof von, before ennoblement Dalin (1708–1763), Swedish author, historian, 5, 118, 134, 138

Dante Alighieri (1265–1321), Italian author, 401

Daumont, Petter (18th century), Swedish barber-surgeon, 393

Dee, John (1527–1608), English mathematician, natural philosopher, mystic, 50

Democritus of Abdera (c. 460–c. 370 BC), Greek natural philosopher, 225, 226, 233, 242, 381, 383

Demophilus (c. 150 AD), Greek author, 310

Derham, William (1657–1735), English theologian, natural philosopher, 39, 246, 264, 354

Desargues, Girard (1591–1661), French mathematician, 29

Descartes, René (1596–1650), French philosopher, mathematician, natural philosopher, 11, 39, 45, 46, 56, 60, 72, 75, 80, 131, 132, 148, 169, 172, 180, 181, 183, 189, 190, 197, 198, 222, 224–226, 232, 237, 242, 246, 249, 288, 296, 300, 304, 306, 313, 314, 317, 319,

325, 326, 335, 341, 357, 359, 360, 366, 367, 379, 380

Diderot, Denis (1713–1784), French author, philosopher, 392

Digges, Thomas (1546–1595), English mathematician, astronomer, 367

Diophantus (c. 250 AD), Greek mathematician, 152

Ditton, Humphry (1675–1715), English mathematician, 58

Döbelius, Johan Jacob, ennobled as von Döbeln (1674–1743), Swedish physician, professor of medicine in Lund, 203

Drake, Anders von (1682–1744), Swedish official, politician, president of the Board of Trade, 105

Dress, Otto (c. 1626–1697), Swedish ironworks owner, 253

Drossander, Andreas (1648–1696), Swedish physician, professor of medicine in Uppsala, 188, 379

Düben, Andreas, the Elder (c. 1597–1662), German-Swedish composer, organist, 162

Duhre, Anders Gabriel (c. 1680–1739), Swedish mathematician, agriculture teacher, 65

Duns Scotus, Johannes (c. 1266–1308), Scottish theologian, philosopher, 361

Dünster, Benjamin (died 1730), Finnish pretender, claiming to be Charles XII, 109

Dupleix, Scipion (1569–1661), French philosopher, 369

Duræus, Samuel (1718–1789), Swedish mathematician, professor of physics in Uppsala, 52, 80

Duseen, Swedish clerk, 393

E

Eenberg, Johan (died 1709), Swedish professor of practical philosophy in Uppsala, 7, 42

Ehrenstrahl, David Klöcker (1628–1698), German-Swedish artist, 95

Ekman, Natanael Olofsson (c. 1682–1746), Swedish inspector of mines, 272

Elvius, Pehr, the Elder (1660–1718), Swedish professor of astronomy in Uppsala, 42, 43, 46–48, 51, 52, 61, 93, 167, 250, 311, 313, 315, 316

Emerson, Ralph Waldo (1803–1882), American author, philosopher, 12

Empedocles (c. 492–c. 432 BC), Greek philosopher, 233

Epicurus (341–c. 270 BC), Greek philosopher, 225, 226, 242, 319, 382

Erasmus of Rotterdam, Desiderius (c. 1469–1536), Dutch humanist, 394

Ercker, Lazarus (died 1593), German metallurgist, 258

Eriksson, Gunnar (born 1931), Swedish professor emeritus of the history of science and ideas in Uppsala, 24, 64, 156, 167, 172, 195, 321

Euclid (c. 300 BC), Greek mathematician, 50, 60, 64, 73, 286, 287, 379

Eudoxus of Cnidos (c. 408–c. 355 BC), Greek astronomer, mathematician, 313

F

Fabricius, Johann Albert (1668–1736), German theologian, philologist, professor in Hamburg, 362

Faggot, Jacob (1699–1777), Swedish surveyor, official, 105, 106

Feif, Casten (1662–1739), Swedish baron, head of the war department, 53, 106, 112

Fermat, Pierre de (1601–1665), French mathematician, 306

Fibonacci, Leonardo (Leonardo of Pisa) (c. 1170–c. 1250), Italian mathematician, 306

Ficino, Marsilio (1433–1499), Italian humanist, philosopher, 388

Flamsteed, John (1646–1719), English astronomer, 48, 53, 54

Fleck, Ludwik (1896–1961), Polish-Israeli theorist of science, physician, 24

Florus, Lucius Annaeus (2nd century AD), Roman author, 7

Fontana, Niccolò (c. 1499–1557), Italian mathematician, 70

Fontenelle, Bernard Le Bovier de (1657–1757), French author, philosopher, secretary of the Académie des Sciences, 49, 70, 73, 212, 224, 313, 315, 368

Forsius, Sigfrid Aron (1550s–1624), Swedish professor of astronomy in Uppsala, 162, 226, 229–231, 254, 270, 341, 355, 388

Frängsmyr, Tore (born 1938), Swedish professor of history of science in

Uppsala, 11, 136, 142, 195, 241, 251, 256, 293, 362

Frederick I (1676–1751), landgrave of Hessen, Swedish king, 115

Frederick II, 'Frederick the Great' (1712–1786), king of Prussia, 113

Frese, Jacob (c. 1690–1729), Swedish poet, 155

Frodbohm, Margaretha (1694–1761), Swedish, married to Michael Hysing, 155

Frölich, David (1595–1648), Hungarian mathematician, 149

G

Galen of Pergamum (Claudius Galenus) (129–199), Greek physician, 189, 201

Galilei, Galileo (1564–1642), Italian astronomer, physicist, mathematician, philosopher, 314

Gassendi, Pierre (1592–1655), French philosopher, mathematician, natural philosopher, 226, 335

Geber. *See* Jabir ibn Hayyan

Geisler, Johan Tobias (1683–1729), Swedish surveyor, 166

Gestrinius, Martinus Erici (1594–1648), Swedish professor of mathematics in Uppsala, 50, 104, 286

Gilbert, William (1544–1603), English physician, physicist, 340, 367

Ginzburg, Carlo (born 1939), Italian historian, professor at the University of Bologna and the University of California, Los Angeles, 5

Gjörwell, Carl Christoffer, the Elder (1731–1811), Swedish publicist, 163

Glauber, Johann Rudolf (1604–1670), German chemist, 250, 258, 274

Goethe, Johann Wolfgang von (1749–1832), German author, 12

Goldbach, Christian (1690–1764), German mathematician, 161

Görtz, Georg Heinrich von (1668–1719), German baron, statesman, head of the purchasing deputation, 100, 106, 108, 117

Graham, George (c. 1675–1751), British clockmaker and instrument maker, 346

Granberg, Lars Bengtson 'Lasse på Jorden' (1678–c. 1718), Swedish mental arithmetician, 111, 112

Gregory, David (1659–1708), Scottish mathematician, astronomer, 319

Grimaldi, Francesco Maria (1618–1663), Italian physicist, astronomer, Jesuit, professor in Bologna, 169

Groot, Huig de (Grotius) (1583–1645), Dutch humanist, philosopher of law, 73

Grundell, Daniel (died 1716), Swedish artillery officer, 216

Gudhemius, Petrus (1679–1751), Swedish vicar of Böne, Västergötland, 196

Guericke, Otto von (1602–1686), German experimental physicist, mayor of Magdeburg, 172, 367

Guglielmini, Domenico (1655–1710), Italian mathematician, natural scientist, medical scholar, 187

Gusmão, Lourenço de (1686–1724), Brazilian Jesuit, 212

Gustavus II Adolphus (1594–1632), Swedish king, 101

Gutenberg, Johann (1394–1468), German printer, 3

Gyllenborg, 150

H

Häll, Jan (born 1955), Swedish Ph.D. in the history of ideas in Uppsala, 12

Halley, Edmond (1656–1742), British astronomer, professor of geometry in Oxford, Astronomer Royal, 53, 54, 110, 246, 368

Hamann, Johann Georg (1730–1788), German philosopher, 375

Hårleman, Carl (1700–1753), Swedish architect, superintendent of public works, 407

Harmens, Lars (died 1760), Swedish auscultator, actuary, inspector of mines, 272

Harrison, John (1693–1776), British clockmaker, instrument maker, 56

Harvey, William (1578–1657), English physician, physiologist, 194, 195

Hasselbom, Nils (1690–1764), Swedish student in Uppsala, professor of mathematics in Åbo (Turku), 62

Hauksbee, Francis (c. 1666–1713), English physicist, instrument maker, 173

Hein, David von (18th century), Hessian court councillor, 112

Heister, Lorenz (1683–1758), German anatomist, surgeon, professor in Altdorf and Helmstedt, 362

Helander, Hans (born 1942), Swedish professor
emeritus of Latin in Uppsala, 1, 3, 7,
23, 24, 91, 277, 312

Hellwig, Christoph von (1663–1721), German
physician, 252, 258, 259, 332

Helmont, Johann Baptist van (1577–1644),
Flemish chemist, physician, natural
philosopher, 195

Helvetius. See Schweitzer

Henckel, Johann Friedrich (1678–1744),
German mineralogist, chemist,
councillor of mines, 259, 262, 274, 275

Henrion, Denis. See Mangin

Heraclitus (c. 500 BC), Greek philosopher,
195, 284, 383

Hermes Trismegistus, Egyptian teacher of
wisdom, 339, 388

Herodotus (c. 480–c. 420 BC), Greek historian,
50

Hesiod (c. 700 BC), Greek poet, 318

Hesselia, Beata, Swedish (died 1759),
married to Johan Kolmodin, sister of
Andreas and Johan Hesselius, cousin of
Swedenborg, 309

Hesselius, Andreas (1677–1733), Swedish
vicar of Christina Parish, Delaware, and
Gagnef, Dalarna, brother of Beata and
Johan Hesselius, cousin of Swedenborg,
209, 210, 312

Hesselius, Johan (1687–1752), Swedish doctor
of medicine, physician, palaeontologist,
brother of Beata and Andreas Hesselius,
cousin of Swedenborg, 145, 163, 215

Hevel, Jan (Johannes Hevelius) (1611–1687),
Polish astronomer, 54

Hiorter, Olof Petrus (1696–1750), Swedish
astronomer, Observator Regius, 346,
347

Hippocrates of Cos (c. 460–c. 370 BC), Greek
physician, 201

Hjärne, Urban, before ennoblement Hiärne
(1641–1724), Swedish physician,
natural philosopher, president of the
Collegium Medicum, vice-president of
the Board of Mines, 90, 101, 132, 133,
139, 142, 143, 145, 149, 155, 186, 230,
237, 249, 253, 275

Hobbes, Thomas (1588–1679), English
philosopher, 126, 242, 301, 368

Hoffmann, Friedrich (1660–1742), German
physician, chemist, 189, 192, 200, 259,
262

Hoffwenius, Petrus (1630–1682), Swedish
physician, professor of medicine and
physics in Uppsala, 45, 169, 226, 230,
232, 244, 246, 288, 314, 326, 335, 379

Hogarth, William (1697–1764), British painter,
engraver, art theorist, 351, 397

Holtzbom, Anders (1660s–1711), Swedish
physician, draughtsman, 150

Homberg, Wilhelm (1652–1715), Dutch
chemist, 230

Homer (8th century BC), Greek author, 318

Hooke, Robert (1635–1703), English natural
scientist, inventor, 49, 50, 53, 169, 238,
287, 296, 297

Hoorn, Johan von (1662–1724), Swedish
physician, assessor in Collegium
Medicum, court physician, 49, 63

Höpken, Anders Johan von (1712–1789),
Swedish count, official, Councillor of
the Realm, 65, 283, 396, 407

Horn, Bengt (1623–1678), Swedish baron,
Councillor of the Realm, warrior, 9, 95

Hugh of Saint Victor (1096–1141), French
theologian, Augustinian friar in
Paris, 161

Hultkrantz, Johan Vilhelm (1862–1938),
Swedish professor of anatomy in
Uppsala, 13

Hume, David (1711–1776), Scottish
philosopher, 369

Hume, James (died c. 1639), Scottish
mathematician, 83

Humerus, Bonde (1659–1727), Swedish
professor of mathematics, Oriental
languages, theology in Lund, cathedral
dean, 38, 39, 90, 321

Huygens, Christiaan (1629–1695), Dutch
mathematician, astronomer, physicist,
51, 67, 158, 169, 179, 193, 240, 319

Hyde, James (died 1910), English
Swedenborgian, 37, 59, 156,
170, 183

Hysing, Michael (1687–1756), Swedish
merchant, married to M. Frodbohm,
155

J

Jabir ibn Hayyan (8th century–c. 815),
Arabian alchemist, physician, natural
philosopher, 267

Jeremiah (c. 626 BC), Old Testament prophet,
100

Jesus Christ of Nazareth (c. 4 BC–c. 30 AD),
son of God, 4, 8, 80, 165, 216, 298, 370,
374, 377, 403

Jesus, son of Sirach, 285
Job, a righteous man in the Old Testament, 273, 285, 399
Johansson, Lars (1638–1674), Swedish poet, 165
John (1st century AD), evangelist, 163, 370, 403
Johnson, Mark (born 1949), American professor of philosophy at the University of Oregon, 21
Jönsdotter, Ester (born 1682), Swedish maid, starvation artist, 100, 203
Jonsson, Inge (born 1928), Swedish professor emeritus of literature in Stockholm, 9, 11, 12, 24, 32, 72, 80, 122, 124, 133, 154, 163, 181, 183, 221, 292, 293, 301, 319, 346, 349, 361, 368, 370, 371, 374, 375, 380, 387–391, 393, 399, 403

K
Kalmeter, Henrik (1693–1750), Swedish auscultator in the Board of Mines, councillor of commerce, 246, 247
Kalsenius, Andreas (1688–1750), Swedish bishop of Västerås, 6, 51
Kant, Immanuel (1724–1804), German philosopher, professor of logic and metaphysics in Königsberg, 9, 12, 347
Kellner, David (1670s–1748), German medical scholar, chemist, 252, 258
Kepler, Johannes (1571–1630), German astronomer, 38, 39, 287, 288, 314, 388
Kertzenmacher, Peter (Petrus) (16th century), German alchemist, 267
Kircher, Athanasius (1602–1680), German Jesuit, natural philosopher, 28, 39, 54, 122, 123, 136, 142, 148, 149, 156, 160, 162, 163, 172, 195, 196, 212, 224, 236, 250, 269, 280, 281, 283, 308, 348, 403, 404
Kirchmaier, Georg Caspar (1635–1700), German professor of rhetoric in Wittenberg, 252
Klingenstierna, Samuel (1698–1765), Swedish professor of geometry, experimental physics, 65, 66, 73, 293, 368
Klopper, Jan (died 1734), Dutch-Swedish painter, drawing master at Uppsala University, 140, 237
Kolmodin, Johan Michaelis (1648–1724), Swedish rural dean in Nysätra, Uppland, married to Beata Hesselia, 309
Komenský. See Comenius

König, Emanuel (1658–1731), Swiss natural historian, medical scholar, 258, 259, 270
Kopernik. See Copernicus.
Kräutermann, Valentin. See Hellwig
Krebs. See Cusanus
Kullin, Lars Julius (1714–1795), Swedish lecturer in mathematics in Gothenburg, vicar of Lundby and Tuve, 65
Kunckel von Löwenstern, Johann (1630–1703), German alchemist, glass engineer, 268, 270, 298, 299
Küster von Rosenberg, Johannes (1614–1685), German physician, 122

L
La Hire, Philippe de (1640–1718), French mathematician, 58, 60
La Mettrie, Julien Offroy de (1709–1751), French physician, philosopher, 382
La Motraye, Aubry de (1674–1743), French merchant, travelogue writer, 208.
Lagerbring, Sven (1707–1787), Swedish professor of history in Lund, 65, 134, 179, 293
Lagercrantz, Olof (1911–2002), Swedish author, critic, publicist, 12
Lakoff, George (born 1941), American linguist, professor at the University of California, Berkeley, 21
Lamm, Martin (1880–1950), Swedish professor of the history of literature in Stockholm, 11, 17, 155, 293, 382, 387, 395, 402, 403
Lana Terzi, Francesco (1631–1687), Italian count, natural scientist, Jesuit, 212
Lavater, Johann Caspar (1741–1801), Swiss author, theologian, priest in Zurich, 407
Leeuwenhoek, Antonie van (1632–1723), Dutch natural scientist, microscopist, 48, 49, 188, 231, 258, 262, 387, 389
Leibniz, Gottfried Wilhelm von (1646–1716), German philosopher, mathematician, 43, 46, 51, 66, 89, 91, 93, 94, 111, 112, 123, 126, 161, 246, 293, 301, 302, 306, 308, 315, 319, 333, 357, 361, 366–369, 373, 375, 380, 381, 390
Lémery, Nicolas (1645–1715), French chemist, 237, 244, 258, 259, 332
Lenhammar, Harry (born 1930), Swedish professor emeritus of ecclesiastical history, 12, 65

Leonardo da Vinci (1452–1519), Italian artist,
    architect, inventor, natural scientist, 39
Lesser, Friedrich Christian (1692–1754),
    German pastor, 362
Leucippus (5th century BC), Greek
    philosopher, atomist, 225
L'Hospital, Guillaume-François-Antoine de
    (1661–1704), French mathematician,
    57, 58
Lidén, Johan Hinric (1741–1793), Swedish
    university librarian, professor of history
    in Uppsala, 163, 415
Linder, Johan, ennobled as Lindestolpe
    (1678–1724), Swedish physician, 180,
    190
Linnaeus, Carl, ennobled as von Linné
    (1707–1778), Swedish physician,
    botanist, professor of medicine in
    Uppsala, 6, 11, 26, 49, 108, 135, 139,
    153, 154, 188, 208, 222, 240, 278, 283,
    284, 297, 307, 311, 318, 362, 363, 391,
    395, 407
Lips, Joest (Justus Lipsius) (1547–1606),
    Flemish philologist, historian,
    philosopher, 310
Lister, Martin (c. 1638–1712), English
    zoologist, 259, 262, 263
Llull, Ramón (c. 1235–1316), Catalan
    theologian, philosopher, poet, 126
Locke, John (1632–1704), English philosopher,
    11, 41, 75, 80, 225, 232, 242, 357, 359,
    392
Lohman, Carl Johan (1694–1759), Swedish
    poet, 283
Löhneyß, Georg Engelhard von (1552–1625),
    German mining engineer, metallurgist,
    258, 262
Lotman, Yuri Mikhailovich (1922–1993),
    Russian literature scholar, semiotician,
    professor in Tartu, 15, 20, 32, 401
Lucidor, Lasse. See Johansson
Lucretius Carus, Titus (c. 95–c. 55 BC),
    Roman poet, philosopher, 41, 132, 225,
    226, 232, 242, 299, 353, 368, 381, 382
Ludenius, Jacob (died 1712), Swedish
    physician, doctor of medicine, 7
Ludwig Rudolf (1671–1735), Duke of
    Brunswick and Lüneburg, 114
Luke, evangelist, 154, 318
Lully, Giovanni Battista (Jean-Baptiste Lully)
    (1632–1687), Italian-French composer,
    224
Lundius, Nicolaus (1656–1726), Swedish
    Saami, Lapland priest, 150

Lundström, Magnus (1687–1720), Swedish
    master builder, master engineer in
    Falun, 107
Luther, Martin (1483–1546), German
    theologian, reformer, 100

M
Maigret, Philippe (18th century), French
    engineer officer, 87
Malebranche, Nicolas (1638–1715), French
    philosopher, theologian, 43, 323, 360,
    361, 380
Mallet, Fredric (1728–1797), Swedish
    astronomer, observator, professor of
    mathematics, 56
Malmström, Johan (1674–1727), Swedish
    vice-librarian, professor of Roman law
    in Uppsala, 48
Malpighi, Marcello (1628–1694), Italian
    physician, anatomist, physiologist, 194,
    386, 387
Manderström, Martin Ludvig (1691–1780),
    Swedish court steward, councillor of
    war, 171
Mandeville, Bernard de (1670–1733),
    Dutch-British philosopher, physician,
    282
Mangin, Clément Cyriaque de (Denis Henrion)
    (1570–1642), French mathematician,
    83
Marcus Aurelius (121–180 AD), Roman
    emperor, Stoic, 195, 353, 382
Marshall, John (c. 1687–1720), English
    instrument maker, optician in London,
    24, 48
Martin, Pehr (Peter) (1687–1727), Swedish
    assistant lecturer in medicine in
    Uppsala, 153, 249
Maupertuis, Pierre Louis Moreau de
    (1698–1759), French natural scientist,
    astronomer, 151
Melanchthon, Philipp, originally Schwarzerd
    (1497–1560), German humanist,
    reformer, 400
Melle, Jacob à (1659–1743), German polymath
    in Lübeck, 143
Menius, Friedrich (died 1659), German
    professor of history in Dorpat (Tartu),
    388
Mersenne, Marin (1588–1648), French
    theologian, mathematician, musical
    theorist, 72, 73, 133, 148, 158, 162,
    167, 242, 353, 354

Methusalah (Methusalem), eighth patriarch in the line from Adam to Noah, 317, 318

Milliet Dechales, Claude François (1621–1678), French mathematician, Jesuit, 51, 169, 230

Miłosz, Czesław (1911–2004), Polish author, Nobel laureate, 12, 402

Milton, John (1608–1674), English poet, 11, 397

Molyneux, William (1656–1698), Irish astronomer, physicist, 39, 392

Moræa, Sara Elisabet (1716–1806), Swedish, daughter of Johan Moræus, married to Carl Linnaeus, 307

Moræus, Johan (1672–1742), Swedish physician, town doctor in Falun, 6, 136, 188, 318

More, Henry (1614–1687), English theologian, 73, 367, 382

More, Sir Thomas (1477–1535), English politician, author, 20

Mört, Johan (died 1743), Swedish tutor, 64

Moses, Old Testament prophet, 228, 253, 319, 341

Musschenbroek, Johan van (1687–1748), Dutch instrument maker, 48

Musschenbroek, Pieter van (1692–1761), Dutch physicist, professor in Leiden, 60, 262, 337

Mylius, Gottlieb Friedrich (1675–1726), German jurist, natural scientist, 308

N

Napier, John (1550–1617), Scottish mathematician, 93

Naumachius, Greek poet, 5

Newcomen, Thomas (1663–1729), British engineer, inventor, 247, 248

Newton, Sir Isaac (1642–1727), English mathematician, physicist, 5, 12, 39, 46, 53, 58, 60, 65, 67, 169, 174, 176, 193, 231, 242, 288, 315–317, 322, 346, 367, 405

Noah, a righteous Jewish man, 138, 143, 319

Nordberg, Jöran Andersson (1677–1744), Swedish court preacher, historian, 83, 84, 88, 109–114, 407

Nordborg, Olof (1681–1745), Swedish master of arts, pastor of the Swedish congregation in London, 37

Nordencrantz, Anders, before ennoblement Bachmanson (1697–1772), Swedish author on economics and philosophy, 283

Nordenskiöld, August (1754–1792), Swedish alchemist, Swedenborgian, 171

Norman, Maria, married to Nils Ahlstedt, Swedenborg's housekeeper, 415

O

Odel, Anders (1718–1773), Swedish politician, author, 114

Odhelius, Erik, ennobled as Odelstierna (1661–1704), Swedish physician, mining scientist, 181, 226, 242, 253

Oetinger, Friedrich Christoph (1702–1782), German theosopher, 12

Olaus Magnus (1490–1557), Swedish churchman and scholar, 142

Origen (c. 185–c. 254), Greek church father, theologian in Alexandria, 399

Örström, Andreas (18th century), Swedish student in Uppsala, 168

Otto, Samuel (17th century), Swedish surveyor, 150, 172, 253

Ovid, Publius Ovidius Naso (43 BC–c. 18 AD), Roman poet, 4, 11, 84, 132, 133, 154, 210, 281, 296, 301, 310, 318, 341, 353

P

Palladio. See Pietro

Palmqvist, Johan (c. 1650–1716), Swedish diplomat, 58, 131

Papke, Jeremias (1672–1755), German professor of mathematics in Greifswald, 116

Paracelsus, born Philippus Aureolus Theophrastus Bombastus von Hohenheim (1493–1541), Swiss physician, natural philosopher, 253, 339

Parmenides (6th–5th century BC), Greek philosopher, 71, 390

Pascal, Blaise (1623–1662), French mathematician, physicist, philosopher of religion, 72, 88, 93, 126, 151, 287, 296, 354

Paul (died 60s AD), Roman apostle, missionary, 403

Peringskiöld, Johan, the Elder (1654–1720), Swedish antiquarian, 4, 111, 277, 317

Perrault, Charles (1628–1703), French author, official, 379

Peter I, 'Peter the Great', Pyotr Alexeyevich (1672–1725), Russian tsar, 7, 118

Philippus, Greek comic playwright, 381

Philo of Alexandria (c. 25 BC–c. 40 AD), Greek-Jewish philosopher, 399

Picinelli, Filippo (c. 1604–c. 1667), Italian emblematist, 154

Pico della Mirandola, Giovanni (1463–1494), Italian humanist, philosopher, 286

Pietro, Andrea di (1508–1580), Italian architect, 307

Pijl, Andreas Samuel (1679–1752), Swedish student in Uppsala, vicar of Nyköping, 196

Pitcairne, Archibald (1652–1713), Scottish professor of medicine in Leiden, 187

Plato (427–347 BC), Greek philosopher, 70, 71, 74, 146, 148, 162, 186, 230, 233, 249, 287, 309, 362, 403

Plautus, Titus Maccius (c. 250–184 BC), Roman comic playwright, 7

Pliny the Elder (23/24–79 AD), Roman author, 283, 340, 353

Plotinus (c. 205–270), Greek philosopher, 11, 364, 387, 388, 397, 403

Polhem, Christopher, before ennoblement Polhammar (1661–1751), Swedish inventor, natural philosopher, 7, 43, 153, 166

Polhem, Emerentia 'Mensa' (1703–1760), Swedish, daughter of Christopher Polhem, married to Reinhold Rücker, 104, 171

Polhem, Gabriel (1700–1772), Swedish mechanic, courtier, son of Christopher Polhem, 171, 217

Polhem, Maria 'Maja' (1698–1754), Swedish, daughter of Christopher Polhem, married to M. L. Manderström, 104, 171

Porphyry, Malchus (c. 234–c. 305), Greek philosopher, 122, 387

Porta, Giambattista della (1535–1615), Italian natural philosopher, 38

Proclus (412–485), Greek philosopher, mathematician, 50, 74, 75, 286–288, 309, 388, 389

Prosperin, Erik (1739–1803), Swedish astronomer, 9, 65

Ptolemy, Claudius (died c. 165 AD), Greek astronomer, geographer, mathematician, 313

Publilius Syrus (105–43 BC), Roman writer of mimes, 6, 115

Pufendorf, Samuel von (1632–1694), German-Swedish baron, historian, jurist, philosopher, professor of natural and international law in Lund, 42, 73, 318

Pyl, Christopher (Pylius) (1678–1739), Pomeranian scholar, 91

Pythagoras (c. 570–c. 497 BC), Greek philosopher, mathematician, 88, 95, 162, 166, 310

Q

Quensel, Conrad (1676–1732), Swedish astronomer, professor of mathematics in Lund, 55

Quinault, Philippe (1635–1688), French dramatist, opera librettist, 224

R

Rålamb, Åke Claesson (1651–1718), Swedish baron, author, 51

Ramée, Pierre de la (Petrus Ramus) (1515–1572), French humanist, educator, 122

Rappe, Niklas (1668–1727), Swedish major general, 233, 235

Ray, John (1627–1705), English botanist, 362

Réaumur, René Antoine Ferchault de (1683–1757), French mining scientist, entomologist, 270, 271

Regnard, Jean François (1655–1709), French author, traveller, 150, 208

Rehoboam (reigned c. 926–c. 910 BC), king of Judah, 100, 101

Retzelius, Olof (1671–1732), Swedish master of arts, secretary, 167

Reuterholm, Hedvig (18th century), Swedish baroness, married to Daniel Tilas, 407

Reyneau, Charles René (1656–1728), French mathematician, 58, 59, 93, 304

Rhyzelius, Andreas Olai (1677–1761), Swedish court preacher, bishop of Linköping, 6, 7, 101, 165

Ribbing, Per (1670–1719), Swedish baron, county governor of Uppland, chief prefect of Stockholm, 99

Riccioli, Giovanni Battista (1598–1671),
    Italian astronomer, Jesuit, 54, 56, 183
Richter, Christian Friedrich (1676–1711),
    German medical scholar, natural
    scientist, 39, 141, 187, 362
Riddermarck, Andreas (Wetterhamn)
    (1651–1707), Swedish professor of
    mathematics in Lund, 160
Ripa, Cesare. See Campani
Roberg, Lars (1664–1742), Swedish professor
    of anatomy and practical medicine in
    Uppsala, 2, 3, 39, 41–43, 45, 63, 69,
    136, 141, 145, 152, 163, 169, 183, 184,
    186, 188, 190, 191, 201, 223, 237, 242,
    254, 260, 262, 275, 285, 308
Roberval, Gilles Personne de (1602–1675)
    French mathematician, 306
Roman, Johann Helmich (1694–1758),
    Swedish composer, court conductor,
    163
Rømer, Ole (1644–1710), Danish astronomer,
    professor in Copenhagen, 169, 179
Rosén von Rosenstein, Nils, before
    ennoblement Rosén (1706–1773),
    Swedish physician, professor of
    medicine in Uppsala, 183, 190
Rosenadler. See Upmarck
Ross, Johan, the Younger (1650s–c. 1716),
    Swedish cathedral organist in Skara,
    162
Rößler, Balthasar (1605–1673), German
    mining engineer, 252, 253
Rubens, Peter Paul (1577–1640), Dutch
    painter, 351
Rücker, Reinhold, ennobled as Rückerschöld
    (1690–1759), Swedish district governor,
    justice of appeal, married to Emerentia
    Polhem, 171
Rudbeck, Olof, the Elder (1630–1702),
    Swedish natural scientist, historian,
    professor of medicine in Uppsala, 26,
    56, 57, 64, 80, 123, 134, 138, 150, 162,
    165–167, 186, 189, 191, 194, 222, 237,
    295, 318, 321
Rudbeck, Olof, the Younger (1660–1740),
    Swedish botanist, philologist, professor
    of medicine in Uppsala, 41, 43, 136,
    149, 150, 152, 163, 167, 242
Rudbeckius, Johannes (1581–1646), Swedish
    university man, bishop of Västerås, 4
Rüdiger, Andreas (1673–1731), German
    philosopher, 246, 382
Rüger (18th century), German secretary in
    Dresden, 292

Runcrantz, Carl August Wilhelm (1750–1828),
    Swedish lecturer in Kalmar, 293
Ruysch, Frederik (1638–1731), Dutch
    anatomist, professor in Amsterdam,
    188
Rydelius, Andreas (1671–1738), Swedish
    philosopher, professor of logic and
    metaphysics, theology, bishop of Lund,
    4, 63, 74, 112, 126, 189, 191, 222, 293,
    357, 403

S
Sault, Richard (died 1702), English
    mathematician, 65
Savery, Thomas (c. 1650–1715), English
    engineer, inventor, 246
Scaliger, Joseph Justus (1540–1609), Italian-
    French philologist, professor in Leiden,
    7
Scapula, Johann (c. 1540–1581), Swiss
    professor of Greek and morals in
    Lausanne, 412
Schefferus, Johannes (1621–1679), German-
    Swedish philologist, Professor
    Skytteanus in Uppsala, 149, 150, 310
Scheffner, Johann Georg (1736–1820), German
    lawyer, author, 375
Scheiner, Christoph (1579–1650), German
    astronomer, 39
Schelling, Friedrich Wilhelm Joseph von
    (1775–1854), German philosopher, 9,
    12
Schenmark, Nils (1720–1788), Swedish
    astronomer, professor of mathematics in
    Lund, 56
Schmidius. See Sebastian Schmidt
Schmidt, Johann Andreas (1652–1728),
    German doctor in Helmstedt, 44
Schmidt, Sebastian (1617–1696), German
    theologian, 398
Schönberg, Anders, the Younger (1737–1811),
    Swedish official, historian, 415
Schönström, Peter, the Younger (1682–1746),
    Swedish army officer, historian, cousin
    of Swedenborg, 272–274
Schott, Gaspar (1608–1666), German
    mathematician, physicist, engineer, 39,
    122, 156, 157, 160, 163, 172, 209, 213,
    283, 404
Schubert, Gotthilf Heinrich von (1780–1860),
    German medical scholar, 9
Schultze, Lars G. (1701–1765), Swedish
    auscultator in the Board of Mines, 266

Schweitzer, Johann Friedrich (1625–1709), Swiss physician, alchemist, 267

Seneca, Lucius Annaeus (*c.* 4 BC–65 e.Kr), Roman Stoic, playwright, statesman, 6, 115, 195, 283, 354

Serenius, Jacob (1700–1776), Swedish dean in Nyköping, bishop of Strängnäs, 379

Shaftesbury, Anthony Ashley Cooper, Earl of (1671–1713), English philosopher, author, 397

Sheldon, Charles (1655–1739), English-Swedish master shipbuilder in Karlskrona, 133

Sheldon, Gilbert (1710–1794), Swedish shipbuilder, 135

Shishak I (10th century BC), Egyptian pharaoh, 100

Sigstedt, Cyriel Odhner (1888–1959), American Swedenborgian, 10, 12

Sinclair, Malcolm (1690–1739), Swedish major, 114, 118

Skutenhielm, Anders (1688–1753), Swedish representative in Copenhagen, court councillor, minister, secretary of state, 346

Snell, Willebrord van Roijen (1580–1626), Dutch astronomer, mathematician, 174

Solomon (reigned *c.* 965–*c.* 926 BC), king of Israel, 94, 155, 285, 307

Spångberg, Jakob, Swedish carpenter's hand, lodging with Swedenborg, 415

Spegel, Haquin (1645–1714), Swedish archbishop, 4, 41, 154–156, 196, 224, 283, 341, 354, 382

Spengler, Oswald (1880–1936), German philosopher, 25, 306

Spinoza, Baruch (1632–1677), Dutch philosopher, 73, 357, 359, 366

Spole, Anders (1630–1699), Swedish mathematician, professor of astronomy in Uppsala, 39, 48, 74, 117, 150, 151, 156, 169, 181, 226, 242, 285, 316, 321

Stahl, Georg Ernst (1660–1734), German chemist, physician, professor of medicine in Halle, 177, 201, 268

Steensen, Niels (Nicolaus Steno) (1638–1686), Danish anatomist, geologist, 145, 187

Steuch, Elof (1687–1772), Swedish professor of Greek in Lund, professor of mathematics in Uppsala, 64

Stevin, Simon (1548–1620), Dutch engineer, physicist, mathematician, 62, 104

Stiernhielm, Georg (1598–1672), Swedish poet, official, natural philosopher, 62,
95, 96, 103, 104, 121, 127, 132, 160, 218, 220, 282, 285, 288, 319, 353, 387, 388

Stroh, Alfred Henry (1878–1922), American Swedenborgian, 10

Ström, Christian (died 1710), Swedish instrument maker, medical scholar, 49

Strömberg, Zacharias (18th century), Swedish merchant in Amsterdam, 108

Sturm, Johann Christopher (1635–1703), German professor of mathematics and physics in Altdorf, 3, 26, 39, 44, 83, 149, 160, 182, 209, 212, 213

Suetonius Tranquillus, Gaius (*c.* 100 AD), Roman author, 99

Swab, Anton Antonsson von (1702–1768), Swedish mining scientist, auscultator, assessor in the Board of Mines, 272

Swammerdam, Jan (1637–1680), Dutch anatomist, entomologist, 282, 387, 396

Swedberg, Albrecht (1684–1696), Swedish, brother of Emanuel Swedenborg, 165

Swedberg, Daniel Danielsson (1648–1733), Swedish inspector of mines, brother of Jesper Swedberg, 150

Swedberg, Jesper (1653–1735), Swedish court preacher, professor of theology in Uppsala, bishop of Skara, father of Emanuel Swedenborg, 2, 41, 88, 90, 109, 155, 163, 165, 310, 412

Swedenborg, Anna (1686–1766), Swedish, sister of Emanuel Swedenborg, married to Eric Benzelius the Younger, 6

Swedenborg, Catharina (1693–1770), Swedish, sister of Emanuel Swedenborg, married to Jonas Unge, 138

Swedenborg, Hedvig (1690–1728), Swedish, sister of Emanuel Swedenborg, married to Lars Benzelstierna, 139, 186

Swedenborg, Jesper (1694–1771), Swedish army officer, brother of Emanuel Swedenborg, 109, 165

Sweelinck, Jan Pieterszoon (1562–1621), Dutch organist, composer, 162

Swift, Jonathan (1667–1745), British author, 49, 282

Sylvius, Franciscus. *See* Boë

T

Tafel, Rudolph Leonard (1831–1893), American Swedenborgian, professor of philology in St Louis, 9

Tartaglia. *See* Fontana

Tatishchev, Vasily Nikitich (1686–1750),
    Russian historian, geographer, mine
    pioneer, 105, 272–274
Tesauro, Emanuele (1592–1675), Italian
    theorist of literature, 23
Thales of Miletus (c. 625–c. 545 BC), Greek
    philosopher, 340
Theobald, James (1688–1759), English natural
    historian, member of the Royal Society,
    346
Thomas Aquinas (c. 1225–1274), Italian
    theologian, philosopher, 361, 362,
    373
Thomasius, Christian (1655–1728), German
    philosopher, professor of law in Halle,
    112
Tilas, Daniel (1712–1772), Swedish baron,
    mining scientist, councillor of mines,
    407
Tiselius, Daniel (1682–1744), Swedish natural
    scientist, vicar of Hammar, Närke, 135,
    141, 142, 196, 241, 249
Törner, Fabian (1666–1731), Swedish
    professor of logic and eloquence in
    Uppsala, 7, 42, 152
Torricelli, Evangelista (1608–1647), Italian
    mathematician, physicist, 173, 306
Touscher, Daniel (1662/63–1733), Swedish
    vicar of Skön, Medelpad, 149
Trier, Johann Wolfgang (1686–1750), German
    councillor of mines, jurist, 274
Triewald, Mårten (1691–1747), Swedish
    inventor, fortification officer, 65, 232,
    241, 248, 249, 285, 312
Troili, Samuel, the Elder (1693–1758),
    Swedish inspector of mines at Stora
    Kopparberg, 74
Troilius, Jacob (1657–1717), Swedish vicar of
    Husby, Dalarna, 42, 122, 228, 235, 236
Tschirnhaus, Ehrenfried Walther von (1651–
    1708), German scientist, inventor,
    156
Tuxen, Christian Friis Møller de (1713–1792),
    Danish general, 65, 171

U
Ulrika Eleonora the Elder (1656–1693),
    Swedish queen, married to Charles XI,
    258
Ulrika Eleonora the Younger (1688–1741),
    Swedish queen, married to Frederick I,
    258

Unge, Andreas Amberni (1662–1736),
    Swedish rural dean, cathedral dean in
    Skara, brother of Jonas Unge, 136, 138
Unge, Jonas (1681–1755), Swedish vicar of
    Lidköping, brother of Andreas Unge,
    married to Catharina Swedenborg, 7,
    141, 156
Upmarck, Johan, ennobled as Rosenadler
    (1664–1743), Swedish Professor
    Skytteanus, censor librorum, 42, 43,
    108, 151, 388

V
Valentini, Michael Bernhard (1657–1729),
    German professor of medicine and
    physics in Gießen, 39, 160, 212, 254,
    259, 262–266
Vallemont, Pierre Le Lorrain de (1649–1721),
    French priest, 38, 83, 251, 253–254
Vallerius, Göran (1683–1744), Swedish
    surveyor, inspector of mines, assessor
    in the Board of Mines, son of Harald
    Vallerius the Elder, 43, 58, 134, 167,
    173, 209, 283
Vallerius, Harald, the Elder (1646–1716),
    Swedish professor of mathematics in
    Uppsala, musician, 20, 30, 38, 41, 43,
    45, 50–52, 64, 73, 93, 131, 139, 149,
    162, 163, 166, 167, 172, 196, 230, 237,
    242, 288, 315, 388
Vallerius, Harald, the Younger (1680–1724),
    Swedish student in Uppsala, secretary
    of court in Stockholm, son of Harald
    Vallerius the Elder, 43
Vallerius, Johan (1677–1718), Swedish
    professor of mathematics, son of Harald
    Vallerius the Elder, 43, 47, 64, 65, 85,
    93, 167, 237, 313
Varignon, Pierre (1654–1722), French
    mathematician, professor at the Collège
    Mazarin and the Collège de France in
    Paris, 58, 60, 349
Vauban, Sébastien le Prestre de (1633–1707),
    French engineer officer, 83, 217
Vermeer van Delft, Jan (1632–1675), Dutch
    painter, 38
Vico, Giambattista (1668–1744), Italian
    philosopher, historian, 23, 24, 310
Viète, François (1540–1603), French
    mathematician, 81
Vieussens, Raymond (c. 1635–1715), French
    medical scholar, anatomist, 184, 191,
    198, 386

Virgil, Publius Vergilius Maro (70–19 BC),
    Roman poet, 208, 266, 267, 269, 318
Vitali, Girolamo (died 1698), Italian monk, 285
Vitruvius Pollio, Marcus (1st century BC),
    Roman architect, engineer, 217
Vivaldi, Antonio (1678–1741), Italian violinist,
    composer, 162
Voltaire, François Marie Arouet de (1694–
    1778), French philosopher, historian,
    author, 87, 93, 113, 392

W

Wallerius, Nils (1706–1764), Swedish
    philosopher, theologian, doctor of
    physics, professor of logic and
    metaphysics and of apologetics in
    Uppsala, 246
Wallis, John (1616–1703), English scientist,
    professor of geometry in Oxford, 111,
    316
Wargentin, Pehr Wilhelm (1717–1783),
    Swedish astronomer, statistician, 56
Warmholtz, Carl Gustaf (1713–1785), Swedish
    bibliographer, 113
Wassenius, Birger Jonas (1687–1771), Swedish
    astronomer, lecturer in mathematics in
    Gothenburg, 115, 141
Werner, Johan Henrik (died 1735), Swedish
    printer in Uppsala, 62, 96
Whitman, Walter (1819–1892), American
    author, 12
Wilhelm (1682–1760), German landgrave of
    Hessen-Kassel, brother of Frederick I,
    266

Wilkins, John (1614–1672), English
    theologian, philosopher, linguist, 122,
    209
Willis, Thomas (1621–1675), English
    anatomist, physician, 39, 187, 189, 191,
    198, 258, 386
Winsløw, Jacob (1669–1760), Danish
    anatomist, 386
Wolf, Johann Christoph (1683–1739), German
    professor of Oriental languages, priest,
    109, 293
Wolff, Christian von (1679–1754), German
    baron, philosopher, professor of
    mathematics and philosophy in Halle
    and Marburg, 39, 51, 60, 74, 109, 123,
    125, 144, 258, 262, 280, 288, 292–294,
    300–302, 307, 332, 339, 357, 358, 361,
    363, 365, 369, 372, 373, 375–377, 380,
    382, 389, 405
Wren, Sir Christopher (1632–1723), English
    architect, scientist, mathematician, 53,
    153
Wretman, Joachim (18th century), Swedish
    merchant in Amsterdam, 414

Z

Zahn, Johann (1641–1707), German
    Premonstratensian canon, 39, 41, 313
Zellinger the Younger, Christian (1660–1718),
    Swedish cathedral organist in Uppsala,
    166
Zeno of Elea (5th century BC), Greek
    philosopher, 301, 368

17908890R00269

Printed in Great Britain
by Amazon